Climate Change, Ecology and Systematics

Climate change has shaped life in the past and will continue to do so in the future. Understanding the interactions between climate and biodiversity is a complex challenge to science. With contributions from 60 key researchers, this book examines the ongoing impact of climate change on the ecology and diversity of life on earth. It discusses the latest research within the fields of ecology and systematics, highlighting the increasing integration of their approaches and methods. Topics covered include the influence of climate change on evolutionary and ecological processes such as adaptation, migration, speciation and extinction, and the role of these processes in determining the diversity and biogeographic distribution of species and their populations. This book ultimately illustrates the necessity for global conservation actions to mitigate the effects of climate change in a world that is already undergoing a biodiversity crisis of unprecedented scale.

TREVOR R. HODKINSON is Senior Lecturer in Botany at the School of Natural Sciences, Trinity College Dublin. He is Head of the Botany Molecular Laboratory and Assistant Curator of the Herbarium. He specialises in the research fields of molecular systematics, genetic resources and taxonomy.

MICHAEL B. JONES holds the Chair of Botany in the School of Natural Sciences, Trinity College Dublin. He is a plant ecophysiologist and his research focuses on the study of climate–plant interactions, particularly on the effects of climate on photosynthesis, growth and primary productivity.

STEPHEN WALDREN is Senior Lecturer in Botany and Curator of the Trinity College Dublin Botanic Garden. His research interests are in the areas of conservation biology and phylogeography.

JOHN A. N. PARNELL, currently Head of the School of Natural Sciences at Trinity College Dublin, is Professor of Systematic Botany and Curator of the Herbarium. His research interests are predominantly in the fields of plant taxonomy and systematics, working mainly on the floras of Ireland and Thailand.

The Systematics Association Special Volume Series

SERIES EDITOR

DAVID J. GOWER

Department of Zoology, The Natural History Museum, London, UK

The Systematics Association promotes all aspects of systematic biology by organising conferences and workshops on key themes in systematics, running annual lecture series, publishing books and a newsletter, and awarding grants in support of systematics research. Membership of the Association is open globally to professionals and amateurs with an interest in any branch of biology, including palaeobiology. Members are entitled to attend conferences at discounted rates, to apply for grants and to receive the newsletter and mailed information; they also receive a generous discount on the purchase of all volumes produced by the Association.

The first of the Systematics Association's publications, *The New Systematics* (1940), was a classic work edited by its then-president Sir Julian Huxley. Since then, more than 70 volumes have been published, often in rapidly expanding areas of science where a modern synthesis is required.

The Association encourages researchers to organise symposia that result in multi-authored volumes. In 1997 the Association organised the first of its international Biennial Conferences. This and subsequent Biennial Conferences, which are designed to provide for systematists of all kinds, included themed symposia that resulted in further publications. The Association also publishes volumes that are not specifically linked to meetings, and encourages new publications (including textbooks) in a broad range of systematics topics.

More information about the Systematics Association and its publications can be found at our website: www.systass.org.

Previous Systematics Association publications are listed after the index for this volume.

THE SYSTEMATICS ASSOCIATION SPECIAL
VOLUME 78

Climate Change, Ecology and Systematics

EDITED BY

TREVOR R. HODKINSON,
MICHAEL B. JONES,
STEPHEN WALDREN
and JOHN A. N. PARNELL

School of Natural Sciences, Trinity College Dublin, Ireland

CAMBRIDGE UNIVERSITY PRESS
Cambridge, New York, Melbourne, Madrid, Cape Town,
Singapore, São Paulo, Delhi, Tokyo, Mexico City

Cambridge University Press
The Edinburgh Building, Cambridge CB2 8RU, UK

Published in the United States of America by Cambridge University Press, New York

www.cambridge.org
Information on this title: www.cambridge.org/9780521766098

First published 2011

Printed in the United Kingdom at the University Press, Cambridge

A catalogue record for this publication is available from the British Library

Library of Congress Cataloguing in Publication data
Climate change, ecology, and systematics / [edited by] Trevor Hodkinson ... [et al.]. – 1st ed.
p. cm. – (Systematics association special volume series)
ISBN 978-0-521-76609-8 (hardback)
1. Climatic changes. 2. Climatology–Mathematical models. 3. Ecology–Environmental aspects.
4. Biology–Classification. I. Hodkinson, Trevor R.
QC981.C628 2011
577.2′2–dc22 2010051697

ISBN 978-0-521-76609-8 Hardback

Contents

Contributors

J. Arroyo, Departamento de Biología Vegetal y Ecología, Universidad de Sevilla, Spain

P. Baas, National Herbarium of the Netherlands, Leiden, the Netherlands

R. M. Bateman, Jodrell Laboratory, Royal Botanic Gardens, Kew *and* School of Geography, Earth and Environmental Sciences, University of Birmingham, UK

J. Bernardo, Department of Natural Resources, Cornell University, NY *and* Southern Appalachian Biodiversity Institute, Roan Mountain, TN, USA

Y. Bouchenak-Khelladi, Department of Botany, University of Cape Town, South Africa

R. Caballero, School of Mathematical Sciences, University College Dublin, Ireland

A. Caffarra, Department of Environmental Sciences, Fondazione E. Mach, Istituto Agrario San Michele all'Adige, Italy

L. W. Chatrou, Nationaal Herbarium Nederland *and* Biosystematics Group, Wageningen University, the Netherlands

C. A. Couch, Royal Botanic Gardens, Kew, UK

T. L. P. Couvreur, Institut de Recherche pour le Développement, Montpellier, France

A. Culham, School of Biological Sciences *and* The Walker Institute for Climate Change, University of Reading, UK

E. Diskin, School of Natural Sciences, Trinity College Dublin, Ireland

A. Donnelly, School of Natural Sciences, Trinity College Dublin, Ireland

G. C. Douglas, Kinsealy Research Centre, Teagasc, Dublin, Ireland

C. J. Ellis, Royal Botanic Garden Edinburgh, UK

F. Fernández-Manjarrés, CNRS *and* Université Paris-Sud XI, Orsay *and* AgroParisTech, Paris, France

N. Frascaria-Lacoste, CNRS *and* Université Paris-Sud XI, Orsay *and* AgroParisTech, Paris, France

M. Hall, Centre for Middle Eastern Plants, Royal Botanic Garden Edinburgh, UK

R. L. Hodd, Botany and Plant Science, National University of Ireland, Galway, Ireland

T. R. Hodkinson, School of Natural Sciences, Trinity College Dublin, Ireland

D. G. Hole, School of Biological and Biomedical Sciences, Durham University, UK *and* Science and Knowledge Division, Conservation International, Arlington, VA, USA

B. Huntley, Ecosystem Science Centre, School of Biological and Biomedical Sciences, Durham University, UK

M. B. Jones, School of Natural Sciences, Trinity College Dublin, Ireland

C. T. Kelleher, National Botanic Gardens, Glasnevin, Dublin, Ireland

S. Lötters, Biogeography Department, Trier University, Germany

P. Lynch, School of Mathematical Sciences, University College Dublin, Ireland

P. J. Mayhew, Department of Biology, University of York, UK

J. C. McElwain, School of Biology and Environmental Science, University College Dublin, Ireland

A. G. Miller, Centre for Middle Eastern Plants, Royal Botanic Garden Edinburgh, UK

A. M. Muasya, Department of Botany, University of Cape Town, South Africa

K. J. Niklas, Department of Plant Biology, Cornell University, NY, USA

J. O'Halloran, Biological, Earth and Environmental Sciences, University College Cork, Ireland

B. F. O'Neill, School of Natural Sciences, Trinity College Dublin, Ireland

J. A. N. Parnell, School of Natural Sciences, Trinity College Dublin, Ireland

J. Peñuelas, Center for Ecological Research and Forestry Applications (CSIC), Campus Universitat Autònoma de Barcelona, Spain

A. Pletsers, School of Natural Sciences, Trinity College Dublin, Ireland

H. Proctor, School of Natural Sciences, Trinity College Dublin, Ireland

F. Rindi, Dipartimento di Scienze del Mare, Università Politecnica delle Marche, Ancona, Italy

F. Rodríguez-Sánchez, Departamento de Biología Vegetal y Ecología, Universidad de Sevilla, Spain

D. Rödder, Herpetology Department, Zoologisches Forschungsmuseum Alexander Koenig, Bonn *and* Biogeography Department, Trier University, Germany

S. Schick, Biogeography Department, Trier University, Germany

S. Schmidtlein, Vegetation Geography Department, Bonn University, Germany

M. J. Sheehy Skeffington, Botany and Plant Science, National University of Ireland, Galway, Ireland

D. A. Simpson, Royal Botanic Gardens, Kew, UK

T. Sparks, Fachgebiet für Ökoklimatologie, Technische Universität München, Germany *and* Institute of Zoology, Poznań University of Life Sciences, Poland *and* Department of Zoology, University of Cambridge, UK

R. Stirnemann, School of Natural Sciences, Trinity College Dublin, Ireland

M. Thomasset, School of Natural Sciences, Trinity College Dublin *and* Kinsealy Research Centre, Teagasc, Dublin, Ireland

S. Waldren, School of Natural Sciences, Trinity College Dublin, Ireland

E. A. Wheeler, Department of Wood and Paper Science, North Carolina State University, NC, USA

J. J. Wieringa, Nationaal Herbarium Nederland *and* Biosystematics Group, Wageningen University, the Netherlands

K. J. Willis, Department of Zoology, University of Oxford, UK

S. G. Willis, Ecosystem Science Centre, School of Biological and Biomedical Sciences, Durham University, UK

R. Yahr, Royal Botanic Garden Edinburgh, UK

C. Yesson, Institute of Zoology, Zoological Society of London, UK

Preface

The 21 chapters of this book are based on the theme of a Special Conference of the Systematics Association and the Linnean Society of London, held at Trinity College Dublin (TCD), Ireland, in September 2008. During the three-day Climate Change and Systematics conference, there were stimulating presentations, posters and discussions covering a broad range of ecological and systematic research relating to climate change; these influenced the shape and content of this volume. Papers were contributed by a number of conference delegates and by others subsequently invited to broaden the book's scope or address particular theoretical issues.

Consideration of the book's theme began when Richard Bateman, the then President of the Systematics Association, invited John Parnell and the School of Natural Sciences, TCD, to host a conference on the topic and to base a Systematics Association volume around its conclusions. The ideas were refined in discussions with Alan Warren, the then Systematics Association Special Volumes series editor. We are grateful to both for their input and encouragement. Two anonymous book proposal reviewers provided valuable content guidance and many anonymous reviewers also helped to improve the chapter contributions. We are particularly grateful for the manuscript preparation input of Sandra Velthuis of Whitebarn Consulting, who has worked long and hard to proofread chapters and standardise their format, to Hugh Brazier, the excellent copy editor, and to the production team at Cambridge University Press, who have been highly supportive and professional. Finally we thank all 57 contributing authors to the book, many of whom also peer-reviewed other chapters. We encourage all readers to support the activities of the Systematics Association (www.systass.org).

T.R. Hodkinson, M.B. Jones,
S. Waldren and J.A.N. Parnell

SECTION 1

Introduction

1

Integrating ecology and systematics in climate change research

T. R. Hodkinson

School of Natural Sciences, Trinity College Dublin, Ireland

Abstract

Interactions between climate and biodiversity are complex and present a serious challenge to scientists who aim to reconstruct the ways in which climate change has shaped life in the past and will contine to do so in the future. This chapter introduces the contributions made to climate change research by the fields of ecology and systematics and outlines how their approaches and methods have, often through necessity, become increasingly integrated. It explores: (1) how climate change has influenced evolutionary and ecological processes such as adaptation, migration, speciation and extinction; (2) how these processes determine the diversity and biogeographic distribution of species and their populations; and (3) how ecological and systematic studies can be applied to conservation and policy planning in our rapidly changing world.

Climate Change, Ecology and Systematics, ed. Trevor R. Hodkinson, Michael B. Jones, Stephen Waldren and John A. N. Parnell. Published by Cambridge University Press.

1.1 Introduction to climate change, ecology and systematics

> Not only does the marvellous structure of each organised being involve the whole past history of the earth, but such apparently unimportant facts as the presence of certain types of plants or animals in one island rather than in another, are now shown to be dependent on the long series of past geological changes, on those marvellous astronomical revolutions which cause a periodic variation of terrestrial climates, on the apparently fortuitous action of storms and currents in the conveyance of germs, and on the endlessly varied actions and reactions of organised beings on each other. (Wallace, 1880)

Climate exerts a strong selective pressure on organisms and their ecosystems and has continuously presented new challenges to life since it began approximately 3.5 billion years ago (Knoll and Lipps, 1993; Fig 1.1). Species and their populations can respond to climate change by adaptive evolution or by migrating geographically to track their favoured climate. If they fail to do either of these, and if they lack sufficient phenotypic plasticity in terms of their climatic tolerances, they face extinction (Thomas et al., 2004, 2006; Pearson, 2006). Evolution has a tremendous capacity to offer solutions to changing selective pressures, but extinctions induced by climate change are inevitable.

Climate change is best seen as a driver for both creating diversity and reducing it. Climate change has left a record in the taxonomic and ecological patterns of the diversity that has existed on earth and the evolutionary history of that diversity (Parmesan and Yohe, 2003; Root et al., 2003). It is important to study these fingerprints to understand the evolutionary and ecological potential of life to adapt to climate change and resist extinction (Woodward and Kelly, 2008). There is no evidence that life has ever gone totally extinct (Benton and Twichett, 2003) and there are, conservatively, an estimated four million species alive today. This is a remarkable resilience given the global change it has endured throughout its history. The fossil record shows that there have been major periods of extinction, but life has never been totally eclipsed and forced to start again. Several megaextinctions are well known, but the role of climate change as a causal agent in these events is far from clear, even though it has been implicated in all of them (Raup and Sepkoski, 1982; Mayhew, Chapter 4).

Interactions between climate, evolution and ecosystems are complex, and their study has become a highly challenging and multidisciplinary research field (Lovejoy and Hannah, 2005; Rosenzweig et al., 2007). Ecological and systematic research is central to these efforts and essential for predicting the impacts of climate change. This chapter introduces the contributions made by these two key research fields and provides supporting information to make the book's content more accessible.

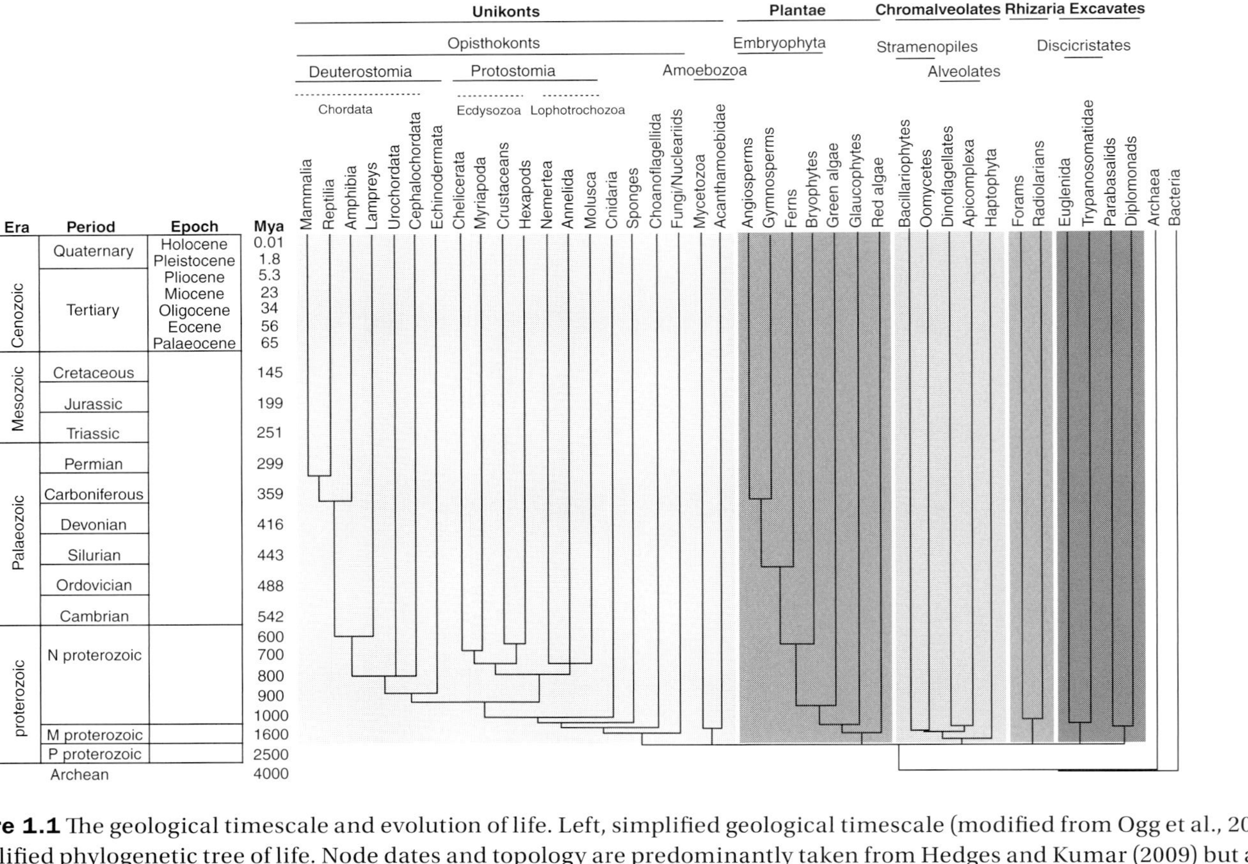

Figure 1.1 The geological timescale and evolution of life. Left, simplified geological timescale (modified from Ogg et al., 2008). Right, simplified phylogenetic tree of life. Node dates and topology are predominantly taken from Hedges and Kumar (2009) but also from Palmer et al. (2004). Note that dates represent times when the taxa diverged from their common ancestor shown on the tree (root node) and do not represent dates for the age of the taxon listed (crown node). Supergroups of Eukaryotes are shown with thick black bars and different shades. Other major groups are shown with thin or dotted bars.

1.1.1 Climate and global change

According to the Intergovernmental Panel on Climate Change (IPCC, 2007a), climate change 'refers to a change in the state of the climate that can be identified (e.g. using statistical tests) by changes in the mean and/or the variability of its properties, and that persists for an extended period, typically decades or longer. It refers to any change in climate over time whether due to natural variability or as a result of human activity.' However, when evaluating the impacts of climate change on life it is necessary to consider atmosphere, geological factors, biogeochemical processes and several other aspects of global change. The term 'global change' is thus often used widely in the scientific community, and it can be defined as 'any consistent trend in the environment - past, present or projected - that affects a substantial part of the globe' (*Global Change Biology* journal homepage at www.wiley.com). The term includes climate change.

Evidence demonstrating the certainty of climate change comes from a range of sources including direct measurements, proxy data, ecological and evolutionary footprints, changes in sea level and changes in the cryosphere. There is unequivocal evidence for recent global warming from such records (IPCC, 2007a; Trenberth et al., 2007). The cryosphere is shrinking, and we have witnessed widespread melting of snow, ice and glaciers and rising global sea level (IPCC, 2007a). At a global scale, we have direct thermometer readings of global air and ocean temperatures dating back about 200 years. The oldest humans alive today, have on average witnessed a global 0.7 °C rise in temperature during their lifetime (100-year linear trend from 1906 to 2005 of 0.74 °C - IPCC, 2007b). The linear warming trend over the 50 years from 1956 to 2005 is nearly twice that for the 100 years from 1906 to 2005 (0.13 °C per decade - IPCC, 2007b). The temperature increase is greater at higher latitudes, where average Arctic temperatures have increased at almost twice the global average rate in the past century. Precipitation increases have been recorded in North and South America, northern Europe and northern central Asia, while precipitation declines have been recorded in the Sahel and Mediterranean, South Africa and parts of Asia. The frequency of extreme weather events, such as hot and cold days and heat waves, also appears to be variable. Climate change therefore shows regional and continental variation.

We must look to palaeoclimatic studies to infer prehistoric climates (Juckes et al., 2007). These studies make use of measurements of change derived from various sources including borehole temperatures, ocean sediment pore-water change and glacier extent changes, as well as proxy measurements involving the changes in chemical, physical and biological parameters that reflect past changes in the environment where the proxy grew or existed (Lowe and Walker, 1997; Stokstad, 2001; IPCC, 2007b). Some biological organisms alter their growth and/or population dynamics in response to changing climate. These climate-induced changes are well recorded in the past growth of living and fossil specimens or assemblages of organisms such as corals or trees. Tree-ring width and tree-ring density chronologies are used to infer past temperature changes based on calibration with temporally overlapping instrumental data (IPCC, 2007a). Plant stomatal densities and size are also useful proxies of past carbon dioxide (CO_2) levels over recent (Woodward, 1987) and geological time (McElwain and Chaloner, 1995; Royer, 2001).

Palaeoclimatic studies have documented, amongst other things, contrasting periods of time that are known as 'icehouse' and 'hothouse' climatic periods (Caballero and Lynch, Chapter 2). Extreme icehouse conditions in the Neoproterozoic have been termed 'snowball earth' (Hoffman et al., 1998). Two periods of intense and long-lived icehouse glaciation periods can be defined in the Phanaerozoici the late Cenozoic (past 30 million years) and the Permo-Carboniferous (330–260 million years ago – mya), with a shorter glaciation in the late Ordovician (440 mya). The current global climate relative to these periods would be described as icehouse, and this has characterised the Quaternary (starting 1.8 mya), with the Last Glacial Maximum occurring *c.* 20 000 years ago (Lowe and Walker, 1997; Hewitt, 2000). Hothouse or warm periods in the earth's history can also be recognised (Huber et al., 2000; Huber and Sloan, 2001; Huber and Caballero, 2003; Caballero and Lynch, Chapter 2) and have, for example, occurred in the Eocene, Cretaceous and Jurassic (Eldrett et al., 2007; Zeebe et al., 2009). The Cretaceous probably witnessed the hottest temperatures of the Phanaerozoic and represented the earth in an extreme greenhouse mode (140–65 mya), a period with substantial polar forests (Beerling and Woodward, 2001).

Evidence from palaeoclimatic studies therefore points to big swings in climate throughout geological time, including the Phanaerozic when 'complex' animals and plants diversified (McElwain and Punyasena, 2007; McElwain et al., Chapter 5). The 'Cambrian explosion', a period of rapid diversification of life at the beginning of the Palaeozoic, occurred during a period of markedly different climate and atmosphere than today (Royer et al., 2004). The remaining Palaeozoic was also an era of significant climate change (Caballero and Lynch, Chapter 2) during which the major radiation of land plants took place (Willis and McElwain, 2002; Palmer et al., 2004; Fig 1.1). The Cenozoic was the main era of diversification of mammals and angiosperms and has witnessed both hothouse and icehouse conditions. Much of life therefore evolved in climates very different from today or over periods of substantial climate change. We are only just beginning to unravel how such great climatic fluctuations influenced evolution.

1.1.2 Causes of climate change and interactions with life

Climate changes because the global energy budget of the earth varies over time (Caballero and Lynch, Chapter 2). The energy budget is mainly determined by changes in solar irradiance and atmosphere and by changes in the properties of the earth's surface. Climate forcing is an imposed, natural or anthropogenic perturbation of the earth's energy balance with space (IPCC, 2007b). The overall response of global climate to forcing factors is complex, however, due to a number of positive and negative feedbacks that can have a strong influence on the climate system (Meehl et al., 2007; Trenberth et al., 2007). Coupled global circulation models used to model climate change, starting with the early attempts of Phillips (1956), attempt to provide detailed description of the dynamics and physics of the atmosphere and ocean (Caballero and Lynch, Chapter 2).

Solar irradiance, including orbital forcing (Milancovitch cycles), and greenhouse gases, especially CO_2, methane (CH_4) and nitrous oxide (N_2O), are known to be major forcing factors (Archer, 2007; Trenberth et al., 2007). Astronomical calculations demonstrate that periodic changes in characteristics of the earth's orbit around the sun, including its tilt, control the seasonal and latitudinal distribution of incoming solar radiation at the top of the atmosphere (insolation). Orbital factors have been linked closely with glaciation cycles of the Quaternary and other climate shifts in the past (EPICA, 2004; Wunsch, 2004; Caballero and Lynch, Chapter 2). Galactic solar ray flux has also been implicated in climate change (Shaviv and Veizer, 2003; Svensmark and Calder, 2007; Trenberth et al., 2007).

The role of atmospheric greenhouse gases in climate change is well accepted, with evidence from both theoretical and empirical studies. Greenhouse gases act primarily to change the atmospheric absorption of outgoing radiation and consequently influence the temperature. The scientific basis of greenhouse theory was developed in the late nineteenth and early twentieth centuries by various researchers such as Tyndall (1865), who discovered the greenhouse properties of gases and water, and Arrhenius (1908), who was the first to model the effects of changes in the concentration of atmospheric CO_2 on climate and develop the 'hothouse theory'. CO_2 has been linked to major extremes of climate in recent history (IPCC, 2007a) and the deep past (Caldeira and Kasting, 1992; Hoffman et al., 1998; Donnadieu et al., 2004; Pierrehumbert, 2004; DeConto et al., 2008). Evidence for the influence of greenhouse gases on climate comes from ancient air and other matter such as dust trapped in ancient ice. Records cover much of the Quaternary (Fig 1.2). Stable water isotopic fractions ($\delta^{18}O$ and deuterium) in snowfall are temperature-dependent and can be used to record past temperature. It is also possible to collect several other sources of data such as greenhouse gases in these cores. Deuterium excess is considered as a proxy for past ocean surface temperature at the moisture source region. The $\delta^{18}O$ record is a proxy for past air temperatures at the ice-core site. The size of ice crystals is also of value (EPICA, 2004).

Impressive data sets from ice cores have been collected in Antarctica and Greenland. An ice-core record from the Russian Vostok station in central east Antarctica reached a depth of 3623 metres, corresponding to a time of

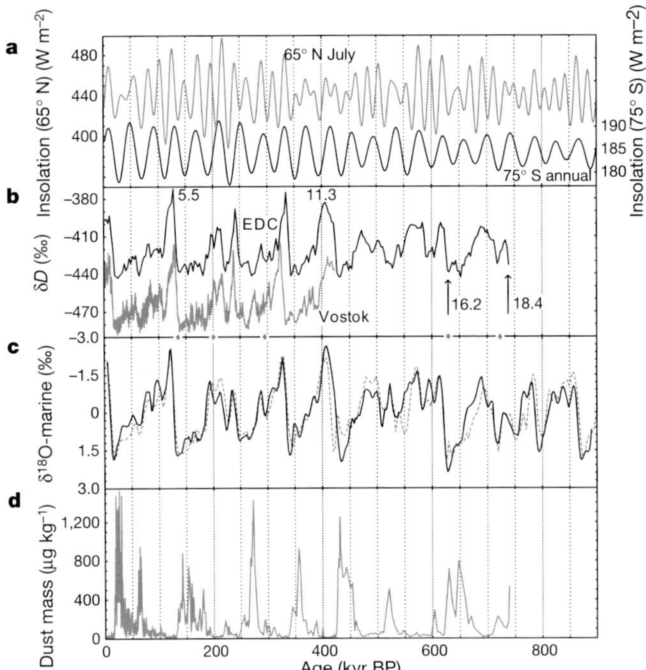

Figure 1.2 Antarctic ice-core data from EPICA Dome C and Vostok with other palaeoclimatic records. (a) Insolation records. Upper curve (left axis), mid-July insolation at 65° N; lower curve (right axis), annual mean insolation at 75° S, the latitude of the ice core at Dome C. (b) Upper curve, δD from EPICA Dome C (EDC; 3000-year averages). Lower curve, Vostok δD; arrows indicate marine isotope stage numbers. (c) Marine oxygen isotope. The solid line is the tuned low latitude stack of site MD900963 and ODP6773. The dashed line is a stack of seven sites for the last 400 kyr but consisting only of ODP site 677 for the earlier period. (d) Dust from EPICA Dome C. Reproduced with permission from EPICA (2004).

approximately 420 000 years before present (Fig 1.2; Petit et al., 1999). More recently, a deuterium record has been extracted from an ice core taken at Dome C in the Antarctic by the European Project for Ice Coring (EPICA, 2004). It gives a 750 000 year record including the last eight glaciations (Fig 1.2). The Vostok cores also show a close association between greenhouse gases and climate. CO_2 and CH_4 measurements closely mirrored temperate proxy data: as temperature rose in the interglacials so too did the greenhouse gases. Ocean sediment cores show similar variation and the same glacial cycles over the Quaternary, indicated by the isotope ratios of buried microfossil shells (such as foraminifera – McManus, 2004).

Evidence for the association of greenhouse gases and Quaternary climate is strong, but a number of publications based on other palaeontological and geochemical proxy data sets (ice cores do not exist at this timescale) indicate that the relationship between climate and atmospheric CO_2 may be less clear for pre-Quaternary time (Pagani et al., 1999, 2005; Pearson and Palmer, 2000; Veizer et al., 2000; McElwain, 2002; Royer et al., 2004). Studies of ocean sediment profiles assessing the $\delta^{18}O$ and $\delta^{13}C$ record of deep-sea benthic taxa (belemnites, brachiopods, conodonts and foraminifera) during the last 80 million years (Zachos et al., 2001, 2008; Cramer et al., 2009) or during the Phaenerozoic (Veizer et al., 2000; Shaviv and Veizer, 2003) have revealed the complexity of factors contributing to climate change including earth boundary conditions/ tectonics, orbital factors, cosmic ray fluxes and greenhouse gases. However, the role of greenhouse gases in palaeoclimates should not be underestimated or dismissed (Royer et al., 2004).

The role of the earth's surface, its geology and biological life (the climate–biogeosphere interaction) is important in regulating climate and atmosphere. Both biophysical (e.g. albedo) and biogeochemical (e.g. weathering and carbon cycle) processes are important (Berner, 1997, 2004; Kump and Pollard, 2008). The carbon cycle and abundance of greenhouse gases are tightly coupled with biological life. CO_2 levels are controlled by geological activity such as supply from volcanoes and metamorphic degassing and removal by chemical weathering of calcium and magnesium silicate rocks (Beerling and Berner, 2005). There are several destabilising positive and stabilising negative feedbacks between life (especially plants) and greenhouse gas levels. Aquatic and terrestrial photosynthetic organisms act as major carbon sinks and regulate excess CO_2 in the atmosphere (Beerling and Woodward, 2001; Beerling and Berner, 2005). CO_2 levels have also greatly influenced biological evolution and diversification, so the coevolutionary feedbacks on CO_2 and climate are complex (Beerling and Berner, 2005; Woodward and Kelly, 2008; Beerling, 2009). The history of oxygen is also complex, tightly coupled not only with biological evolution but also with the geosphere via recycling of the earth's crust (Copley, 2001; Berner, 2004; Lenton, 2004).

Climate change can occur relatively quickly and be stimulated by various tipping elements that may push the earth system past critical states into qualitatively different modes of operation and climate phases (Archer, 2007; Lenton et al., 2008). For example, changes in global ocean currents are believed to be important (Masson-Delmotte et al., 2005; Steffensen et al., 2008). The global ocean circulation responsible for large interhemispheric and interocean exchanges of mass, heat and fresh water is known as the meridional overturning circulation (MOC; also related to the thermohaline circulation). Rapid and large climate changes (2–4 °C from one year to the next) have been linked to abrupt circulation changes in the Atlantic component of the MOC (NGRIP, 2004; Steffensen et al., 2008). Such warming can also result in massive iceberg discharges known as Heinrich events (Heinrich, 1988; Broecker, 1994), and these may themselves trigger global climate change (Broecker, 1994). Another tipping element involves critical changes in regulatory carbon sinks (Beerling and Woodward, 2001; Beerling and Berner, 2005) that may accelerate the rate of climate change (Cox et al., 2000; Cramer et al., 2001). The role of forests might decline over this century as their ability to absorb CO_2 and synthesise biomass saturates is overtaken by the release of CO_2 by respiration in a hotter, drier future climate (Cramer et al., 2001).

1.2 Adaptation, speciation and extinction

> We shall best understand the probable course of natural selection by taking the case of a country undergoing some slight physical change, for instance, of climate. The proportional numbers of its inhabitants will almost immediately undergo a change, and some species will probably become extinct. (Darwin, 1859)

Darwin and his influential contemporary Wallace laid the foundations for our understanding of adaptation, speciation, extinction and biogeography (Darwin and Wallace, 1858; Darwin, 1859; Wallace, 1869, 1876). The diversity of life is influenced by a complex interaction of abiotic (temperature, water, light, nutrients/poison, wind, fire), biotic (competition, herbivory, predation) and historical (plate tectonics) factors. These factors structure species and populations into forms that we recognise in several ways. They can for example be classified ecologically into biomes or niches, or taxonomically into realms or other units of species diversity (Wallace, 1876, 1880; Breckle, 2002; Walther et al., 2002; Engelbrecht et al., 2007; Cox et al., 2010).

1.2.1 Biomes and niches

At a global scale, species are adapted to their biome. Biomes are ecological species formations, usually on an intercontinental scale, defined by distinct life forms, characteristic physiognomy (stature, habitat, etc.) and similar taxonomic diversity (Campbell et al., 2008). Aquatic biomes include coral reefs, oceanic pelagic and benthic zones, estuaries and intertidal zones. They are determined mainly by chemical and physical differences such as salt concentration and temperature. For example, coral reefs are sensitive to temperatures below *c.* 18–20 °C and above 30 °C (Campbell et al., 2008) and are therefore found primarily in the tropics. However, global zonation of aquatic biome types is not as obvious as vegetation type. This is because terrestrial biomes (Fig 1.3) are influenced primarily by climate, and especially water availability and temperature (Breckle, 2002). Vegetation biomes therefore show latitudinal zones, because there are latitudinal patterns to climate (Breckle, 2002). Biome zones have also shifted through geological time with fluctuating climate and global positioning of the continental plates (Fig 1.4; Ziegler et al., 2003).

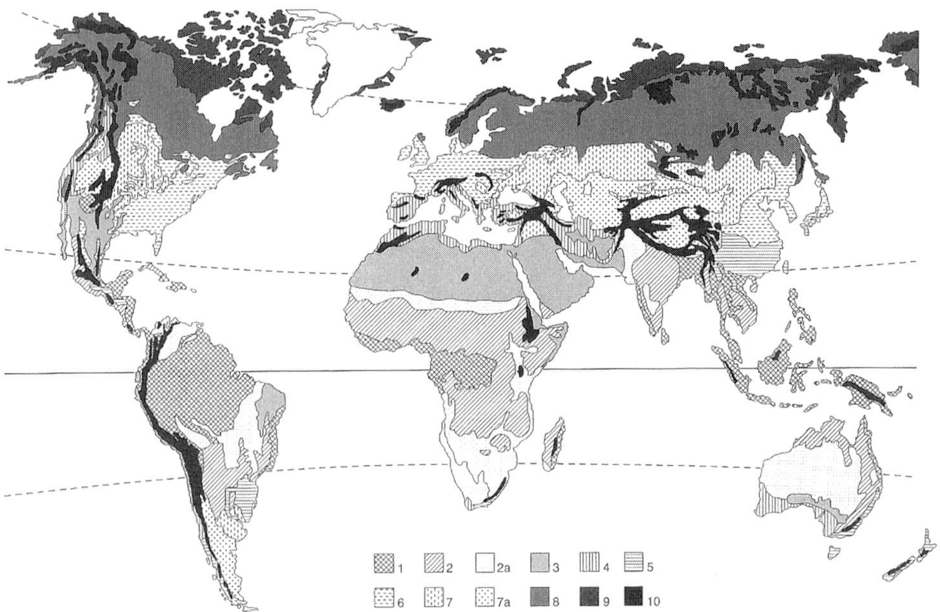

Figure 1.3 Major biomes of the world. Map shows simplified vegetation zones. 1, evergreen rainforests; 2, semi-evergreen and wet season green forests; 2a, savannas, grasslands, dry woodlands; 3, hot deserts and semi-deserts; 4, sclerophyllic woodlands and winter rain; 5, moist warm temperate woodlands; 6, deciduous forests; 7, steppes; 7a, semi-deserts and deserts with cold winters; 8, boreal coniferous zone; 9, tundras, 10, mountains. Reproduced with permission from Breckle (2002).

At a finer geographical scale, organisms are adapted to their ecological niche. The niche is often more informative than the biome for ecological study of individual species, because organisms do not respond to the approximated global averages that are used to define biomes (Walther et al., 2002). Regional changes that are highly spatially heterogeneous are thus of critical importance. Furthermore, biotic as well as abiotic factors determine the ecological niche of an organism (Grinnell, 1917). Because of this, Hutchinson (1957, 1978) made the distinction between the fundamental and realised niche (Rödder et al., Chapter 11). The fundamental niche is defined by the abiotic conditions in which a species is able to persist, whereas the realised niche describes the conditions in which a species persists given the presence of other species.

Some groups of organisms have evolved to occupy highly different ecological niches. In other groups ecological niches have remained more or less conserved. The tendency of species to retain ancestral ecological characteristics is known as 'niche conservatism' (Peterson et al., 1999; Wiens and Graham, 2005). This can be contrasted with 'niche lability', which describes the tendency of species to change their ancestral ecological niche characteristics (Hardy, 2006; Pearman et al.,

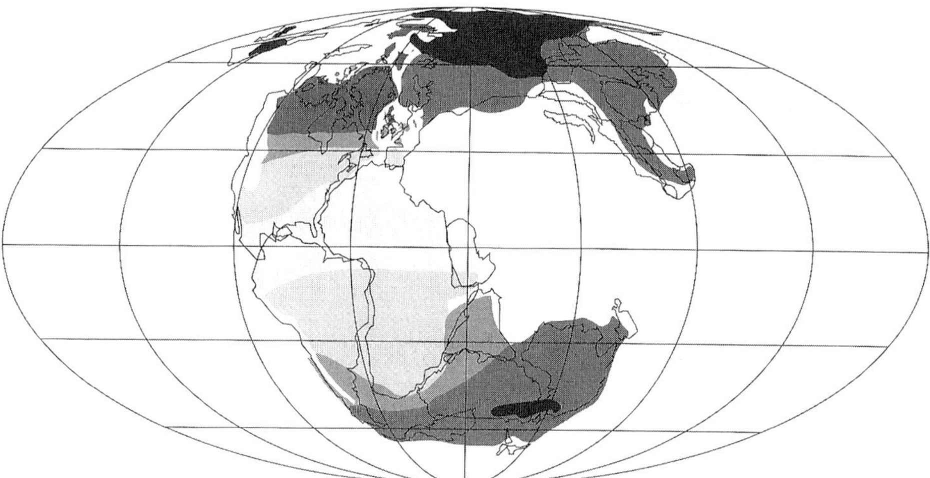

Figure 1.4 Geological distribution of the continents in the early Jurassic, with global vegetation biomes. White, tropical summerwet; light grey, subtropical desert; grey, winterwet; dark grey, warm temperate; black, cool temperate. Reproduced with permission from Willis and McElwain (2002).

2007). Examples of niche conservation in relation to climate include Tethyan laurels (Rodríguez-Sánchez and Arroyo, Chapter 13), *Cyclamen* (Yesson and Culham, Chapter 12), amphibians (Rödder et al., Chapter 11) and grasses (Bouchenak-Khelladi and Hodkinson, Chapter 7).

1.2.2 Realms and species diversity

It has been long known that species diversity and taxonomic groupings are non-randomly distributed across the globe and, as a consequence, taxonomic realms have been described (Wallace, 1876; Schmidt, 1954; Wnuk, 1996). Realms are areas with distinct suites of taxa, each area including a significant proportion of endemic families. Realms are strongly influenced by geographic and tectonic history, climate and dispersal ability of species. The major floristic realms broadly match the major faunal zones (Fig 1.5). The breakup of Pangea during the past 200 million years (Mesozoic and Cenozoic – Fig 1.4) resulted in the isolation of continents and has left an evolutionary/taxonomic footprint. The unique terrestrial faunas and floras of South America and Australia exist because these continents have been islands for much of the last 100 million years and because many species have limited dispersal ability. There are large differences between the tropical floras and faunas of the Old and New World. These have been termed the Palaeotropics and Neotropics, respectively (Fig 1.5). Thus although suitable climate exists for many species in different areas of the world, they may not occupy those areas because

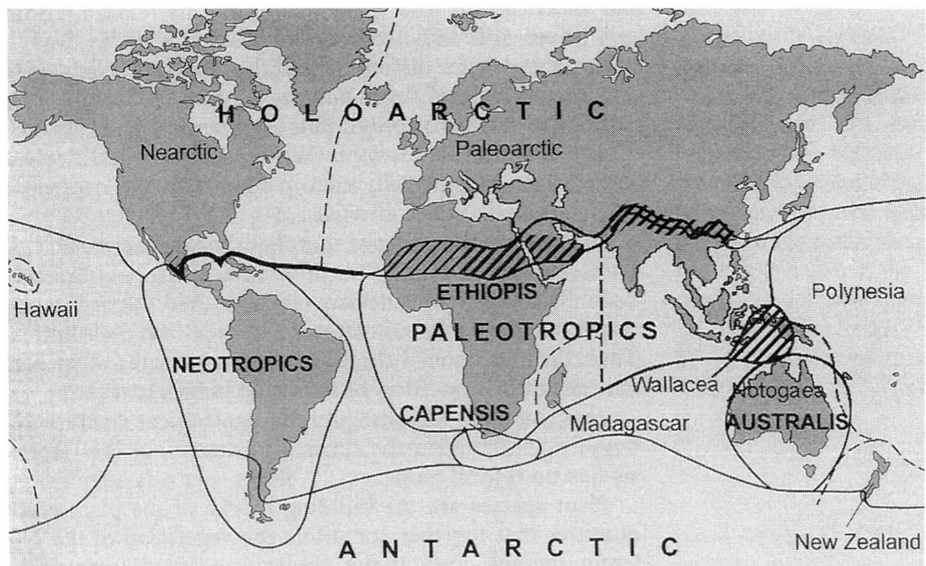

Figure 1.5 Floristic and faunal realms of the world: floristic realms shown in uppercase letters, faunal realms in lowercase. Reproduced with permission from Breckle (2002).

they are restricted by dispersal ability. Where there has been more opportunity for dispersal, larger latitudinal realms prevail. For example, Greenland and North America separated from Eurasia relatively recently, and hence floristic differences between them are smaller than between the tropical continents. Therefore, the Holarctic is often considered as a realm instead of the Nearctic and Palaearctic that are recognised for fauna (Fig 1.5; Schmidt, 1954). Clear patterns can also be seen between hemispheres, such as amongst gymnosperm families. Taxodiaceae and Pinaceae have a largely northern hemisphere distribution, and many Podocarpaceae such as *Araucaria* have a largely southern hemisphere distribution (Breckle, 2002).

The geographical distribution of species diversity is also closely linked to climate and latitude (Jablonski et al., 2006). The tropics have much higher species diversity than other areas (Woodward and Kelly, 2008). There are two major explanations for the latitudinal patterns in species diversity, known as the 'time area hypothesis' and 'diversification rate hypothesis' (Willig et al., 2003; Benton, 2009). In the former, time and area interact to influence diversification. For example, the tropical clades have had longer to speciate and have done so in a larger geographical area because the tropical belt is larger and older than comparatively large belts elsewhere. In the diversification rate hypothesis, it is assumed that there are higher rates of speciation and lower rates of extinction in the tropics than elsewhere. Studies have also shown a positive correlation between primary productivity and species diversity (Woodward and Kelly, 2008). Primary productivity shows a climatic and latitudinal pattern, being higher in the tropics than in temperate regions. Biodiversity hotspots have recently received much attention (Myers et al., 2000; Crawford, 2008), and climate seems to be critical to their existence. In some cases, such as the South African fynbos and the southwestern Australia hotspot, they have remained relatively climatically stable for long periods of time, allowing the accumulation of high biodiversity. Given the evidence for the diversification rate hypothesis, we would also expect rates of speciation to vary with time, because climate has changed over time. The fossil record should, therefore, be able to offer clues to the interaction of diversity and climate (Mayhew, Chapter 4; McElwain et al., Chapter 5).

The fossil record shows several periods of global extinction and origination of higher taxonomic groups at rates and magnitudes greatly exceeding background levels (Alroy et al., 2008; Mayhew et al., 2008). There have been major cycles in fossil diversity (Rohde and Muller, 2005; Mayhew, Chapter 4; McElwain et al., Chapter 5). The 'big five' recorded megaextinctions, based largely on marine records, include the end-Ordovician (446 mya), the Frasnian–Famennian (late Devonian – 371 mya), the Permian–Triassic (251 mya), the Triassic–Jurassic (200 mya) and the Cretaceous–Palaeogene (65 mya) (Raup and Sepkoski, 1982; Sheehan, 2001; Sepkoski, 2002; Benton and Twichett, 2003). Although there is no clear consensus on the precipitating causes of such extinctions, there is evidence they may have been forced by abiotic change including meteorites, volcanism, euxinia and greenhouse gases (Peters and Foote, 2002; Benton, 2003; Benton and Twichett, 2003; Wignall, 2005; Meyer and Kump, 2008; Peters, 2008). Large-scale extinctions of plants, for example, were all characterised by large excursions in stable carbon isotopic composition, indicating major disturbance of the global carbon cycle between atmosphere, biosphere and rock reservoirs (McElwain et al., 1999; McElwain and Punyasena, 2007; van de Schootbrugge et al., 2009; McElwain et al., Chapter 5). See Mayhew (Chapter 4) for a perspective on fauna. Global-scale analyses demonstrate that extinction rates are generally elevated during hothouse phases and biodiversity is depressed. There is also evidence for delayed biological recovery in terms of diversity origination rates after extinctions (Kirchner and Weil, 1990; Cornette et al., 2002; Alroy, 2008).

To understand the cycles of diversity over time it is useful to again consider abiotic (Court Jester model – Barnosky, 2001) and biotic factors (Red Queen model – van Valen, 1973) and a mixture of the two (Mayhew, Chapter 4). They operate over different geographical and temporal scales. For example, biotic factors involved in the Red Queen model, such as competition, herbivory and predation, occur over short time spans and are relatively predictable. Abiotic factors involved in the Court Jester model operate over thousands and millions of years (Benton, 2009), as they involve climate and tectonic events that shape larger-scale patterns regionally and globally, and are less predictable. Mayhew (Chapter 4) argues that the fossil evidence suggests that abiotic factors such as climate are a major influence on biodiversity through time, but relatively predictably so, unlike the paradigm of the Court Jester. To accommodate this he suggests a third paradigm, the Ace of Spades, where there is abiotic but predictable extinction.

1.2.3 Phylogeny and key innovations

Phylogenetic approaches are becoming increasingly useful for the study of biological diversification and its link with climate change (Donoghue, 2005, 2008; Edwards et al., 2007). The tree of life is not balanced, and some groups are much more species-rich than others (Hodkinson and Parnell, 2007). For example, the largest five orders of insects, representing just 6% of all insect orders, contain 83% of their species. One order, Coleoptera, contains a staggering *c.* 350 000 species (which is 35% of all insect species). The same can be seen in the flowering plants, where five families, representing just 1% of all angiosperm families, contain 32% of its species (Hodkinson and Parnell, 2007). Some groups of organisms have therefore been winners and others losers in terms of the evolution of high numbers of species. Many lineages have gone extinct (Hodkinson and Parnell, 2007). It is a challenge to discover the biological attributes of these groups and their key innovations that have enabled them to become so successful. Phylogenetic approaches can highlight shifts in diversification of clades and date when they occurred (Hodkinson et al., 2007; Bouchenak-Khelladi and Hodkinson, Chapter 7). Sister clade comparisons can also be used for assessing the significance of contrasting biological traits on the success of those clades (Purvis, 1996).

Several key innovations, facilitating extensive species diversification, have been linked to climate change. In plants, these innovations need to be interpreted with care (Donoghue, 2005) but notwithstanding they include the initial evolution of their leaves (Beerling et al., 2001), their photosynthetic abilities under climatic stress (Bouchenak-Khelladi and Hodkinson, Chapter 7) and their wood anatomy (Baas and Wheeler, Chapter 6). These innovations have, in turn, accelerated the diversification of animals such as insects (Kenrick and Crane, 1997) and mammals (Bouchenak-Khelladi et al., 2009).

The evolution of leaves is closely linked to climatic and atmospheric change (Beerling, 2005). Leaves have evolved at least twice (microphylls in lycophytes and megaphylls in ferns, gymnosperms and angiosperms). Megaphylls originated from the developmental modification of lateral branches, and this innovation greatly increased the ability of the terrestrial vegetation to fix CO_2 at a global scale (Beerling, 2005). Megaphylls became widespread at the close of the Devonian (360 mya) but their rise seems tightly linked to climate and CO_2. They began to dominate during the late Palaeozoic, which witnessed a 90% drop in CO_2 (Berner, 2004). This corresponded with a rise in stomatal density. Increased stomatal density allowed the evolution of larger leaves by permitting greater evaporative cooling and alleviating the requirement for convective heat loss. The large leaves gradually appeared as CO_2 levels declined and stomatal numbers rose to increase evaporative cooling and ease the thermal burden of observed solar energy (Beerling et al., 2001).

Leaves are held above the ground to capture light, sometimes higher than 100 metres in the tallest trees such as the coastal redwoods of California (Pakenham, 2003), but increased height produces a hydraulic problem, as they need to transport water to their leaves against gravity (Baas and Wheeler, Chapter 6). Dixon and Joly (1895) proposed a solution to this problem: they recognised the chemical properties of water (the cohesion–tension theory), and that transpiration (evaporation at the leaf) allows water to travel up the water-conducting elements (xylem tracheids and vessels) to the leaf. Wood clearly provides the support needed for a tree to reach the required height to compete with other trees for light. There is growing evidence that xylem evolution has been driven by functional adaptations to climate change (Baas and Wheeler, Chapter 6). Climate change has contributed to multiple parallelisms and reversals in vessel, fibre, parenchyma and ray modifications. For example, scalariform perforations add to the flow resistance in vessels, putting selective pressure on their elimination in environments demanding high conductive efficiency such as lowland tropics or drought-stressed environments, but can be of high value in trapping gas bubbles in thawing xylem sap at frost-prone latitudes and altitudes (Baas and Wheeler, Chapter 6).

There is also strong evidence that the evolution of photosynthesis and its variants such as C_3 and C_4 metabolism are linked to climate change. C_4 photosynthesis is found in several taxonomic groups but is particularly common in the grass family Poaceae (Bouchenak-Khelladi et al., 2008, 2009; Bouchenak-Khelladi and Hodkinson, Chapter 7) and the sedge family Cyperaceae (Besnard et al., 2009; Simpson et al., Chapter 19). Temperature and precipitation largely determine the geographical distribution of the predominantly tropical C_4 and predominantly temperate C_3 grasses (Poorter and Navas, 2003; Sage, 2004). There is good evidence from dated phylogenetic trees and fossils to show that C_4 photosynthesis in grasses first evolved in the hot conditions of the Eocene (at *c.* 30 mya) during periods of declining CO_2, and that it has done so independently on approximately 20–30 occasions depending on the analysis performed (Christin et al., 2008; Bouchenak-Khelladi et al., 2009). Although there is an association of C_4 evolution with declining CO_2, and good physiological evidence to show C_4 plants are more efficient under hot dry conditions with limited CO_2 than their C_3 counterparts (the CO_2 starvation hypothesis), we must be careful not to confuse correlation with cause (Roalson, 2008). Grasses did not rise to dominance until the Miocene, and not earlier than 10 million years ago. It is clear that a whole system of factors has contributed to their success, including climatic and atmospheric tolerance, fire resistance, competitive ability and nutrient efficiency. The evolution of C_4 photosynthesis undoubtedly facilitated the expansion of tropical grasslands and savannas (Edwards and Still, 2008; Osborne and Freckleton, 2009; Bouchenak-Khelladi et al., 2010; Bouchenak-Khelladi and Hodkinson, Chapter 7). In these ecosystems, grasses were coevolving with ungulate herbivores, and some evidence points to an evolutionary arms race between the two (Bouchenak-Khelladi et al., 2009; Bouchenak-Khelladi and Hodkinson, Chapter 7).

1.2.4 Adaptation

Species have limits to their distribution and possess populations that occupy marginal habitats. These populations demarcate an endpoint in adaptation to a changing environment (Crawford, 2008). The margins of a species can sometimes be abrupt and easily visible, such as the interface between one vegetation type and another (e.g. *Nothofagus* forest in the Andes). However, they can also be diffuse, such as the interface (ecotones) between southern limits of the boreal forest and northern limits of the deciduous broadleaved forest (Crawford, 2008). When the limits of distribution are reached, populations must adapt or migrate to avoid extinction. Different species will respond in different ways, and in practice there is an interaction between adaptation, migration and plasticity in response to climate change (Bradshaw, 1965; Davis and Shaw, 2001). Evidence of widespread niche conservation would indicate that there might be more potential to migrate than to evolve to cope with short- or long-term climatic change. Forest trees provide a good example of this interaction and are therefore discussed in some detail below.

As trees tracked the shifting climate during the Holocene, evidence suggests that it was easier for species to disperse seed and establish in new environments than to evolve a new range of climatic preferences. Such niche conservation is common in other groups of organisms (Wiens and Graham, 2005). However, this does not mean that that the populations are undifferentiated. Migration is not an alternative to adaptation.

Widespread temperate and boreal trees that shifted latitudes during the Holocene display much genetic variability in their nuclear DNA, but little geographical differentiation (Davis and Shaw, 2001). This indicates a high degree of gene flow among populations. Much of this occurs through pollen, as the species are generally wind-pollinated outbreeders. Chloroplast DNA (cpDNA) markers (maternally inherited and useful for tracking seed dispersal) show greater differentiation. For example, Petit et al. (2002), in a wide-ranging European-scale study of neutral oak cpDNA haplotypes, showed that populations are genetically differentiated in geographical space. The haplotypes also reflect the postglacial history when populations migrated from glacial refugia. One lineage expanded from an Iberian refugium throughout northwest Europe in an oceanic distribution (Kelleher et al., 2004) while others were continental, expanding from other refugia such as in Italy or the Balkans. Some forest trees in the northern hemisphere show a decline in genetic variability from south to north due to several factors including stochastic loss through repeated founder effects (Davis and Shaw, 2001). These postglacial migration and diversity patterns can also be seen with animal populations such as brown bears in Europe (Hewitt, 2000; Lowe et al., 2004).

Even though diversity analyses can indicate the potential for evolution, studies of adaptation or studies on the adaptive genes themselves are required to directly study adaptive variation. Common garden experiments (provenance trials) have long been used in forest research to help growers select populations with particular growth traits. These provenance trials have also offered a powerful tool for studying climate tolerance and response. The studies have demonstrated that modern populations that have shifted ranges in the past are adapted to the climatic conditions where they now grow (Davis and Shaw, 2001; Jump et al., 2009). For example, Scots pine (*Pinus sylvestris* L.) migrated across central Europe from the south as temperatures warmed about 15 000 years ago. Transplant trials throughout northern Sweden showed striking differentiation with respect to survival (Eriksson et al., 1980). At each site, mortality was higher for trees transplanted there from lower latitudes than for trees native to the site.

Selection and gene flow are important processes in range shifts (Excoffier et al., 2009). Natural selection will sieve out genotypes less suited to local conditions. For northward range shifts (in the northern hemisphere), the arrival of seed (e.g. from southern populations) may contribute to adaptation, and selection will also promote new genetic combinations. Studies are teasing apart the population dynamics and genetics of advancing and retracting ranges (Pauli et al., 2006; Sexton et al., 2009). For example, populations at the leading edge of the migrating front may be enhanced by gene transfer from the middle of the range. In contrast, populations at the trailing edge of the range receive less pollen from better adapted populations (Davis and Shaw, 2001; Excoffier et al., 2009). Adaptation at the trailing edge of a species' range depends largely on variation in the local population and may be slower than at the advancing edge (Davis and Shaw, 2001).

Range shifts also offer new conditions for hybridisation (Rieseberg and Carney, 1998; Rieseberg et al., 2003; Soltis and Soltis, 2009). Major ecological transitions can be facilitated by hydridisation. In the tree genus *Fraxinus*, for example, hybrid zones occur in various parts of Europe and facilitate gene flow between otherwise largely climatically separated species. Ecological niche modelling has shown that climate change will cause a change in the potential size and distribution of hybrid-zone populations and their parental species (Thomasset et al., Chapter 15). Similar studies on the influence of climate on the fate of hybridising species have been undertaken on hybrid zones between eastern and western species of North American birds (Swenson, 2006). The hybrid zones are geographically clustered and associated with the sharp temperature gradient of the Great Plains to the Rocky Mountains ecotone. The range borders of the parental species are thus maintained by strong selection against the hybrids; the distinction of the parental species is therefore likely to be maintained.

Recently, phylochronological studies using ancient DNA have been used to study the genetic response of species to climate change (Hadly et al., 2004). Advances in sequencing, especially emulsion-based clonal amplification, assist in producing sequence information from small fragments of DNA that it is possible to retrieve from fossil material of differing ages (Ramakrishnan and Hadly, 2009). This approach allows us to link population genetics and evolution, changes in phenotypic traits, the genes that govern them and changes in the environment. This is made possible by serially taking DNA genotype data from fossils. Serially sampled genetic data permit the chronologic testing of the reconstructed population. Using this phylochronological approach, Hadly et al. (2004) investigated the effect of changing population size of the montane vole and northern pocket gopher during times of climatic change (including the little ice age and the medieval warm period) in mountain habitats of western North America. Their studies could track gene diversity during that time and showed that the two species responded differently to climate change; their response was linked to life-history strategy and gene flow.

Organisms may deal with a variable environment by evolving the capacity for phenotypic plasticity. Plasticity is the ability of an organism to change its phenotype in response to environmental change. It is, in itself, an adaptive trait (Bradshaw, 1965). Phenotypic plasticity can be seen in the phenology of plants and animals. Phenology refers to the timing of recurring life-cycle events (Donnelly et al., Chapter 8). Climate change has caused several phenological changes in a wide range of organisms (Hall et al., 2007). Warmer spring temperatures in recent years have influenced bud burst in trees (Fitter and Fitter, 2002; Cleland et al., 2007), appearance and abundance of insects (Roy and Sparks, 2000) and migration of birds (Cotton, 2003; Menzel et al., 2006; Huntley et al., Chapter 16). Phenotrends have also been observed in marine pelagic environments (Edwards and Richardson, 2004). Some of these changes can be attributed to phenotypic plasticity. For example, bud burst in trees has changed in response to climate change within the same individual trees (Donnelly et al., Chapter 8). Furthermore, long-term genetic studies of American red squirrels in the Arctic indicated that a high proportion (62%) of the change in breeding dates occurring over a 10-year period was the result of phenotypic plasticity and 13% due to population genetic change (Berteaux et al., 2004). However, sometimes changes in phenology require adaptation. Donnelly et al. (Chapter 8) discuss the relationship between adaptation and plasticity for phenological traits in plants, insects and birds.

Studies of genes involved in adaptation are required for the study of adaptation at a molecular level. Some attempts have been made with neutral markers (Jump and Peñuelas, 2005; Jump et al., 2006, 2008) or candidate genes of adaptive traits (Neale and Ingvarsson, 2008). Genes controlling complex adaptive traits are usually discovered via approaches such as quantitative trait loci or association mapping (Neale and Savolainen, 2004). Once discovered, it is possible, via resequencing of different populations, to assess variation in these target genes and test the presence of selection using modern population genetic approaches (mutation rates and departures from neutrality to indicate they might be under some form of natural selection). For example, roughly 20% of the *c.* 300 forest tree genes that have been tested showed departure from neutrality (but demographic processes may also account for such a departure – Neale and Ingvarsson, 2008). Genes for biotic (e.g. disease resistance) and abiotic (water-use efficiency, cold tolerance, bud set) factors have been investigated (Neale and Ingvarsson, 2008; Donnelly et al., Chapter 8).

One of the earliest studies on genotypic and phenotypic variation in forest trees using the candidate gene approach was by Ingvarsson et al. (2006). They studied a phytochrome gene (*phyB2*) involved in dormancy induced by shortening of the photoperiod (short day induced bud set) in *Populus*. A resequencing study in European aspen (*P. tremula* L.) found several DNA sequence variants (single nucleotide polymorphisms, SNPs) that showed significant clinal variation with latitude (Donnelly et al., Chapter 8). Hall et al. (2007) also report adaptive population differentiation across a latitudinal gradient in aspen. Other species and candidate genes have been investigated in the study of genes of ecological and adaptive significance, including bud burst genes in oak (Derory et al., 2010), drought stress genes in pine (Eveno et al., 2008) and stem cold hardiness and wood formation genes in Douglas fir (Krutovsky and Neale, 2005). Studies on other organisms such as corals and algae are also finding evidence for genetic adaptation of populations to climate change (Bradshaw and Holzapfel, 2006; Parmesan, 2006). Rindi (Chapter 9) discusses adaptation in terrestrial algae in response to climate change and outlines evidence of genetic adaptation at the molecular level. These studies are therefore confirming the conclusions of common garden experiments that suggest strong adaptive differentiation within species. They are also identifying the specific genes involved in those traits, and characterising their variation.

1.3 Biogeography, migration and ecological niche modelling

> It is notorious that each species is adapted to the climate of its own home: species from an arctic or even from a temperate region cannot endure a tropical climate, or conversely. (Darwin, 1859)

It is essential to know how the geographic distribution of species has been affected by climate change and how it will be influenced in the future. Range shifts often entail niche shifts (Parmesan and Yohe, 2003; Wiens and Graham, 2005). Palaeoecological studies on the glacial history of habitat types, especially in Europe and North America (Lowe and Walker, 1997), have produced thousands of pollen and macrofossil diagrams compiled in databases. These have provided records of species' abundances as they changed through space and time (at regional and continental scales) and at high taxonomic detail. These records of species distribution therefore provide a palaeoecological way of tracking how species have migrated in response to climate change and how their population sizes have varied. This evidence has been supported by recent DNA studies of extant and ancient samples (Hewitt, 2000; Petit et al., 2002).

Range shifts associated with recent climate change are clearly seen with altitude. Empirical studies have shown how changing temperatures can drive upward or downward range shifts of species in mountains. Lowland birds have started breeding in montane cloud forest in Costa Rica (Pounds et al., 1999). The range of Edith's checkerspot butterfly has shifted upward by 105 m in Mexico and North America (Parmesan, 2006). Similarly, the lower altitudinal limits of 16 Spanish butterfly species have increased by an average of 212 m in 30 years (concurrent with a 1.3 °C rise in temperature – Wilson et al., 2005). Several shifts have been recorded for the elevation of mountain tree lines (Parmesan and Yohe, 2003; Crawford, 2008; Jump et al., 2009).

Species in mountain regions are typically restricted to relatively narrow altitudinal bands, and change in their distribution is easy to observe (Crawford, 2008). In contrast, latitudinal range shifts are less easy to document because of the geographical scale involved (Jump et al., 2009). For recent global change, there is less evidence for range shifts with latitude than with altitude. There is an approximate 1 °C change in temperature with a 167 m change in altitude (5–6.5 °C per 1000 m) or a 145 km change in latitude (6.9 °C per 1000 km at 45° N). Thus stratification of ecosystems can be much more readily seen in mountain regions than across latitudes (Jump et al., 2009). Recent changes in plant distribution in mountains have been reported for a range of species (tropical epiphytes, alpine plants and forest species), and mountain tree lines have shifted upwards (forwards) by up to 130 m over the past 50 years (Jump et al., 2009). The shift seems to be biased to the leading edge, but some studies also show similar-magnitude retractions of the trailing edge (due to elevated mortality and reproductive decline – Jump et al., 2009).

Given the strong evidence of range shifts in mountains, we can predict what should occur over latitude (Jump et al., 2009). We would expect latitudinal shifts of tens to hundreds of kilometres (based on the altitude-to-latitude model). Reports of latitudinal range shifts are rare, because of lack of research and methodological issues, but some reports have been made in several groups of organisms. For example, in plants, there has been an expansion of forest over the Alaskan tundra over the last 50 years (and increase in the latitudinal position and density of trees and shrubs such as white spruce and green alder). For butterflies, Parmesan et al. (1999) found that nearly two-thirds of 35 non-migratory European species had shifted their ranges north by 35–240 km and only two species had shifted south. In the most extreme cases, such as the sooty copper, the southern edge contracted concurrent with the northern-edge expansion. For birds, a mean northward shift of *c.* 20 km over a 20-year period has been recorded in UK populations (Thomas and Lennon, 1999). For lichens, 77 new epiphytic lichens colonised the Netherlands from the south between 1979 and 2001 (van Herk et al., 2002; Ellis and Yahr, Chapter 20). Finally, for mammals, range expansion has been recorded for the red fox, with a simultaneous range retreat of the arctic fox (Hersteinsson and MacDonald, 1992).

At lower latitudes, shifts are associated with movements of tropical species into more temperate areas (Parmesan, 2006). North African species are moving into Spain and France, and Mediterranean species are moving up into the continental interior. For example, large populations of the African plain tiger butterfly have established in Spain (Haeger, 1999). In the Americas, the rufous hummingbird has shown a large shift in its winter range, expanding from Mexico into southern USA (Parmesan, 2006). Over this time winter temperatures rose by approximately 1 °C.

1.3.1 Ecological niche and distribution modelling

It is possible to use modelling approaches to understand the distribution of species and their ecological niches in relation to climate change, and to make inferences about past distributions (hindcasting) and future distributions (forecasting). These approaches use a variety of statistical methods to model species' geographic distributions and their ecological niches in relation to climatic and topographic variables. They are variously known as climate envelope models, climatic niche models, species distribution models, ecological niche models, or statistical niche models (Peterson, 2006; Wake et al., 2009; Rödder et al., Chapter 11). Niche modelling applies powerful computational tools to species' locality data assembled through fieldwork and specimens in herbaria and museum collections. The modelling approach involves: (1) georeferenced localities for the study species (collection localities and their latitude/longitude coordinates); (2) climatic variable data (e.g. temperature and precipitation variables) at the collection sites and surrounding areas; and (3) algorithms for estimating the climatic niche envelope of those species, based on the distribution of climatic variables where they occur and do not occur within a region (Wiens and Graham, 2005; Rödder et al., Chapter 11).

A range of modelling resources are available (Soberón and Peterson, 2004; Culham and Yesson, Chapter 10; Rödder et al., Chapter 11). Georeferenced localities are available on the internet. The Global Biodiversity Information Facility provides such records for free (www.gbif.org). Fine-scale climatic data sets such as WorldClim are also freely available, based on information from a large number of weather stations and statistical and modelling extrapolations to locations without weather stations (Beaumont et al., 2005; Hijmans and Graham, 2006). There is also a diverse range of methods to construct ecological niche models and model species distributions, including environmental envelopes, ordination approaches, generalised regression models, genetic algorithms and Bayesian methods (Guisan and Thuiller, 2005; Araújo and Guisan, 2006; Elith et al., 2006; Heikkinen et al., 2006; Peterson, 2006; Phillips et al., 2006; Elith and Leathwick, 2009). In general, the statistical model establishes a relationship between point locality data and environmental layers that describe the variation in a climatic variable over space (Peterson, 2006).

The model is then used to create a map of predicted species distribution, given these environmental variables (Wiens and Graham, 2005). When this map of predicted range is overlaid on the actual species distribution map it is possible to see how well the climatic variables predict the species' range limits. Matching distributions support the hypothesis that the specialised climatic tolerances of a species may limit its geographical spread. If a range is overpredicted, the climate variables indicate that the species should have a wider distribution than its distribution data suggest. Overprediction may indicate that climate is not the primary factor that limits the species' geographical distribution (Wiens and Graham, 2005).

It is also possible with different approaches to determine which climatic variables are the most important for limiting geographical distribution of the species (Wiens and Graham, 2005). These data can be used in various ways, such as in conservation biology. They are commonly used for forecasting and hindcasting (Pearman et al., 2008). Several studies have incorporated distribution models and geographical information systems (GIS) in phylogenetic studies, an approach known as phyloclimatic modelling (Kozak et al., 2008; Culham and Yesson, Chapter 10).

1.3.2 Phyloclimatic modelling

Evolutionary processes, such as speciation and genetic divergence of populations, are heavily influenced by environmental variables. Therefore, phyloclimatic modelling approaches can help establish the ecological causes of evolutionary processes. A growing number of evolutionary and systematic studies are making use of climate envelope models and the extensive data sets available in GIS (Kozak et al., 2008; Rödder et al., Chapter 11). This is an improvement over previous studies that would have used only a limited set of environmental variables because of lack of data availability. For example, phylogeographic studies often only considered geographic distance to explain their results and did not analyse complex environmental data. GIS-based data tools are now making it relatively simple to include such information in evolutionary research (Kozak et al., 2008; Culham and Yesson, Chapter 10). Applications are numerous and include speciation (Graham et al.,

2004b; Rödder et al., Chapter 11), phylogeography (Yesson and Culham, Chapter 12) and character evolution (Kozak et al., 2008).

Rödder et al. (Chapter 11) use climatic envelope models to delimit the potential palaeodistribution (sympatry and allopatry) of taxa belong to the Afrotropical reed frog genus *Hyperolius*, and hence use these distribution data to help interpret speciation and species delimitation in the group. In another approach of phyloclimatic modelling, the limiting climatic variables obtained from ecological niche modelling can be assessed phylogenetically (e.g. mapped onto a phylogenetic tree or compared to genetic distances) to see how they change during the evolutionary history of the group (Graham et al., 2004a, 2004b; Davis, 2005; Yesson and Culham, 2006a, 2006b; Edwards et al., 2007; Kozak et al., 2008; Jakob et al., 2009). This will determine if they are conserved over the evolutionary history of the group or if they are labile. Examples include Yesson and Culham (Chapter 12) for *Cyclamen* evolution and Chatrou et al. (Chapter 14) for the study of the impact of climate change on the origin and future of East African rainforest trees. Chatrou et al. (Chapter 14) use a phyloclimatic approach to identify the environmental variables associated with the origin of East African endemics of the genus *Monodora*. Yesson and Culham (Chapter 12) review phyloclimatic work undertaken on the Mediterranean *Cyclamen* protected by CITES legislation and provide new results on the palaeogeography and palaeoclimate of the genus.

Many of the phyloclimatic approaches use GIS data sets and tools (e.g. DIVA GIS) to obtain environmental data for each locality and then analyse these statistically. They can test for correlations between environmental variables and spatial patterns of phenotypic divergence. Ruegg et al. (2006) used GIS to study the relationship between genetic differentiation and climatic differentiation on song evolution in Swainson's thrush. They quantified temperature and precipitation across the species' range and found that the acoustic divergence of the populations was not correlated with genetic distance but was correlated with geographical distances in climate. Climate was shown to be an indirect cause because song was closely related to the acoustic differences in climatically determined forest type (rainforest versus coniferous). It is clear that there is great potential to use GIS approaches to help establish an evolutionary framework to estimate how species respond to climate change.

1.4 Conservation

> There can be no purpose more enspiriting than to begin the age of restoration, reweaving the wondrous diversity of life that still surrounds us. (Wilson, 1992)

1.4.1 Impacts

Despite uncertainty in climate predictions there is compelling evidence for rapid recent human-driven climate change (Ruddiman, 2003; Juckes et al., 2007; Meehl et al., 2007). Even if the cause is not anthropogenic, there is serious reason to be concerned about the impacts of climate change, because of escalating global population growth and habitat loss and fragmentation. Several studies have reported the impacts of climate change on individual species and habitats. Declining populations have already been recorded in several groups of organisms including amphibians (McMenamin et al., 2008), birds (Both et al., 2006; Møller et al., 2008), mammals (Moritz et al., 2008) and several other groups (Parmesan and Yohe, 2003; Parmesan, 2006). Climate change can also influence species' interactions by, for example, introducing potentially invasive species (Vila et al., 2000; Broennimann et al., 2007; Broennimann and Guisan, 2008; Rödder and Lötters, 2009; Thomasset et al., Chapter 15; Simpson et al., Chapter 19), and can influence interactions with pests and diseases (Parmesan, 2006). However, meta-analyses have been necessary to show broad-scale patterns of the impacts of recent climate change on populations and ecosystems. These studies have shown a 'coherent fingerprint' of climate on life (Parmesan and Yohe, 2003). In their meta-analyses of 1700 species, Parmesan and Yohe (2003) defined a diagnostic biological fingerprint of temporal and spatial 'sign switching' responses uniquely predicted by climate trends of the last century. Meta-analyses therefore confirmed predictions based on individual species or habitats.

1.4.2 Prediction

Predicting the response of organisms to future climates is a serious challenge to conservationists and ecologists (Pearson and Dawson, 2003; Parmesan, 2006; Sutherland, 2006; Huntley, 2007a). There is a requirement to forecast many things including range shifts, extinction risks, population sizes, biome shifts, disturbance regimes and biogeochemical weathering (Williams et al., 2007). There is also a pressing need to be able to do this at local as well as international scales (Walther et al., 2002). Progress with climate-change predictions at regional and local levels is discussed by Meehl et al. (2007) and Caballero and Lynch (Chapter 2).

The risk of population decline and extinction is not even across taxonomic groups or across geographical space. Species' responses to climate change are likely to depend on interactions between population processes, between species, and between demographic and landscape dynamics (Keith et al., 2008). Dispersal ability and habitat preference of species are crucial. Studies predict that risks of extinctions are particularly high in mountainous regions, high-latitude areas and in areas with high endemism (IUCN, 2001; Hannah et al., 2008). Species become increasingly threatened with extinction as suitable habitat becomes reduced and new areas remain unreachable due to natural and anthropogenic barriers to dispersal (Hannah et al., 2007). Conservation strategies need to consider the results of such studies (Hannah et al., 2008). Hodd and Sheehy Skeffington (Chapter 21) discuss the threat of climate change to mountain plant communities. They provide a case study of mountain bryophyte species that occur in scarce communities of hyperoceanic montane heath in Ireland. These communities are clearly vulnerable to climate change and require careful monitoring and management. A further case study is provided by Ellis and Yahr (Chapter 20), who discuss lichens in arctic–alpine ecosystems.

The first extinctions of species attributed to recent global warming were of mountain- or ocean-restricted species. Many cloud-forest-dependent amphibians have declined or gone extinct on a mountain in Costa Rica (Pounds et al., 2006) due to the indirect impact of climate change. An estimated 67% of harlequin frogs in Central and South American tropics have disappeared over the last 20–30 years due to climate shifts, especially in precipitation, favouring the pathogenic chytrid fungus *Batrachochytrium dendrobatidis* Longcore, Pessier and D. K. Nichols, 1999 (Pounds et al., 2006). Sharp population declines of frogs have also been linked to this epidemic disease in western USA (Walther et al., 2002). Similar temperature factors are significant for extinctions in range-restricted coral reefs. Several periods of mass coral bleaching have occurred since 1979, and these events are increasing in frequency and intensity. Up to 16% of the world's reef-building corals died during the most severe bleaching and extreme temperate period in the El Niño event of 1997–98 (Parmesan, 2006). The problem was exasperated by poor dispersal of symbionts between reefs.

Making predictions about extinction probability is shrouded in uncertainty, but some approaches have been useful. Thomas et al. (2004, 2006) used an approach combining climate envelope modelling and statistics that estimated the probability of extinction. Using projections of species' distributions for future climate scenarios, they assessed extinction risks for sample regions covering some 20% of the earth's terrestrial surface. Methods to estimate extinction were based on the species–area relationship, which is a well-established empirical power-law relationship describing how the number of species relates to area (MacArthur and Wilson, 1967; Rosenzweig, 1995). This relationship predicts adequately the numbers of species that become extinct or threatened when the area available to them is reduced. They predicted, on the basis of mid-range climate warming scenarios for 2050, that 15–37% of species in their sample of regions and taxa will be 'committed to extinction'.

Conservationists have established networks of protected areas and other designated sites. These are static, fixed geographical locations, and there is concern amongst both governmental and non-governmental conservation bodies that such sites may, in the future, fail to provide suitable conditions for the species and ecosystems for whose protection they were established (Araújo et al., 2004; Hannah, 2008). Huntley et al. (Chapter 16) used *c.* 1700 bird species breeding in sub-Saharan Africa and a network of 803 important bird areas (IBAs) identified and designated in that region as a model to explore this issue. They used climate envelope models fitted to current climate to predict the future occurrences of the species in the IBAs. They showed that the networks have the potential to maintain most species under future climate-change scenarios, but the outcome depends on substantial species turnover. This and other studies have shown that the effectiveness of the networks will depend on their connectivity and sympathetic management of the wider landscape (Hannah et al.. 2008; Hole et al., 2009, in press).

Range shifts are limited by habitat fragmentation (Hijmans and Graham, 2006). In Madagascar, for example, there is 90% endemism amongst plants, mammals, reptiles and amphibians. The impacts of current and future climate change are likely to be high because suitable habitat for most of these is much reduced and fragmented due to deforestation that has claimed approximately 90% of the island's natural forest (Hannah et al., 2008), providing a poor environment for large-scale range shifts of species. Changes in atmospheric gases also need to be considered. Future life will potentially experience novel atmospheres and novel environments (Williams and Jackson, 2007, Williams et al., 2007). We need to make predictions about the impact of future climate change on communities, but no analogue communities may exist. It will be a challenge for ecologists to study ecosystems they have never seen (Williams et al., 2007). Models need to incorporate net primary productivity (NPP) that is predicted to increase with increased CO_2 (Korner, 2006). Woodward and Kelly (2008) show that rising NPP should potentially increase species diversity but suggest that the effect will not be enough to offset the damage caused by habitat loss.

Models can also be limited because they are often based on the assumption of niche conservation (Huntley, 2007b). It seems that some species may have undergone rapid niche shifts while others experienced long periods of niche stability. This casts doubt on the prediction of climate-change impacts by species distribution models that do not consider adaptation. There is a need to know how the potential for niche change varies amongst species, and to take this into account in the modelling, or it may be necessary to choose species more carefully for modelling (Pearman et al., 2008). There is also a need for models to make predictions about population size (Green et al., 2008). It is clear that ecological niche models will need to become more complex (Thuiller, 2004; Pearson et al., 2006; Thuiller et al., 2008; Huntley et al., 2010; Rödder et al., Chapter 11).

1.4.3 Conservation and primary taxonomic data

The ability to study the response of life to climate change, to make predictive conclusions about its fate, and to provide strategies for its conservation, is entirely dependent on the quality of primary data. For many studies the primary data are taxonomic. Most studies are dependent on accurate species identification, some rely on accurate taxon inventories (species lists), and some rely on location data of predefined taxa (species occurrence data). The taxonomic and systematic community therefore has the responsibility to provide accurate and informative data to all biologists. However, progress in the world's oldest science, taxonomy, is unacceptably slow given the biodiversity crisis (Hodkinson and Parnell, 2007). Only *c.* 1.7 million species, from the aforementioned conservative estimate of four million species, have been described, and at current rates the process could take several more centuries to near completion (Hodkinson and Parnell, 2007). There is thus a serious taxonomic impediment to progress in ecological and climatic change research (Hall and Miller, Chapter 17). Taxonomy is an enabling science, and there is an urgent need to better characterise species when they are first described. For example, ecological information is generally lacking when species are described. To make taxonomy predictive, Bateman (Chapter 3) argues that much stricter requirements should be placed on species descriptions. The taxonomic community is probably far from meeting that logical aspiration because of ideological, historical and practical constraints. Furthermore, systematic studies are required to better define species. It is clear that we have underestimated species diversity because many species are cryptic and have thus far gone unnoticed. Bernardo (Chapter 18) provides a thorough review of species concepts and cryptic diversity and gives examples of where ecological studies such as ecological niche modelling can fail to reach accurate conclusions because of such shortcomings. There is also a pressing need to speed up taxonomic discovery and provide more useful taxonomic resources to end users such as ecologists (Hall and Miller, Chapter 17). Several innovations offer huge potential to reduce the taxonomic impediment, including digital literature, DNA barcoding and improved field guides (Bateman, Chapter 3; Hall and Miller, Chapter 17).

Ecologists and systematists need to communicate the implications of their science for improved conservation planning, policy, research and management. Conservation biology is a complex multidisciplinary field, and it will be important to develop robust principles and detailed recommendations (Barnard and Thuiller, 2008; Carnaval and Moritz, 2008). Significant care will be necessary to test the predictive science using theoretical and empirical studies.

1.5 Conclusions: crisis upon crisis

It is alarming to note that the current climate crisis has coincided with a biodiversity crisis, which in turn has been met by a taxonomic crisis. There is, therefore, an urgent need to deal with all these issues, and ecological and systematic research are central to these efforts. Ecological and systematic research interact powerfully to explain the impact of climate and other aspects of environmental change on ecological niches and species diversity through space and time. They help us understand the past and allow us to make meaningful predictions about the future. It is the aim of this book to demonstrate how these research fields are at their most useful for climatic change research when fully integrated. Results from such research are vital for policy and conservation work at a time when life on earth faces an unprecedented threat from many factors, including climate change, which must not be ignored.

Acknowledgements

I would like to thank Alison Donnelly, Michael Jones, John Parnell and Stephen Waldren for reviewing the contents of this chapter.

References

Alroy, J. (2008). Dynamics of origination and extinction in the marine fossil record. *Proceedings of the National Academy of Sciences of the USA*, **105**, 11536–11542.

Alroy, J., Aberhan, M., Bottjer, D. J. et al. (2008). Phanerozoic trends in the global diversity of marine invertebrates. *Science*, **321**, 97–100.

Araújo, M. B. and Guisan, A. (2006). Five (or so) challenges for species distribution modelling. *Journal of Biogeography*, **33**, 1677–1688.

Araújo, M. B., Cabeza, M., Thuiller, W., Hannah, L. and Williams, P. H. (2004). Would climate change drive species out of reserves? An assessment of existing reserve-selection methods. *Global Change Biology*, **10**, 1618–1626.

Archer, D. (2007). Methane hydrate stability and anthropogenic climate change. *Biogeosciences*, **4**, 521–544.

Arrhenius, S. (1908). *Worlds in the Making: the Evolution of the Universe*. New York, NY: Harper.

Barnard, P. and Thuiller, W. (2008). Global change and biodiversity: future challenges. *Biology Letters*, **4**, 553–555.

Barnosky, A. D. (2001). Distinguishing the effects of the red queen and court jester on Miocene mammal evolution in the northern rocky mountains. *Journal of Vertebrate Paleontology*, **21**, 172–185.

Beaumont, L. J., Hughes, L. and Poulsen, M. (2005). Predicting species' distributions: use of climatic parameters in BIOCLIM and its impact on predictions of species' current and future distributions. *Ecological Modelling*, **186**, 250–269.

Beerling, D. J. (2005). Leaf evolution: gases, genes and geochemistry. *Annals of Botany*, **96**, 345–352.

Beerling, D. J. (2009). *The Emerald Planet: How Plants Changed Earth's History*. Oxford: Oxford University Press.

Beerling, D. J. and Berner, R. A. (2005). Feedbacks and the coevolution of plants and atmospheric CO_2.

Proceedings of the National Academy of Sciences of the USA, **102**, 1302–1305.

Beerling, D. J. and Woodward, F. I. (2001). *Vegetation and the Terrestrial Carbon Cycle: Modelling the First 400 Million Years*. Cambridge: Cambridge University Press.

Beerling, D. J., Osborne, C. P. and Chaloner, W. G. (2001). Evolution of leaf-form in land plants linked to atmospheric CO_2 decline in the Late Palaeozoic era. *Nature*, **410**, 352–354.

Benton, M. J. (2003). *When Life Nearly Died: the Greatest Mass Extinction of All Time*. London: Thames and Hudson.

Benton, M. J. (2009). The Red Queen and the Court Jester: species diversity and the role of biotic and abiotic factors through time. *Science*, **323**, 728–732.

Benton, M. J. and Twitchett, R. J. (2003). How to kill (almost) all life: the end-Permian extinction event. *Trends in Ecology and Evolution*, **18**, 358–365.

Berner, R. A. (1997). The rise of plants and their effect on weathering and atmospheric CO_2. *Science*, **276**, 544–546.

Berner, R. A. (2004). *The Phanerozoic Carbon Cycle: CO_2 and O_2*. Oxford: Oxford University Press.

Berteaux, D., Reale, D., McAdam, A. G. and Boutin, S. (2004). Keeping pace with fast climate change: can Arctic life count on evolution? *Integrative and Comparative Biology*, **44**, 140–151.

Besnard, G., Muasya, A. M., Russier, F. et al. (2009). Phylogenomics of C_4 photosynthesis in sedges (Cyperaceae): multiple appearances and genetic convergence. *Molecular Biology and Evolution*, **26**, 1909–1919.

Both, C., Bouwhuis, S., Lessells, C. M. and Visser, M. E. (2006). Climate change and population declines in a long-distance migratory bird. *Nature*, **441**, 81–83.

Bouchenak-Khelladi, Y., Salamin, N., Savolainen, V. et al. (2008). Large multi-gene phylogenetic trees of the grasses (Poaceae): progress towards complete tribal and generic level sampling. *Molecular Phylogenetics and Evolution*, **47**, 488–505.

Bouchenak-Khelladi, Y., Verboom, G. A., Hodkinson, T. R. et al. (2009). The origins and diversification of C4 grasses and savanna-adapted ungulates. *Global Change Biology*, **15**, 2397–2417.

Bouchenak-Khelladi, Verboom, G. A., Savolainen, V. and Hodkinson, T. R. (2010). Biogeography of the grasses (Poaceae): a phylogenetic approach to reveal evolutionary history in geographical space and geological time. *Botanical Journal of the Linnean Society*, **162**, 543–557.

Bradshaw, A. D. (1965). Evolutionary significance of phenotypic plasticity in plants. *Advances in Genetics*, **13**, 115–155.

Bradshaw, W. E. and Holzapfel, C. M. (2006). Evolutionary response to rapid climate change. *Science*, **312**, 1477–1478.

Breckle, S. W. (2002). *Walter's Vegetation of the Earth: the Ecological Systems of the GeoBiosphere*, 4th edn. Berlin: Springer-Verlag.

Broecker, W. S. (1994). Massive iceberg discharges as triggers for global climate change. *Nature*, **372**, 421–424.

Broennimann, O. and Guisan, A. (2008). Predicting current and future biological invasions: both native and invaded ranges matter. *Biological Letters*, **4**, 585–589.

Broennimann, O., Treier, U. A., Müller-Schärer, H. et al. (2007). Evidence of climatic niche shift during biological invasion. *Ecology Letters*, **10**, 701–709.

Caldeira, K. and Kasting, J. F. (1992). Susceptibility of the early Earth to irreversible glaciation caused by carbon dioxide clouds. *Nature*, **359**, 226–228.

Campbell, N. A., Reece, J. B., Urry, L. A. et al. (2008). *Biology*, 8th edn. London: Pearson Benjamin Cummings.

Carnaval, A. C. and Moritz, C. (2008). Historical climate modelling predicts patterns of current biodiversity in the Brazilian Atlantic forest. *Journal of Biogeography*, **25**, 1187–1201.

Christin, P. A., Besnard, G., Samaritani, E. et al. (2008). Oligocene CO_2 decline promoted C4 photosynthesis in grasses. *Current Biology*, **18**, 37–43.

Cleland, E. E., Chuine, I., Menzel, A., Mooney, H. A. and Schwartz, M. D. (2007). Shifting plant phenology in response to global change. *Trends in Ecology and Evolution*, **22**, 357–365.

Copley, J. (2001). The story of O. *Nature*, **410**, 862–864.

Cornette, J. L., Lieberman, B. S. and Goldstein, R. H. (2002). Documenting a significant relationship between macroevolutionary origination rates and Phanerozoic pCO_2 levels. *Proceedings of the National Academy of Sciences of the USA*, **99**, 7832–7835.

Cotton, P. A. (2003). Avian migration phenology and global climate change. *Proceedings of the National Academy of Sciences of the USA*, **100**, 12219–12222.

Cox, M., Cox, C. B. and Moore, P. D. (2010). *Biogeography: an Ecological and Evolutionary Approach*, 8th edn. Oxford: Wiley.

Cox, P. M., Betts, R. A., Jones, C. D., Spall, S. A. and Totterdell, I. J. (2000). Acceleration of global warming due to carbon-cycle feedbacks in a coupled climate model. *Nature*, **408**, 184–187.

Cramer, B. S., Toggweiler, J. R., Wright, J. D., Katz, M. E. and Miller, K. G. (2009). Ocean overturning since the Late Cretaceous: inferences from a new benthic foraminiferal isotope compilation. *Paleoceanography*, **24**, PA4216, 1–14.

Cramer, W., Bondeau, A., Woodward, F. I. et al. (2001). Global response of terrestrial ecosystem structure and function to CO_2 and climate change: results from six dynamic global vegetation models. *Global Change Biology*, **7**, 357–373.

Crawford, R. M. M. (2008). *Plants at the Margin: Ecological Limits and Climate Change*. Cambridge: Cambridge University Press.

Darwin, C. (1859). *On the Origin of Species by Means of Natural Selection, or the Preservation of Favoured Races in the Struggle for Life*. London: John Murray.

Darwin, C. R. and Wallace, A. R. (1858). On the tendency of species to form varieties; and on the perpetuation of varieties and species by natural means of selection. *Journal of the Proceedings of the Linnean Society of London, Zoology*, **3**, 46–50.

Davis, M. B. (2005). Comparison of climate space and phylogeny of *Marmota* (Mammalia: Rodentia) indicates a connection between evolutionary history and climate preference. *Proceedings of the Royal Society of London B*, **272**, 519–526.

Davis, M. B. and Shaw, R. G. (2001). Range shifts and adaptive responses to quaternary climate change. *Science*, **292**, 673–679.

DeConto, R. M., Pollard, D., Wilson, P. et al. (2008). Thresholds for Cenozoic bipolar glaciation. *Nature*, **455**, 652–656.

Derory, J., Scotti-Saintagne, C. Bertocchi, E. et al. (2010). Contrasting relations between diversity of candidate genes and variation of bud burst in natural and segregating populations of European oaks. *Heredity*, **104**, 438–448.

Dixon, H. and Joly, J. (1895). On the ascent of sap. *Philosophical Transactions of the Royal Society of London B*, **186**, 563–576.

Donnadieu, Y., Goddéris, Y., Ramstein, G., Nédélec, A. and Meert, J. (2004). A snowball Earth climate triggered by continental break-up through changes in runoff. *Nature*, **428**, 303–306.

Donoghue, M. J. (2005). Key innovations, convergence, and success: macroevolutionary lessons from plant phylogeny. *Paleobiology*, **31**, 77–93

Donoghue, M. J. (2008). A phylogenetic perspective on the distribution of plant diversity. *Proceedings of the National Academy of Sciences of the USA*, **105**, 11549–11555.

Edwards, E. J. and Still, C. J. (2008). Climate, phylogeny and the ecological distribution of C4 grasses. *Ecology Letters*, **11**, 266–276.

Edwards, E. J., Still, C. J. and Donoghue, M. J. (2007). The relevance of phylogeny to studies of global climate change. *Trends in Ecology and Evolution*, **22**, 243–249.

Edwards, M. and Richardson, A. J. (2004). Impact of climate change on marine pelagic phenology and trophic mismatch. *Nature*, **430**, 881–884.

Eldrett, J. S., Harding, I. C., Wilson, P. A., Butler, E. and Roberts, A. P. (2007). Continental ice in Greenland during the Eocene and Oligocene. *Nature*, **446**, 176–179.

Elith, J. and Leathwick, J. R. (2009). Species distribution models: ecological explanation and prediction across space and time. *Annual Review of Ecology, Evolution, and Systematics*, **40**, 677–697.

Elith, J., Graham, C. H., Anderson, R. P. et al. (2006). Novel methods improve prediction of species' distributions from occurrence data. *Ecography*, **29**, 129–151.

Engelbrecht, B. M. J., Comita, L. S., Condit, R. et al. (2007). Drought sensitivity shapes species distribution patterns in tropical forests. *Nature*, **447**, 80.

EPICA (2004). Eight glacial cycles from an Antarctic ice core. *Nature*, **429**, 623–628.

Eriksson, G., Andersson, S., Eiche, V., Ifver, J. and Persson, A. (1980). Severity index and transfer effects on survival and volume production of *Pinus sylvestris* in Northern Sweden. *Studia Forestalia Suecica*, **156**, 1–31.

Eveno, E., Collada, C., Guevara, M. A. et al. (2008). Contrasting patterns of selection at *Pinus pinaster* Ait. Drought stress candidate genes as revealed by genetic differentiation analysis. *Molecular Biology and Evolution*, **25**, 417–437.

Excoffier, L., Foll, M. and Petit, R. J. (2009). Genetic consequences of range expansions. *Annual Review in Ecology, Evolution, and Systematics*, **40**, 481–501.

Fitter, A. H. and Fitter, R. S. R. (2002). Rapid changes in flowering time in British plants. *Science,* **296**, 1689–1691.

Graham, C. H., Ferrier, S., Huettman, F., Moritz, C. and Peterson, A. T. (2004a). New developments in museum-based informatics and applications in biodiversity analysis. *Trends in Ecology and Evolution,* **19**, 497–503.

Graham, C. H., Ron, S. R., Santos, J. C., Schneider, C. J. and Moritz, C. (2004b). Integrating phylogenetics and environmental niche models to explore speciation mechanisms in dendrobatid frogs. *Evolution,* **58**, 1781–1793.

Green, R. E., Collingham, Y. C., Wills, S. G. et al. (2008). Performance of climate envelope models in retrodicting recent changes in bird population size from observed climatic change. *Biological Letters,* **4**, 599–602.

Grinnell, J. (1917). The niche-relationships of the California Thrasher. *Auk,* **34**, 427–433.

Guisan, A. and Thuiller, W. (2005). Predicting species' distributions: offering more than simple habitat models. *Ecology Letters,* **8**, 993–1009.

Hadly, E. A., Ramakrishnan, U., Chan, Y. L. et al. (2004). Genetic response to climate change: insights from ancient DNA and phylochronology. *PLOS Biology,* **2**, 1–10.

Haeger, J. F. (1999). *Danaus chrysippus* (Linnaeus 1758) en la Península Ibérica: migraciones o Dinámica de metapoblaciones? *Shilap,* **27**, 423–430.

Hall, D., Luquez, V., Garcia, V. M. et al. (2007). Adaptive population differentiation in phenology across a latitudinal gradient in European aspen (*Populus tremula,* L.): a comparison of neutral markers, candidate genes and phenotypic traits. *Evolution,* **61**, 2849–2860.

Hannah, L. (2008). Protected areas and climate change. In *Year in Ecology and Conservation Biology 2008,* ed. R. S. Ostfeld and W. H. Schlesinger. Oxford: Blackwell, pp. 201–212.

Hannah, L., Midgley, G., Andelman, S. et al. (2007). Protected area needs in a changing climate. *Frontiers in Ecology and the Environment,* **5**, 131–138.

Hannah, L., Dave, R., Lowry, P. P. et al. (2008). Climate change adaptation for conservation in Madagascar. *Biology Letters,* **4**, 590–594.

Hardy, C. R. (2006). Reconstructing ancestral ecologies: challenges and possible solutions. *Diversity and Distributions,* **12**, 7–19.

Hedges, S. B. and Kumar, S. (2009). Discovering the timetree of life. In *The Timetree of Life,* ed. S. B. Hedges and S. Kumar. Oxford: Oxford University Press, pp. 3–18.

Heikkinen, R. K., Luoto, M., Araújo, M. B. et al. (2006). Methods and uncertainties in bioclimatic envelope modeling under climate change. *Progress in Physical Geography,* **30**, 751–777.

Heinrich, H. (1988). Origin and consequences of cyclic ice rafting in the Northeast Atlantic Ocean during the past 130,000 years. *Quaternary Research,* **29**, 142–152.

Hersteinsson, P. and MacDonald, D. W. (1992). Interspecific competition and the geographical distribution of red and arctic foxes *Vulpes vulpes* and *Alopex lagopus. Oikos,* **64**, 505–515.

Hewitt, G. (2000). The genetic legacy of the Quaternary ice ages. *Nature,* **405**, 907–913.

Hijmans, R. J. and Graham, C. H. (2006). The ability of climate envelope models to predict the effect of climate change on species' distributions. *Global Change Biology,* **12**, 2272–2281.

Hodkinson, T. R. and Parnell, J. A. N. (2007). Introduction to the systematics of species rich groups. In *Towards the Tree of Life: Systematics of Species Rich Groups*, ed. T. R. Hodkinson and J. A. N. Parnell. Boca Raton, FL: CRC Press, pp. 3–20.

Hodkinson, T. R., Savolainen, V., Jacobs, S. W. et al. (2007). Supersizing: progress in documenting and understanding grass richness. In *Towards the Tree of Life: Systematics of Species Rich Groups*, ed. T. R. Hodkinson and J. A. N. Parnell. Boca Raton, FL: CRC Press, pp. 279–298.

Hoffman, P. F., Kaufman, A. J., Halverson, G. P. and Schrag, D. P. (1998). A Neoproterozoic snowball earth. *Science,* **281**, 1342–1346.

Hole, D. G., Willis, S. G., Pain, D. J. et al. (2009). Projected impacts of climate change on a continent-wide protected area network. *Ecology Letters,* **12**, 420–431.

Hole, D. G., Huntley, B., Arinaitwe, J. et al. (in press). Towards a management framework for key biodiversity networks in the face of climatic change. *Conservation Biology.*

Huber, B. T., MacLeod, K. G. and Wing, S. L., eds. (2000). *Warm Climates in Earth History.* Cambridge: Cambridge University Press.

Huber, M. and Caballero, R. (2003). Eocene El Niño: evidence for robust tropical dynamics in the 'hothouse'. *Science,* **299**, 877–881.

Huber, M. and Sloan, L. C. (2001). Heat transport, deep waters and thermal gradients: coupled climate simulation of an Eocene greenhouse climate. *Geophysical Research Letters,* **28**, 3841–3884.

Huntley, B. (2007a). *Climatic Change and the Conservation of European Biodiversity: Towards the Development of Adaptation Strategies.* Strasbourg: Council of Europe, Convention of the Conservation of European Wildlife and Natural Habitats.

Huntley, B. (2007b). Evolutionary response to climate change? *Heredity,* **98**, 247–248.

Huntley, B., Barnard, P., Altwegg, R. et al. (2010). Beyond bioclimatic envelopes: dynamic species' range and abundance modelling in the context of climatic change. *Ecography,* **33**, 11–16.

Hutchinson, G. E. (1957). Concluding remarks. *Cold Spring Harbor Symposia on Quantitative Biology,* **22**, 415–427.

Hutchinson, G. E. (1978). *An Introduction to Population Ecology.* New Haven, CT: Yale University Press.

Ingvarsson, P. K., Garcia, M. V., Hall, D., Luquez, V. and Jansson, S. (2006). Clinal variation in phyB2, a candidate gene for day-length-induced growth cessation and bud set, across a latitudinal gradient in European aspen (*Populus tremula*). *Genetics,* **172**, 1845–1853.

Intergovernmental Panel on Climate Change (IPCC) (2007a). *Climate Change 2007: Synthesis Report. Contribution of Working Groups I, II and III to the Fourth Assessment Report of the Intergovernmental Panel on Climate Change,* ed. R. K. Pachauri and A. Reisinger, Geneva: IPCC.

Intergovernmental Panel on Climate Change (IPCC) (2007b). *Climate Change 2007: the Physical Science Basis. Contribution of Working Group*

I to the Fourth Assessment Report of the Intergovernmental Panel on Climate Change, ed. S. Solomon, D. Qin, M. Manning et al. Cambridge: Cambridge University Press.

International Union for Conservation of Nature and Natural Resources (IUCN) (2001). *IUCN Red List Categories and Criteria: Version 3.1.* Gland and Cambridge: IUCN.

Jablonski, D., Roy, K. and Valentine, J. W. (2006). Out of the tropics: evolutionary dynamics of the latitudinal diversity gradient. *Science*, **314**, 102–106.

Jakob, S. S., Martinez-Meyer, E. and Blattner, F. R. (2009). Phylogeographic analyses and paleodistribution modeling indicate pleistocene in situ survival of *Hordeum* species (Poaceae) in southern Patagonia without genetic or spatial restriction. *Molecular Biology and Evolution*, **26**, 907–923.

Juckes, M. N., Allen, M. R., Briffa, K. R. et al. (2007). Millennial temperature reconstruction intercomparison and evaluation. *Climate of the Past*, **3**, 591–609.

Jump, A. S. and Peñuelas, J. (2005). Running to stand still: adaptation and the response of plants to rapid climate change. *Ecology Letters*, **8**, 1010–1020.

Jump, A. S., Hunt, J. M., Martinez-Izquierdo, J. A. and Peñ uelas, J. (2006). Natural selection and climate change: temperature-linked spatial and temporal trends in gene frequency in *Fagus sylvatica*. *Molecular Ecology*, **15**, 3469–3480.

Jump, A. S., Peñuelas, J., Rico, L. et al. (2008). Simulated climate change provokes rapid genetic change in the Mediterranean shrub *Fumana thymifolia*. *Global Change Biology*, **14**, 637–643.

Jump, A. S., Matyas, C. and Peñuelas, J. (2009). The altitude-for-latitude disparity in the range retractions of woody species. *Trends in Ecology and Evolution*, **24**, 694–701.

Keith, D. A., Akçakaya, H. R., Thuiller W. et al. (2008). Predicting extinction risks under climate change: coupling stochastic population models with dynamic bioclimatic habitat models. *Biology Letters*, **4**, 560–563.

Kelleher, C. T, Hodkinson, T. R., Kelly, D. L. and Douglas, G. C. (2004). Characterisation of chloroplast DNA haplotypes to reveal the provenance and genetic structure of oaks in Ireland. *Forest Ecology and Management*, **189**, 123–131.

Kenrick, P. and Crane, P. R. (1997). The origin and early evolution of plants on land. *Nature*, **389**, 33–39.

Kirchner, J. W. and Weil, A. (1990). Delayed biological recovery from extinctions throughout the fossil record. *Nature*, **404**, 177–180.

Knoll, A. H. and Lipps, J. H. (1993). Evolutionary history of prokaryotes and protists. In *Fossil Prokaryotes and Protists*, ed. J. H. Lipps. Boston, MA: Blackwell Scientific Publications, pp. 19–93.

Korner, C. (2006). Plant CO_2 responses: an issue of definition, time and resource supply. *New Phytologist*, **172**, 393–411.

Kozak, K. H., Graham, C. H. and Wiens, J. J. (2008). Integrating GIS-based environmental data into evolutionary biology. *Trends in Ecology and Evolution*, **23**, 141–148.

Krutovsky, K. V. and Neale, D. B. (2005). Nucleotide diversity and linkage disequilibrium in cold-hardiness- and wood quality-related candidate genes in Douglas fir. *Genetics*, **171**, 2029–2041.

Kump, L. R. and Pollard, D. (2008). Amplification of Cretaceous warmth by biological cloud feedbacks. *Science*, **320**, 195.

Lenton, T. M. (2004). The coupled evolution of life and atmospheric oxygen. In *Evolution on Planet Earth*, ed. L. J. Rothschild and A. Lister. Boston, MA: Academic Press.

Lenton, T. M., Held, H., Kriegler, E. et al. (2008). Tipping elements in the earth's climate system. *Proceedings of the National Academy of Sciences of the USA*, **105**, 1786–1793.

Lovejoy, T. E. and Hannah, L., eds. (2005). *Climate Change and Biodiversity*. New Haven, CT: Yale University Press.

Lowe, A., Harris, S. and Ashton, P. (2004). *Ecological Genetics: Design, Analysis, and Application*. Oxford: Blackwell.

Lowe, J. J. and Walker, M. J. C. (1997). *Reconstructing Quaternary Environments*. Harlow: Longman.

MacArthur, R. H. and Wilson, E. O. (1967). *The Theory of Island Biogeography*. Princeton, NJ: Princeton University Press.

Masson-Delmotte, V., Jouzel, J., Landais, A. et al. (2005). GRIP deuterium excess reveals rapid and orbital-scale changes in Greenland moisture origin. *Science*, **309**, 118–121.

Mayhew, P. J., Jenkins, G. B. and Benton, T. G. (2008). A long-term association between global temperature and biodiversity, origination and extinction in the fossil record. *Proceedings of the Royal Society of London B*, **275**, 47–53.

McElwain, J. C. (2002). Is the greenhouse theory a fallacy? A paleontological paradox. *Palaios*, **17**, 417–418.

McElwain, J. C. and Chaloner, W. G. (1995). Stomatal density and index of fossil plants track atmospheric carbon dioxide in the Paleozoic. *Annals of Botany*, **76**, 389–395.

McElwain, J. C. and Punyasena, S. (2007). Mass extinction events and the plant fossil record. *Trends in Ecology and Evolution*, **22**, 548–557.

McElwain, J. C., Beerling, D. J. and Woodward, F. I. (1999). Fossil plants and global warming at the Triassic–Jurassic boundary. *Science*, **285**, 1386–1390.

McManus, J. F. (2004). A great grand-daddy of ice cores. *Nature*, **429**, 611–612.

McMenamin, S. K., Hadly, E. A. and Wright, C. K. (2008). Climatic change and wetland desiccation cause amphibian decline in Yellowstone National Park. *Proceedings of the National Academy of Sciences of the USA*, **105**, 16988–16993.

Meehl, G. A., Stocker, T. F., Collins, W. D. et al. (2007). Global climate projections. In *Climate Change 2007: the Physical Science Basis. Contribution of Working Group I to the Fourth Assessment Report of the Intergovernmental Panel on Climate Change*, ed. S. Solomon, D. Qin, M. Manning et al. Cambridge: Cambridge University Press, pp. 747–845.

Menzel, A., Sparks, T. H., Estrella, N. et al. (2006). European phenological response to climate change matches the warming pattern. *Global Change Biology*, **12**, 1969–1976.

Meyer, K. M. and Kump, L. R. (2008). Oceanic euxinia in Earth history: causes and consequences. *Annual Review of Earth and Planetary Sciences*, **36**, 251–288.

Møller, A. P., Rubolini, D. and Lehikoinen, E. (2008). Populations of migratory bird species that did not show a phenological response to climate change are declining. *Proceedings of the National Academy of Sciences of the USA*, **105**, 16195–16200.

Moritz, C., Patton, J. L., Conroy, C. J. et al. (2008). Impact of a century of climate change on small-mammal communities in Yosemite National Park, USA. *Science*, **322**, 261–264.

Myers, N., Mittermeier, R. A., Mittermeier, C. G., da Fonseca, G. A. B. and Kent, J. (2000). Biodiversity hotspots for conservation priorities. *Nature*, **403**, 853–858.

Neale, D. B. and Ingvarsson, P. K. (2008). Population, quantitative and comparative genomics of adaptation in forest trees. *Current Opinion in Plant Biology*, **11**, 149–155.

Neale, D. B. and Savolainen, O. (2004). Association genetics of complex traits in conifers. *Trends in Plant Science*, **9**, 325–330.

North Greenland Ice Core Project (NGRIP) (2004). High-resolution record of Northern Hemisphere climate extending into the last interglacial period. *Nature*, **431**, 147–151.

Ogg, J. G., Ogg, G. and Gradstein, F. (2008). *Geologic Time Scale*. Cambridge: Cambridge University Press.

Osborne, C. P. and Freckleton, R. P. (2009). Ecological selection pressures for C4 photosynthesis. *Proceedings of the Royal Society of London B*, **276**, 1753–1760.

Pagani, M., Arthur, M. A. and Freeman, K. H. (1999). Miocene evolution of atmospheric carbon dioxide. *Paleoceanography*, **14**, 273–292.

Pagani, M., Zachos, J. C., Freeman, K. H., Tipple, B. and Bohaty, S. (2005). Marked decline in atmospheric carbon dioxide concentrations during the Paleogene. *Science*, **309**, 600–603.

Pakenham, T. (2003). *Remarkable Trees of the World*. London: Weidenfeld and Nicolson.

Palmer, J. D., Soltis, D. E. and Chase, M. W. (2004). The plant tree of life: an overview and some points of view. *American Journal of Botany*, **91**, 1437–1445.

Parmesan, C. (2006). Ecological and evolutionary responses to recent climate change. *Annual Review of Ecology, Evolution, and Systematics*, **37**, 637–669.

Parmesan, C. and Yohe, G. (2003). A globally coherent fingerprint of climate change impacts across natural systems. *Nature*, **421**, 37–42.

Parmesan, C., Ryrholm, N., Stefanescu, C. et al. (1999). Poleward shifts in geographical ranges of butterfly species associated with regional warming. *Nature*, **399**, 579–583.

Pauli, H., Gottfried, M., Reiter, K., Klettner, C. and Grabherr, G. (2006). Signals of range expansions and contractions of vascular plants in the high Alps: observations (1994–2004) at the GLORIA master site Schrankogel, Tyrol, Austria. *Global Change Biology*, **13**, 147–156.

Pearman, P. B., Guisan, A., Broennimann, O. and Randin, C. F. (2007). Niche dynamics in space and time. *Trends in Ecology and Evolution*, **23**, 149–158.

Pearman, P. B., Randin, C. F., Broennimann, O. et al. (2008). Prediction of plant species' distributions across six millennia. *Ecology Letters*, **11**, 357–369.

Pearson, P. N. and Palmer, M. R. (2000). Atmospheric carbon dioxide concentrations over the past 60 million years. *Nature*, **406**, 695–699.

Pearson, R. G. (2006). Climate change and the migration capacity of species. *Trends in Ecology and Evolution*, **21**, 111–113.

Pearson, R. G. and Dawson, T. P. (2003). Predicting the impacts of climate change on the distribution of species: are bioclimate envelope models useful? *Global Ecology and Biogeography*, **12**, 361–371.

Pearson, R. G., Thuiller, W., Araújo, M. B. et al. (2006). Model-based uncertainty in species range prediction. *Journal of Biogeography*, **33**, 1704–1711.

Peters, S. E. (2008). Environmental determinants of extinction selectivity in the fossil record. *Nature*, **454**, 626–629.

Peters, S. E. and Foote, M. (2002). Determinants of extinction in the fossil record. *Nature*, **416**, 420–424.

Peterson, A. T. (2006). Uses and requirements of ecological niche models and related distributional models. *Biodiversity Informatics*, **3**, 59–72.

Peterson, A. T., Soberón, J. and Sánchez-Cordero, V. (1999). Conservation of ecological niches in evolutionary time. *Science*, **285**, 1265–1267.

Petit, J. R., Basile, I., Leruyuet, A. et al. (1999). Four climate cycles in Vostok ice core. *Nature*, **387**, 359–360.

Petit, R. J., Csaikl, U. M., Bordács, S. et al. (2002). Chloroplast DNA variation in European white oaks: phylogeography and patterns of diversity based on data from over 2600 populations. *Forest Ecology and Management*, **156**, 5–26.

Phillips, N. (1956). The general circulation of the atmosphere: a numerical experiment. *Quarterly Journal of the Royal Meteorological Society*, **82**, 123–164.

Phillips, S. J., Anderson, R. P. and Schapire, R. E. (2006). Maximum entropy modeling of species geographic distributions. *Ecological Modelling*, **190**, 231–259.

Pierrehumbert, R. T. (2004). High levels of atmospheric carbon dioxide necessary for the termination of global glaciation. *Nature*, **429**, 646–649.

Poorter, H. and Navas, M. L. (2003). Plant growth and competition at elevated CO_2: on winners, losers and functional groups. *New Phytologist*, **157**, 175–198.

Pounds, J. A., Fogden, M. P. L. and Campbell, J. H. (1999). Biological responses to climate change on a tropical mountain. *Nature*, **398**, 611–615.

Pounds, J. A., Bustamente, M. R., Coloma, L. A. et al. (2006). Widespread amphibian extinctions from epidemic disease driven by global warming. *Nature*, **439**, 161–167.

Purvis, A. (1996). Using interspecies phylogenies to test macroevolutionary hypotheses. In *New Uses for New Phylogenies*, ed. P. H. Harvey, A. J. Leigh Brown, J. Maynard Smith and S. Nee. Oxford: Oxford University Press, pp. 153–168.

Ramakrishnan, U. and Hadly, E. A. (2009). Using phylochronology to reveal cryptic population histories: review and synthesis of 29 ancient DNA studies. *Molecular Ecology*, **18**, 1310–1330.

Raup, D. M. and Sepkoski, J. J. (1982). Mass extinctions in the marine fossil record. *Science*, **215**, 1501–1503.

Rieseberg, L. H. and Carney, S. E. (1998). Plant hybridization. *New Phytologist*, **140**, 599–624.

Rieseberg, L. H., Raymond, O., Rosenthal, D. M. et al. (2003). Major ecological transitions in wild sunflowers facilitated by hybridization. *Science*, **301**, 1211–1216.

Roalson, E. H. (2008). C_4 photosynthesis: differentiating causation and coincidence. *Current Biology*, **18**, 167–168.

Rödder, D. and Lötters, S. (2009). Niche shift or niche conservatism? Climatic properties of the native and invasive range of the Mediterranean Housegecko *Hemidactylus turcicus*. *Global Ecology and Biogeography*, **18**, 674–687.

Rohde, R. A. and Muller, R. A. (2005). Cycles in fossil diversity. *Nature*, **434**, 208–210.

Root, T. L., Price, J. T., Hall, K. R., Schneider, S. H. and Rosenzweig, C. (2003). Fingerprints of global warming on wild animals and plants. *Nature*, **421**, 57–60.

Rosenzweig, C., Casassa, G., Karoly, D. J. et al. (2007). Assessment of observed changes and responses in natural and managed systems. In *Climate Change 2007: Impacts, Adaptation and Vulnerability. Contribution of Working Group II to the Fourth Assessment Report of the Intergovernmental Panel on Climate Change*, ed. M. Parry, O. Canziani, J. Palutikof, P. van der Linden and C. Hanson. Cambridge: Cambridge University Press, pp. 79–131.

Rosenzweig, M. L. (1995). *Species Diversity in Space and Time*. Cambridge: Cambridge University Press.

Roy, D. B. and Sparks, T. H. (2000). Phenology of British butterflies and climate change. *Global Change Biology*, **6**, 407–416.

Royer, D. L. (2001). Stomatal density and stomatal index as indicators of paleoatmospheric CO_2 concentration. *Review of Palaeobotany and Palynology*, **114**, 1–28.

Royer, D. L., Berner, R. A., Montañez, I. P., Tibor, N. J. and Beerling, D. J. (2004). CO_2 as a primary driver of Phanerozoic climate. *Geological Society of America Today*, **14**, 4–10.

Ruddiman, W. F. (2003). The anthropogenic greenhouse era began thousands of years ago. *Climatic Change*, **61**, 261–293.

Ruegg, K., Slabbekoorn, H., Clegg, S. and Smith, T. B. (2006). Divergence in mating signals correlates with ecological variation in the migratory songbird, Swainson's thrush (*Catharus ustulatus*). *Molecular Ecology*, **15**, 3147–3156.

Sage, R. F. (2004). The evolution of C-4 photosynthesis. *New Phytologist*, **161**, 341–370.

Schmidt, K. P. (1954). Faunal realms, regions, and provinces. *Quarterly Review of Biology*, **29**, 322–331.

Sepkoski, J. J. (2002). A compendium of fossil marine animal genera. *Bulletin of American Paleontology*, **363**, 1–560.

Sexton, J. P., McIntyre, P. J., Angert, A. L. and Rice, K. J. (2009). Evolution and ecology of species range limits. *Annual Review of Ecology, Evolution, and Systematics*, **40**, 415–436.

Shaviv, N. J. and Veizer, J. (2003). Celestial driver of Phanerozoic climate? *Geological Society of America Today*, **13**, 4–10.

Sheehan, P. M. (2001). The Late Ordovician mass extinction. *Annual Review of Earth and Planetary Sciences*, **29**, 331–364.

Soberón, J. and Peterson, A. T. (2004). Biodiversity informatics: managing and applying primary biodiversity data.

Philosophical Transactions of the Royal Society of London B, **359**, 689–698.

Soltis, P. S. and Soltis, D. E. (2009). The role of hybridization in plant speciation. *Annual Review of Plant Biology*, **60**, 561–588.

Steffensen, J. P., Andersen, K. K., Bigler, M. et al. (2008). High-resolution Greenland ice core data show abrupt climate change happens in few years. *Science*, **321**, 680–684.

Stokstad, E. (2001). Myriad ways to reconstruct past climate. *Science*, **292**, 658–659.

Sutherland, W. J. (2006). Predicting the ecological consequences of environmental change: a review of the methods. *Journal of Applied Ecology*, **43**, 599–616.

Svensmark, H. and Calder, N. (2007). *The Chilling Stars: a New Theory of Climate Change*. Cambridge: Icon Books.

Swenson, N. G. (2006). GIS-based niche models reveal unifying climatic mechanisms that maintain the location of avian hybrid zones in a North American suture zone. *Journal of Evolutionary Biology*, **19**, 717–725.

Thomas, C. D. and Lennon, J. J. (1999). Birds extend their ranges northwards. *Nature*, **399**, 213.

Thomas, C. D., Cameron, A., Green, R. E. et al. (2004). Extinction risk from climate change. *Nature*, **427**, 145–148.

Thomas, C. D., Franco, A. M. A. and Hill, J. K. (2006). Range retractions and extinction in the face of climate warming. *Trends in Ecology and Evolution*, **21**, 415–416.

Thuiller, W. (2004). Patterns and uncertainties of species' range shifts under climate change. *Global Change Biology*, **10**, 2020–2027.

Thuiller, W., Albert, C., Araú jo, M. B. et al. (2008). Predicting global change impacts on plant species' distributions: future challenges. *Perspectives in Plant Ecology, Evolution and Systematics*, **9**, 137–152.

Trenberth, K. E., Jones, P. D., Ambenje, P. et al. (2007). Observations: surface and atmospheric climate change. In *Climate Change 2007: the Physical Science Basis. Contribution of Working Group I to the Fourth Assessment Report of the Intergovernmental Panel on Climate Change*, ed. S. Solomon, D. Qin, M. Manning et al. Cambridge: Cambridge University Press, pp. 235–336.

Tyndall, J. (1865). *Heat Considered as a Mode of Motion*, 2nd edn. London: Longman Green.

van de Schootbrugge, B., Quan, T. M., Lindström, S. et al. (2009). Floral changes across the Triassic/Jurassic boundary linked to flood basalt volcanism. *Nature Geoscience*, **2**, 589–594.

van Herk, C. M., Aptroot, A. and van Dobben, H. F. (2002). Long-term monitoring in the Netherlands suggests that lichens respond to global warming. *Lichenologist*, **34**, 141–154.

van Valen, L. (1973). A new evolutionary law. *Evolutionary Theory*, **1**, 1–30.

Veizer, J., Godderis, Y. and François, L. M. (2000). Evidence for decoupling of atmospheric CO_2 and global climate during the Phanerozoic eon. *Nature*, **408**, 698–701.

Vila, M., Weber, E. and Antonio, C. M. D. (2000). Conservation implications of invasion by plant hybridization. *Biological Invasions*, **2**, 207–217.

Wake, D. B., Hadley, E. A., and Ackerly, D. D. (2009). Biogeography, changing climates, and niche evolution. *Proceedings of the National*

Academy of Sciences of the USA, **106**, 19631-19636.

Wallace, A. R. (1869). *The Malay Archipelago; The Land of the Orang-utan and the Bird of Paradise; A Narrative of Travel With Studies of Man and Nature*. London: Macmillan.

Wallace, A. R. (1876). *The Geographical Distribution of Animals; With A Study of the Relations of Living and Extinct Faunas as Elucidating the Past Changes of the Earth's Surface*. London: Macmillan.

Wallace, A. R. (1880). *Island Life: Or, The Phenomena and Causes of Insular Faunas and Floras, Including a Revision and Attempted Solution of the Problem of Geological Climates*. London: Macmillan.

Walther, G. R., Post, E., Convey, P. et al. (2002). Ecological responses to recent climate change. *Nature*, **416**, 389–395.

Wiens, J. J. and Graham, C. H. (2005). Niche conservatism: integrating evolution, ecology, and conservation biology. *Annual Review of Ecology and Systematics*, **36**, 519–539.

Wignall, P. B. (2005). The link between large igneous province eruptions and mass extinctions. *Elements*, **1**, 293–297.

Williams, J. W. and Jackson, S. T. (2007). Novel climates, no-analog communities, and ecological surprises. *Frontiers in Ecology and the Environment*, **5**, 475–482.

Williams, J. W., Jackson, S. T. and Kutzbach, J. E. (2007). Projected distributions of novel and disappearing climates by 2100 AD. *Proceedings of the National Academy of Sciences of the USA*, **104**, 5738–5742.

Willig, M. R., Kaufmann, D. M. and Stevens, R. D. (2003). Latitudinal gradients of biodiversity: pattern, process, scale and synthesis. *Annual Review of Ecology, Evolution, and Systematics*, **34**, 273–309.

Willis, K. J. and McElwain, J. C. (2002). *The Evolution of Plants*. Oxford: Oxford University Press.

Wilson, E. O. (1992). *The Diversity of Life*. Cambridge, MA: Harvard University Press.

Wilson, J. W, Gutiérrez, D., Martinez, D., Agudo, R. and Monserrat, V. J. (2005). Changes to the elevational limits and extent of species ranges associated with climate change. *Ecological Letters*, **8**, 1138–1146.

Wnuk, C. (1996). The development of floristic provinciality during the Middle and Late Paleozoic. *Review of Palaeobotany and Palynology*, **90**, 6–40.

Woodward, F. I. (1987). Stomatal numbers are sensitive to increase in CO_2 from pre-industrial levels. *Nature*, **327**, 617–618.

Woodward, F. I. and Kelly, C. K. (2008). Responses of global plant biodiversity capacity to changes in carbon dioxide concentration and climate. *Ecology Letters*, **11**, 1229–1237.

Wunsch, C. (2004). Quantitative estimate of the Milankovitch-forced contribution to observed Quaternary climate change. *Quarterly Science Reviews*, **23**, 1001–1012.

Yesson, C. and Culham, A. (2006a). A phyloclimatic study of Cyclamen. *BMC Evolutionary Biology*, **6**, 72.

Yesson, C. and Culham, A. (2006b). Phyloclimatic modelling: combining phylogenetics and bioclimatic modelling. *Systematic Biology*, **55**, 785–802.

Zachos, J. C., Pagani, M., Sloan, L., Thomas, E. and Billups, K. (2001). Trends, rythms and aberrations in global climate 65 Ma to present. *Science*, **292**, 686–693.

Zachos, J. C., Dickens, G. R. and Zeebe, R. E. (2008). An early Cenozoic perspective on greenhouse warming and carbon-cycle dynamics. *Nature*, **451**, 279–283.

Zeebe, R. E., Zachos, J. C. and Dickens, G. R. (2009). Carbon dioxide forcing alone insufficient to explain Palaeocene–Eocene Thermal Maximum warming. *Nature Geoscience*, **2**, 576–580.

Ziegler, A.M., Eshel, G., McAllister, R. et al. (2003). Tracing the tropics across land and sea: Permian to present. *Lethaia*, **36**, 227–254.

2

Climate modelling and deep-time climate change

R. Caballero and P. Lynch

School of Mathematical Sciences, University College Dublin, Ireland

Abstract

Detailed and reliable understanding of past climate change is a key ingredient in unravelling how climate has influenced life on earth and will continue to do so in the future. Palaeoclimatology and climate modelling have both made rapid strides over the past decades, and there has been fruitful two-way interaction between the two fields. The application of climate models to palaeoclimates has proved useful both in interpreting palaeoclimate proxy data and in testing the robustness and generality of climate models. Here, we give an overview of the current state of climate modelling and review recent progress in understanding deep-time climate change, with emphasis on problems where climate models have played a salient role. By suitably adjusting the concentration of atmospheric greenhouse gases, climate models can be made to replicate many key climatic transitions in the earth's history. However, important discrepancies remain between modelled climates and proxy reconstructions, particularly on the warm end of the spectrum.

Climate Change, Ecology and Systematics, ed. Trevor R. Hodkinson, Michael B. Jones, Stephen Waldren and John A. N. Parnell. Published by Cambridge University Press.

2.1 Introduction

Climate science deals with reconstructing and explaining the long-term mean and variability of physical conditions in the earth's envelope. A striking feature emerging from such analysis is the vast range of timescales on which there is significant variability. Part of this variability, including the diurnal and annual cycles, is periodic and predictable, but mostly it is random and unpredictable. We know from direct experience that the weather changes from hour to hour, from day to day, and from year to year. Analysis of long-term instrumental records and of palaeoclimatological proxy data shows that the same is true on longer timescales; conditions change randomly from century to century, millennium to millennium, and from aeon to aeon. Figure 2.1 shows a composite spectrum of surface temperature and temperature proxies spanning timescales from a few days up to about 1 million years (Huybers and Curry, 2006). There are sharp peaks at the annual frequency and subannual harmonics, as expected. There are also some more poorly defined peaks at the 41 000-year and 100 000-year orbital variability frequencies (see Wunsch, 2003, 2004, for a discussion of the statistical significance of these peaks). However, most of the spectrum shows very little structure, and is best approximated by straight lines of constant slope. Thus, at least on timescales up to around 1 million years, climate variability is mostly red noise: fully stochastic, unpredictable variability skewed towards low frequencies.

Time series whose spectra are red up to the lowest frequencies, such as those shown in Fig 2.1, are said to possess 'long memory' or 'long-range correlation' (Beran, 1994). A feature of long-memory processes is that any time-series segment of finite length will typically show a trend, because the segment can be thought of as being 'embedded' in a longer-term fluctuation. Figures 2.2 and 2.3 show temperature reconstructions over the Cenozoic (up to 65 million years ago – mya – Zachos et al., 2001) and the Phanaerozoic (up to 545 mya – Royer et al., 2004). Interestingly, both these time series show clear trends (for the Phanaerozoic time series in Fig 2.3, this is only the case if the oxygen isotope data are appropriately corrected for pH effects – Royer et al., 2004). Although this trend by itself is not enough to infer long memory, the important implication is that the climate system appears never to have been in a steady state, even over time spans comparable to the age of the earth. Life has coevolved with climate over this time, and much remains to be understood about the way in which life and climate have mutually conditioned each other's evolution. This chapter will focus on climatic aspects, providing a backdrop for the more biologically oriented chapters in the rest of the book.

Mitchell (1976) and Kutzbach (1976) provide classic accounts of the mechanisms underlying long-term climate variability. The climate system contains two fluid components, the atmosphere and ocean, which are strongly heated in the tropics and more weakly heated in the high latitudes. This heating gradient

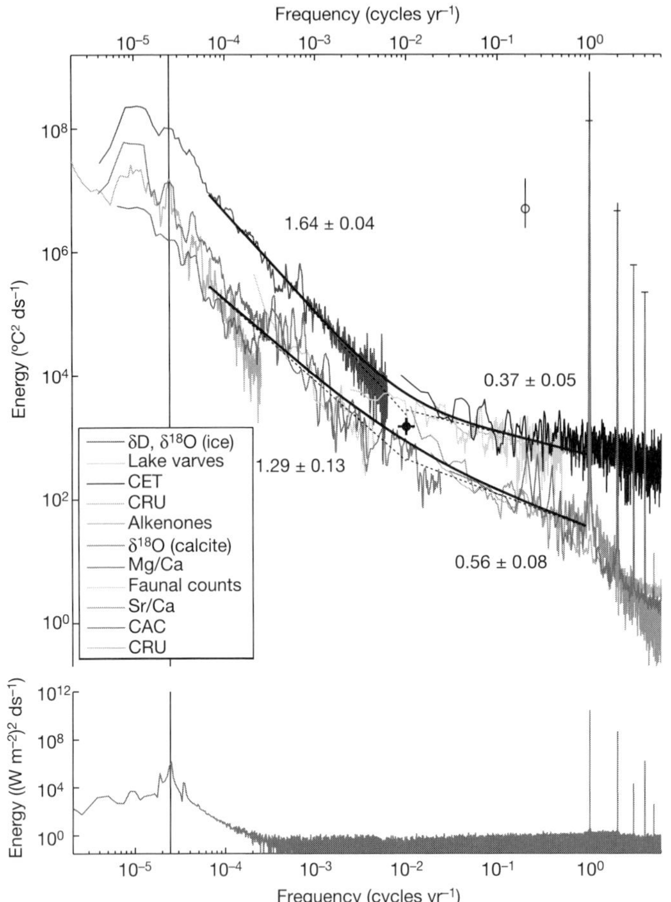

Figure 2.1 Patchwork spectral estimate using instrumental and proxy records of surface temperature variability. The more energetic spectral estimate is from high-latitude continental records and the less energetic estimate from tropical sea-surface temperatures. Power-law estimates for 1.1–100-year and 100–15 000-year periods are listed along with standard errors, and are indicated by the dashed lines. The sum of the power laws fitted to the long and short period continuum is indicated by the black curve. The vertical line segment indicates the approximate 95% confidence interval, where the circle indicates the background level. The mark at 1/100 yr indicates the region midway between the annual and Milankovitch periods. At the bottom is the spectrum of insolation at 65° N sampled monthly over the past million years plus a small amount of white noise. The vertical black line indicates the 41-kyr obliquity period. Reproduced with permission from Huybers and Curry (2006). *See colour plate section.*

triggers internal instabilities leading to ceaseless, chaotic motion in the ocean–atmosphere system with timescales from days to a few thousand years and spatial scales from hundreds to many thousands of kilometres (the fact that heating is

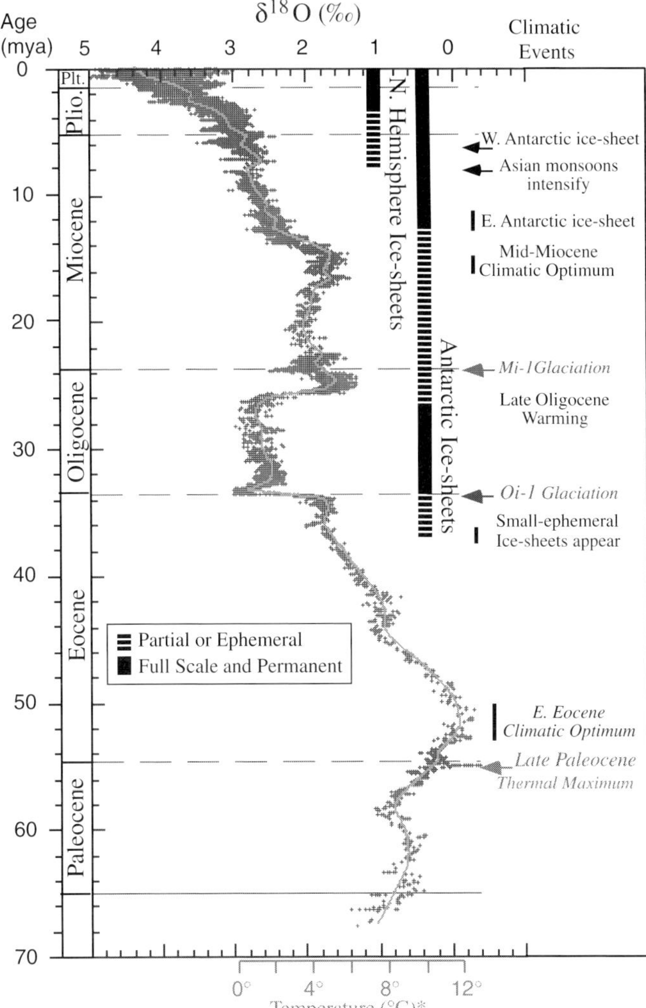

Figure 2.2 Global deep-sea oxygen isotope records for the Cenozoic era, based on data compiled from more than 40 ocean drilling sites. The raw data were smoothed using a five-point running mean and curve fitted with a locally weighted mean. The oxygen isotope temperature scale was computed for an ice-free ocean, and thus only applies to the time preceding the onset of large-scale glaciation on Antarctica (35 mya). From the early Oligocene to the present, much of the variability in the oxygen isotope record reflects changes in Antarctica and northern hemisphere ice volume. The vertical bars provide a rough qualitative representation of ice volume in each hemisphere relative to the Last Glacial Maximum (21 000 years ago), with the dashed bar representing periods of minimal ice coverage (less than 50%), and the full bar representing close to maximum ice coverage (more than 50% of present). Some key tectonic and biotic events are listed as well. Adapted from Zachos et al. (2001). *See colour plate section.*

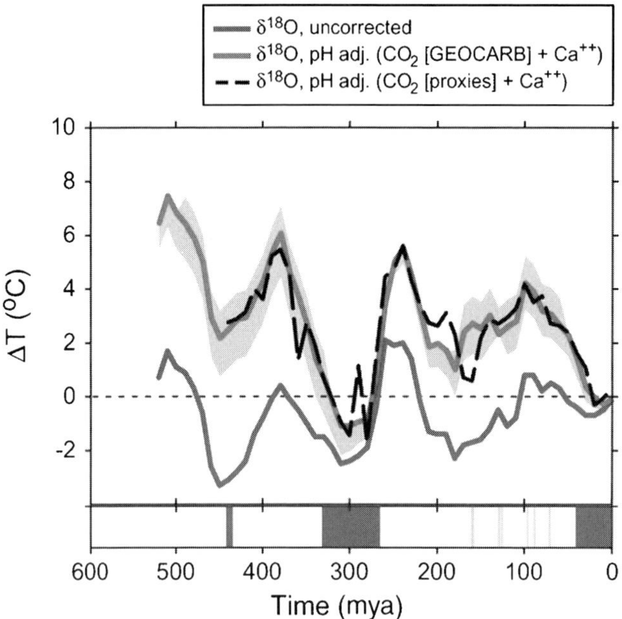

Figure 2.3 Shallow marine carbonate oxygen isotope record over the Phanaerozoic aeon. The blue curve corresponds to temperature deviations relative to today calculated by Shaviv and Veizer (2003). In the two remaining curves, data from the blue curve have been adjusted for pH effects due to changes in seawater calcium ion concentration and CO_2 based either on model (GEOCARB) or proxy reconstructions. Blue bands in the strip along the bottom indicate icehouse intervals, with extensive, permanent continental ice sheets (dark blue) or cool climates, with modest, ephemeral ice sheets (light blue). Adapted from Royer et al. (2004). *See colour plate section.*

stronger near the surface than aloft also triggers convective motions in the atmosphere with much shorter space and time scales). The atmosphere and ocean also interact with non-fluid components having much longer timescales. The build-up of large continental ice sheets requires in the order of 10^5 years. The long-term carbon cycle, which controls atmospheric carbon dioxide (CO_2) concentrations and thus the greenhouse effect and surface temperature, involves the slow weathering of silicate rocks (Walker et al., 1981), which sets a timescale in the order of 10^5 years for the drawdown of atmospheric CO_2. Tectonic processes, including the creation of large mountain ranges, the opening/closing of ocean seaways and the establishment of major igneous provinces, have timescales of 10^6–10^7 years. The formation and breakup of supercontinents, loosely speaking the Wilson cycle (Wilson, 1966), have timescales of 500 million years. At the very longest timescales, the sun's luminosity has been steadily increasing, by perhaps 40% over geologic time. The 'faint young sun' paradox – the problem of reconciling evidence for liquid water early in the earth's history with the low insolation then prevalent, first pointed out by Sagan and Mullen (1972) – has yet to be fully resolved. Finally, the appearance and evolution of life on earth has also had a major impact on climate, the details of which have been only partially unravelled.

By comparison with the earth's previous history, the past 10 000 years (the Holocene) have been relatively quiescent, with no major swings in ice volume or sea level and relatively mild changes in global and regional temperatures. This is also the period in which agriculture and settled human society first arose. Human population and its economic activities have now reached a scale such that they may strongly impact climate at the regional and global scales. Since human society has evolved and is adapted to a particular set of climatic conditions, any change is likely to result in widespread disruption and hardship. Perhaps the most alarming possibility is of a major melting of the Greenland and/or Antarctic ice sheets. Total collapse of the Greenland ice sheet alone would cause sea-level rise of around 7 m (IPCC, 2007), potentially submerging many of the world's most populous cities and driving hundreds of millions of people into destitution.

The problem of understanding the climate system and predicting its evolution in the near future has gained much prominence in scientific and public debate. The study of palaeoclimates helps, amongst other things, to answer questions about the extent and speed of climate change. It also serves as a test bed for theories of climate. Quantitative assessment of future climate change relies very heavily on global climate models, which contain a detailed description of the physics governing climate. Application of climate models to palaeoclimates helps solve some of the puzzles posed by these climates, and provides a particularly stringent test of model robustness and reliability. In the following sections, we provide a brief review of climate modelling and of long-term climate change. We emphasise time periods where the interplay of models and palaeoclimate proxy data has been particularly illuminating; these fall mostly (but not entirely) in the Cenozoic, for which proxy data are most abundant.

2.2 Climate modelling

2.2.1 The physical basis of climate modelling

The earth's climate is governed by fundamental and well-understood principles of physics. In weather prediction and climate models, these principles are expressed as mathematical equations which are then solved by numerical means. Washington and Parkinson (2005) provide an excellent introduction to climate modelling. For an account of the evolution of numerical weather prediction models, see Lynch (2006).

The motions of the atmosphere and oceans evolve according to the laws of dynamics first established by Newton in the seventeenth century. These laws relate changes in the motion to the forces applied to the system: if we can identify and specify the forces, we can use the laws of motion to deduce how the system will evolve. Thus, the dynamics can be described in precise terms. However, dynamics alone are insufficient. We also need to consider the effects of heating and cooling. These effects are described by the laws of thermodynamics and electromagnetic radiation, which were brought to light in the nineteenth century. Climate is driven by solar energy. Because of the spherical shape of the earth and the configuration of its orbit around the sun, the amount of incoming solar energy is highly variable, depending strongly on the time of day, the season and the latitude. The climate system can be viewed as an enormous thermodynamic engine, with the atmosphere and oceans playing the role of the working fluid, transporting heat from source to sink regions and performing work in the process.

In addition to dynamics and thermodynamics, there is the principle of conservation of mass, which provides a quantitative relationship called the continuity equation. The physical variables are linked through a relationship known as the equation of state. Finally, the physical principles of how matter and radiation interact, and how water changes between the solid, liquid and gaseous phases, complete the picture. We can specify the state of the ocean–atmosphere system at a particular time by providing the values of the physical variables everywhere throughout the system. Thus, if we give the pressure, temperature, density and (for the ocean) salinity, and the velocity in each direction and at every point, the configuration of the system is determined. Such data are called the initial conditions. Given these conditions, the physical principles may now be used to deduce the future evolution of the system.

Each physical law has a precise mathematical expression, and the assemblage of laws provides us with a closed set of equations. Mathematicians describe these as non-linear partial differential equations. They cannot be solved by analytical means; the solution in each case must be calculated by numerical approximations. The complete system of equations was first assembled around 1900 by scientists interested in forecasting weather by rational means. However, 50 years had to pass before the goal of numerical weather prediction began to become a reality.

In 1950 a drastically simplified mathematical model of the atmosphere was used to simulate the evolution of the weather over a one-day period (Charney et al., 1950). A few years later, Phillips carried out the first long-range simulation of the general circulation of the atmosphere. He used a two-level model with simplified equations and geometry, and with rudimentary physics. The computation used a spatial grid of 16×17 points, and the simulation was for a period of about one month. Idealised initial conditions were used and, during the simulated month, a state developed having some of the key features of the observed atmosphere. Starting from a zonal (westerly) flow with small random perturbations, a wave disturbance with wavelength of 6000 km developed. It had the characteristic westward tilt with height found in baroclinic waves, the large-scale unstable waves found in the mid-latitude atmosphere, and the disturbance moved eastward at about 20 m s^{-1}, which is comparable to observed atmospheric waves. Phillips examined the energy exchanges of the developing wave and found good qualitative agreement with observations of wave systems in the atmosphere. He also examined the flow averaged around parallels of latitude, the mean meridional flow, and found circulations corresponding to the three atmospheric cells.

Phillips completed his experiment in 1955 and communicated the results to John von Neumann, who immediately recognised their significance and arranged a conference in Princeton in October 1955: Application of Numerical Integration Techniques to the Problem of the General Circulation (see Lewis, 1998). Following this conference, Phillips entered the research for the first Napier Shaw Memorial Prize. On 20 June 1956, the adjudicators recommended that the prize be given to Phillips for his essay 'The general circulation of the atmosphere: a numerical experiment', which had been published in the *Quarterly Journal of the Royal Meteorological Society* (Phillips, 1956).

Phillips's work had a galvanising effect on the meteorological community. Within 10 years, there were several major research groups modelling the general circulation of the atmosphere, some of the leading ones being at the Geophysical Fluid Dynamics Laboratory in Princeton, New Jersey, USA, the National Center for Atmospheric Research in Boulder, Colorado, USA, the Meteorological Office's Hadley Centre in Exeter, UK, and the Max Planck Institute for Climate Research in Hamburg, Germany.

2.2.2 Current global models

Following the seminal work of Phillips, the development of general circulation models (GCMs) advanced rapidly, hand in hand with increasing computer power. Standard current-generation GCMs include a detailed description of the dynamics and physics of both atmosphere and ocean (these are referred to as 'coupled' GCMs), and also incorporate sophisticated treatment of the hydrological cycle over land, and of the formation and dynamics of sea ice. For palaeoclimate applications, a standard GCM can be coupled to a dynamic ice-sheet model. Ice sheets evolve over timescales of 10^4–10^5 years, far in excess of what is feasible for a continuous GCM simulation given current computer capabilities, so in this case the coupling is performed asynchronously, i.e. in a sequence of quasi-equilibrium steps. Much current effort is being devoted to incorporating aspects of the carbon cycle, including interaction with land and ocean biota, to more refined atmospheric chemistry and aerosol schemes, and to incorporation of oxygen isotope fractionation.

As an example of a current climate model, we consider HadCM3, developed at the Hadley Centre. Many earlier coupled models needed a flux adjustment (additional artificial heat and moisture fluxes at the ocean surface) to produce good simulations. The higher ocean resolution of HadCM3 was a major factor in removing this requirement. To test its stability, HadCM3 has been run for over 1000 years' simulated time and shows minimal drift in its surface climate. The atmospheric component of HadCM3 has 19 levels and a latitude/longitude resolution of 2.5 × 3.75 degrees, with grid of 96 × 73 points covering the globe. The resolution is about 417 × 278 km at the equator. The physical parameterisation package of the model is very sophisticated. The radiative effects of minor greenhouse gases as well as CO_2, water vapour and ozone are explicitly represented. A parameterisation of background aerosol is included. The land surface scheme includes freezing and melting of soil moisture, surface runoff and soil drainage. The convective scheme includes explicit down draughts. Orographic and gravity wave drag are modelled. Cloud water is an explicit variable in the large-scale precipitation and cloud scheme. The atmospheric component of the model allows the emission, transport, oxidation and deposition of sulphur compounds to be simulated interactively. The oceanic component of HadCM3 has 20 levels with a horizontal resolution of 1.25 × 1.25 degrees, permitting important details in the oceanic current structure to be represented. The model is initialised directly from the observed ocean state at rest, with a suitable atmospheric and sea-ice state. HadCM3 is being used for a wide range of climate studies, which will form crucial inputs to the forthcoming Fifth Assessment Report (AR5) of the Intergovernmental Panel on Climate Change (IPCC), which will be finalised in 2014.

2.2.3 Downscaling and regional climate models

Global GCMs are run at coarse spatial resolution, typically using a 100 km grid. They are unable to resolve many important subgrid-scale features such as clouds and topography. As a result, GCMs are not suitable for regional impact studies. To address this problem, various downscaling methods are employed to study local-scale climate impacts. The outputs of the global models are used as inputs to the regional models. The values generated during the execution of the global model are saved, and are used to specify the values around the boundaries of the limited domain, that is, the lateral boundary conditions. Wilby and Wigley (1997) identified four categories of downscaling: regression methods, weather pattern-based approaches, stochastic weather generators, and limited-area modelling.

In the regression method, the fine details of the past climate are related to the coarse representation provided by the global model. Assuming the relationship holds under a different climate regime, the regional details of the future climate can be deduced from coarse-grain predictions generated by the global model. This statistical downscaling has been used with some success. However, the hypothesis that the relationship between coarse- and fine-grain structures remains unchanged under a changing climate is open to serious question. We therefore seek alternative approaches.

A regional climate model (RCM) uses the same physical principles and mathematical equations as global models, but covers only a limited geographical domain. This allows the use of a much finer computational grid. Whereas a global model might have points separated by 100 km, a regional model would typically have a 10 km grid. Thus, it is capable of resolving a large range of atmospheric phenomena that 'fall between the gaps' of a GCM. In particular, the representation of mountains is much better in the RCM. Since temperature and precipitation are so strongly coupled with elevation and slope, this is a major advantage.

The climate scientists at University College Dublin and Met Éireann have been collaborating for several years on a project called the Community Climate Change Consortium for Ireland (C_4I). The objective is to develop regional climate models and apply them to the question of Ireland's future climate. The main results to date are summarised in the report *Ireland in a Warmer World* (McGrath and Lynch, 2008). The work has employed a range of regional models.

2.2.4 Uncertainty of the predictions

Climate models are the best means we have for predicting future changes in our climate. They have a solid scientific basis in the principles of physics and are increasing rapidly in sophistication and in the accuracy with which they can simulate the climate. However, the atmosphere is chaotic, that is, highly sensitive to very small changes, so that if the starting conditions are changed in a small way, the subsequent evolution will be completely different. Thus, no matter how good the models are, there will always be a degree of uncertainty in the predictions. There are many sources of uncertainty. Firstly, the initial state of the system is not known precisely; this is particularly the case for the deep ocean. Secondly, the external forcings cannot be known exactly; for example, the level of CO_2 20 or 40 years hence depends on population growth and on energy use, which may be strongly influenced by technological developments. Thirdly, the models themselves have imperfections, and many approximations must be made in constructing practical models. Fourthly, the physical processes in the atmosphere and ocean are complex and not completely understood. Finally, there are many feedbacks in the climate system that make it very sensitive to small changes.

Among the processes that can act as positive feedbacks, we may consider the level of water vapour, the ice albedo effect, cloud amount, carbon uptake by forest, and emission of methane from melting tundra. An increase of cloudiness may have either a cooling or a warming effect on climate, depending on the cloud height and structure. In general, low clouds act to cool the climate and high clouds have a warming effect. But in reality things are much more complicated. For climate predictions over the next century, cloud feedbacks remain the greatest source of uncertainty (e.g. Soden and Held, 2006).

2.2.5 Probabilistic prediction: the ensemble approach

In view of the many uncertainties in predicting future climate, scientists now approach the problem from a stochastic point of view. For example, instead of trying to predict the increase of temperature by 2100 in a deterministic manner, we aim to predict both the most probable change and also the probable range of changes that may be observed. In essence, we try to forecast the probability density function rather than just a single deterministic value. To do this, we need to assess the sensitivity of the predictions to various errors in the initial state, in the model itself and in the specified external forcings. Considering a range of values of initial conditions and other parameters, we perform a collection of numerical predictions, called an ensemble. Depending on the available computer power, the ensemble may contain anything from a few predictions to as many as a hundred.

Using the ensemble outputs, we can examine how the predictions differ, and the extent to which they spread out from each other. If they form a compact set of closely related values, we can be confident that the average of these values is a reliable guide to what may happen. If, however, they are widely dispersed, we may have relatively little confidence in the prediction. The IPCC predictions are generally based on ensembles of model runs, and are given in a probabilistic form. For example, the mean temperature increase under the assumptions of a particular emission scenario (A1b) is given as a 'best estimate' of 2.8 K and a 'likely range' of between 1.7 K and 4.4 K.

2.3 Long-term climate change

2.3.1 Greenhouse and icehouse climates

A useful classification of the earth's climate types is into icehouse climates, which feature large-scale continental ice sheets and widespread perennial sea ice at high latitudes, and greenhouse climates, which do not. We are currently in an icehouse world, but the earth has switched between the two states several times during its history. Based mostly on the presence or absence of geological evidence for large-scale glaciation, Frakes et al. (1992) identify five icehouse periods separated by four greenhouse periods during the past 600 million years, each interval lasting 50–100 million years. However, not all the cool periods identified were of the same intensity, as indicated in Fig 2.3. There is abundant evidence for two full-blown glaciations, during the early Carboniferous to mid-Permian (340–250 mya) and the late Cenozoic (34 mya to present), while the late Jurassic to early Cretaceous (183–105 mya) and late Ordovician to early Silurian (458–421 mya) cool periods may have seen only ephemeral ice sheets and seasonal sea ice.

Understanding how earth shifts between greenhouse and icehouse is a central question in palaeoclimatology. Changing levels of atmospheric greenhouse gas concentrations, especially CO_2, appear to play an important role in these transitions. CO_2 reconstructions over the Phanaerozoic are subject to much uncertainty, but they generally show reduced CO_2 levels during the Carboniferous–Permian and the late Cenozoic, qualitatively matching the glaciation record (Royer et al., 2004; Royer, 2006). There is also considerable modelling evidence that continental glaciation is controlled by CO_2 levels. The association between climate conditions and CO_2 is somewhat problematic for the end-Ordovician glaciation (440 mya, see Fig 2.3), which occurred during a period of apparently high CO_2. The discrepancy may be due to the short duration of this glacial interval, which is possibly not well resolved by the CO_2 data. It should also be recalled that insolation at that time was around 3–4% lower than today.

The question of what ultimately caused the CO_2 fluctuations over geological time is less settled. Long-term atmospheric CO_2 levels are controlled by the balance between the rate of volcanic outgassing and the rate of drawdown due to chemical weathering of silicate rocks (Walker et al., 1981; Berner et al., 1983). Plate tectonics can affect both source and sink: increased seafloor spreading rates or the subduction of carbonate-rich sediments will lead to increased outgassing, while the breakup of a supercontinent, continental drift from a temperate to a tropical zone and mountain uplift will all increase weathering rates and lead to reduced CO_2 levels. As a specific example, Kent and Muttoni (2008) suggest that high CO_2 during the early Cenozoic was due to the closing of Tethys (see Rodríguez-Sánchez and Arroyo, Chapter 13), a relatively shallow tropical ocean with a deep carpet of carbonate sediments, whose subduction led to a period of increased outgassing. This period ended with the collision of India and Asia, which also brought the highly weatherable Deccan Traps into the equatorial humid belt, explaining the subsequent decline in CO_2 through the Eocene and Oligocene. The evolution of land plants over geological time also had a major impact on the long-term carbon cycle, notably though the effect of root systems on weathering rates (Berner, 2004) and potentially through a host of other feedback loops (Beerling and Berner, 2005).

An important related question concerns the role of climate change in driving mass extinction events, particularly the 'big five' extinction events at the end-Ordovician (446 mya), late Devonian (371 mya), Permian–Triassic (PT) boundary (251 mya), Triassic–Jurassic (TJ) boundary (200 mya) and Cretaceous–Tertiary (KT) boundary (65 mya) (Sepkoski, 1982). The PT, TJ and KT extinctions coincided with intense volcanism events, forming the Siberian Traps, Central Atlantic Magmatic Province and the Deccan Traps respectively, and the PT and TJ were also periods of rapidly rising CO_2 and temperature (McElwain et al., 1999; Benton and Twitchett, 2003). It is not clear, however, whether global warming by itself is sufficient to explain the mass extinctions (Mayhew et al., 2008), or whether other 'kill mechanisms' need to be invoked. It is widely accepted that the KT extinction was associated with a large bolide impact (Alvarez et al., 1980), but there is scant evidence for such impacts during the other extinctions. Global warming can lead to sluggish oceanic circulation and widespread anoxia, which could explain extinction in the oceanic realm (Kiehl and Shields, 2005). It has also been hypothesised that the emission of sulphur dioxide (SO_2) and other pollutants from flood basalts may have had a directly noxious effect on biota, especially plants (van de Schootbrugge et al., 2009).

2.3.2 Neoproterozoic snowball earth

In the most extreme case, the earth can exist in a stable equilibrium where the oceans are entirely covered in ice. Whether this snowball earth scenario has ever actually occurred remains contentious. A brief paper by Kirschvink (1992), noting evidence for low-latitude glaciation during the Neoproterozoic (*c.* 750 mya), suggested a snowball earth could have existed at that time. The idea was promoted powerfully by Hoffman et al. (1998), who provided more extensive evidence, as well as a coherent narrative, of how the earth could enter and exit such a state. The paper triggered a spurt of publications on the topic. While there is general consensus that the Neoproterozoic was an extreme icehouse period featuring low-latitude glaciation, the community is divided as to whether the planet was truly covered in ice, or whether the sea-ice line reached only to the subtropics, leaving a broad swath of open ocean in the tropics (so-called 'slushball earth'). The two hypotheses are qualitatively different as regards their impacts on life: a true snowball would have killed off most marine life, since no light would have reached the underlying ocean through the thick layer of ice, while in a slushball scenario life could have continued relatively unperturbed in the tropics.

Climate model simulations suggest that a transition to a snowball earth can indeed occur under Neoproterozoic conditions, though this is somewhat sensitive to the precise level of atmospheric CO_2 (e.g. Donnadieu et al., 2004). A greater difficulty with the snowball hypothesis concerns the exit mechanism. The basic idea is that, since the hydrological cycle comes to a halt during a snowball episode and the oceans are isolated from the atmosphere, there are no sinks for atmospheric CO_2, so given enough time subaerial volcanism will gradually build up large atmospheric CO_2 concentrations. Eventually, a threshold is crossed where tropical temperatures are high enough to melt the ice, at which point a catastrophic meltback ensues as the surface albedo feedback works in reverse (Caldeira and Kasting, 1992; Hoffman et al., 1998). However, attempts to reproduce this process using a detailed climate model (Pierrehumbert, 2004, 2005) show that the system remains far short of deglaciation even at extremely elevated CO_2, in the order of 20% of the total mass of the atmosphere. Thus, either the earth was never in a snowball state, or other as yet unknown mechanisms explain the transition out of that state.

2.3.3 The early Cenozoic greenhouse

The early Eocene climate optimum

The early Cenozoic was a time of extreme global warmth, culminating in the early Eocene climate optimum (EECO) around 50 mya. Oxygen isotope proxies indicate that benthic temperatures (which reflect high-latitude annual mean ocean surface conditions) could have been as high as 12 °C (Zachos et al., 2001). Direct reconstruction of Arctic ocean surface temperatures using the TEX86 proxy suggest values as high as 19 °C (Sluijs et al., 2006). High-latitude continental interiors were also much warmer than today (Greenwood and Wing, 1995). In the tropics, recent sea-surface temperature reconstructions using $\delta^{18}O$ in well-preserved foraminifera indicate temperatures in excess of 30 °C, independently confirmed by TEX86 estimates (Pearson et al., 2007). Overall, these reconstructions suggest a global mean temperature during the EECO of perhaps 25 °C, around 10 °C higher than today.

Simulating the EECO poses a major challenge to climate models. Atmospheric CO_2 concentration during the EECO is known to have been much higher than today, with estimates ranging from around 1000 to 4000 ppm (compared to 280 ppm in modern pre-industrial times). Setting model CO_2 to values within this range, it is possible to match the 10 °C increase in global mean temperature (Shellito et al., 2003), but it has proved impossible to capture the very weak equator–pole temperature gradient indicated by the proxy reconstructions. This 'low-gradient paradox' suggests that current climate models may be missing some key physical ingredient responsible for the reduction in meridional temperature gradient.

Two types of explanation are possible: either some unknown or poorly represented radiative feedback acts to limit temperatures in the tropics or boost them at the poles, or dynamical atmosphere/ocean poleward heat transport increases much more rapidly with temperature than predicted by the models. Clouds play a major role in most explanations of the first type, either cooling the tropics (Lindzen et al., 2001) or warming the high latitudes (Sloan and Pollard, 1998; Abbot and Tziperman, 2008). As for the second category, it appears that poleward atmospheric heat transport in models cannot increase above a certain limit, for dynamical reasons that remain unclear (Caballero and Langen, 2005; O'Gorman and Schneider, 2008). Model ocean transports also do not increase much in coupled Eocene simulations (Huber and Sloan, 2001). Recent work has focused on the role of tropical cyclones, which are unresolved in current climate models and are therefore an ideal candidate for a missing process. Tropical cyclone frequency and intensity are predicted to increase in a warmer world; by increasing vertical mixing in the ocean, this could lead to stronger poleward heat transport (Emanuel, 2001, 2002). However, direct implementation of a temperature-dependent ocean mixing rate in a climate model, designed to mimic the effect of tropical cyclones, failed to show much increase in heat transport to the high latitudes (Korty et al., 2008).

More recently, it has become apparent that much of the model data discrepancy may actually be due to deficiencies in the data. While older temperature reconstructions from pelagic foraminifera (e.g. Zachos et al., 1994) showed Eocene tropical temperatures comparable to or even lower than today, more recent studies using exceptionally well-preserved forams indicate much higher temperatures (Pearson et al., 2007), suggesting that the colder estimates are biased low due to diagenesis. With these new tropical temperature estimates, the low-gradient paradox becomes much less severe.

Hyperthermals

The early Eocene was also exceptional in that it was punctuated by a series of hyperthermals, short-lived events of extreme global warmth. Several such events have been documented (Lourens et al., 2005). The best-known hyperthermal, the Palaeocene–Eocene thermal maximum (PETM), saw global temperature soar by over 5 °C in less than 10 000 years (Zachos et al., 2001), and then revert to pre-event levels over a longer period of about 150 000 years. The warming was accompanied by a sharp negative carbon isotope excursion (a reduction of the $^{12}C/^{13}C$ ratio) of global extent, and rapid shoaling of the calcite compensation depth indicative of ocean acidification (Zachos et al., 2005). The generally accepted interpretation of these facts is that a rapid injection of isotopically light carbon into the atmosphere caused a large, transient increase in the greenhouse effect, producing the rapid warming. The carbon anomaly was subsequently reabsorbed on the typical timescale associated with silicate rock weathering and oceanic deposition.

The origin of the requisite large, rapid carbon injection remains enigmatic. Several hypotheses have been put forward. The carbon could have been 'baked' out of deep rocks by volcanic intrusions (Svensen et al., 2004), although this is hard to reconcile with the short-lived and recurrent nature of the events (Lourens et al., 2005). Catastrophic methane release from clathrate deposits has also been invoked (Dickens, 2003), though building large deposits under warm early Eocene conditions is difficult (Buffett and Archer, 2004). Finally, the isotopically light carbon could have come from the large-scale respiration of organic matter, possibly through the dessication of epicontinental seas (Higgins and Schrag, 2006).

Whatever its origin, the rapid carbon injection during hyperthermals is the closest known analogue to the ongoing anthropogenic carbon release. In this context, hyperthermals provide a perspective on the kind of climate changes we may expect in future, as well as a unique opportunity to test the climate models used to predict these changes. The most recent estimate puts the total carbon release during the PETM at no more than 3000 Gt (Zeebe et al., 2009), which would have resulted is somewhat less than a doubling of the pre-existing atmospheric CO_2

concentration. This is insufficient to explain the estimated global warming of 5 °C during the PETM, given the typical sensitivity of current climate models of around 3 °C per doubling of CO_2. Thus, if we believe these estimates to be correct, either some unknown positive feedbacks acted during the PETM to amplify the warming, or the sensitivity of current climate models is biased towards low values.

2.3.4 Late Cenozoic glaciation

Glaciating Antarctica

Through the late Cretaceous and early Palaeogene, the earth was in a greenhouse state free of permanent continental ice sheets. Then, at the Eocene–Oligocene (EO) boundary (*c.* 34 mya), a large ice sheet abruptly formed covering Antarctica. The transition is marked by a step-like rise in the benthic $\delta^{18}O$ record (Fig 2.2). Early work conjectured that the transition was tectonically driven, coinciding with the separation of Australia from Antarctica (Kennett, 1977). The opening of the Tasman Gateway created a continuous ocean belt around Antarctica, bringing the Antarctic Circumpolar Current into existence and supposedly interrupting the southward flow of warm water from the subtropics that existed previously. The resulting thermal isolation of Antarctica would then have led to its glaciation. A study using a simplified ocean–atmosphere model showed that opening a circumpolar seaway would indeed give a modest cooling over Antarctica (Sijp and England, 2004). On the other hand, simulations using a full physics coupled ocean–atmosphere climate model under late Eocene conditions showed that Antarctica was in fact thermally isolated even before the opening of the Tasman Gateway (Huber et al., 2004); no warm subtropical currents reached high southern latitudes even with Tasmania attached to Antarctica.

DeConto and Pollard (2003) proposed an entirely different scenario for the onset of Antarctic glaciation. They ran a global atmospheric model with late Eocene topography coupled to an ice-sheet model, and gradually reduced the level of atmospheric CO_2. They found that once the concentration dropped below a threshold level, an ice sheet abruptly and spontaneously appeared over Antarctica. Although the change in CO_2 was gradual, the build-up of the ice sheet was rapid due to the strong feedbacks at play. The most obvious feedback involves surface albedo: ice and snow are much more reflective than most other types of surface cover, so their presence reduces net absorbed insolation and leads to lower temperatures, promoting further ice formation. Additionally, increasing ice-sheet height enhances mass gain during winter and reduces mass loss during summer (because a high ice-sheet top is colder and therefore less subject to melting), so that the ice sheet's mass balance becomes increasingly positive the larger it grows. This positive feedback is finally halted by downhill ice flow once the ice sheet becomes sufficiently massive. Antarctica, having high orography near the pole, is particularly sensitive to this mechanism, while the Arctic, with an ocean around the pole and no high orography on the surrounding continents, requires much lower levels of CO_2 to form an ice sheet.

Glaciating the Arctic

While the early Oligocene emplacement of a continental-scale Antarctic ice sheet is well established, the time of first appearance of ice sheets in the northern hemisphere remains controversial. Recent work shows evidence for ice-rafted debris in the northern hemisphere during the Eocene and early Oligocene (Eldrett et al., 2007; Tripati et al., 2008). Moreover, palaeothermometry studies using magnesium/calcium isotopes suggest little benthic temperature change across the EO boundary, implying that the entire 1.5% shift in $\delta^{18}O$ is attributable to ice-sheet growth (Billups and Schrag, 2003; Coxall et al., 2005). A shift this large would require more ice growth than can be accommodated on Antarctica. Overall, this new evidence suggests that northern hemisphere ice sheets appeared in the early Oligo-cene and that the EO boundary glaciation was in fact bipolar. However, DeConto et al. (2008) found that Arctic glaciation requires CO_2 concentration to drop below 280 ppm in their atmosphere–ice-sheet coupled model, a level far below that which prevailed during the early Oligocene according to the palaeo-CO_2 reconstruction of Pagani et al. (2005). In addition, palaeotemperature reconstruction using the TEX86 proxy shows that high latitude sea-surface temperatures dropped by about 5 °C across the EO boundary (Liu et al., 2009). This cooling can account for much of the observed oxygen isotope excursion, and the remaining fraction can be explained by an ice volume that can be comfortably accommodated on Antarctica alone. The evidence for ice-rafted debris can be explained by localised, ephemeral ice caps on the higher orography.

The final transition to the modern icehouse world, including orbitally paced waxing and waning of continental-scale northern hemisphere ice sheets, is thought to have occurred in the late Pliocene, *c.* 2.7 mya (Shackleton et al., 1984). Several mechanisms have been proposed for this transition, and there is as yet no clear consensus on which was the dominant one. The Panamanian hypothesis notes the rough coincidence between glacial onset and final closure of the Panamanian seaway (Keigwin, 1982). By preventing mixing of Atlantic and Pacific waters, this closure could have increased the strength of the Atlantic conveyor-belt circulation, increasing the supply of moisture to Greenland and ultimately feeding a larger ice sheet. Atlantic temperature proxy data provide some support for this scenario (Bartoli et al., 2005). The uplift hypothesis suggests that the rise of the Rockies and Himalayan Plateau during the Cenozoic would have altered atmospheric circulation patterns, making Greenland's climate cooler, moister and more conducive to ice-sheet formation (Ruddiman and Kutzbach, 1989). A more recent hypothesis points to evidence that during the late Miocene and early Pliocene the tropical Pacific was in a permanent El Niño state, i.e. one with permanently warm conditions in the eastern basin and no east–west equatorial temperature gradient (Wara et al., 2005). Transition to the modern regime, with cooler average conditions in the east and interannual fluctuations between El Niño (warm) and La Niña (cool) events, roughly coincides with glacial onset. Noting the modern observed impacts of El Niño events on high-latitude climate, some authors have drawn a causal connection between the two transitions (Molnar and Cane, 2002; Philander and Fedorov, 2003). Finally, it is possible that CO_2 concentrations decreased through the Pliocene (Lunt et al., 2008), bringing the system across the threshold for Arctic glaciation (DeConto et al., 2008). A test of all these hypotheses using a comprehensive atmosphere–ocean ice-sheet model found that only a drop in CO_2 could fully account for glacial onset (Lunt et al., 2008). Furthermore, model date comparison by Huber and Caballero (2003) showed that an active El Niño–Southern Oscillation (ENSO) cycle probably existed during the Eocene, when all conditions for a permanent El Niño were satisfied, so it is difficult to see what could have induced such a state in the Miocene.

2.4 Conclusions

Earth's climate system fluctuates on all timescales, from hours to billions of years. Over its history, earth has gone through a succession of greenhouse and icehouse states, which correlate well with periods of high and low atmospheric CO_2 concentration. Climate modelling is playing an increasingly important role in unravelling the intricate details of these transitions. At the same time, palaeoclimate applications expose some of the greatest limitations of current models. In particular, palaeoclimate proxy data from the very warm climates of the early Eocene suggest that climate models may substantially underestimate the sensitivity of surface temperature to CO_2, particularly at high latitudes. While the long-standing low-gradient paradox may be nearing resolution due to improved palaeotemperature estimates, the emerging challenge is to explain how palaeoclimates may have become so warm with relatively modest levels of CO_2. Changes in clouds, perhaps linked to changes in ocean biota (Kump and Pollard, 2008), may provide part of the answer. This provides an exciting avenue for future research in bioclimate interactions.

References

Abbot, D. S. and Tziperman, E. (2008). Sea ice, high-latitude convection, and equable climates. *Geophysical Research Letters*, **35**, L03702.

Alvarez, L. W., Alvarez, W., Asaro, F. and Michel, H. V. (1980). Extraterrestrial cause for the Cretaceous–Tertiary extinction. *Science*, **208**, 1095–1108.

Bartoli, G., Sarnthein, M., Weinelt, M. et al. (2005). Final closure of Panama and the onset of northern hemisphere glaciation. *Earth and Planetary Science Letters*, **237**, 33–44.

Beerling, D. J. and Berner, R. A. (2005). Feedbacks and the coevolution of plants and atmospheric CO_2. *Proceedings of the National Academy of Sciences of the USA*, **102**, 1302–1305.

Benton, M. J. and Twitchett, R. J. (2003). How to kill (almost) all life: the end-Permian extinction event. *Trends in Ecology and Evolution*, **18**, 358–365.

Beran, J. (1994). *Statistics for Long-Memory Processes*. Monographs on Statistics and Applied Probability, Boca Raton, FL: Chapman and Hall/CRC.

Berner, R. A. (2004). *The Phanerozoic Carbon Cycle: CO_2 and O_2*. Oxford: Oxford University Press.

Berner, R. A., Lasaga, A. and Garrels, R. (1983). The carbonate-silicate geochemical cycle and its effect on atmospheric carbon dioxide over the past 100 million years. *American Journal of Science*, **283**, 641–683.

Billups, K. and Schrag, D. P. (2003). Application of benthic foraminiferal Mg/Ca ratios to questions of Cenozoic climate change. *Earth and Planetary Science Letters*, **209**, 181–195.

Buffett, B. and Archer, D. (2004). Global inventory of methane clathrate: sensitivity to changes in the deep ocean. *Earth and Planetary Science Letters*, **227**, 185–199.

Caballero, R. and Langen, P. L. (2005). The dynamic range of poleward energy transport in an atmospheric general circulation model. *Geophysical Research Letters*, **32**, L02705.

Caldeira, K. and Kasting, J. F. (1992). Susceptibility of the early Earth to irreversible glaciation caused by carbon dioxide clouds. *Nature*, **359**, 226–228.

Charney, J. G., Fjørtoft, R. and von Neumann, J. (1950). Numerical integration of the barotropic vorticity equation. *Tellus*, **2**, 237–254.

Coxall, H. K., Wilson, P. A., Pälike, H., Lear, C. H. and Backman, J. (2005). Rapid stepwise onset of Antarctic glaciation and deeper calcite compensation in the Pacific Ocean. *Nature*, **433**, 53–57.

DeConto, R. M. and Pollard, D. (2003). Rapid Cenozoic glaciation of Antarctica induced by declining atmospheric CO_2. *Nature*, **421**, 245–249.

DeConto, R. M., Pollard, D., Wilson, P. et al. (2008). Thresholds for Cenozoic bipolar glaciation. *Nature*, **455**, 652–656.

Dickens, G. R. (2003). Rethinking the global carbon cycle with a large, dynamic and microbially mediated gas hydrate capacitor. *Earth and Planetary Science Letters*, **213**, 169–183.

Donnadieu, Y., Goddéris, Y., Ramstein, G., Nédélec, A. and Meert, J. (2004). A snowball Earth climate triggered by continental break-up through changes in runoff. *Nature*, **428**, 303–306.

Eldrett, J. S., Harding, I. C., Wilson, P. A., Butler, E. and Roberts, A. P. (2007). Continental ice in Greenland during the Eocene and Oligocene. *Nature*, **446**, 176–179.

Emanuel, K. (2001). Contribution of tropical cyclones to meridional heat transport by the oceans. *Journal of Geophysical Research*, **106**, 14771–14781.

Emanuel, K. (2002). A simple model for multiple climate regimes. *Journal of Geophysical Research*, **107** (D9), 4077.

Frakes, L. A., Francis, J. E. and Syktus, J. I. (1992). *Climate Modes of the Phanerozoic*. Cambridge: Cambridge University Press.

Greenwood, D. R. and Wing, S. L. (1995). Eocene continental climates and latitudinal temperature gradients. *Geology*, **23**, 1044–1048.

Higgins, J. A. and Schrag, D. P. (2006). Beyond methane: towards a theory for the Paleocene-Eocene thermal maximum. *Earth and Planetary Science Letters*, **245**, 523–537.

Hoffman, P. F., Kaufman, A. J., Halverson, G. P. and Schrag, D. P. (1998). A Neoproterozoic snowball Earth. *Science*, **281**, 1342–1346.

Huber, M. and Caballero, R. (2003). Eocene El Niño: evidence for robust tropical dynamics in the 'hothouse'. *Science*, **299**, 877–881.

Huber, M. and Sloan, L. C. (2001). Heat transport, deep waters and thermal gradients: coupled climate simulation of an Eocene greenhouse climate. *Geophysical Research Letters*, **28**, 3841–3884.

Huber, M., Brinkhuis, H., Stickley, C. E. et al. (2004). Eocene circulation of the Southern Ocean: was Antarctica kept warm by subtropical waters? *Paleoceanography*, **19**, PA4026.

Huybers, P. and Curry, W. (2006). Links between annual, Milankovitch and continuum temperature variability. *Nature*, **441**, 329–332.

Intergovernmental Panel on Climate Change (IPCC) (2007). *Climate Change 2007: the Physical Science Basis. Contribution of Working Group I to the Fourth Assessment Report of the Intergovernmental Panel on Climate Change*, ed. S. Solomon, D. Qin, M. Manning et al. Cambridge: Cambridge University Press.

Keigwin, L. (1982). Isotopic paleoceanography of the Caribbean and East Pacific: role of Panama uplift in late Neogene time. *Science*, **217**, 350–353.

Kennett, J. P. (1977). Cenozoic evolution of Antarctic glaciation, the circum-Antarctic Ocean, and their impact on global paleoceanography. *Journal of Geophysical Research*, **82**, 3843–3860.

Kent, D. V. and Muttoni, G. (2008). Equatorial convergence of India and early Cenozoic climate trends. *Proceedings of the National Academy of Sciences of the USA*, **105**, 16065–16070.

Kiehl, J. T. and Shields, C. A. (2005). Climate simulation of the latest Permian: implications for mass extinction. *Geology*, **33**, 757–760.

Kirschvink, J. L. (1992). Late Proterozoic low-latitude global glaciation: the snowball earth. In *The Proterozoic Biosphere: a Multidisciplinary Study*, ed. J. W. Schopf and C. Klein. Cambridge: Cambridge University Press, pp. 51–52.

Korty, R. L., Emanuel, K. A. and Scott, J. R. (2008). Tropical cyclone-induced upper-ocean mixing and climate: application to equable climates. *Journal of Climate*, **21**, 638–654.

Kump, L. R. and Pollard, D. (2008). Amplification of Cretaceous warmth by biological cloud feedbacks. *Science*, **320**, 195.

Kutzbach, J. E. (1976). The nature of climate and climatic variations. *Quaternary Research*, **6**, 471–480.

Lewis, J. M. (1998). Clarifying the dynamics of the general circulation: Phillips's 1956 experiment. *Bulletin of the American Meteorological Society*, **79**, 39–60.

Lindzen, R. S., Chou, M. D. and Hou, A. Y. (2001). Does the earth have an adaptive infrared iris? *Bulletin of the American Meteorological Society*, **82**, 417–432.

Liu, Z., Pagani, M., Zinniker, D. et al. (2009). Global cooling during the Eocene-Oligocene climate transition. *Science*, **323**, 1187–1190.

Lourens, L. J., Sluijs, A., Kroon, D. et al. (2005). Astronomical pacing of late Palaeocene to early Eocene global warming events. *Nature*, **435**, 1083–1087.

Lunt, D. J., Foster, G. L., Haywood, A. M. and Stone, E. J. (2008). Late Pliocene Greenland glaciation controlled by a decline in atmospheric CO_2 levels. *Nature*, **454**, 1102–1105.

Lynch, P. (2006). *The Emergence of Numerical Weather Prediction: Richardson's Dream*. Cambridge: Cambridge University Press.

Mayhew, P. J., Jenkins, G. B. and Benton, T. G. (2008). A long-term association between global temperature and biodiversity, origination and extinction in the fossil record. *Proceedings of the Royal Society of London B*, **275**, 47–53.

McElwain, J. C., Beerling, D. J. and Woodward, F. I. (1999). Fossil plants and global warming at the Triassic–Jurassic boundary. *Science*, **285**, 1386–1390.

McGrath, R. and Lynch, P., eds. (2008). *Ireland in a Warmer World: Scientific Predictions of the Irish Climate in the Twenty-First Century*. Dublin: Met Éireann.

Mitchell, J. M. (1976). An overview of climatic variability and its causal mechanisms. *Quaternary Research*, **6**, 481–493.

Molnar, P. and Cane, M. A. (2002). El Niño's tropical climate and teleconnections as a blueprint for pre-Ice Age climates. *Paleoceanography*, **17**, 1021.

O'Gorman, P. A. and Schneider, T. (2008). The hydrological cycle over a wide range of climates simulated with an idealized GCM. *Journal of Climate*, **21**, 3815–3832.

Pagani, M., Zachos, J. C., Freeman, K. H., Tipple, B. and Bohaty, S. (2005). Marked decline in atmospheric carbon dioxide concentrations during the Paleogene. *Science*, **309**, 600–603.

Pearson, P. N., van Dongen, B. E., Nicholas, C. J. et al. (2007). Stable warm tropical climate through the Eocene Epoch. *Geology*, **35**, 211–214.

Philander, S. G. and Fedorov, A. V. (2003). Role of tropics in changing the response to Milankovich forcing some three million years ago. *Paleoceanography*, **18**, 1045.

Phillips, N. (1956). The general circulation of the atmosphere: a numerical experiment. *Quarterly Journal of the Royal Meteorological Society*, **82**, 123–164.

Pierrehumbert, R. T. (2004). High levels of atmospheric carbon dioxide necessary for the termination of global glaciation. *Nature*, **429**, 646–649.

Pierrehumbert, R. T. (2005). Climate dynamics of a hard snowball earth. *Journal of Geophysical Research*, **110**, D1111.

Royer, D. L. (2006). CO_2-forced climate thresholds during the Phanerozoic. *Geochimica et Cosmochimica Acta*, **70**, 5665–5675.

Royer, D. L., Berner, R. A., Montañez, I. P., Tabor, N. J. and Beerling, D. J. (2004). CO_2 as a primary driver of Phanerozoic climate. *Geological Society of America Today*, **14**, 5.

Ruddiman, W. F. and Kutzbach, J. E. (1989). Forcing of late Cenozoic Northern Hemisphere climate by plateau uplift in southern Asia and the American West. *Journal of Geophysical Research*, **94**, 18, 409–418, 427.

Sagan, C. and Mullen, G. (1972). Earth and Mars: evolution of atmospheres and surface temperatures. *Science*, **177**, 52–56.

Sepkoski, J. J. (1982). Mass extinctions in the Phanerozoic oceans: a review. *Geological Society of America Special Papers*, **191**, 283–289.

Shackleton, N., Backman, J., Zimmerman, H. et al. (1984). Oxygen isotope calibration of the onset of ice-rafting and history of glaciation in the North Atlantic region. *Nature*, **307**, 620–623.

Shaviv, N. J. and Veizer, J. (2003). Celestial driver of Phanerozoic climate? *Geological Society of America Today*, **13**, 4–10.

Shellito, C. J., Sloan, L. C. and Huber, M. (2003). Climate sensititvity to atmospheric CO_2 levels in the Early-Middle Paleogene. *Paleogeography, Paleoclimatology, Paleoecology*, **193**, 113–123.

Sijp, W. P. and England, M. H. (2004). Effect of the Drake Passage throughflow on global climate. *Journal of Physical Oceanography*, **34**, 1254–1266.

Sloan, L. C. and Pollard, D. (1998). Polar stratospheric clouds: a high latitude warming mechanism in an ancient greenhouse world. *Geophysical Research Letters*, **25**, 3517–3520.

Sluijs, A., Schouten, S., Pagani, M. et al. (2006). Subtropical Arctic Ocean temperatures during the Palaeocene/Eocene thermal maximum. *Nature*, **441**, 610–613.

Soden, B. J. and Held, I. M. (2006). An assessment of climate feedbacks in coupled ocean–atmosphere models. *Journal of Climate*, **19**, 3354–3360.

Svensen, H., Planke, S., Malthe-Sørenssen, A. et al. (2004). Release of methane from a volcanic basin as a mechanism for initial Eocene global warming. *Nature*, **429**, 542–545.

Tripati, A. K., Eagle, R. A., Morton, A. et al. (2008). Evidence for glaciation in the Northern Hemisphere back to 44 Ma from ice-rafted debris in the Greenland Sea. *Earth and Planetary Science Letters*, **265**, 112–122.

van de Schootbrugge, B., Quan, T. M., Lindström, S. et al. (2009). Floral changes across the Triassic/Jurassic boundary linked to flood basalt volcanism. *Nature Geoscience*, **2**, 589–594.

Walker, J. C. G., Hays, P. B. and Kasting, J. F. (1981). A negative feedback mechanism for the long-term stabilization of the earth's surface temperature. *Journal of Geophysical Research*, **86**, 9776–9782.

Wara, M. W., Ravelo, A. C. and Delaney, M. L. (2005). Permanent El Niño-like conditions during the Pliocene warm period. *Science*, **309**, 758–761.

Washington, W. M. and Parkinson, C. L. (2005). *An Introduction to Three-Dimensional Climate Modeling*, 2nd edn. Sausalito, CA: University Science Books.

Wilby, R. L. and Wigley, T. M. L. (1997). Downscaling general circulation model output: a review of methods and limitations. *Progress in Physical Geography*, **21**, 530–548.

Wilson, J. T. (1966). Did the Atlantic close and then re-open? *Nature*, **211**, 676–681.

Wunsch, C. (2003). The spectral description of climate change including the 100 ky energy. *Climate Dynamics*, **20**, 353–363.

Wunsch, C. (2004). Quantitative estimate of the Milankovitch-forced contribution to observed Quaternary climate change. *Quarterly Science Reviews*, **23**, 1001–1012.

Zachos, J. C., Stott, L. D. and Lohmann, K. C. (1994). Evolution of early Cenozoic marine temperatures. *Paleoceanography*, **9**, 353–387.

Zachos, J. C., Pagani, M., Sloan, L., Thomas, E. and Billups, K. (2001). Trends, rythms and aberrations in global climate 65 Ma to present. *Science*, **292**, 686–693.

Zachos, J. C., Röhl, U., Schellenberg, S. A. et al. (2005). Rapid acidification of the ocean during the Paleocene–Eocene thermal maximum. *Science*, **308**, 1611–1615.

Zeebe, R. E., Zachos, J. C. and Dickens, G. R. (2009). Carbon dioxide forcing alone insufficient to explain Palaeocene–Eocene Thermal Maximum warming. *Nature Geoscience*, **2**, 576–580.

3

The perils of addressing long-term challenges in a short-term world: making descriptive taxonomy predictive

R. M. Bateman

Jodrell Laboratory, Royal Botanic Gardens, Kew and *School of Geography, Earth and Environmental Sciences, University of Birmingham, UK*

Abstract

Increased political interest in addressing environmental issues, notably climate change, conservation and landscape restoration, has the potential to strengthen the focus, integration and profile of systematic biology. In particular, it could rescue descriptive taxonomy from its current state of near-extirpation in the developed world. However, exploiting this opportunity will require greater consensus than the systematics community has previously achieved, together with the determination to resist exaggerating the value of existing systematic data and of technological advances such as DNA barcoding and web-based identification. Descriptive taxonomy erects hypotheses of species existence that must be tested using other categories of data if systematics is to become a genuinely predictive enterprise. Prediction followed by recommendations for adaptation and/or mitigation, each essential to address the consequences of climate change, are possible only with

Climate Change, Ecology and Systematics, ed. Trevor R. Hodkinson, Michael B. Jones, Stephen Waldren and John A. N. Parnell. Published by Cambridge University Press.

good knowledge of the species and ecosystems under scrutiny. Taxonomic data alone are of little value, but, equally, non-taxonomic data are rarely of value in the absence of a taxonomic framework. Instead of seeking shortcuts to, or even substitutes for, taxonomy in the hope of accelerating the rate of superficial species description (and redescription), the climate change challenge is best addressed by obligatorily increasing the rigour required in taxonomic descriptions. This especially requires: (1) escaping from traditional typology by prescribing minimum levels of both morphological and molecular data via obligatory online registration of species; (2) requiring taxonomists to state the species concept(s) employed in each study; (3) improving feedback to taxonomy from identifications performed by non-systematists; and (4) prioritising groups for taxonomic study according to the importance of the questions that the study group can address. More broadly, an effective response to the challenge of climate change requires that enabling disciplines that are inherently long-term, notably descriptive taxonomy and ecological/environmental monitoring, are integrated more effectively and then adequately insulated from the vagaries of traditional three-year funding cycles.

3.1 Introduction

3.1.1 Background

Although the scientific community has long paid close attention to present-day and fossil evidence of profound climate change, it did not impinge greatly on the minds of either public or policy makers for several decades. It is rare that one can point to a single day as marking a major threshold of interest in, and commitment to, a particular research topic by a national government, but such an event occurred on 30 October, 2006. On that day, the Stern report, commissioned by the UK government, persuasively linked climate change to macroeconomic projections, thereby immediately capturing the attention of both policy makers and the media worldwide (Stern, 2006). For example, the *Independent* newspaper (UK) of 31 October devoted not only its entire cover but also 15 interior pages to the report.

In recent years, where policy has led, funding has rapidly followed. Predictably, a significant proportion of the UK's biological and environmental research funding was soon redirected towards climate change. For example, the 2007–12 strategy of the UK's Natural Environment Research Council (NERC) permits only a single strategic goal: 'To deliver the scientific evidence needed for governments, business and society to respond urgently to the increasing pressures on natural resources and global climate' (NERC, 2007). Many research disciplines have reoriented themselves rapidly in response to this remarkable shift in socially responsible science. In contrast, it is my contention that the rapid rise of climate change studies has exposed the damagingly poor integration and waning political influence of the systematics community. Moreover, I believe that this weakness is rooted as much in the way that we pursue our science as in the way that we pursue political influence.

This chapter was inspired most strongly by two texts that were both published in 2008. The first was *Systematics and Taxonomy*, the proceedings and recommendations of the fourth review of the UK's systematics base that had been conducted since 1992 (House of Lords, 2008). In my judgement, the main objectives of this review were to explain and remedy the continued inability of the UK's taxonomic base to obtain adequate funding, integrate efficiently with other biological disciplines, and contribute effectively to high-priority policy objectives. The second publication was *The New Taxonomy*, a Systematics Association volume edited by Wheeler (2008a). The book sought to present an integrated and (in part) high-tech vision for the future of taxonomy, to prioritise taxonomy above other disciplines within systematic biology, and to distinguish taxonomy from the process-driven 'Modern Synthesis' (e.g. Huxley, 1940 – a document that effectively constituted the founding manifesto of the Systematics Association). In addition, I have drawn upon certain articles published recently in *The Systematist* and from several prescriptions for taxonomy derived primarily from the UK and North America (Systematics Agenda 2000, 1994; Blackmore et al., 1998; Futuyma, 1998; Cracraft et al., 2002; House of Lords, 2002, 2008; Crane, 2003; Godfray and Knapp, 2004; Wheeler, 2008b).

3.1.2 Precepts

This chapter is based on the recognition that taxonomy, evolution and climate change share one characteristic that is of paramount importance: they all operate on timescales far longer than the period covered by a typical research grant (or a typical term in government). I make no claims for the objectivity of this text. Rather, I have drawn on three decades operating as a palaeo-environmentalist, palaeontological and neontological systematist, and on a dozen years interacting with politicians and policy makers. This experience, gained in the UK and in the USA, has led me to the following precepts regarding neontological (as opposed to palaeontological) taxonomy:

(1) The main challenges presented by climate change and biodiversity conservation are ecological, falling under the broad headings of monitoring, remediation and adaptation.

(2) There is currently a damaging imbalance between efforts made to mine and synthesise existing data relating to climate change compared with the greater effort needed to generate genuinely novel, 'bespoke' data.

(3) Taxonomy should be defined narrowly to cover only formal description and naming, whereas systematic biology should be defined broadly to encompass

all areas of comparative biology. The two terms (which cause much confusion, both within and outwith the discipline) are far from being synonymous.

(4) Taxonomy *s.s.* makes a vital but largely indirect contribution to applied challenges such as climate change; contrary to some recent statements, taxonomy erects hypotheses rather than testing them, and so is not a predictive science. Thus, taxonomy can feed directly into biodiversity monitoring via identification but not into attempts at remediation or adaptation.

(5) Taxonomic activity should be better integrated into, and driven by, other biological disciplines. In contrast, taxonomic funding should be separated out and linked with other fundamentally long-term 'enabling' disciplines such as ecological and environmental monitoring.

3.2 Clarifying the relationship between taxonomy and systematics

There is perhaps no better illustration of the stubborn individualism of systematists than our inability to reach consensus on the definitions of, and relationship between, taxonomy and systematics. This perennial ambiguity has both conceptual and practical consequences. For example, the introduction of the House of Lords (2008) review offered the following definitions:

> *Taxonomy* is the scientific discipline of describing, delimiting and naming organisms, both living and fossil; *systematics* is the process of organising taxonomic information about organisms into a logical classification that provides the framework for all comparative studies [implying a phylogenetic framework]. Systematics and taxonomy are referred to collectively as *systematic biology*.

However, when drafting the Systematics Association's evidence to the Lords review (House of Lords, 2008), I preferred a significantly more inclusive definition of systematic biology that extended well beyond classification, encompassing any activity that employs the comparative method. This broad definition incorporates many studies in evolutionary biology (a phrase startlingly absent from the Lords review), crucially including population genetic data and the derived subdiscipline of biogeography that has been termed phylogeography (e.g. Avise, 2000).

Further developing these definitions, the core sister disciplines within systematic (comparative) biology are taxonomy and phylogenetics. Taxonomy concerns the delimitation, description and identification of all species and other taxa, both extant and extinct. This has traditionally been pursued via global treatments of a relatively narrow range of species (monographs) or regional treatments of a much wider range of species (inventories leading to floras/faunas), typically based primarily on reference collections of specimens. Taxonomy is increasingly assimilating information from various nucleic-acid-based approaches such as population genetics and DNA 'barcoding'. Phylogenetics explores the evolutionary relationships among the species and other taxa that have been generated by taxonomists, using various comparative approaches to explore both morphological and DNA-based information. Its ultimate aim is to reconstruct and interpret the 'tree of life'. Both taxonomy and phylogenetics are integrated into the naming of taxa and their organisation into hierarchical systems of classification.

Taxonomy provides the essential framework for any biological study, by providing the formal classifications and names – and also the standardised terminology – to reference, describe and identify the organisms that constitute the earth's biota. It can justly be argued that any biological study, and certainly any comparative biological study, is ultimately rooted in taxonomy. Taxonomic activities can usefully be categorised in two subdisciplines. Descriptive taxonomy is relatively creative and proactive, inevitably requiring specialist knowledge and involving the highly prescribed formal description of new taxa, preferably followed by their redescription as further relevant data are gathered. Applied taxonomy is generally more reactive, encompassing the subsequent use of those classifications to identify organisms, and the dissemination of the resulting data on species recognition and distribution. It is these definitions of systematics, phylogenetics, and descriptive and applied taxonomy that are employed in the present text. Mistakenly synonymising taxonomy into systematics (e.g. Enghoff and Seberg, 2006) is more than a mere semantic error. If taxonomy is substantially different from most other scientific disciplines (including the remainder of systematic biology), then we obscure those differences at our peril.

So, just what does differentiate taxonomy *s.s.* from other scientific disciplines, including other aspects of systematic biology? Taxonomy is a classic enabling (or 'foundation') science, here defined as disciplines that erect rather than test hypotheses. As described by May (1999),

> The task of inventorying is sometimes mistaken for 'stamp collecting' by thoughtless colleagues ... but such information is a prerequisite to the proper formulation of evolutionary and ecological questions, and essential for rational assignment of priorities in conservation biology.

Taxonomy is anchored directly and firmly in collections in general, and type specimens in particular. It is dictated by complex nomenclatural legislation that requires expert knowledge, and thus far it has gained only limited benefits from the application of technological 'shortcuts' (e.g. Enghoff and Seberg, 2006; Flowers, 2007); taxonomy instead relies on the hard-won experience of specialists. Taxonomic outputs are diverse and tend to attract exceptionally low scores on criteria such as the Institute for Scientific Information (ISI) impact factor (the dominant measure of how many times a published paper is cited in other published papers); indeed, if published in specialist journals, edited volumes or monographs, the outputs lie outside the realm of impact factors. This perceived weakness is partly compensated for by their exceptionally long half-life; old outputs can even trump recent outputs, notwithstanding technological advances. However, the overarching feature of taxonomy is the need for long-term stability of both activities and funding. The uses made of taxonomic data may change in emphasis through time, as exemplified by the recent elevation of climate change research, but the primary goal of taxonomy – documenting life on earth – does not.

Taxonomy is most commonly promoted to non-systematists on a platform of practical applications that have direct and obvious impacts on society. Although climate change has become pre-eminent in the post-Stern world, the list of applications of taxonomy is long: crop development (including forestry and fisheries) and biofuels (House of Commons, 2007); sustainability (Duraiappah and Naeem, 2005); pests, diseases and health; bioprospecting; bioremediation; conservation of biodiversity (Crane, 2003); recreation and education. Linking systematics disciplines to applications gives the most informative perspective on their relationships (Fig 3.1). Before considering further the relationship of taxonomy with its many user communities, I will critically review some basic assumptions inherent in standard taxonomic practice.

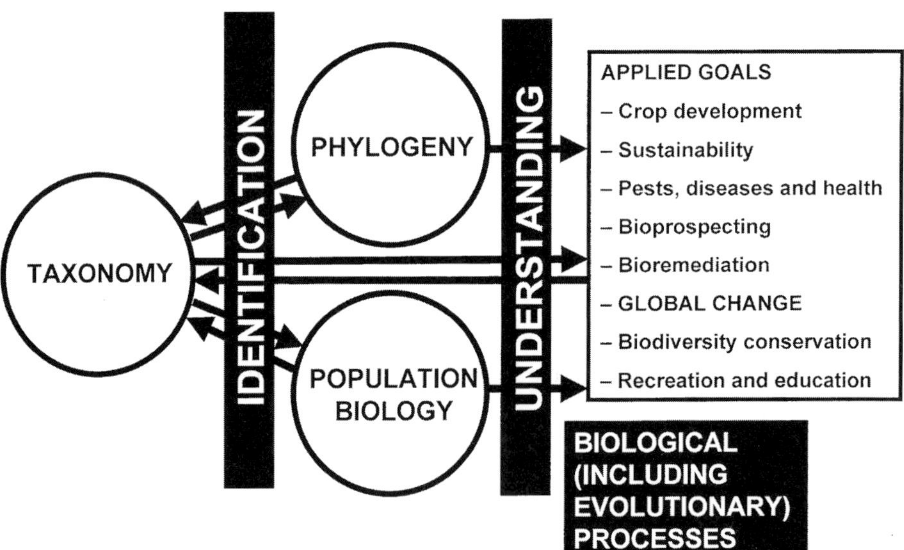

Figure 3.1 Flow diagram illustrating how taxonomic information must pass (in part via phylogenetics and population biology) through a filter of biological understanding in order to adequately inform applied questions such as climate change. Modified after Bateman in House of Lords (2008).

3.3 Escaping from the typological straightjacket

Systematics may appear overly anarchic compared with other scientific disciplines. However, there is one notable exception, where order is ruthlessly imposed. Approaches to erecting and legally validating a taxonomic species (i.e. formal nomenclature) are heavily prescribed by the International Codes of Botanical, Zoological and Microbial Nomenclature. These volumes are so complex and legalistic that they would not look out of place on a barrister's bookshelf. More importantly, the rules are both obligatory and applied worldwide, representing a rare example of genuinely global unification. However, the formal name is only one-third of the traditional description of a new species (termed the protologue); at least one representative specimen (the type) and a formal description are also required. The species epithet (e.g. *perennis*) is not unique or usable until expanded into a Latin binomial of genus and species names, thus immediately necessitating placement in a hierarchical classification (e.g. *Bellis perennis*), and it is not valid unless succeeded by the author of the resulting Linnean binomial (e.g. *Bellis perennis* L.). Assignment to a family is also routine (Asteraceae).

In the case of eukaryotic organisms, the type specimen is often only a portion of a single individual that represents just one stage of its life history. Indeed, under duress, it can even be substituted by an image and can lack all details of its geographical origin and source habitat. Even if multiple specimens are available, one is typically prioritised as the selected holotype, which then operates as an immutable reference point for all subsequent taxonomic consideration of the species in question. The formal description of most novel species is exclusively morphological, and at least implicitly couched in terms of characters divided into contrasting states (e.g. petals red versus white).

I find it astonishing that the International Code of Botanical Nomenclature dictates that the formal description must be given in Latin but makes no requirement whatsoever regarding the content of the description (Bateman, 2009b). As long as we can satisfactorily translate 'Resembles Mickey Mouse' into Latin, that single phrase will constitute a perfectly valid (if utterly useless) description of *Organismus michaelis-mus*! I have long toyed with the idea of describing a new taxon using only DNA characters – for example, 'possesses the nucleotide cytosine at a location 156 nucleotides from the 5' end of the ITS region of the nuclear ribosomal DNA'. This description would be valid; no rule in the Code stipulates that the formal description must include morphological characters. Nor, sadly, does any rule state that a new taxon cannot be described until genetic and/or ecological as well as morphological data have been acquired from at least one specimen, and preferably from a representative range of populations. If such a rule did exist, taxonomic 'divination' would at last be superseded by basic science, and far fewer frivolous names would be coined.

Thus, as currently practised, erecting and legally validating a taxonomic species is fundamentally a typological process. We have insufficient information at our disposal to adequately circumscribe that species, in terms of either intrinsic properties such as phenotype and genotype or emergent properties such as its distribution and ecology. In short, the lack of internationally agreed minimum requirements for the nature, quantity and quality of data underpinning a formal description weakens the scientific credentials of taxonomy. Moreover, morphological phylogenetic studies, and until recently most molecular phylogenetic studies, routinely represented an entire species as a single row of data (often presenting only a minute portion of the genome of a single organism) and so are similarly typological.

This routine reliance on typology means that taxonomy *s.s.* plays an essential role in erecting species hypotheses but cannot adequately test those hypotheses. Arguably, a weak test of the biological validity of taxonomically erected species rests in the discovery of further specimens that can be compared with the type for the same characters. Fortunately, as discussed below, much stronger tests can be applied by gathering relevant data via other branches of comparative biology.

Crucially, the type and the epithet are unique data points that cannot in practice be modified; they can only be accepted or rejected (i.e. they are fundamentally typological). Only the description is subject to modification as data accumulate, allowing progression from the typological origin (point of first description) toward the unattainable goal of complete knowledge (Fig 3.2). Refining the protologue is most simply achieved by testing it against other putatively conspecific individuals using the same category of data – a process better known to biologists as identification! More individuals and/or populations are added, often involving expansion of sampling into new habitats and/or geographical regions. A stronger test is to

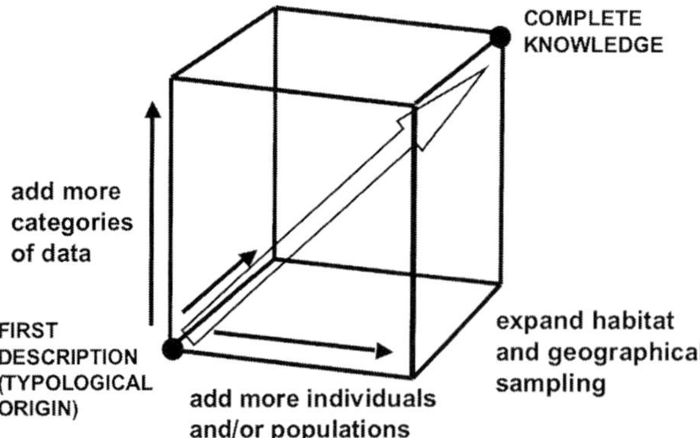

Figure 3.2 Hypothetical cube illustrating how improved sampling of individuals and categories of descriptive characters can eventually propel a species from initial tentative recognition to deep and potentially predictive biological knowledge.

analyse putatively conspecific individuals for a different category of data, seeking congruent patterns in contrasting categories of biological data (e.g. testing genetically a circumscription that was based initially on morphology). And the strongest test is to broaden the taxonomic spectrum under consideration to other putatively non-conspecific individuals, seeking discontinuities in the data that are likely to represent species boundaries; a level of comparison best achieved using quantitative phenetic and/or cladistic methods (e.g. Bateman, 2001).

Whatever approach is taken to systematics, it is clear that we need considerably more information about a species than just an epithet and a type specimen before that species becomes useful, not least in climate change studies. Until adequately tested, a hypothesised species is of no practical value. Before exploring this viewpoint further, I will consider in greater detail the significance of: (1) analysing a wider range of individuals within a hierarchical context; and (2) gathering multiple categories of taxonomically relevant data.

3.4 The value of a demographic approach to species delimitation employing both morphological and molecular taxonomic data

Once species have been formally erected, their biological validity can only be tested by employing an approach to species delimitation that is 'demographic', as defined by Bateman (2001) (Fig 3.3). This requires extensive quantitative description of individuals in multiple populations per putative species. Individuals are aggregated into populations in search of the discontinuities that separate the sets of populations that together constitute bona fide species. In contrast, overlap between populations of putatively different species is assumed to indicate conspecificity.

Admittedly, both discontinuities and overlaps between populations are by definition 'metastable'; they are vulnerable to elimination by further sampling. For example, in the hypothetical morphological ordination of eight populations shown in Fig 3.3a, a clear discontinuity exists between two clusters of six populations; thus, each of the two clusters is circumscribed as a distinct species. However, subsequently sampling a further four intermediate populations (Fig 3.3b) bridges the former discontinuity, so that the spread of organisms is now continuous and the definition of a species employed here requires that the two former species should be merged into only one revised species. We might then choose to greatly expand the range of characters that we measure, either by adding other morphological characters or, if our original analysis has already thoroughly described the morphology of the organisms, by adding molecular data. In this case, these additional data reveal a newly recognised discontinuity between two groups of eight

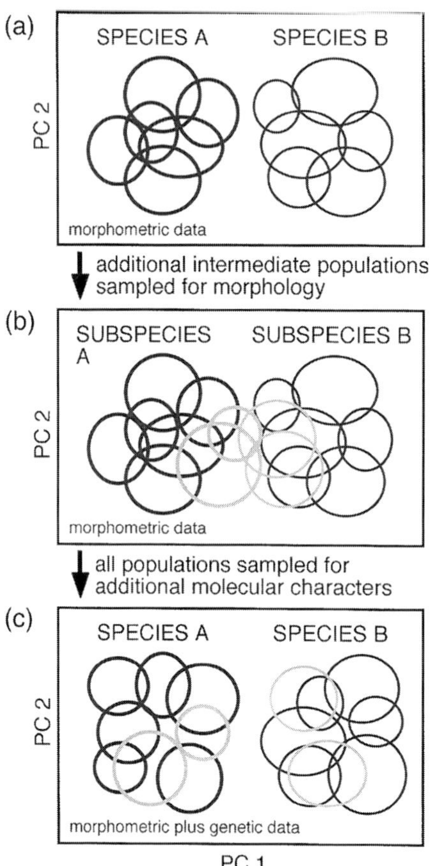

Figure 3.3 Species delimitation via population aggregation. Hypothetical ordinations illustrating the 'metastability' of perceived discontinuities among putative species. (a) Multivariate ordination of morphometric data from 12 orchid populations (each shown as an oval that represents the range of variation circumscribed by individuals within the population) reveals a morphological discontinuity that delimits two species, A and B. (b) Sampling of morphometric data from four additional intermediate populations (lighter ovals) bridges the perceived discontinuity, requiring that the two species now be merged into a single aggregate species. (c) Sampling of DNA sequence data from all 16 populations reveals a new, genetically based discontinuity between the former species A and B, arguably requiring the restoration of the two taxa to full species status. PC, principal coordinate. Reproduced with permission from Bateman (2001).

populations, most likely reflecting reproductive isolation. This insight obliges us to once again recognise two distinct species (Fig 3.3c).

Thus, our all-important perception of either discontinuity or overlap between groups of populations is heavily dependent upon sampling. We benefit from increasing the range of putative species analysed, the number of populations studied within each species, and the number of individuals studied within each population. Perhaps even more important is maximising the number of characters that we use to describe those organisms, and especially the number of categories of character that we quantify. The most crucial test of species boundaries is whether the discontinuities found using different categories of data coincide. Employing a combination of morphometrics, cytology and DNA sequence data from both the biparentally inherited nuclear genome and uniparentally inherited plastid genome (plants) or mitochondrial genome (animals and fungi, less often plants), all derived from the same individuals, is especially powerful. We need data on phenotype and genotype to adequately delimit (rather than merely erect) species, because phenotype confers crucial information regarding the functionality and ecology of the species, whereas genetic data should reveal whether our putative species show the required level of reproductive isolation.

This approach emphasises a particular brand of congruence test. The concept of congruence will be familiar to any phylogenetically aware systematist, but the term is traditionally used to indicate congruence between different characters of the same kind within the context of a single phylogeny-building exercise. Here, the term is used in the sense of topological congruence, between trees built of different kinds of character but analysed for the same spectrum of taxa (and, ideally, of individuals representing those taxa). This kind of congruence test based on multiple data sets is most powerful when a genotypic discontinuity coincides with a phenotypic discontinuity, thereby implying both genetic cohesion (gene flow at least heavily restricted) and functional constraints arising from that cohesion that influence ecological 'behaviour'. Similar multivariate circumscriptions (at or below the species level) or topologies (above the species level) between contrasting kinds of data set, greatly reinforce the analysts' belief that the species being recognised conform to both a phenotypic definition of a species based on morphology and a genotypic definition of a species based on a high degree of reproductive isolation (Table 3.1). Species that are both phenotypically and genotypically cohesive when thoroughly sampled are beyond reproach.

Rieseberg et al. (2006) conducted a meta-analysis of 218 studies of 400 plant and animal genera to assess the levels of incongruence observed between traditional names, morphometric delimitation and genetic delimitation. The good news was that 83% of plant studies and 88% of animal studies revealed discrete phenotypic clusters and thus offered the potential for rigorous species delimitation. The bad news was that correspondence between phenetic clusters and the boundaries of the species as previously described using traditional descriptive taxonomic methods was only 53% in plants and 52% in animals. Interestingly, the authors attributed 87% of the extensive incongruence between the two data sets to oversplitting at the species level by traditional taxonomists.

Table 3.1 Comparison of morphological and molecular taxonomy.

Criterion	Morphological taxonomy	Molecular taxonomy
Main cost	Specialist taxonomist's salary	Technician's salary (decreasing); laboratory equipment and consumables
Experience	Accrued gradually over long period	Accrued rapidly
Species comparison	Slow, and in practice limited to presumed close relatives	Rapid, largely automated comparison with all previously sequenced species
Results in the field	Often obtained	Not yet, but (very?) soon
Information on function	Available to varying degrees	Not yet, but eventually

Another valuable test of the accuracy of traditional species erection will be provided by burgeoning data provided via attempts at DNA-based identification known as DNA barcoding (Savolainen et al., 2005). Most current estimates of the accuracy of DNA barcoding lie between 60% and 95% (e.g. Quicke, 2004; Meier, 2008), but it is questionable whether we presently possess enough case studies based on large bodies of both phenotypic and genotypic data to estimate its likely accuracy and reliability. My own experiences (e.g. Devey et al., 2008; Bateman et al., in press) suggest that, in temperate orchids at least, the figure could lie at or even below the lower threshold of the quoted range – in other words, close to the *c.* 50% accuracy provided by classical morphological treatments according to Rieseberg et al. (2006).

In practice, most named species have not yet been adequately tested and so remain in biological limbo, trapped at the lowest (typological) level in the systematic hierarchy. I believe that the typological level lies below the threshold of utility when facing practical challenges such as monitoring and countering climate change. Before explaining my concerns, I will briefly consider the seriously undervalued relationship between taxonomy and identification.

3.5 The overlooked feedback loop between identification and taxonomy

Much of the concern expressed in the House of Lords (2008) review focused on the diminishing and ageing nature of the UK's remaining professional descriptive taxonomists. Similar trends are evident in most developed countries. There is no doubt that the discipline is genuinely in crisis. Indeed, I suspect that the total amount of descriptive taxonomic activity is even less than is generally believed; in addition to the decrease in the number of active professional practitioners, the percentage of the average practitioner's time spent on hands-on descriptive taxonomy has progressively declined as practitioners have been obliged by circumstance to become their own publicists and fundraisers. I also believe that taxonomists spend substantially less time than previously identifying organisms for others.

My impression is that the bulk of identification work now conducted worldwide is actually performed by what I would term parataxonomists. This term has most commonly been used as part of a prescription for progressing taxonomy and identification skills in developing countries, by educating individuals to a level where basic field identification can be achieved with a reasonable level of accuracy (e.g. Hall and Miller, Chapter 17). However, I would argue that the bulk of organismal identifications made in the UK are now also conducted by parataxonomists; specifically, by an informal alliance of retired professional taxonomists, part-time amateur natural historians (often operating under the auspices of specialised natural history societies), and professional practitioners of disciplines related to systematics such as ecology and environmental science.

If 'traditional' morphological and molecular identification are indeed now largely conducted by non-systematists, this is not necessarily a negative outcome; it implies both an ongoing need for broad identification skills (and the taxonomies that underlie them) and for dissemination of those skills across disciplines. Unfortunately, systems have not yet been widely introduced to allow feedback to taxonomy from identifications made by parataxonomists.

The need for such systems will be reinforced by the imminent sea change in the nature of parataxonomy that will be prompted by the development of hand-held DNA sequencers (Roach, 2005; Blasej et al., 2006). At present, such devices appear to be the preserve of the American military (Gutierrez, in press), but they are likely to follow the same pathway to popularisation as their previous invention, the Global Positioning System (GPS). DNA barcoding will then become universally available to natural historians (Bateman, 2009a). Specimens will be sequenced in the field within minutes, the resulting sequences broadcast via satellite to pass through Blast-type searches at GenBank or the European Molecular Biology Organization (EMBO), and a list of progressively decreasing sequence similarities will return via satellite to the fieldworker in seconds, offering a prioritised series of potential identities. The same palmtop device will also be able to record images of the organisms in question, and to retrieve comparable images and interactive files of diagnostic morphological characters.

This empowerment of parataxonomists will represent a serious challenge to a professional taxonomic community that thus far seems to have singularly failed to consider the consequences. Firstly, taxonomic expertise is likely to be more in demand, particularly to guide and educate the parataxonomists. Secondly, and more importantly, parataxonomy will provide a golden opportunity to remedy one of the two great weaknesses of DNA barcoding: the paucity of adequately vouchered reference sequences from a wide range of species. It is essential that systematists consider how best to incorporate parataxonomists' identifications into taxonomic revisions; at present, the required positive feedback loop is shamefully neglected.

3.6 Why predictivity is essential to address practical challenges such as climate change

Returning to the question of the utility of species descriptions, I should justify my deliberately provocative statement that basic taxonomic descriptions are inadequate for most practical purposes. My basic thesis is that a great deal more knowledge about the biology, ecology and distribution of a species is required before its responses to environmental change can be predicted (e.g. Barnard and Thuiller, 2008). And without that predictivity, the environment cannot be managed in order to conserve the species nor, conversely, can that species be used to monitor (and where necessary, attempt to remediate) environmental change. Efforts to infer underlying evolutionary processes are also undermined. Three case studies will serve to illustrate this point.

3.6.1 Monitoring the evolution of Darwin's finches on Daphne Major

Arguably the most impressive field study of the dynamics of evolution ever attempted is the 30-year monitoring of Darwin's finches on the Galapagan island of Daphne Major by Grant and Grant (e.g. Grant and Grant, 2002, 2008; Grant et al., 2004). During that period, the two main study species – the narrow-beaked *Geospiza scandens* (Gould, 1837) and broad-beaked *G. fortis* Gould, 1837 – have responded in parallel to climatic perturbations for most supposedly adaptive characters, including body size and beak size. However, variation in beak shape exhibited a more complex pattern of variation through time that is dominated by years when the island experienced unusual climatic conditions. For example, the serious drought of 1977 caused preferential survival of large-seeded food plants, advantaging large-beaked finches, whereas the El Niño rains of 1983 caused small-seeded food plants to smother larger-seeded food plants, advantaging pointed-beaked finches (Grant and Grant, 2002). More interestingly from the viewpoint of inferring speciation mechanisms, it became evident that the modest amount of genuine phenotypic divergence that apparently separates the two finch species is easily negated by a low level (< 2%) of introgressive hybridisation caused by these transient climatic shifts, largely because insufficient numbers of male *G. fortis* survive droughts (Grant et al., 2004; Grant and Grant, 2008; for a broader discussion of hybridisation and introgression see Thomasset et al., Chapter 15). In addition, the developmental genetic controls of beak shape and size have been elucidated, thus linking genetics, development and ecology (Abzhanov et al., 2004). This depth of phenotypic, genotypic and ecological understanding permits accurate predictions of the responses of these finches to climate change, enabling us both to use them as indicators of climate change and to conserve them more effectively.

3.6.2 Unravelling the complex life history of the large blue butterfly

The life history of the large blue butterfly (*Maculinea arion* (L., 1758)), an obligatory parasite on both thyme and social ants (Als et al., 2004), is the stuff of science fiction. Larval instars 1–4 feed on *Thymus serpyllum* L. foliage in the conventional butterfly manner, but they are then carried by *Myrmica* ants into their nests, where they proceed to feed voraciously on the ant larvae, pupate, and eventually emerge to repeat the entire unintuitive cycle. Perhaps partly as a consequence of its complex life history, the large blue became extirpated in the British Isles, and attempts to reintroduce it were unsuccessful until its life history had been fully documented. Moreover, as noted by Als et al. (2004), all three species involved in this extraordinary symbiosis are relatively sensitive to climate change, raising severe doubts about the long-term prognosis for the continued presence in the British Isles of the large blue and other vulnerable Lepidoptera. Although a major concern for conservationists, this symbiotic system should prove to be an unusually sensitive indicator of climate change.

3.6.3 Explaining the failure of Corsican pine as a UK timber crop

The fast-growing Corsican pine (*Pinus nigra* J. F. Arnold) presently constitutes 21% of the UK's standing timber crop. An equally alien fungal pathogen, *Dothistroma septosporum* (Dorog.) M. Morelet, which arrived in the UK in 1954, initially had little effect on the Corsican pine. However, its population exploded in 1995, as increased spring and summer rainfall optimised spore dispersal and infection (Brown and Webber, 2008). Wood formation is radically reduced in all infected trees and many ultimately die. The negative effects have been sufficiently profound that a five-year moratorium has been imposed on commercial planting of Corsican pine in the UK. Significantly, the pathogen has a far less drastic effect on the related radiata pine (*Pinus radiata* D. Don) within the bounds of its native distribution in California, where climatic conditions allow a more balanced relationship between host and pathogen, than in plantations located further north in the USA. Given this depth of biological knowledge, it should now be possible to predict the climatic conditions under which this ailing but potentially valuable timber tree could prove viable in the timber industries of both the UK and the USA. Nonetheless, further research is desirable to demonstrate conclusively that the negative effect of the pathogen on the host is not merely correlative but also causal.

3.7 Why long-term research is essential to address practical challenges such as climate change

Although they were selected primarily to illustrate the value of in-depth understanding of the autecology of species, the three preceding examples (especially the finches) also highlight the value of long-term observation. The three case studies summarised below further illustrate the importance of long-term data collection, at the same time emphasising the vagaries of funding such projects.

3.7.1 Permanent tropical forest plots

Since the Smithsonian Institution established a forest dynamics plot at Barro Colorado Island, Panama, in 1980, a worldwide network of permanent 50 ha plots has gradually been developed (Plotkin et al., 2000; Condit et al., 2005). The plots were established primarily to allow long-term monitoring, and thus deeper understanding, of local changes in the ecology of tropical forests. Exceptionally detailed taxonomic inventories have been carried out on the best of these plots, including monitoring the relative performance of each individual tree and shrub. More recently, it became apparent that the plot network also had the capability to monitor the ecological effects of global climate change. Unfortunately, likely sensitivity to climate change had not been a primary criterion when selecting the locations of the plots. It was therefore fortuitous that the 50 ha plot established in Belize by the Natural History Museum, London, in 1999 was located on thin limestone-derived soils and was therefore likely to be especially sensitive to fluctuations in rainfall (Garwood, 1999). Unfortunately, there was insufficient financial or political support, either within or outside the institution, to maintain the project, which was abandoned in 2004. The scientific importance of the remaining plots has increased greatly in recent years, but this has not been matched by increased resourcing. Even though these plots epitomise the effective integration of taxonomy and ecology to monitor climate change, they are not prioritised by funding bodies.

3.7.2 Permanent agricultural plots

A similarly chequered history is evident in the two 'classical' experiments maintained at the UK's Biotechnology and Biological Sciences Research Council-funded institute at Rothamsted, near London: Broadbalk (arable) and Park Grass (pasture). Park Grass is the world's oldest ecological experiment; the complex of plots was established in 1856 to determine the effect of different fertiliser treatments on hay meadows. However, in recent years it has gained a new lease of life in the study of environmental and climatic influences on species richness and of the genetics of local adaptation and drift (Silvertown et al., 2006). Although the time and other resources invested in this 150-year experiment have been considerable, mirroring the 50 ha tropical forest plots in requiring full and frequently repeated biodiversity inventories, the investment is modest when assessed on an annual basis, and the results have been invaluable. It has proven possible to track climate change using not only changes in the presence/absence and relative frequency of species but also in the genetic composition of populations and their estimated fitness (Silvertown et al., 2006). Nonetheless, historical review of the accumulated data shows that different categories of data obtained from Park Grass vary greatly in quality and frequency through time, reflecting both the transience of scientific fashion and the pragmatism of funding allocation. Astonishingly, the entire invaluable experiment was almost terminated in the 1980s on the grounds of presumed lack of relevance to modern science.

3.7.3 Permanent marine monitoring stations

Located several kilometres offshore, Plymouth Marine Laboratory's Oceanic Station L4 is part of the Marine Environmental Change Network, which coordinates long-term marine time series in the UK. Water samples are taken from the site weekly by ship, documenting both nutrient levels and biodiversity spectra through detailed identifications of many groups of microorganisms (Frost et al., 2006; Frost, 2007). Comparison of species frequencies and nutrient levels across the entire network of oceanic sampling stations has revealed cyclicity at several different scales, most notably annual (Worm et al., 2008). These stations effectively constitute an early warning system for climate-induced changes in sea temperature. Of particular concern is the diminution and potential cessation of the North Atlantic Conveyor, which carries warm currents from the Gulf of Mexico to western Europe and is largely responsible for the present equable climate. There is increasing evidence that, during the Pleistocene period, the Conveyor either ceased or diverted far to the south repeatedly, causing the periodic imposition of a periglacial climate that most strongly affected the British Isles. The most recent cold period, which lasted from *c.* 12 800 to 11 500 calibrated years before present (bp), appears to have involved transitions between temperate and periglacial and back to temperate conditions, requiring as little as a decade to complete both cooling and warming transitions (Alley et al., 1993; Kobashi et al., 2008).

Sampling at Station L4 began in about 1910, but understandably was interrupted during the Second World War by the laying of German mines. However, it was also interrupted, less understandably, during the period 1988–99 by withdrawal of funding from NERC – the resulting crucial break in the time-series data is consequently referred to as the 'NERC gap'. At the time of writing, funding for the project will last for only a further two years, possibly extendable by an additional three years. This is a classic example of a false economy, given that weakening of the North Atlantic Conveyor would most likely be detected first in marine microorganisms and that cessation of the Conveyor would literally return western Europe to the 'stone age'.

3.8 The bottom line: core funding from central government

Research infrastructures in most countries of the developed world have, in recent years, been increasingly influenced by political (or at least policy-driven) pressures. These have in turn created funding environments that are ever more short-term and competitive, and that emphasise projects that are believed to be capable of bringing immediate socioeconomic benefits. Under such academically perilous circumstances, studies of evolution and especially of the relatively uncharismatic discipline of taxonomy have been seriously disadvantaged, even before the global 'credit crunch' forced fiscal reprioritisation from late 2008. I can best illustrate these points through the system with which I am most familiar – that of the UK.

Of the three long-term monitoring projects described earlier, one failed completely and the other two barely survived periods of indefensible indifference on the part of their sponsoring government departments and funding bodies. Neither of the survivors has yet gained an assured future. Together, they eloquently illustrate the uniformly short-term nature of science planning and support within the British research system. My impression is that the level of commitment to similar projects is little better in most other countries. The spread of chronic 'short-termism' into scientific research from its increasingly prescriptive political sponsors has left the scientific community ill-equipped to deal with the recent, uncharacteristically persistent interest shown by politicians in climate change. Indeed, once again, it is the dominantly voluntary/membership organisations that have made the most impressive specific contributions to documenting the biota of the British Isles (e.g. the plant atlas edited by Preston et al., 2002) and have together provided the framework of the UK's National Biodiversity Network (NBN – www.nbn.org.uk).

Reviewing the evidence given to, and recommendations emanating from, the recent House of Lords (2008) overview of the UK's systematics base reveals a majority view that the UK's research councils currently offer (barely) adequate support to phylogenetics, molecular evolution and bioinformatics. However, despite assertions to the contrary, proposals related to both descriptive taxonomy and interdisciplinary systematics routinely fall through the gaps between the remits of the three relevant research councils. In my opinion, there are two ostensibly contrasting potential responses to this conundrum of funding descriptive taxonomy: (1) inextricably link taxonomy programmes to other research disciplines that are compatible but more fundable; and/or (2) separate taxonomy from the rest of systematic biology and promote taxonomic research as a special case of a long-term 'enabling' science that deserves unequivocal support. There also exists a potential compromise solution that deserves greater consideration: link taxonomy to its user groups when planning research programmes, but link taxonomy to other enabling disciplines, notably ecological and environmental monitoring, when organising its funding.

In order to determine whether such a compromise course could be viable, I will first consider the major trends in government funding of science since I first entered the UK's research environment 30 years ago. As in most developed countries, research in the UK can be divided into two government-sponsored arenas: the universities and the research institutes. Research in British universities is consistently benchmarked as being internationally competitive, a level of success that has been achieved by placing greater emphasis on obtaining funding for specific research projects and then concentrating those limited resources in ever smaller numbers of universities (a trend that may finally have reached its long-overdue acme – Corbyn, 2009). Inevitably, in such a competitive funding environment, long-term research has been stripped to supply more charismatic short-term research projects. University research has become ever more firmly tied to hypothesis-testing science projects funded for periods of three years (or at most five years). It was equally inevitable that such selection pressure, when applied over decades, would eventually eliminate from the university sector 'foundation/enabling' sciences such as descriptive taxonomy. Only recently has this trend led to expression of serious concern by policy makers and funders.

Fortunately, the traditional refuge of long-term non-medical research in the UK – and, to a lesser degree, in many other countries – has not been the universities but rather networks of permanent research institutes, many established by Victorian philanthropists, that specialised in the study of agriculture, forestry and fisheries, thereby encompassing the terrestrial, freshwater and marine environments. Some of these institutes, notably the natural history museums, botanic gardens and some zoological gardens, focused on systematics. Through the last 30 years, these networks of institutes have followed the university sector by switching to research goals that have been shorter-term and at least partially externally funded. However, despite this evident responsiveness to government policy, they suffered profound reductions in direct funding from central government and radical reduction in number, even before the global economic downturn took place.

It is ironic that the sector arguably least affected hitherto by these depredations in the UK has been that of the major systematics institutes; these have undoubtedly suffered vicissitudes, but their continued existence was not seriously challenged until very recently. To some extent, these institutes have been protected collectively by the remarkable fact that each is answerable to a different UK government department. However, this devolution of responsibility has also guaranteed the absence of 'joined-up' biodiversity policy, and responses to policy, across the UK (Bateman, 2008). It is also ironic that the wellbeing of the systematics institutes has been rather more frequently debated than the fate of other networks of research institutes – institutes that should be the natural users of the information generated by systematists in general and taxonomists in particular, most notably the institutes that comprise the NERC Centre for Ecology and Hydrology (CEH). As recently as 2006, the CEH institutes suffered radical rationalisation (BBC, 2006). These environmental institutes are the very bodies that should be the primary users of the products of the taxonomic institutes, applying identification skills in order to monitor biodiversity, predict change and develop adaptive and remedial strategies.

Thus, if one were organising from first principles an infrastructure to coordinate policy and research relating to biodiversity and climate change in the UK, it would not even vaguely resemble the status quo at the time of writing. If the UK government is genuinely committed to researching, monitoring, adapting to and, where possible, remediating climate change, it needs to take far more radical decisions than merely making climate change the sole strategic goal of NERC. I have noted similar trends, albeit less extreme, in many other developed countries. In the USA, a series of systematics initiatives instituted by the National Science Foundation brought at least temporary relief. The Assembling the Tree of Life programme provided a framework for phylogeny reconstruction while the Planetary Biodiversity Inventory and Partnerships for Enhancing Expertise in Taxonomy programmes gave a much needed fillip to monographic research. These programmes attracted multimillion-dollar investment (e.g. Wheeler, 2008b). A further, even more ambitious initiative, the Legacy Infrastructure Network for Natural Environments, has been proposed (Page et al., 2005).

However, the inherent weakness of such programmes is that they are inevitably of finite duration, typically being rapidly surpassed in the continual flux of political priorities and exigencies. The transience of funding fashion is now adversely affecting both the North American systematics programmes and their European equivalents, such as the European Distributed Institute for Taxonomy (EDIT; e.g. Scoble, 2008). Longer-term, genuinely sustainable solutions are required. These should include:

(1) strengthening links between systematics institutes, ecological/environmental institutes and universities in order to rebuild the inevitably prolonged 'biodiversity apprenticeship' needed to educate a new generation of multiskilled taxonomists;
(2) accruing and coordinating further project-specific funding, perhaps by linking multiple funding sources in successive tranches in order to encompass periods of 10 or more years rather than the current two or three (maximum five) years;
(3) developing performance criteria specifically for long-term 'enabling' sciences that reflect the quantity, quality and policy relevance of the data generated and the efficiency with which they are disseminated to users;
(4) explicitly linking descriptive taxonomy to other enabling sciences, notably ecological and environmental monitoring.

3.9 Prioritisation of goals within taxonomy

3.9.1 'Don't panic': responding effectively to the biodiversity crisis

If major changes are required in the policy and funding arenas, they are equally desirable within the taxonomic community. Much depends on how taxonomists respond to the increased concern expressed by politicians with regard to the challenges posed by increases in the rates of biodiversity loss and especially climate change. It has become routine to refer to both of these interconnected challenges as 'crises'. Should we therefore be indulging in crisis management?

Most observers accept that it is taking undesirably long to document global biodiversity. For example, Wheeler (2003) noted that 'Using traditional methods, taxonomists have documented about 1.7 million species in 245 years. Assuming 8.3 million species remain to be described, and a rate of progress equal to the post-Linnaean period, we will need 1196 years to complete the job.' Ergo, 'The urgency of the biodiversity crisis dictates that the taxonomic ... community can no longer accept business as usual.' Moreover, it is evident that the modest taxonomy initiatives developed over the last decade in Europe and the less modest but now dwindling taxonomic initiatives developed in North America have made at best an equally modest impact on the global rate of species description. For example, Flowers (2007) presented quantitative data to support his assertion that '15 years of initiatives, programs, speeches, white papers, and some substantial financial support have resulted in no significant change in the rate of species discovery in ... Coleoptera and Diptera' (beetles and flies). Furthermore Loebl and Leschen (2005) reported that 89% of the dwindling number of Coleoptera taxonomists still active in Europe were unfunded in 2005 and 91% operated primarily from home, emphasising the ever-increasing reliance for descriptive taxonomy on the goodwill of an ageing and largely unwaged community.

I have limited sympathy with the various general remedies that have recently been suggested by many of my colleagues. For example, I disagree with Schram's (2004) opinion, driven by his by no means unique enthusiasm for 'terascale bioinformatics', that 'Systematics as a realist, truth-seeking activity is doomed ... As the science of systematics speeds up [while] shifting to the strong inference [= relativist] mode, there will be no time nor tolerance for isolated researchers defending to the death their island redoubts.' Given such provocative statements, I consider it essential to resolve the tension that exists between those like Schram who argue that increasing the speed of taxonomic research is paramount and those like myself who believe that it is more important to increase the rigour of taxonomic description (in other words, to ensure that the species and associated descriptions that are the basic currency of biology have been circumscribed using multiple, well-sampled data sets). Of course, both speed and rigour are desirable, and both should in theory be feasible, but in practice the resources available to what remains of the global taxonomic community were woefully inadequate even before the present global economic downturn began to bite.

How can we greatly accelerate the rate of species description without substantially increased resourcing? Although there have been many attempts to make efficiency gains, modest 'savings' might still be made if the systematics community was better organised. Ironically, one possible route to simplification would be to develop *fewer* taxonomy initiatives. The community has become even more fragmented as a result of attempting to service large numbers of small initiatives. This suboptimal pattern is a direct consequence of short-termism in funding; it is far easier to obtain funds to initiate a project than to subsequently maintain it.

For example, I have lost count of the number of biodiversity database initiatives that began in a blaze of publicity but soon either failed completely or gradually ossified. Both outcomes are unfortunate, as databases are of little value until well populated with high-quality data. These databases illustrate another important point: they are often populated largely with recycled biodiversity data, rather than catalysing the generation of genuinely new data. Much recent biodiversity-related activity has involved reorganising existing data rather than pushing forward the frontiers of science. Reorganisation sounds like a good idea (and probably is), but even this remedy should not be sanctioned uncritically. Many recent international initiatives in taxonomy have focused on collating taxonomic information in 'open access' electronic environments. Although I make regular and grateful use of some such websites, I also recognise that the capability and versatility of non-specialist search engines such as Google, combined with the increasing availability of scientific information online, arguably reduce the need to physically aggregate such information in single databases. And many regional floras and faunas – the ultimate long-term taxonomic projects – also rely heavily on recycling traditional information. Schram (2004) made a related uncompromising point: 'I never cease to be amazed at how much alpha-taxonomic science has been sloppily done – an earmark perhaps of its cottage-industry tradition.' If the taxonomic community succeeded in generating global monographs for all groups of organisms, floras and faunas could be derived from those aggregated monographs, thereby reducing the need for further primary research at a regional scale.

Where research initiatives do plan to generate novel data, they increasingly seek to cut corners by developing technological fixes. It is important that each such supposed innovation is evaluated critically before it receives substantial resourcing. For example, Humphries (2003) rightly noted that '[Recent systematic] investigations have erred on the technological and algorithmic sides of achievement rather than on the epistemological and theoretical.' Of these schemes, the most interesting are those that involve automation, characterising either morphology through visual recognition systems (e.g. MacLeod, 2007) or DNA through automated sequencing (Savolainen et al., 2005). However, the use of these technologies by the most obvious constituencies – amateur natural historians and researchers operating in biological disciplines other than systematics – has rarely been considered at the policy level. Unless carefully managed, these innovations could conspire to make typical species descriptions (and redescriptions) even more superficial.

Also highlighted previously was the general failure of the positive feedback loop from species identification back to species delimitation during taxonomic revision. Most taxonomic identifications are performed for a wide variety of applied purposes, where practicality (notably speed and cost) is generally prioritised over accuracy and repeatability. The understandable desire among user communities for rapid, repeatable and confident identifications – in other words, for stability of species description and classification – can threaten the credibility of taxonomy as a genuine science, which should be held to the same strictures of verification and falsifiability as any other bona fide science. As eloquently expressed by Seberg (2004), 'Hypotheses about relationships are always subject to revision as new information becomes available, or existing data are reinterpreted. The taxonomy of species is not fixed. In taxonomy, "stability is ignorance", and the mere idea behind creating a [prescribed] unitary taxonomy runs counter to scientific practice.'

The standing of taxonomy as a mainstream science is further undermined by resource limitation, which inevitably encourages prioritisation of taxonomic lacunae over previously studied groups. The emphasis of taxonomic terra incognita appears particularly counterproductive in the light of my thesis that basic taxonomic descriptions are essential but not of practical value in isolation.

The reputation of descriptive taxonomy as a science suffers further avoidable damage as a result of the fact that, despite the extensive literature on species concepts (e.g. Mayden, 1997; Rieseberg et al., 2006), most taxonomic studies fail to explicitly specify which (if any) particular species concept has been applied. Any erstwhile users of that taxonomic product are thus left unsure whether the species (or, more likely, group of species) that is described is suitable for addressing their particular questions, not least those relating to climate change.

Documenting the earth's biota remains a compelling goal for many systematists, but it is orders of magnitude too large and too nebulous to convince most potential funding bodies to provide tangible support. As there exists an almost limitless number of taxonomic groups still awaiting detailed examination, prioritisation among them is essential. Taxonomists should work with evolutionary biologists and other users of systematic data to better prioritise taxonomic groups for detailed systematic study, using as their primary criterion the range and importance of scientific and/or social questions that could be answered by each group. The next logical step is to select the most useful analytical techniques and species concept(s) to answer those questions using that group. This approach allows us to proceed beyond documentation of biodiversity to a deeper level of biological understanding. In addition, it exploits the deep affection held by taxonomists for their chosen group(s), while allowing us to maintain better contact with the many disciplines, both practical and academic, that ultimately depend on taxonomic outputs.

Question-driven taxonomy should ultimately result in a 'postmodern synthesis' capable of revealing the 'modern synthesis' for what it truly is – a myopic sketch of a biological reality that is far more complex and intriguing. Our understanding of the diversity of both evolutionary and ecological processes has advanced greatly during the last half-century, but taxonomic concepts have failed to keep pace. Indeed, this crucial mismatch has not even been adequately debated yet.

3.9.2 Depth versus breadth of knowledge

I believe strongly that gaining depth of knowledge regarding species is more important than gaining breadth when addressing climate change issues (see also Bernardo, Chapter 18). In this context, I have increasingly come to share the opinion of many colleagues that taxonomy as currently practised is damagingly anarchic. However, I disagree with the more commonly suggested remedial constraints, notably agreeing lists of recommended Linnean binomials based on traditional taxonomy, or prioritising alternative, binomial-free approaches such as automated identification via DNA barcoding. No one should doubt the pressures placed on the community to offer spurious stability; for example, 'in a world demanding certainty (e.g. lists of species known by 2010), admitting the inbuilt trials and tribulations of species delimitation seems somewhat dangerous, if applaudably honest' (Flann, 2006). Submitting to such pressures would simply make taxonomy even more prescriptive and even less scientific. As noted by Seberg (2004),

> To relegate taxonomy to a high-tech service industry centred around ... DNA sequences will deprive evolutionary biology of its most important function: the testing of evolutionary hypotheses at all levels, from the evolution of characters, over the evolution of species, to the evolution of clades ... There are better solutions to the 'taxonomic impediment' than to become ignorant of the organisms we work with.

There exists one key aspect of descriptive taxonomy where universally respected compliance with complex rules routinely replaces anarchy. Specifically, taxonomists have long been dictated to by the obligatory authoritarianism of the International Codes of Nomenclature (Bateman, 2009b). I would certainly support recent arguments (expressed most vociferously by the zoological community and epitomised by the establishment of ZooBank – Polaszek et al., 2008) that we should expand this authoritarian ethos by developing a centralised registry for new formal taxon names and combinations, thus mirroring the undeniable success of GenBank/EMBO as near-universal repositories for genetic data. However, I would extend taxonomic authoritarianism even further, in recognition of my belief that predictive biology in an area such as climate change needs far more than a name – it also requires substantial knowledge of morphology, genetics, habitat, distribution and autecology (preferably also synecology). I firmly believe that *the systematics community* should set an obligatory baseline of biological information necessary to register a new name for an extant species.

Determining the precise nature of these baseline data would obviously require detailed discussion within the community. My initial negotiating position would be that, for eukaryotes at least, the threshold of acceptance would require both morphological and DNA data, plus the deposition of at least one high-quality type specimen accompanied by at least basic data on its geographical location and habitat. A putative species falling short of this threshold could be registered only as a parataxon – that is, as a hypothesis of species existence, awaiting future testing of that hypothesis. Alternatively, we could develop an agreed hierarchy of species notation, to reflect radical differences in the breadth and depth of the underpinning data.

Such strictures would more likely slow than accelerate the rate of species discovery (contravening the exhortations of Schram, 2004), but at least the resulting species would have a good chance of being usefully robust and predictive. In addition, taxonomists, phylogeneticists, population biologists, ecologists and environmentalists would be encouraged to work more closely with each other and to develop more integrated research proposals and policy advice. This increased integration would negate the obvious retort that any requirement for sequence data in the protologue would undermine taxonomy in developing countries, where molecular technologies are less readily available. Instead, better links would be formed between morphologically experienced taxonomists in the developing countries and molecularly experienced taxonomists in developed countries, to the considerable benefit of both.

The resulting joined-up science, linking long-term enabling research to shorter-term, hypothesis-driven projects, should encourage both funders and policy makers to put their own houses in order, by developing joined-up resourcing that takes adequate account of long-term as well as short-term research. I do not believe that mankind will ever truly complete either the Linnean taxonomic 'project' or the Darwinian evolutionary-phylogenetic 'project', but we could at least increase their efficiency by running them in parallel and in stronger synergy.

Acknowledgements

I wrote this contribution while salaried via NERC grant NE/E004369/1 (PI Jason Hilton). Andrew Moffat helpfully drew my attention to the sad case of the Corsican pine. Paula Rudall kindly critiqued, and three anonymous reviewers unkindly criticised, an early draft of the manuscript. I thank the Biosciences Federation and the Systematics Association for providing me with temporary platforms that gave me valuable insights into the workings of the British parliamentary system and their consequences for biodiversity-related policy. And I thank the conference organisers for demonstrating to me that a (junior) minister of the environment can actually be persuaded to remain at a scientific meeting on climate change beyond the opening ceremony.

References

Abzhanov, A., Protas, M., Grant, P. R., Grant, B. R. and Tabin, C. J. (2004). *Bmp4* and morphological variation of beaks in Darwin's finches. *Science*, **305**, 1462–1465.

Alley, R. B., Meese, D. A., Shuma, C. A. et al. (1993). Abrupt accumulation increase of the Younger Dryas termination in the GISP2 ice core. *Nature*, **262**, 527–529.

Als, T. D., Vila, R., Kandul, P. et al. (2004). The evolution of alternative parasitic life histories in large blue butterflies. *Nature*, **432**, 386–390.

Avise, J. C. (2000). *Phylogeography: the History and Formation of Species.* Cambridge, MA: Harvard University Press.

Barnard, P. and Thuiller, W. (2008). Global change and biodiversity: future challenges. *Biology Letters*, **4**, 553–555.

Bateman, R. M. (2001). Evolution and classification of European orchids: insights from molecular and morphological characters. *Journal Europäischer Orchideen*, **33**, 33–119.

Bateman, R. M. (2008). When diagnosis is easier than cure. *Research Fortnight*, **10**, 16.

Bateman, R. M. (2009a). Field-based molecular identification within the next decade? *Journal of the Hardy Orchid Society*, **6**, 31–34.

Bateman, R. M. (2009b). What's in a name? *Journal of the Hardy Orchid Society*, **6**, 53–63 and 88–99.

Bateman, R. M., James, K. E., Rudall, P. J. and Cozzolino, S. (in press). Contrast in morphological versus molecular divergence between two closely related Eurasian species of *Platanthera* (Orchidaceae) suggests recent evolution with a strong allometric component. *Biological Journal of the Linnean Society*.

BBC (2006). Research lab closures to go ahead. BBC News, 13 March 2006. news.bbc.co.uk/1/hi/england/4801814.stm.

Blackmore, S., Watson, E., Bateman, R. M. et al. (1998). *The Web of Life: A Strategy for Systematic Biology in the United Kingdom*. London: UK Systematics Forum/Office of Science and Technology.

Blasej, R. G., Kumareson, P. and Mathies, R. A. (2006). Microfabricated bioprocessor for integrated nanoliter-scale Sanger DNA sequencing. *Proceedings of the National Academy of Sciences of the USA*, **103**, 7240–7245.

Brown, A. and Webber, J. (2008). *Red Band Needle Blight of Conifers in Britain*. Research Note 0002, Forestry Commission.

Condit, R., Ashton, P., Balslev, H. et al. (2005). Tropical tree alpha-diversity: results from a worldwide network of large plots. *Biologiske Skrifter Kongelige Danske Videnskabernes Selskab*, **55**, 565–582.

Corbyn, Z. (2009). Elite v-cs fear 'end of road' for concentration of research. *Times Higher Education* (1 January), 4–5.

Cracraft, J., Donoghue, M., Dragoo, J., Hillis, D. and Yates, T., eds. (2002). *Assembling the Tree of Life*. National Science Foundation.

Crane, P. R., ed. (2003). *Measuring Biodiversity for Conservation*, Policy document 11/03. London: The Royal Society.

Devey, D. S., Bateman, R. M., Fay, M. F. and Hawkins, J. A. (2008). Friends or relatives? Phylogenetics and species delimitation in the controversial European orchid genus *Ophrys*. *Annals of Botany*, **101**, 385–402.

Duraiappah, A. K. and Naeem, S., eds. (2005). *Ecosystems and Human Well-Being: Biodiversity Synthesis*. Washington, DC: World Resources Institute.

Enghoff, H. and Seberg, O. (2006). A taxonomy of taxonomy and taxonomists. *The Systematist*, **27**, 13–15.

Flann, C. (2006). Why taxonomy is fundamental to conserving the plants of this world: a review of E. Leadlay and S. Jury. *Taxonomy and plant conservation*. *The Systematist*, **27**, 19–21.

Flowers, R. W. (2007). Taxonomy's unexamined impediment. *The Systematist*, **28**, 3–7.

Frost, M. T., ed. (2007). *The Marine Environmental Change Network – Strategy Meeting Report*. London: Department for Environment, Food and Rural Affairs.

Frost, M. T., Jefferson, R. and Hawkins, S. J., eds. (2006). *The Evaluation of Time Series: Their Scientific Value and Contribution to Policy Needs*. Plymouth: Marine Biology Association.

Futuyma, D. J., ed. (1998). *Evolution, Science, and Society: Evolutionary Biology and the National Research Agenda*. American Society of

Naturalists [etc.]. www.rci.rutgers.edu/~ecolevol/fulldoc.pdf.

Garwood, N. C. (1999). A new forest dynamics plot in Belize. *Inside the Center of Forest Dynamics*, Summer, 2.

Godfray, H. C. J. and Knapp, S., eds. (2004). *Taxonomy for the Twenty-First Century.* London: The Royal Society.

Grant, P. R. and Grant, B. R. (2002). Unpredictable evolution in a 30-year study of Darwin's finches. *Science*, **296**, 707–711.

Grant, P. R. and Grant, B. R. (2008). *How and Why Species Multiply: The Radiation of Darwin's Finches.* Princeton, NJ: Princeton University Press.

Grant, P. R., Grant, B. R., Markert, J. A., Keller, L. F. and Petren, K. (2004). Convergent evolution of Darwin's finches caused by introgressive hybridization and selection. *Evolution*, **58**, 1588–1599.

Gutierrez, T. (in press). Field identification of vectors and pathogens of military significance. In *Descriptive Taxonomy Serving Biodiversity*, ed. M. F. Watson and C. Lyal. Systematics Association Special Volume 79. Cambridge: Cambridge University Press.

House of Commons (2007). *Are Biofuels Sustainable?* House of Commons Environmental Audit Committee 1st Report of Session 2007–08, HC76-11. London: Stationery Office.

House of Lords (2002). *What on Earth? The Threat to the Science Underpinning Conservation*, House of Lords Select Committee on Science and Technology Paper 118. London: Stationery Office.

House of Lords (2008). *Systematics and Taxonomy: Follow-up* (S. Sutherland, chair). 5th report of Session 2007–08, House of Lords Science and Technology Committee, HL162. London: Stationery Office.

Humphries, C. J. (2003). Biennial meeting: a presidential review. *The Systematist*, **22**, 10–14.

Huxley, J. (1940). *The New Systematics.* Oxford: Oxford University Press.

Kobashi, T., Severinghaus, J. P. and Barnota, J. M. (2008). 4 ± 1.5 °C abrupt warming 11,270 years ago identified from trapped air in Greenland ice. *Earth and Planetary Science Letters*, **268**, 397–407.

Loebl, I. and Leschen, R. A. B. (2005). Demography of coleopterists and their thoughts on DNA barcoding and the Phylocode, with commentary. *Coleopterists Bulletin*, **59**, 284–292.

MacLeod, N., ed. (2007). *Automated Taxon Identification in Systematics: Theory, Approaches and Applications.* Systematics Association Special Volume 74. Boca Raton, FL: CRC Press/Taylor and Francis.

May, R. M. (1999). The dimensions of life on earth. In *Nature and Human Society: The Quest for a Sustainable World*, ed. P. H. Raven and T. Williams. Washington DC: National Academy of Sciences, pp. 30–45.

Mayden, R. L. (1997). A hierarchy of species concepts: the denouement in the saga of the species problem. In *Species: the Units of Biodiversity*, ed. M. F. Claridge, H. A. Dawah and M. R. Wilson. London: Chapman and Hall, pp. 381–421.

Meier, R. (2008). DNA sequences in taxonomy: opportunities and challenges. In *The New Taxonomy*, ed. Q. D. Wheeler. Systematics Association Special Volume 76. Boca Raton, FL: CRC Press, pp. 95–127.

Natural Environment Research Council (NERC) (2007). *Next Generation Science for Planet Earth.* Swindon: NERC.

Page, L. M., Bart, H. L., Beaman, R. et al. (2005). *LINNE: Legacy Infrastructure Network for Natural Environments.* Champaign, IL: Illinois Natural History Survey.

Plotkin, J. B., Potts, M. D., Yu, D. W. et al. (2000). Predicting species diversity in tropical forests. *Proceedings of the National Academy of Sciences of the USA*, **97**, 10850–10854.

Polaszek, A., Pyle, R. and Yanega, D. (2008). Animal names for all: ICZN, ZooBank and the New Taxonomy. In *The New Taxonomy*, ed. Q. D. Wheeler. Systematics Association Special Volume 76. Boca Raton, FL: CRC Press, pp. 129–141.

Preston, C. D., Pearman, D. A. and Dines, T. D. (2002). *New Atlas of the British and Irish Flora.* Oxford: Oxford University Press.

Quicke, D. L. (2004). The world of DNA barcoding and morphology. *The Systematist*, **23**, 8–12.

Rieseberg, L. H., Wood, T. E. and Baack, E. J. (2006). The nature of plant species. *Nature*, **440**, 524–527.

Roach, J. (2005). Handheld DNA scanners to ID instantly? *National Geographic News*, 26 January.

Savolainen, V., Cowan, R. S., Vogler, A. P., Roderick, G. K. and Lane, R., eds. (2005). *DNA Barcoding of Life.* London: Royal Society.

Schram, F. R. (2004). The truly new systematics: megascience in the information age. *Hydrobiologia*, **519**, 1–7.

Scoble, M. J. (2008). Networks and their role in e-taxonomy. In *The New Taxonomy*, ed. Q. D. Wheeler. Systematics Association Special Volume 76. Boca Raton, FL: CRC Press, pp. 19–31.

Seberg, O. (2004). The future of systematics: assembling the tree of life. *The Systematist*, **23**, 2–8.

Silvertown, J., Poulton, P., Johnston, E. et al. (2006). The Park Grass experiment 1856–2006. *Journal of Ecology*, **94**, 801–814.

Stern, N. H. (2006). *Stern Review of the Economics of Climate Change.* London: HM Government.

Systematics Agenda 2000 (1994). *Systematics Agenda 2000: Charting the Biosphere.* New York, NY: Society of Systematic Biologists, American Society of Plant Taxonomists, Willi Hennig Society and Association of Systematics Collections.

Wheeler, Q. D. (2003). Transforming taxonomy. *The Systematist*, **22**, 3–5.

Wheeler, Q. D., ed. (2008a). *The New Taxonomy.* Systematics Association Special Volume 76. Boca Raton, FL: CRC Press.

Wheeler, Q. D. (2008b). Taxonomic shock and awe. In *The New Taxonomy*, ed. Q. D. Wheeler. Systematics Association Special Volume 76. Boca Raton, FL: CRC Press, pp. 211–226.

Worm, B., Barbier, E. B., Beaumont, N. et al. (2008). Impacts of biodiversity loss on ocean ecosystem services. *Science*, **314**, 787–790.

SECTION 2

Adaptation, speciation and extinction

4

Global climate and extinction: evidence from the fossil record

P. J. MAYHEW

Department of Biology, University of York, UK

Abstract

The Intergovernmental Panel on Climate Change (IPCC) has acknowledged that climate change represents a tangible threat to species richness, based largely on the evidence from climate envelope modelling and the shifts in ranges of current species. However, because of uncertainty in the accuracy of extinction forecasts from climate envelope modelling, it would be useful to have an alternative source of information. The evidence from the fossil record is less widely discussed, but supports the view that a warmer global climate will increase extinction rates even without other associated human impacts such as habitat loss. Fine-scale studies show heterogeneity in results, but global-scale analyses demonstrate that extinction rates are generally elevated during greenhouse phases and that biodiversity is depressed. These trends are consistent with studies of extinction events that have implicated global warming as a consistent cause, triggered by carbon dioxide (CO_2) release from large igneous province eruptions. They suggest that abiotic factors such as climate are a major influence on biodiversity through time, but relatively predictably so (unlike the paradigm of the Court Jester). They indicate that there are perils of a warm climate distinct from those of climate change alone;

Climate Change, Ecology and Systematics, ed. Trevor R. Hodkinson, Michael B. Jones, Stephen Waldren and John A. N. Parnell. Published by Cambridge University Press.

that global biodiversity loss through climate change will only be reversed on geologic timescales; and that any reduction in global warming will bring some benefit to global biodiversity.

4.1 Introduction

What will be the consequences of anthropogenic climate change during the current and next centuries? This question dominates much of the current research in the biological and earth sciences. Within the evolutionary and ecological sciences, major questions relate to whether and how climate change will cause extinction, and to what extent this might affect biodiversity (Lovejoy and Hannah, 2005). There are three sources of relevant evidence, representing the past, the present and the future.

The evidence from the present represents our observations of extant, or recently extinct, organisms and the changes that they have experienced as a result of current climate change. Although there have been few documented recent extinctions as a result of climate change (e.g. Pounds et al., 2006; Thomas et al., 2006), the effects of climate change on individuals and species are nonetheless pervasive (e.g. Parmesan and Yohe, 2003). Another source of evidence uses these observations, and particularly those on species ranges and range shifts, to project changes in organismic distribution into the future with the aid of climate models (e.g. Thomas et al., 2004). These two sources of evidence form the backbone of the claim of the IPCC (2007) that large-scale extinction can be expected in the current or next century.

The extent of extinction risk from climate change, implied by the climate envelope modelling approach, depends on the simplifying assumptions that must be made, about which there are great uncertainties (Thuiller, 2004; Araújo and Guisan, 2006; Ibanez et al., 2006; Pearson, 2006). It is therefore desirable to have indications of the severity of extinction and biodiversity loss from other sources, and evidence from the past can provide that. In general, the past provides an important context within which current and expected future climate change is viewed. Earth history, for example, contains records of the magnitude of climate fluctuation that has occurred since life began and serves to show how unusual the magnitude and rate of anthropogenic climate change has been (Frakes et al., 1992; Huber et al., 2000; IPCC, 2007). Through anthropogenic climate change, the earth system is expected to enter a state which has not been seen for millions of years (IPCC, 2007), in terms of both temperature (e.g. Royer et al., 2004) and particularly atmospheric CO_2 concentrations (e.g. Berner and Kothavala, 2001). This means that projections based on present-day biotas rely on ever greater extrapolation. Earth history, however, contains the record of how biotas responded to such changes in the past and can thus potentially inform the future, if only the 'rules of response' can be extracted over long time periods. As well as past fluctuations in climate, there has been enormous variation in biodiversity since life began, measured as changes in the number of fossil families and genera across geological strata (e.g. Benton, 1995; Alroy et al., 2008). This should provide the necessary data to identify statistical relationships, if they exist.

In this chapter I review recent findings on the relationship between biodiversity and climate over earth history. I ask if there are observable rules of response, and, finding them, ask what the causes may be and how they may be used to inform the future in ways that complement the lessons of climate envelope modelling. I then discuss the implications of the Phanaerozoic associations for our general understanding of the evolution of biodiversity. Finally, I provide a set of research questions, raised by this discussion, to guide the next generation of studies.

4.2 Global climate and biodiversity over earth history

By far the majority of studies aimed at understanding the link between biodiversity and climate are in some sense selective: that is, they focus on particular taxa, within particular geographic areas, and they cover a relatively small part of the fossil record. While the information gained so far is definitely useful, it is somewhat optimistic to hope that such studies will reveal any consistent rules of response. Indeed, by far the most consistent message that such studies reveal is how heterogeneous the biotic responses to climate variation can be. Three examples serve to illustrate this.

Alroy et al. (2000) examined the evolution of North American mammal communities in response to temperature fluctuations revealed by oxygen isotope ratios during the Cenozoic. The study is a model of careful analysis; not only are the measures of diversity and taxonomic rates carefully standardised for sampling effort, but the analyses take into account the problems of time-series analysis, such as serial trends leading to meaningless correlations, and temporal autocorrelation amongst the data, which are ignored in many studies. However, neither standing biodiversity nor origination or extinction rates show any consistent relationships with the oxygen isotope record.

Gibbs et al. (2006) studied the evolutionary responses of marine nanoplankton at sites in North America across the Eocene thermal maximum. They found that peaks in extinction and origination coincided with the warming, although these opposing forces paralleled each other so closely that biodiversity was hardly affected in the short term. In the longer term, however, extinction remained higher than origination, leading to a drop in diversity.

In contrast, Jaramillo et al. (2006) studied neotropical plant diversity over a large part of the Cenozoic. Overall, increasing biodiversity was positively correlated with temperature, although there was no significant association between temperature and either origination or extinction.

Thus, three different studies give three completely contrasting impressions. It may be that, in the future, meta-analysis can be applied to a large collation of such studies that will reveal underlying consistencies in the responses across groups, regions or time periods. Indeed, the large-scale patterns I discuss below suggest that an underlying consistency will eventually be discovered. However, the most that we can currently infer from the finer-scale studies is that evolutionary processes and biodiversity can sometimes be associated with temperature, although we cannot even suggest what the general direction of the response will be. Given this lack of consensus, therefore, we might be wise to ask if larger-scale studies, in terms of taxonomic representation, geographic scale and timescale, can suggest general rules of response where other studies have failed. There is reason to suspect that they might, because larger-scale analyses should be less susceptible to the idiosyncrasies of particular regions, time periods or taxa. Put simply, if you want to discover general rules, make your data as general as possible.

Fortuitously, large-scale studies support the idea that there are general rules about how global climate regulates biodiversity. In what was possibly the first study of this sort, Denness (1989) correlated marine animal diversity in the fossil record with temperature estimates from a global climate model, and found a positive association (i.e. higher diversity in greenhouse climates). Although the estimates of global temperature used are no longer well supported in the face of new data (and the associations are now thought to be the opposite of those suggested by the paper), such a clear correlation should have stimulated at least further research into whether such long-term associations exist and how they might be regulated. However, the academic community ignored the paper; it has never been cited by another scientific paper, and all the palaeobiologists I have spoken to were ignorant of it.

Fortunately the ideas were revived more recently by a pair of studies linking atmospheric CO_2 concentrations to evolutionary processes. Rothman (2001) correlated a CO_2 proxy against two large data sets across the Phanaerozoic: the number of marine animal genera (Fig 4.1) and terrestrial plant families (not shown). Both were negatively correlated with CO_2 concentrations (i.e. diversity was lower under high CO_2 conditions and vice versa). Cornette et al. (2002) showed that predicted CO_2 concentrations (predicted from the GEOCARB III model) positively correlated with origination rates and, more weakly, with extinction rates in marine animal genera. Given the strong correlation between atmospheric CO_2 concentrations and global temperature during the Quaternary, it could be thought that strong links might be inferred between global biodiversity and global climate

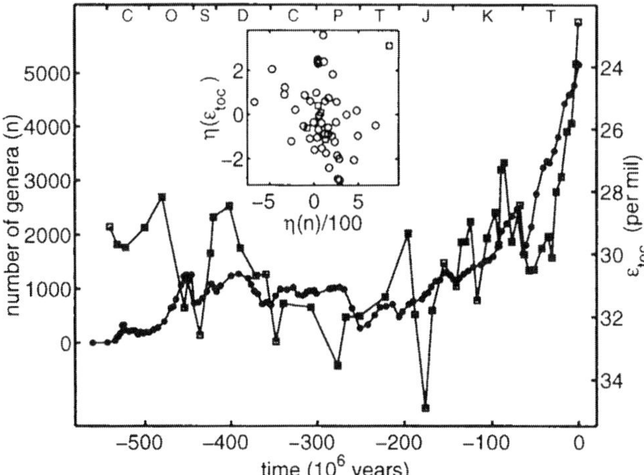

Figure 4.1 Number of marine animal genera and the isotopic fractionation between inorganic and organic carbon against time. Circles show the number of marine animal genera. Squares show ε_{toc}, a proxy of atmospheric CO_2 concentrations (note the reversed scale). The insert shows the overall negative association after removing temporal autocorrelations. A similar association exists for terrestrial plant families (Rothman, 2001). Reproduced from Rothman (2001) with permission of the National Academy of Sciences of the USA.

change. In fact, this was not the case, possibly because at the time it was not well established that CO_2 concentrations and global temperature were strongly linked over the Phanaerozoic as a whole (e.g. Veizer et al., 2000).

The suggestion that a link between temperature and biodiversity existed also came from a study on cycles in the marine animal fossil record by Rohde and Muller (2005; Fig 4.2). Their paper reported a 62-million-year (myr) cycle in the number of genera. However, a lesser-known feature of this paper is that it also reports a longer-term 140 myr cycle in the numbers of genera (Fig 4.2). Rohde and Muller noted that a 140 myr cycle was consistent with the periods of other cycles reported in climate and cosmic rays, and therefore that it might merit further investigation. The climatic cycles in question here are the long-term changes between greenhouse and icehouse modes during the Phanaerozoic (Frakes et al., 1992; Veizer et al., 2000). Frakes et al. (1992) reported the greenhouse phases as early Cambrian to late Ordovican, early Silurian to early Carboniferous, late Permian to mid-Jurassic, and early Cretaceous to early Eocene. The ice-house phases are sandwiched in between: respectively, late Ordovician to early Silurian, early Carboniferous to late Permian, late Jurassic to early Cretaceous, and early Eocene to Present, and recent results also confirm this (Royer et al., 2004 – Fig 4.3a).

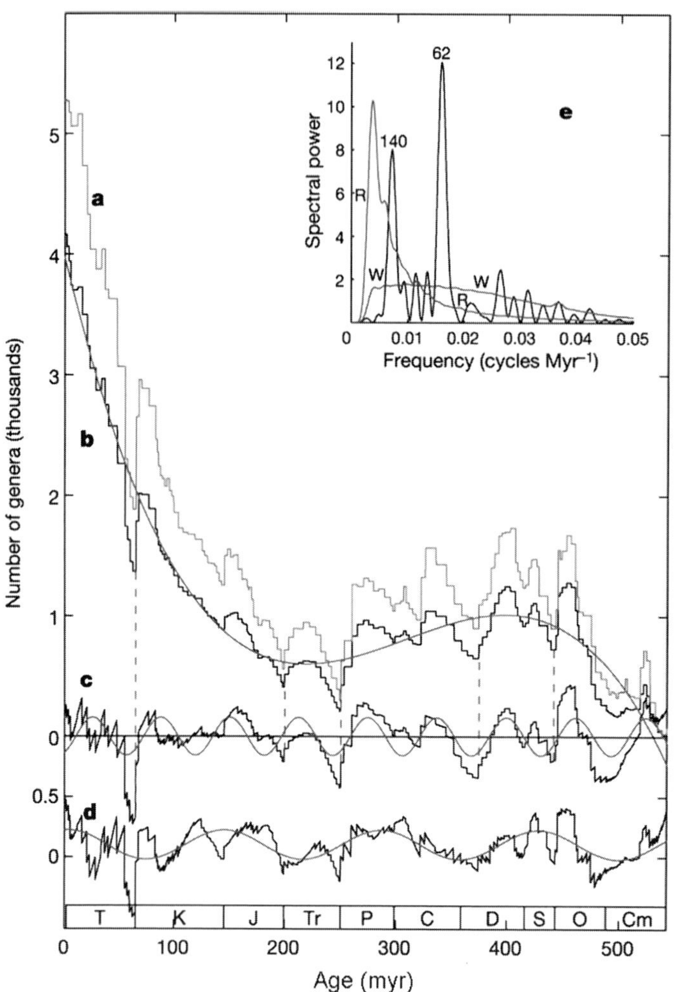

Figure 4.2 Number of marine animal genera against age. (a) All data; (b) singletons and poorly dated genera removed; (c) detrended to show the 62-million-year (myr) cycle; (d) detrended to show the 140 myr cycle; and (e) Fourier spectrum of (c), showing the 62 myr and 140 myr cycles rising beyond the spectral background (W and R). Reproduced from Rohde and Muller (2005) with permission of Macmillan Publishers Ltd. *See colour plate section.*

Further investigation of the trends suggested by Rohde and Muller (2005) was provided by Mayhew et al. (2008), who analysed the fossil record from two data sets: the marine animal genera data set of Sepkoski (2002) and the global family-level data set of Benton (1993) (Fig 4.3b–d), analysed as a whole and separately for marine and terrestrial taxa. Temperature estimates came from Royer et al. (2004), who used the low-latitude sea-surface oxygen isotope data of Veizer et al.

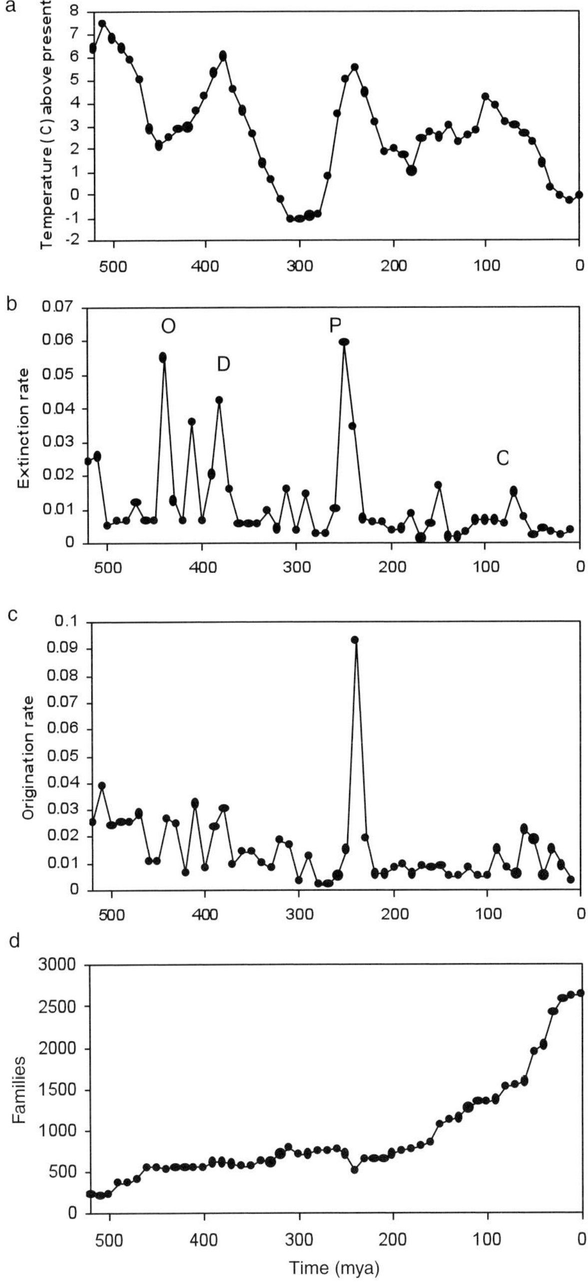

Figure 4.3 Time series (10 myr intervals) of estimated low-latitude sea surface temperature, per capita extinction rates, per capita origination rates and standing family diversity. (a) Estimated low-latitude sea surface temperature from Royer et al. (2004); (b) per capita extinction rates (myr^{-1}); (c) per capita origination rates (myr^{-1}); and (d) standing diversity of all families in Benton (1993) using the maximum dating assumption. Four mass extinctions indicated in (b) are the end-Ordovician (O), late Devonian (D), end-Permian (P) and end-Cretaceous (C).

(2000) to estimate temperature, after accounting for changes to seawater pH (Fig 4.3a) caused partly by changes in atmospheric CO_2 concentrations predicted by GEOCARB III (Berner and Kothavala, 2001). CO_2 concentrations from GEOCARB III and the number of sedimentary rock formations (Peters and Foote, 2001, 2002; Peters, 2005) were used as alternative predictor (control) variables. All data were detrended prior to analysis so the associations refer to the variability around the long-term trends (Fig 4.4), which in the case of the biodiversity data may strongly reflect collection bias (Alroy et al., 2008).

In all data sets, significant negative correlations between biodiversity and temperature were revealed, such biodiversity being relatively low during greenhouse phases (Mayhew et al., 2008; Fig 4.4c). Associations were the opposite for origination and extinction rates, which were positively associated with temperature (Fig 4.4a, b). Temperature during the previous (10 myr) time-step often correlated most strongly with biodiversity and origination, suggesting a lag effect, but this was never the case for extinction. Although CO_2 concentrations were commonly also a significant predictor of biodiversity, origination and extinction, as expected from previous studies, it was generally a weaker predictor than temperature and eliminated from general linear models (Mayhew et al., 2008). This was also expected, since CO_2 concentrations are not known to show a 140 myr cycle coincident with the cycles in biodiversity, but temperature is. Although the sedimentary rock record (in terms of numbers of formations) strongly predicted biodiversity and taxonomic rates, the effect of temperature was retained despite this, and this was also expected, because of the absence of a known 140 myr cycle (Mayhew et al., 2008).

In summary, therefore, current evidence supports the suggestion that the fossil record has distinctive characteristics during greenhouse and icehouse modes of the Phanaerozoic. Three further questions arise, namely: (1) why do the associations exist; (2) what do they imply about the consequences of anthropogenic climate change for biodiversity; and (3) what do they imply in general about the evolution of biodiversity?

4.3 Why do the associations exist?

Because the fossil record, taken at face value, contains a number of inherent biases, it is wise to retain some caution when interpreting the above associations. In particular we should first ask whether the associations imply the operation of evolutionary-ecological processes or whether they reflect geophysical processes such as the relationship between climate and sedimentary rock formation, hence sampling or preservation bias. Given the previous discovery of cycles of biodiversity that are predicted by cycles of sedimentation (Smith and McGowan, 2005, 2007), this problem is a realistic one (but see Melott, 2008).

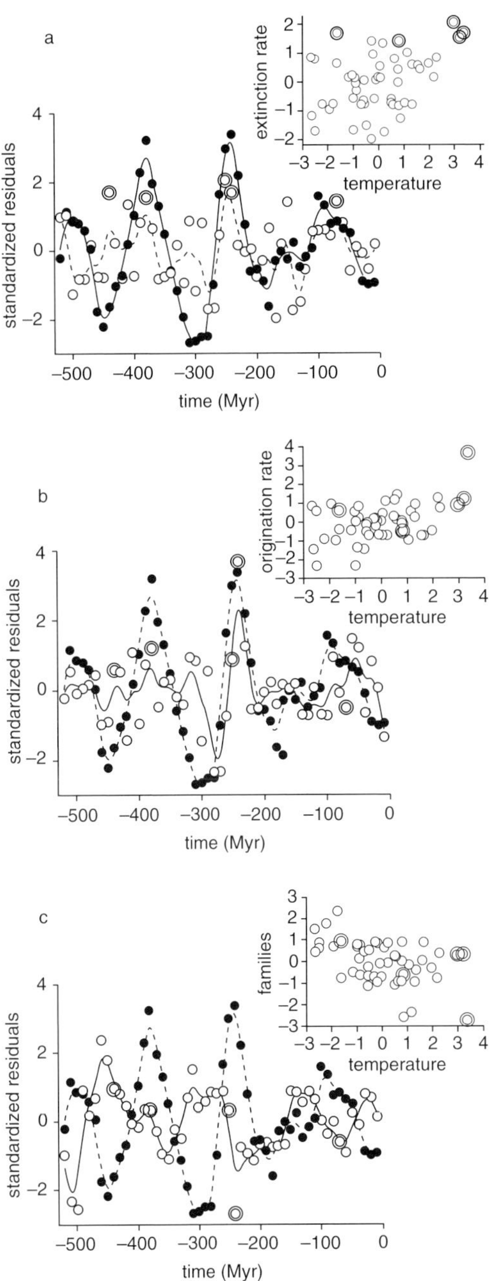

Figure 4.4 Associations between the time series in Fig 4.3. In each case the raw data sets have been transformed to stabilise variance, detrended, and then mean standardised for plotting on the same scales. Inserts show the correlations between the plots. Closed circles are temperature and open circles are (a) extinction rate, (b) origination rate, (c) standing diversity of families. Double open circles indicate the intervals with the five largest extinction rate residuals, corresponding to well-known mass extinction events, which from left to right in the time series are: end-Ordovician, late Devonian, end-Permian (× 2), end-Cretaceous. Lines are fitted curves using 25 d.f. splines. Reproduced from Mayhew et al. (2008) with permission of the Royal Society of London and Blackwell Scientific.

One argument in favour of an evolutionary-ecological explanation comes from the independent evidence concerning the causes of mass extinctions. Detailed geochemical and geophysical evidence from several mass extinctions, discussed in more detail below, supports the idea that high global temperatures have played a role (e.g. Wignall, 2001, 2005; Ward, 2006), and these are consistent in timing with the greenhouse climatic phases analysed above (Mayhew et al., 2008).

It is also encouraging that the associations between temperature and the fossil record are retained even when controlling for the number (or turnover as appropriate) of sedimentary formations (Mayhew et al., 2008). Furthermore, a periodicity of biodiversity of the same order is retained in sample standardised data from the Palaeobiology database (Melott, 2008), suggesting that the biodiversity signal is robust to sampling rate. In addition, some associations become stronger (Mayhew et al., 2008) when measures of diversity or taxonomic rate are used that ignore singletons and are therefore less susceptible to bias (Foote, 2000a). Finally, while it is true that the fossil record can be misleading if taken at face value, it also clearly contains some valid biological signals (e.g. Alroy, 2008; Alroy et al., 2008). Correcting for preservation bias is still a field under development, and while it is sensible to revisit these analyses as new 'corrected' data sets become available, it is also wise to remember that such data sets are not a panacea and come with their own assumptions and challenges (Kerr, 2008).

An alternative approach to 'check' the biological causation might come from studies of the timing of diversification in phylogenetic trees of extant taxa, or of gene genealogies, although these would have to be deep-rooted, and fairly robustly dated. A study along these lines by Ding et al. (2006) found weak evidence of periodic evolution of new transmembrane gene families at 62 myr cycles (corresponding with a 62 myr cycle in biodiversity), but not at 140 myr cycles. More studies of a similar nature would be interesting.

Taking the gambit then that there are evolutionary-ecological explanations to be found, what might they be? A major question is how the correlations revealed might reflect causation. There are two parts to this question. Firstly, how do the three response variables of standing biodiversity, origination and extinction relate to each other? Secondly, what is the action of the explanatory variables?

It is likely that the regulation of biodiversity is via changes in extinction, with origination responding to reductions in diversity via a rebound. Foote (2000b) has shown that over the marine record extinction better predicts biodiversity than origination, probably because of its greater variance. Delayed rise in origination after extinction is also commonly documented (e.g. Kirchner and Weil, 1990; Sepkoski, 1998; Lu et al., 2006; Alroy, 2008). Supporting this notion, the correlations between temperature and biodiversity and origination are often lagged (both lagging behind temperature), but this is never the case for extinction, suggesting that extinction is the more immediate agent of change (Mayhew et al., 2008).

What might be the causal mechanisms underlying associations between global temperature and extinction? An important consideration is that temperature is not the only environmental variable that has varied over geological time, nor the only variable that can affect biodiversity. Atmospheric CO_2 concentrations have already been mentioned above, but other variables that are commonly linked to the fossil record include sea-level variation, large igneous province eruptions and extraterrestrial impacts. The first has long been linked to variation in fossil diversity. A recent study to address this topic is the analysis of Purdy (2008) which shows a strong positive correlation between sea level and the number of marine genera (even without detrending) over the Phanaerozoic. More mechanistically, another study, by Peters (2008), showed that changes in the turnover of different sediment types predict the turnover from Palaeozoic to modern marine faunas, and suggested that sea-level fluctuations are a likely causative link to both. Sea-level fluctuations are, of course, not without relevance to the immediate future of life on earth, since sea-level rises are expected in response to climate warming (IPCC, 2007).

Large igneous province eruptions and extraterrestrial impacts are two of the most widely invoked potential causes of mass extinctions (Arens and West, 2008). The evidence for the influence of impacts is best at the end-Cretaceous extinction (Fig 4.3b), where the coincidental evidence for a large impact crater, pervasive worldwide geochemical evidence and sudden disappearance of many taxa make it a likely contributor, though perhaps not the sole cause (Keller, 2001). There is also pervasive evidence that impacts can increase the severity of extinctions through other causes (Arens and West, 2008). Large igneous province eruptions have been invoked in several mass (and other minor) extinctions (Wignall, 2001). One study has suggested that the combination of volcanism and impact best predicts the severity of the extinction event (Arens and West, 2008).

There is then the potential, not yet fulfilled, to include a greater range of environmental variables such as these into statistical modelling of the fossil record, and also to infer causation between the variables via path-analysis approaches. Clearly causative interactions between many of the variables are likely (see above references). For example, tectonic activity may release CO_2 to the atmosphere; atmospheric CO_2 may increase temperature; tectonic activity may influence sea level; temperature may influence sea level; and sea level and tectonic activity may both influence the types and quantity of sedimentation processes. So far, however, only temperature is known to show the 140 myr cycles that are likely to predict those seen in the fossil record, although fluctuations in sea level, which are not obviously periodic, clearly accord with similar ones in marine genera (Purdy, 2008). Why climate cyclicity might exist in the Phanaerozoic remains an open question; the subject of ongoing debate amongst palaeoclimatologists and astrophysicists (e.g. Shaviv and Veizer, 2003; see also Caballero and Lynch, Chapter 2).

Given, then, that temperature has some direct and pervasive causative effect on extinction, what are the possible mechanisms? There are, again, two major sources of inspiration: mechanistic studies of mass extinctions and studies of how global temperature change affects modern species.

Mechanistic studies of mass extinction, notably the end-Permian (Fig 4.3b) but increasingly others, have recently invoked marine crises, termed 'euxinia', in which large parts of the ocean surface waters become anoxic (Meyer and Kump, 2008). Involved is a web of forces, including CO_2 eruptions from large igneous provinces, such as the Siberian traps at the end-Permian, warming of the oceans, and release of submarine methane hydrates and hydrogen sulphide (H_2S) by sulphur bacteria in anoxic waters (Ward, 2006). On land the H_2S released may have depleted the stratospheric ozone layer (Ward, 2006), and warming, combined with the disposition of the continents in a single large land mass, Pangea, may have combined to lead to widespread desertification (Benton, 2005). Although such forces have only been studied in a few extinction events, the possibility that they may be more pervasive in earth history and correlated with extinction rates generally deserves attention. As well as empirical documentation of the distribution and extent of euxinia, coupled earth–ocean–atmosphere models capable of predicting their incidence through the Phanaerozoic are required to see if euxinia are predicted to be associated with greenhouse phases more commonly than icehouse phases (as seems likely – Meyer and Kump, 2008) and also with extinction severity. Similarly, the extent of low-diversity biomes such as deserts might be higher in greenhouse than in icehouse phases, but statistical studies are warranted.

Studies of the impact of current climate warming on present-day biodiversity emphasise how climate change may cause mismatch between species' current distributions and the expected future distribution of their current climatic envelopes (Thomas et al., 2004). Although much publicity is focused on the changes taking place to cold environments, such as the Antarctic and high Arctic, where the rate of climate change is highest and the effects most visible, most biodiversity is, of course, tropical. Some studies have suggested that tropical species may suffer just as much as, perhaps more than, high-latitude species, because although they may not experience high rates of temperature change, thermal tolerances of tropical species may not be high (Deutsch et al., 2008), and the climatic space that is expected in tropical environments is outside the current norms (Williams et al., 2007). Another potential threat to tropical species is that many have small geographic ranges (Gaston and Blackburn, 2001), and this means that the entire population of a species is vulnerable to local extinction threats. There is some excellent evidence from the fossil record that small geographic ranges can show increased extinction rates in comparison to large ones (e.g. Jablonski and Lutz, 1983), and the phenomenon is also predicted by the equilibrium theory of island biogeography (MacArthur and Wilson, 1967). It is not currently known, but if species in greenhouse climate modes evolve these 'tropical' characteristics more generally, then that may make biodiversity in general more vulnerable to change than would be the case in icehouse modes.

Collectively, then, these two approaches suggest two broad mechanisms by which greenhouse climate modes might elevate extinction. Firstly, warm worlds may be more likely to experience certain types of threat, such as euxinia (the perils of a warm world). Secondly, when threats do occur, such as short-term climatic fluctuations, the severity of the extinction that results from them might be elevated (the perils of change in a warm world).

4.4 Implications for current climate warming

Are Phanaerozoic-scale studies useful for predicting the consequences of anthropogenic climate change? It might be argued that they are not. Firstly, the Phanaerozoic climate measures are at relatively crude temporal scales, and cannot therefore adequately address the effects of short-term climate fluctuations, which we know have occurred in the Tertiary, Quaternary and Holocene. Thus the whole issue of how rapid changes in climate, rather than climates per se, affect biodiversity, simply cannot be easily addressed at those long temporal and global scales. Secondly, short-term processes might be different from long-term processes, and it is a leap of faith to extrapolate from one to the other. Thirdly, confidence in both the climate and biodiversity data is likely to be reduced at longer temporal scales because the data become sparser and more prone to biases.

These three objections are all valid points, but should not in themselves deter us from seriously considering the long-term picture alongside other data (Table 4.1). With regard to timescales, it is known that many of the statistically influential extinctions over the Phanaerozoic occurred over much shorter timescales than is implied by the interval between data points, so short-term changes do contribute to the patterns observed over long timescales. This should not simply be dismissed. Furthermore, the long-term patterns allow us to consider variation that cannot be considered over shorter intervals, hence it adds to our empirical knowledge. Suppose we were only to consider Cenozoic climate fluctuations; we would have finer-scale data and have greater confidence in them, but we would not be able to consider major greenhouse-to-icehouse fluctuations in the same way, reducing the climatic variability we can consider. During the Cenozoic there was a gradual long-term reduction in planetary surface temperatures (Fig 4.3a), representing the transition from a greenhouse to icehouse mode, with shorter-term fluctuations. However, that transition is unreplicated, and would, as standard, be detrended and removed in a time-series analysis (e.g. Alroy et al., 2000; Mayhew et al., 2008; Melott, 2008). Hence a Cenozoic study would consider short-term fluctuations, but of lesser extent.

Table 4.1 Complementary attributes of global Phanaerozoic and finer-timescale studies.

Attribute	Global Phanaerozoic studies	Finer-scale studies
Confidence in raw data	Lower, but high confidence not necessarily needed for associations to hold	High
Spatial scales considered	Large, so can detect general patterns	Small, so idiosyncrasies may mask general patterns
Temporal intervals considered	Long, so need other studies to infer short-term processes, but short-term processes can contribute to patterns	Short, so obviously more relevant to current change, and can consider issues of rate of change
Temporal durations considered	Very long, so can consider larger range of environmental conditions	Shorter, so idiosyncrasies may mask general patterns
Number of taxa considered	Can be all taxa with a fossil record, so general rules may be more apparent	Fewer, studies tend to be taxonomically focused, so idiosyncrasies may mask general patterns

Furthermore, a study over short durations would necessarily have to consider fewer taxa, because the temporal detail required would not be available for all. This may be problematic because of the large idiosyncrasy that is suggested by existing studies; consistent rules may easily go undetected. Finally, the Phanaerozoic-level associations between biodiversity and climate do not rely on the modelled data being accurate in every sense; they simply rely on the underlying pattern of greenhouse and icehouse fluctuations of roughly the estimated duration, and this is known with greater certainty than the precise temperatures. It is perhaps better, therefore, to view the global Phanaerozoic-level studies as complementary to finer-scale studies (Table 4.1).

Once it has been decided that Phanaerozoic-scale studies can in principle provide a context for the effects of current climate change, it is important to consider the level of predictability that such studies show. The associations might be described as moderate in strength, but they do not, prima facie, appear highly predictive (Fig 4.4). In the extinction time series (Fig 4.4a), the largest positive outlier is the end-Ordovician mass extinction (see also Fig 4.3b), which took place during an icehouse mode. At this time, life was not established on land, and the continents were disposed rather southerly. It is conjectured that glaciation and concomitant effects on sea level were a major factor in the extinctions at this time (Sheehan, 2001). However, such glacially caused mass extinctions appear to be the exception rather than the rule. Clearly, then, greenhouse effects are not the only cause of extinction, and other abiotic extremes are also partly or wholly responsible (Arens and West, 2008). This adds error variance to the association with temperature.

The next mass extinction that least conforms to the temperature trend is the end-Cretaceous extinction. Presumably the effects of meteor impact were important here. The fact that the associated residual is not even larger may be just coincidence (the end-Cretaceous was still a relatively warm period). However, temperature may also have played a role in accentuating the scale of the extinction. In general, variance around the trends sometimes appears greater in extinction compared to origination or biodiversity (Fig 4.4b, c; see also Cornette et al., 2002). Likewise there are episodes of high extinction during the Carboniferous icehouse mode and Jurassic cool mode. This suggests that temperature also has tighter, more direct effects on the origination rate. This might be caused, for example, by allowing larger rebounds after extinction during greenhouse modes than in icehouse modes. This idea would fit well with the accord that evolutionary rates may be higher in low-latitude than in high-latitude regions (Cardillo, 1999; Davies et al., 2004).

In general, then, the moderate coefficients of determination of the temperature–biodiversity associations are not necessarily cause for pessimism in the underlying predictability between changes in temperature and biodiversity; they may simply represent the fact that in an observational, uncontrolled data set, many environmental variables are fluctuating at the same time. It is encouraging that the fossil data are in broad agreement with the climate envelope modelling data in suggesting that extensive extinction and biodiversity loss will result from global warming, although they may imply different mechanisms (see below).

The relationship between temperature and extinction rates is much more predictive at high temperatures than at low temperatures (Fig 4.4a). At low temperatures, extinction rates are highly variable across strata. At high temperatures they are always high. This is notable because of the current direction of change in global temperatures (i.e. upwards). Because the long-term background temperatures in tropical seas in the end-Permian were of the order of 6 °C higher than today (Fig 4.3a), and because global surface temperature rises of the order of 6 °C are possible for the current century (IPCC, 2007), this justifies some concern over the possibility of catastrophic changes to global ecosystems on very short timescales. Therefore, of the causative mechanisms outlined by studies that invoke climate change in mass extinctions, we should ask if such mechanisms could operate and become more likely in the current or next century. Methane hydrate stores in the deep ocean are sufficient to cause a catastrophic change in climate if they were released suddenly (Archer, 2007). Worryingly, such stores are sensitive to seafloor temperatures (Buffet and Archer, 2004), but comfortingly, present data indicate that on the timescale of a century or so, the seafloor is relatively buffered against changes to surface temperature (Archer, 2007). However, methane hydrates could still be a significant contributor to future climate warming, and release of methane from permafrost environments is tangible and a contributor to current climate change. On a longer timescale of hundreds of millennia, hydrate release is likely to have a greater effect, perhaps as large as that of fossil-fuel combustion, but this should give time for the anthropogenic warming to be combatted, and would in any case be a chronic rather than a catastrophic change (Archer, 2007).

The incidence of euxinia today is very low, and euxinia are thought to be promoted by periods of global warmth primarily through reduced solubility of oxygen rather than oceanic stagnation (Meyer and Kump, 2008). Anoxia leads to phosphate release from sediments; critical to the widespread incidence of euxinia is a continental configuration that encourages nutrient trapping. Warmth combined with biological productivity can enhance oxygen depletion and sulphide build-up. Models of the nutrient-trapping efficiency of palaeogeographic reconstructions suggest that the end-Permian continental configuration, with Pangea enclosing the Tethys sea, would have led to widespread conditions that favour euxinia. In contrast, the modern world has a much lower nutrient-trapping efficiency, and thus it appears that the likely future incidence of euxinia, while increasing due to temperature, is less likely to be catastrophic (Meyer and Kump, 2008).

There are qualifiers to add to the above summary. Reported incidents of hypoxia in coastal waters have shown a recent rise, and have been described as a major threat to coastal ecosystems globally (Vaquer-Sunyer and Duarte, 2008). While hypoxia is therefore unlikely to cause catastrophic effects on global biodiversity, similar to a mass extinction, it is nonetheless likely to cause increasing depletion of fisheries in those affected areas and some loss of biodiversity (Bakun and Weeks, 2004). It may also contribute to greenhouse emissions, another positive feedback on the climate system.

We might also ask whether the distribution of world biomes is expected to see catastrophic change. Once again, the answer seems to be: some change, but not catastrophic. In general, biome shifts on a scale of centuries will be detectable and significant for biodiversity but not globally catastrophic (Cramer et al., 2001). Biodiversity capacity in the absence of land-use change should actually show a net increase due to the positive effects of CO_2 increases trumping the negative effects of warming (Woodward and Kelly, 2008).

Therefore, while the inferred processes at the end-Permian and other extinctions can be seen to be operating on smaller scales today, current research does not indicate that we will see catastrophic changes on the timescales of centuries, though changes could nonetheless be chronic. However, given the severity of the threat, we need higher certainty. We need much greater confidence in our understanding of the underlying processes than we currently have, and we also need to know the extent of 'breathing space' that we have. In particular, we should try to push global ocean–earth–atmosphere models into an end-Permian environmental state to understand exactly what anthropogenic changes would pose a globally catastrophic threat.

The Phanaerozoic associations bear two wider-ranging messages for our understanding of the effects of future climate change. The first relates to the form of the relationship. This is essentially gradual in nature (Fig 4.4) rather than a threshold effect, with warmer climates leading to larger effects. Climatologists often refer to tipping points in the climate system (Lenton et al., 2008), such as the absence of Arctic ice in summer which will have a significant additional warming effect. The presence of tipping points, which may inevitably be passed, may engender helplessness and apathy amongst policy makers. In contrast, the fossil record suggests that any kind of mitigation action that will reduce global warming will actually have beneficial effects; any decrease in global temperature, however small, will lower extinction risk. This suggests that, far from apathy, what is needed is the resolution to carry through beneficial changes in the knowledge that, however slight, they will have some mitigating effect. The second message relates to the persistence of the effects. Rises in origination after extinctions, at least amongst higher taxa, are likely to be delayed by millions of years, and therefore the biodiversity loss may be very long-term, measurable on geological timescales. This should again encourage immediate action, since any losses of biodiversity are unlikely to be replaced during the lifetime of our species.

4.5 Implications for the evolution of biodiversity

Aside from their potential for predicting the future, associations between global climate and the fossil record are of interest simply because the variability in the history of life is something to understand and explain. The above studies suggest that abiotic factors play a major role in regulating biodiversity through time, and also, because of the taxonomic turnover that occurs through extinction and origination rebounds, in the taxonomic composition of that diversity.

Studies of the fossil record have produced two alternative paradigms related to abiotic and biotic effects on extinction (Benton, 2009). The Red Queen hypothesis holds that biotic forces produce a continual and predictable extinction risk, while the Court Jester hypothesis holds that extinction is abiotic, episodic and unpredictable (Benton, 2009). The results detailed above are not fully consistent with either. Like the Court Jester, extinction is inferred to be mediated through abiotic causes, but like the Red Queen this extinction is relatively predictable. The results therefore demand a third paradigm, of abiotic but predictable extinction, which I propose here to call the 'Ace of Spades' because of the consistent associations of that symbol with death and destruction, and in keeping with the playing-card connotations of the other two paradigms.

Some qualification is needed; the associations relate, of necessity, to fluctuations around the long-term trends. One expects the long-term trends themselves to be regulated by the general extent to which origination outweighs extinction, and how both change over time, and in particular whether growth is exponential or more logistic (Benton, 1995, 1997; Alroy, 2008). However, abiotic factors may influence the long-term trends as well, for example, if the species richness is affected by sea level for marine organisms (Purdy, 2008) or by continental configurations (Valentine and Moores, 1970).

An interesting puzzle that emerges from the Phanaerozoic studies is that the relationship between taxonomic diversity and temperature is negative over time, but is positive across space, as seen in the latitudinal gradient in diversity. To find the opposite effect in time compared to space seems contradictory, prima facie, but a deeper consideration shows a possible solution. The spatial pattern of latitudinal diversity depends on regional processes, while the temporal pattern depends on processes that are presumably global. Essentially, then, the spatial diversity gradients could remain over time while global diversity fluctuates over time. It is perfectly plausible that the global and regional processes might be different: hypothetically, for example, the waxing and waning of high-latitude ice sheets may generally increase extinction rates in high latitudes compared to low latitudes (Willig et al., 2003), while the temporal incidence of marine euxinia may cause temporal variability in global extinction across all latitudes.

In addition, processes or events can occur that cause the build-up of latitudinal gradients in diversity, but that do not contribute towards temporal global diversity trends. Time spent in the tropics and limited dispersal to high-latitude regions are thought to be influential in latitudinal diversity patterns (e.g. Böhm and Mayhew, 2005; Jablonski et al., 2006), but neither can contribute to a global temporal pattern that by definition ignores them.

The spatial and temporal patterns are not contradictory in one respect, and that is in origination rates (positively related to temperature in space and over time). Hence the Phanaerozoic patterns support the idea, derived from spatial diversity patterns, of an association between environmental energy and speciation rates, and generally of evolutionary rates.

4.6 Future research

Like many scientific studies, analysis of the relationship between global temperature and the fossil record poses many more questions than it answers. Because of the significance of the topic, these new questions need urgent attention.

- Do the associations between global climate and the fossil record hold using shorter-term variation?
- Do these associations hold in the face of our increasing knowledge of sampling and preservation biases?
- Do deeply rooted phylogenetic inferences or genealogies provide any support for higher diversification rates during greenhouse modes?
- Is the association between origination and extinction dependent on temperature?
- Does temperature retain its significance in a larger multivariate analysis of many environmental explanatory variables, and what are the causal relationships between the variables?
- How are the temporal associations between temperature and biodiversity consistent with spatial associations of the opposite form?
- Does the incidence of marine euxinia correlate with global temperature?
- Are species' geographic ranges smaller during greenhouse modes?
- Does variation in the extent of biomes over geological time predict extinction intensity, and does it accord with climate modes?
- What increases in temperature in the near future will be required to recreate conditions in the Permian–Triassic mass extinction, if at all?
- What threats to current biodiversity from current climate change result from high temperatures per se as opposed to changes in temperature?

Acknowledgements

I thank Trevor Hodkinson, Mike Jones, Steve Waldren and John Parnell for the invitation to the conference on which this book is based, and for the invitation to contribute a chapter, as well as the many biologists and palaeobiologists who have discussed with me the topics included in it. Two anonymous referees made useful comments on an earlier draft.

References

Alroy, J. (2008). Dynamics of origination and extinction in the marine fossil record. *Proceedings of the National Academy of Sciences of the USA*, **105** (Suppl. 1), 11536–11542.

Alroy, J., Koch, P. L. and Zachos, J. C. (2000). Global climate change and north American mammalian evolution. *Paleobiology*, **26** (Suppl.), 259–288.

Alroy, J., Aberhan, M., Bottjer, D. J. et al. (2008). Phanerozoic trends in the global diversity of marine invertebrates. *Science*, **321**, 97–100.

Araújo, M. B. and Guisan, A. (2006). Five (or so) challenges for species distribution modelling. *Journal of Biogeography*, **33**, 1677–1688.

Archer, D. (2007). Methane hydrate stability and anthropogenic climate change. *Biogeosciences*, **4**, 521–544.

Arens, N. C. and West, I. D. (2008). Press-pulse: a general theory of mass extinction? *Paleobiology*, **34**, 456–471.

Bakun, A. and Weeks, S. J. (2004). Greenhouse gas buildup, sardines, submarine eruptions and the possibility of abrupt degradation of intense marine upwelling ecosystems. *Ecology Letters*, **7**, 1015–1023.

Benton, M. J., ed. (1993). *The Fossil Record 2*. London: Chapman and Hall.

Benton, M. J. (1995). Diversification and extinction in the history of life. *Science*, **268**, 52–58.

Benton, M. J. (1997). Models for the diversification of life. *Trends in Ecology and Evolution*, **12**, 490–495.

Benton, M. J. (2005). *When Life Nearly Died*. London: Thames and Hudson.

Benton, M. J. (2009). The Red Queen and the Court Jester: species diversity and the role of biotic and abiotic factors through time. *Science*, **323**, 728–732.

Berner, R. A. and Kothavala, Z. (2001). GEOCARB III: a revised model of atmospheric CO_2 over Phanerozoic time. *American Journal of Science*, **301**, 182–204.

Böhm, M. and Mayhew, P. J. (2005). Historical biogeography and the evolution of the latitudinal gradient of species richness in the Papionini (Primata: Cercpithicidae). *Biological Journal of the Linnean Society*, **85**, 235–246.

Buffet, B. and Archer, D. (2004). Global inventory of methane clathrate: sensitivity to changes in the deep ocean. *Earth and Planetary Science Letters*, **227**, 185–199.

Cardillo, M. (1999). Latitude and rates of diversification in birds and butterflies. *Proceedings of the Royal Society of London B*, **266**, 1221–1225.

Cornette, J. L., Lieberman, B. S. and Goldstein, R. H. (2002). Documenting a significant relationship between macroevolutionary origination rates and Phanerozoic pCO_2 levels. *Proceedings of the National Academy of Sciences of the USA*, **99**, 7832–7835.

Cramer, W., Bondeau, A., Woodward, F. I. et al. (2001). Global response of terrestrial ecosystem structure and function to CO_2 and climate change: results from six dynamic global vegetation models. *Global Change Biology*, **7**, 357–373.

Davies, T. J., Savolainen, V., Chase, M. W., Moat, J., and Barraclough, T. G. (2004). Environmental energy and evolutionary rates in flowering plants. *Proceedings of the Royal Society of London B*, **271**, 2195–2200.

Denness, B. (1989). Evolution and environmental determinism. *Modern Geology*, **13**, 283–286.

Deutsch, C. A, Tewksbury, J. J, Huey, R. B. et al. (2008). Impacts of climate warming on terrestrial ectotherms across latitude. *Proceedings of the National Academy of Sciences of the USA*, **105**, 6668–6672.

Ding, G., Kang, J., Lui, Q. et al. (2006). Insights into the coupling of duplication events and macroevolution from an age profile of animal transmembrane gene families.

PLoS Computational Biology, **2**, 918–924.

Foote, M. (2000a). Origination and extinction components of taxonomic diversity: general problems. *Paleobiology*, **26** (Suppl.), 74–102.

Foote, M. (2000b). Origination and extinction components of taxonomic diversity: Paleozoic and post-Paleozoic dynamics. *Paleobiology*, **26**, 578–605.

Frakes, L. A., Francis, J. E. and Syktus, J. I. (1992). *Climate Modes of the Phanerozoic*. Cambridge: Cambridge University Press.

Gaston, K. J. and Blackburn, T. M. (2001). *Pattern and Process in Macroecology*. Oxford: Blackwell.

Gibbs, S. J., Bown, P. R., Sessa, J. A., Bralower, T. J. and Wilson, P. A. (2006). Nannoplankton extinction and origination across the Paleocene–Eocene thermal maximum. *Science*, **314**, 1770–1773.

Huber, B. T., MacLeod, K. G. and Wing, S. L., eds. (2000). *Warm Climates in Earth History*. Cambridge: Cambridge University Press.

Ibanez, I., Clark, J. S., Dietze, M. C. et al. (2006). Predicting biodiversity change: outside the climate envelope, beyond the species-area curve. *Ecology*, **87**, 1896–1906.

Intergovernmental Panel on Climate Change (IPCC) (2007). *Climate Change 2007: the Physical Science Basis. Contribution of Working Group I to the Fourth Assessment Report of the Intergovernmental Panel on Climate Change*, ed. S. Solomon, D. Qin, M. Manning et al. Cambridge: Cambridge University Press.

Jablonski, D. and Lutz, R. A. (1983). Larval ecology of marine invertebrates: paleobiological implications. *Biological Reviews*, **58**, 21–89.

Jablonski, D., Roy, K. and Valentine, J. W. (2006). Out of the tropics: evolutionary dynamics of the latitudinal diversity gradient. *Science*, **314**, 102–106.

Jaramillo, C., Rueda, M. J. and Mora, G. (2006). Cenozoic plant diversity in the neotropics. *Science*, **311**, 1893–1896.

Keller, G. (2001). The end-Cretaceous mass-extinction in the marine realm: year 2000 assessment. *Planetary and Space Science*, **49**, 817–830.

Kerr, R. A. (2008). Life's innovations let it diversify, at least up to a point. *Science*, **321**, 24–25.

Kirchner, J. W. and Weil, A. (1990). Delayed biological recovery from extinctions throughout the fossil record. *Nature*, **404**, 177–180.

Lenton, T. M., Held, H., Kriegler, E. et al. (2008). Tipping elements in the earth's climate system. *Proceedings of the National Academy of Sciences of the USA*, **105**, 1786–1793.

Lovejoy, T. E. and Hannah, L., eds. (2005). *Climate Change and Biodiversity*. New Haven: Yale University Press.

Lu, P. J., Yogo, M. and Marshall, C. R. (2006). Phanerozoic marine biodiversity dynamics in light of the incompleteness of the fossil record. *Proceedings of the National Academy of Sciences of the USA*, **103**, 2736–2739.

MacArthur, R. H. and Wilson, E. O. (1967). *The Theory of Island Biogeography*. Princeton, NJ: Princeton University Press.

Mayhew, P. J., Jenkins, G. B. and Benton, T. G. (2008). A long-term association between global temperature and biodiversity, origination and extinction in the fossil record.

Proceedings of the Royal Society of London B, **275**, 47–53.

Melott, A. L. (2008). Long-term cycles in the history of life: periodic biodiversity in the Paleobiology database. *PLoS ONE*, **3**, e4044.

Meyer, K. M. and Kump, L. R. (2008). Oceanic euxinia in Earth history: causes and consequences. *Annual Review of Earth and Planetary Sciences*, **36**, 251–288.

Parmesan, C. and Yohe, G. (2003). A globally coherent fingerprint of climate change impacts across natural systems. *Nature*, **421**, 37–42.

Pearson, R. G. (2006). Climate change and the migration capacity of species. *Trends in Ecology and Evolution*, **21**, 111–113.

Peters, S. E. (2005). Geologic constraints on the macroevolutionary history of marine animals. *Proceedings of the National Academy of Sciences of the USA*, **102**, 1236–1331.

Peters, S. E. (2008). Environmental determinants of extinction selectivity in the fossil record. *Nature*, **454**, 626–629.

Peters, S. E. and Foote, M. (2001). Biodiversity in the Phanerozoic: a reinterpretation. *Paleobiology*, **27**, 583–601.

Peters, S. E. and Foote, M. (2002). Determinants of extinction in the fossil record. *Nature*, **416**, 420–424.

Pounds, J. A., Bustamante, M. R., Coloma, L. A. et al. (2006). Widespread amphibian extinctions from epidemic disease driven by global warming. *Nature*, **439**, 161–167.

Purdy, E. G. (2008). Comparison of taxonomic diversity, strontium isotope and sea-level patterns. *International Journal of Earth Science*, **97**, 651–664.

Rohde, R. A. and Muller, R. A. (2005). Cycles in fossil diversity. *Nature*, **434**, 208–210.

Rothman, D. H. (2001). Global biodiversity and the ancient carbon cycle. *Proceedings of the National Academy of Sciences of the USA*, **98**, 4305–4310.

Royer, D. L., Berner, R. A., Montañez, I. P., Tibor, N. J. and Beerling, D. J. (2004). CO_2 as a primary driver of Phanerozoic climate. *Geological Society of America Today*, **14**, 4–10.

Sepkoski, J. J. (1998). Rates of speciation in the fossil record. *Philosophical Transactions of the Royal Society of London B*, **353**, 315–326.

Sepkoski, J. J. (2002). A compendium of fossil marine animal genera. *Bulletin of American Paleontology*, **363**, 1–560.

Shaviv, N. J. and Veizer, J. (2003). Celestial driver of Phanerozoic climate? *Geological Society of America Today*, **13**, 4–10.

Sheehan, P. M. (2001). The Late Ordovician mass extinction. *Annual Review of Earth and Planetary Sciences*, **29**, 331–364.

Smith, A. B. and McGowan, A. J. (2005). Cyclicity in the fossil record mirrors rock outcrop area. *Biology Letters*, **1**, 443–445.

Smith, A. B. and McGowan, A. J. (2007). The shape of the Phanerozoic marine diversity curve: how much can be predicted from the sedimentary rock record of Western Europe? *Palaentology*, **50**, 765–774.

Thomas, C. D., Cameron, A., Green, R. E. et al. (2004). Extinction risk from climate change. *Nature*, **427**, 145–148.

Thomas, C. D., Franco, A. M. A. and Hill, J. K. (2006). Range retractions and extinction in the face of climate warming. *Trends in Ecology and Evolution*, **21**, 415–416.

Thuiller, W. (2004). Patterns and uncertainties of species' range shifts under climate change. *Global Change Biology*, **10**, 2020–2027.

Valentine, J. W. and Moores, E. M. (1970). Plate-tectonic regulation of faunal diversity and sea level: a model. *Nature*, **228**, 657–659.

Vaquer-Sunyer, R. and Duarte, C. M. (2008). Thresholds of hypoxia for marine biodiversity. *Proceedings of the National Academy of Sciences of the USA*, **105**, 15452–15457.

Veizer, J., Godderis, Y. and François, L. M. (2000). Evidence for decoupling of atmospheric CO_2 and global climate during the Phanerozoic eon. *Nature*, **408**, 698–701.

Ward, P. D. (2006). Impact from the deep. *Scientific American*, October 2006, 64–71.

Wignall, P. B. (2001). Large igneous provinces and mass extinctions. *Earth-Science Reviews*, **53**, 1–33.

Wignall, P. B. (2005). The link between large igneous province eruptions and mass extinctions. *Elements*, **1**, 293–297.

Williams, J. W., Jackson, S. T. and Kutzbach, J. E. (2007). Projected distributions of novel and disappearing climates by 2100 AD. *Proceedings of the National Academy of Sciences of the USA*, **104**, 5738–5742.

Willig, M. R., Kaufmann, D. M. and Stevens, R. D. (2003). Latitudinal gradients of biodiversity: pattern, process, scale and synthesis. *Annual Review of Ecology, Evolution, and Systematics*, **34**, 273–309.

Woodward, F. I. and Kelly, C. K. (2008). Responses of global plant biodiversity capacity to changes in carbon dioxide concentration and climate. *Ecology Letters*, **11**, 1229–1237.

5

Long-term fluctuations in atmospheric CO_2 concentration influence plant speciation rates

J. C. McElwain

School of Biology and Environmental Science, University College Dublin, Ireland

K. J. Willis

Department of Zoology, University of Oxford, UK

K. J. Niklas

Department of Plant Biology, Cornell University, NY, USA

Abstract

The influence of vascular land-plant evolution on long-term fluctuations in atmospheric carbon dioxide (CO_2) concentration has been much discussed. However, the direct evolutionary consequences of changes in past atmospheric CO_2 concentration have not been widely explored despite experimental evidence showing that elevated CO_2 can intensify interplant competition, alter plant reproductive biology, and thus potentially influence plant speciation. In this chapter,

Climate Change, Ecology and Systematics, ed. Trevor R. Hodkinson, Michael B. Jones, Stephen Waldren and John A. N. Parnell. Published by Cambridge University Press.

we put forward the hypothesis that changes in atmospheric CO_2 may have directly enhanced vascular plant speciation rates in the course of land-plant evolution, particularly in the late Palaeozoic when numerous pteridophyte and gymnosperm lineages were radiating rapidly (between approximately 400 and 240 million years ago – mya). Our hypothesis is based on the observation of significant correlations between land-plant speciation rates and long-term records on fluctuations in atmospheric CO_2 levels over the past 410 million years. The relationship is complex, however, with a strong positive correlation between gymnosperm and pteridophyte speciation rates and CO_2 concentration above 1000 ppmv, but a weak negative correlation between angiosperm speciation rates and atmospheric CO_2 levels. These results suggest that fundamental plant evolutionary responses to atmospheric CO_2 may be dependent on evolutionary grade. Model estimates of palaeo-atmospheric CO_2 values are subject to considerable uncertainties, as are large-scale literature compilations of fossil plant speciation rates through geological time. Future collections of plant speciation data and palaeo-CO_2 estimates are therefore required to test this hypothesis further.

5.1 Introduction

One of the many challenges facing the global change research community is predicting what impact future increases in atmospheric CO_2 and global temperature will have on the ecological composition and biodiversity of ecosystems. Despite over three decades of research, numerous reviews and meta-analyses conclude that it is difficult to scale up predictions on natural ecosystem responses to elevated CO_2 from individual pot experiments. This is partly because they are poor predictors of species' responses grown in more 'natural' conditions such as in competition experiments (Poorter and Navas, 2003). A further complication is the paucity of data available on the transgenerational affects of elevated CO_2 (Ward and Kelly, 2004; Lau et al., 2008). For instance, there is much uncertainty as to whether short-term plastic responses by plants to elevated CO_2 will be enhanced or dampened by evolutionary processes. Ward and Kelly (2004) argue that a doubling of atmospheric CO_2 concentration to 700 ppm by the year 2100 has the potential to impose strong selective pressure on plants, because such high CO_2 levels will be novel with respect to the plant's recent evolutionary history (having experienced CO_2 levels of between 170 and 350 ppm for the past *c.* 10 million years) and because large variations in genotypic responses have been observed in experimental studies under high CO_2 conditions. Elegant transgenerational, genotype and environment factor studies are beginning to shed light on some of these issues (Poorter and Navas, 2003). However, a complementary approach is to mine the plant fossil record for underlying relationships between long-term CO_2 change and plant macroevolutionary patterns.

The plant fossil record offers an opportunity to investigate the ecological and evolutionary responses of taxa to atmospheric CO_2 and climatic change at large spatial and long temporal scales, and of a similar or perhaps greater magnitude compared to global changes anticipated by the year 2100. This chapter aims to explore these relationships by specifically investigating the effect of atmospheric CO_2 concentration on plant speciation rates. As speciation and extinction are key determinants of future biodiversity, it is essential that we begin to understand the potential effects of elevated CO_2 on plant macroevolutionary processes.

5.2 A macroscale view of plant evolution

Three major reproductive grades of land plants have successively dominated terrestrial floras over the past 410 million years (Niklas et al., 1983; Knoll, 1985; Niklas, 1997; Willis and McElwain, 2002). Free-sporing vascular species (pteridophytes) dominated Palaeozoic (543–248 mya) floras, whereas gymnosperms and angiosperms dominated Mesozoic (248–65 mya) and Cenozoic (65–0 mya) floras, respectively (Fig 5.1). The adaptive radiation of each plant group was marked by an increase in species diversity and/or relative abundance, and resulted in, or was coincident with, a gradual decline in the diversity and/or relative abundance of the previously dominant plant group (Niklas et al., 1983; Lidgard and Crane, 1988, 1990; Lupia et al., 1999). The early stages of each radiation are characterised by high speciation rates in all three reproductive grades (Niklas et al., 1983; Niklas, 1997).

These patterns of speciation, adaptive radiation and subsequent replacement have been much discussed, yet the potential driving mechanism(s) that may explain the timing of major evolutionary events in the plant fossil record remain elusive. Indeed, to date, neither climatic nor biogeographical (Niklas et al., 1983) nor biological factors appear fully sufficient to explain the timing of each of the three major evolutionary revolutions in terrestrial floristic composition (Barrett and Willis, 2001). Furthermore, in contrast to animal mass extinction events (Gould, 1985; Jablonski, 2001), which are considered by some to reset the evolutionary 'clock' for faunal recompositions, major innovations in floral evolution do not coincide with the recovery stages following mass extinctions (Traverse, 1988; Willis and Bennett, 1995; Niklas, 1997; Willis and McElwain, 2002).

The role of atmospheric CO_2 in both animal (Cornette et al., 2002) and plant evolution (Sage, 2004; McElwain et al., 2005; Lau et al., 2008; Franks and Beerling, 2009) has emerged as an increasingly important area of investigation, particularly in light of current concerns surrounding future climatic and biological responses

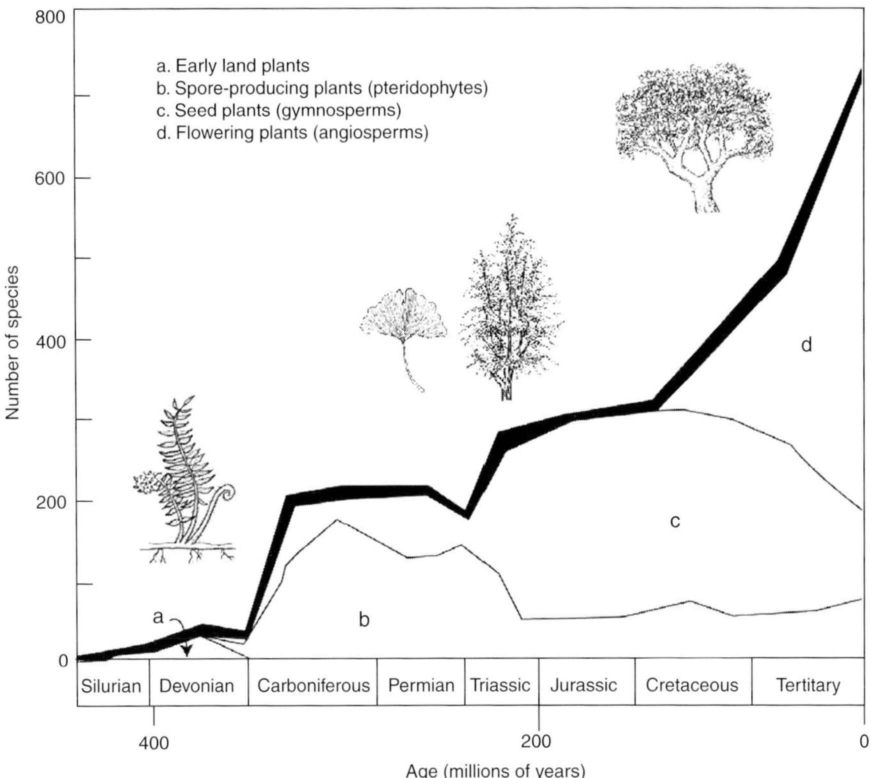

Figure 5.1 Changes in diversity of early land plants, pteridophytes, gymnosperms and angiosperms over geological time. The figure illustrates shifts in the dominance of the different evolutionary grades. Pteridophytes were dominant in the Palaeozoic, gymnosperms in the Mesozoic and angiosperms in the Cenozoic. Modified from Niklas (1997).

to ever-rising CO_2 levels. Marked reductions in atmospheric CO_2 have been invoked as a primary driver for the evolution of true leaves in the late Palaeozoic (Beerling et al., 2001), the diversification and radiation of angiosperms in the Cretaceous (Robinson, 1994; McElwain et al., 2005; Boyce et al., 2009) and the ecological expansion of C_4 plants in the Miocene (Ehleringer et al., 1991; Christin et al., 2008; Bouchenak-Khelladi and Hodkinson, Chapter 7). However, it seems apparent that innovations in plant morphology, anatomy and reproductive biology are not always associated with or restricted to intervals of relatively reduced CO_2 levels.

Recent examination of the timing of major speciation events in the plant fossil record have demonstrated close temporal correlations with times of rising or relatively elevated atmospheric CO_2 (Barrett and Willis, 2001; Willis and McElwain, 2002). We specifically explore the role of atmospheric CO_2 in large-scale patterns of plant evolution (macroevolution) and investigate whether the timing of the three major reproductive innovations in the plant fossil record (that is, the evolution of pteridophytes and gymnosperms in the Palaeozoic and angiosperms in the early Cretaceous) can be attributed in part to a single or group of abiotic causal factors.

5.3 Atmospheric CO_2 and plant macroevolution

Long-term fluctuations in atmospheric CO_2 (on multimillion-year timescales) are controlled by the influence of tectonic and biological factors on two components of the long-term carbon cycles: the silicate–carbonate cycle and the organic-matter cycle (Berner and Kothavala, 2001; Berner, 2006). The main sources of CO_2 within these cycles include thermal decomposition of carbon-containing rocks at mid-ocean ridges and subduction zones, volcanic outgassing and uplift, and the weathering of organic-rich sediments (Berner, 2006). The most important long-term sinks for atmospheric CO_2 are chemical weathering of silicate rocks and the subsequent precipitation of silicates as carbonates in the oceans and the uptake and sequestration of CO_2 in plant cell walls. Terrestrial plants, in particular those with vascular systems, are believed to play an important role in the silicate geochemical cycle by accelerating the rate of silicate weathering through the action of organic acids on primary minerals, by increasing the amount of mineral–water contact time, and by anchoring soils and preventing physical erosion, thereby enabling further mineral dissolution (Berner and Canfield, 1989; Berner, 1997; Algeo and Scheckler, 1998; Moulton and Berner, 1998). The coevolution of fungal associations with plant roots, such as arbuscular mycorrhizal and ectomycorrhizal fungi, are also believed to have greatly enhanced the rates of weathering by vegetation (Taylor et al., 2009).

In considering the role of atmospheric CO_2 concentration on large-scale patterns in plant evolution, it is important to take into account the feedback effect of vascular land-plant evolution on palaeo-atmospheric CO_2 levels via the silicate weathering cycle (Berner, 1997, 2006; Berner and Kothavala, 2001) and to be aware that these feedback effects have been built into estimates of how CO_2 has fluctuated throughout geological time. For instance, the GEOCARB III (Berner and Kothavala, 2001) and GEOCARBSULF (Berner, 2006) models assume that vascular plants have a greater weathering potential than non-vascular plants (e.g. bryophytes) or lichens. These assumptions are based on weathering studies in Iceland, which have shown that the rate of calcium and magnesium ion release from areas vegetated by vascular plants are two to five times that of equivalent areas vegetated by their non-vascular counterparts (Moulton and Berner, 1998). Similarly, although the evidence is less equivocal (Huh and Edmond, 1999), an assumption is made by both models that angiosperms have a greater weathering potential than gymnosperms (Moulton et al., 2000). In order to avoid circularity we have examined relationships between plant speciation rate and long-term CO_2 change using models that exclude (i.e. GEOCARB I – Berner, 1991) as well as include (i.e. GEOCARB III and GEOCARBSULF) the effects of plant evolution on relative silicate weathering rates.

Numerous proxy methods, which are independent of the carbon-cycle model estimates, have also been employed to document fluctuations in atmospheric CO_2 over the Phanaerozoic (Royer et al., 2001, 2004). Briefly these include: (1) analysis of the stable carbon isotopic composition ($\delta^{13}C$) of pedogenic carbonate (Cerling, 1991), known as the palaeosol method; (2) $\delta^{11}B$ analysis of calcite from planktonic foraminifera (Pearson and Palmer, 2000), known as the boron method; (3) analysis of $\delta^{13}C$ of marine phytoplankton (Pagani et al., 1999); (4) stomatal density and index of fossil leaves (Woodward, 1987; McElwain and Chaloner, 1995; Royer, 2001), known as the stomatal method; and (5) $\delta^{13}C$ of fossil liverworts (Fletcher et al., 2005), now known as BRYOCARB (Fletcher et al., 2006). Estimated errors in palaeo-CO_2 prediction using the GEOCARB model, based on sensitivity analysis of all of the processes influencing the long-term carbon cycle, range from ± 75–200 ppm for the Tertiary to as high as ± 3000 ppm for the early Palaeozoic (Royer et al., 2001). Despite these uncertainties, there is good general agreement among the various proxy and modelling estimates that CO_2 concentrations were high in the early Palaeozoic followed by a marked decline in concentrations into the Carboniferous and Permian, with levels somewhere in between these two extremes for the Mesozoic and Cenozoic, respectively (Fig 5.2).

5.4 Data sources and analysis

For the purposes of this analysis we selected long-term palaeo-CO_2 records spanning the past 410 million years from three related geochemical models (GEOCARB I, GEOCARB III and GEOCARBSULF). All GEOCARB models compute palaeo-CO_2 estimates averaged over 10 million years using mass balance calculations of all the long-term inputs and outputs of carbon to and from the atmosphere (Fig 5.2). As the temporal resolution of the GEOCARB models was lower than that of the speciation database, linear interpolation was used to obtain atmospheric CO_2 values for all times represented by plant evolution data. GEOCARBSULF models the long-term evolution of both atmospheric O_2 and CO_2 concentrations and was used here to investigate the effects of $CO_2 : O_2$ ratios on plant speciation rate.

To test for any relationship between atmospheric CO_2 concentration and long-term patterns in plant evolution, simple linear correlations were performed between each CO_2 record and a database of vascular land-plant speciation rates from Niklas et al. (1985). The plant speciation database is based on a large

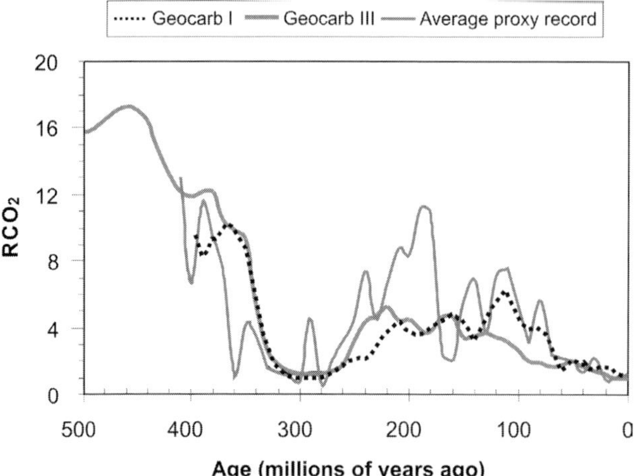

Figure 5.2 Changes in atmospheric CO_2 concentration (RCO_2) over the last 500 million years of earth history. Estimations based on two models of the long-term carbon cycle, GEOCARB I (Berner, 1991) and GEOCARB III (Berner and Kothavala, 2001) and a compilation of proxy-CO_2 records from Royer et al. (2004). RCO_2 is a ratio scale of estimated palaeo-CO_2 to the pre-industrial CO_2 concentration of 300 ppm. Note that GEOCARB I does not take into account the influence of plant evolution on chemical weathering processes, whereas GEOCARB III does.

compilation of published records encompassing over 18 000 fossil plant species, 80% of which are vascular plant taxa (Niklas et al., 1985). Where possible, the database has been corrected for synonymies, and palaeobotanical 'form genera' are not included. As with all palaeontological databases which are based on large compilations of data from published records, the plant speciation database is subject to considerable uncertainties and biases such as incompleteness in the fossil record, variable taxonomic practices amongst researchers and a bias towards data sources from well-studied North America, Europe and Russia. Speciation rates (r_s) were calculated according to the formula:

$$r_S = \frac{1}{D} \cdot \frac{S}{\Delta t}$$

where D is species diversity per interval of time and S is the number of speciation events per unit time t (Niklas et al., 1985).

5.5 Results

Our results demonstrate that atmospheric CO_2 change is a potentially important driving mechanism for land-plant macroevolution. We find statistically significant but complex relationships between palaeo-CO_2 concentrations and plant species origination rates for all carbon-cycle models tested, irrespective of their treatment of the chemical weathering potential of vascular plants (Figs 5.3–5.5). Specifically, when all evolutionary grades were included in the analysis, speciation rates increased linearly with CO_2 concentration above *c.* 1000 ppm (Figs 5.3a, 5.4a) and $CO_2 : O_2$ ratios above 0.01 (Fig 5.5a). A very slight increase in speciation rates was also observed at CO_2 values at the lower end of the scale (< 500 ppm – Figs 5.3a, 5.4a) and at $CO_2 : O_2$ ratios < 0.01 (Fig 5.5a). These complex relationships can be explained in part by different responses of the three evolutionary grades of plant reproduction and morphology to fluctuations in atmospheric CO_2. Palaeozoic taxa consisting exclusively of pteridophytes and gymnosperms show a strong positive relationship between speciation rates and atmospheric CO_2 concentration (Figs 5.3b, 5.4b, 5.5b). In contrast, Cenozoic taxa consisting predominantly of angiosperms show an inverse (but statistically non-significant) relationship with CO_2 (Figs 5.3c, 5.4c, 5.5c), suggesting that factors other than CO_2 were equally or more important in driving angiosperm speciation following their attainment of ecological dominance by the early Cenozoic such as plant/animal coevolution or pollination syndromes (Crepet and Niklas, 2009) or fluctuations in UVB radiation (Willis et al., 2009).

It is unlikely that the observed positive correlation between fossil plant speciation rates and atmospheric CO_2 concentration is due to autocorrelation, as we have tested the relationship using palaeo-CO_2 models which both incorporate (Fig 5.3) and exclude (Fig 5.4) the effects of plant evolution on chemical weathering, and found little or no statistically significant difference in the strength of the correlations. It is also unlikely that our findings can be explained by invoking a simple CO_2 fertilisation effect on plant photosynthetic productivity, as the greatest enhancement of fossil plant speciation rate is evident at times in the past when, based on our understanding of modern plants, CO_2 concentrations exceeded saturation levels for C_3 photosynthesis (i.e. > 1000 ppmv – Korner, 2006). Our findings of a positive correlation between CO_2 concentration and plant speciation rates are more consistent with observations from short-term experimental studies showing a positive correlation between elevated CO_2 and decreased plant generation times (Ward et al., 2000), increased turnover (Phillips and Gentry, 1994), enhanced plant reproductive success (increased seed vitality/biomass and germination – Ward et al., 2000; Hussain et al., 2001) and significantly elevated selection pressure (Bazzaz et al., 1995; Lau et al., 2008) all of which would serve to enhance plant speciation rates in the long term.

CO_2 is a key ingredient for plant growth and survival. It is not surprising, therefore, that elevated levels have been repeatedly shown to stimulate above- and below-ground plant growth and biomass in studies where nutrient and water supplies are kept at optimum levels (DeLucia et al., 1999), i.e. the 'fertilisation effect'. However, in more natural experimental systems (such as those using free-air

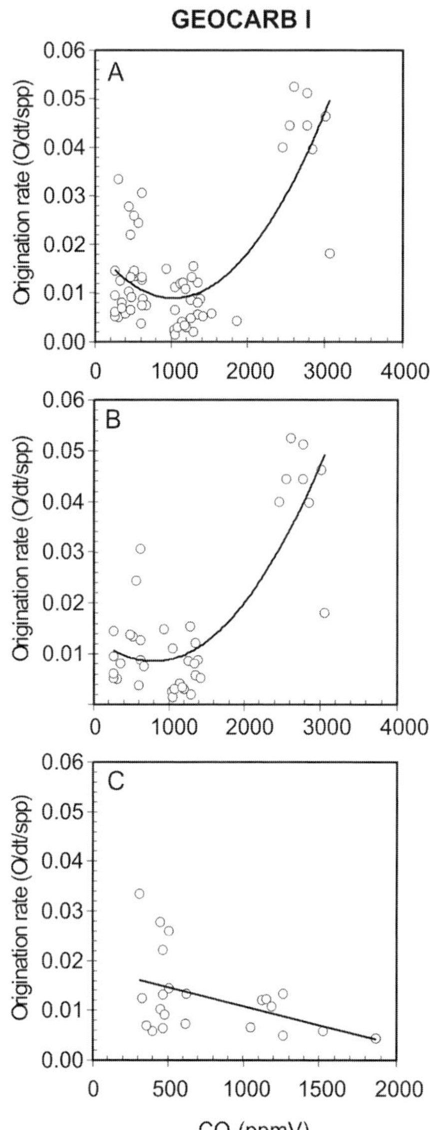

Figure 5.3 Relationship between fossil plant species speciation rate and GEOCARB I modelled palaeo-CO_2 for the last 400 million years for all major plant grades. Speciation rate is based on dS/dt/standing species diversity (Niklas et al., 1983; Niklas, 1997). Modelled palaeo-CO_2 concentration is from GEOCARB I (Berner, 1991). (A) Pteridophytes, gymnosperms and angiosperms ($y = 1^{-08x2} - 2^{-05x} + 0.0194$, $r^2 = 0.5658$). (B) Pteridophytes and gymnosperms only, for the period 400–140 mya ($y = 8^{-09x2} - 1^{-05x} + 0.0135$, $r^2 = 0.6561$). (C) Predominantly angiosperms, for the last 140 million years ($y = -8^{-06x} + 0.0186$, $r^2 = 0.1918$).

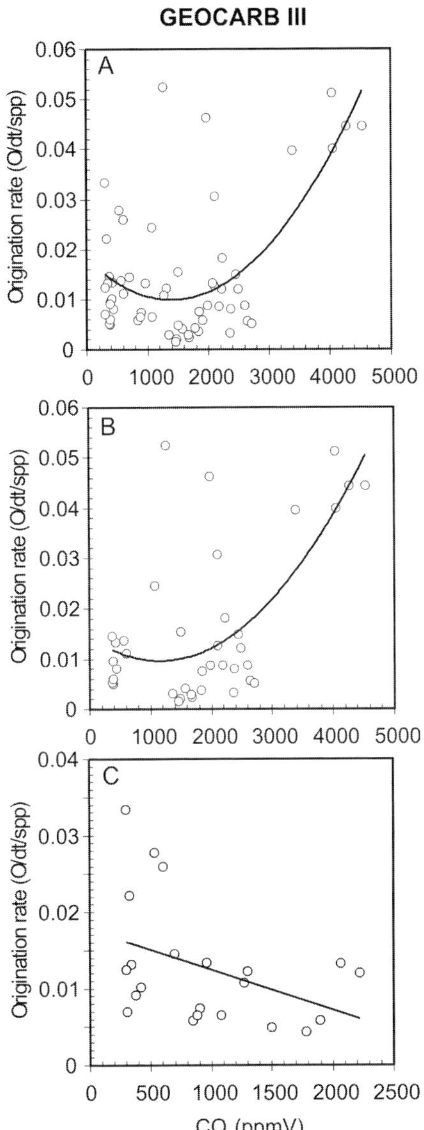

Figure 5.4 Relationship between fossil plant species speciation rate and GEOCARB III modelled palaeo-CO_2 for the last 400 million years for all major plant grades. Speciation rate is based on d*S*/d*t*/standing species diversity (Niklas et al., 1983; Niklas, 1997). Modelled palaeo-CO_2 concentration is from GEOCARB III (Berner and Kothavala, 2001). (A) Pteridophytes, gymnosperms and angiosperms ($y = 4^{-09x2} - 1^{-05x} + 0.0182$, $r^2 = 0.3995$). (B) Pteridophytes and gymnosperms only, for the period 400–140 mya ($y = 4^{-09x2} - 8^{-06x} + 0.0144$, $r^2 = 0.4297$). (C) Predominantly angiosperms, for the last 140 million years ($y = -5^{-06x} + 0.0177$, $r^2 = 0.1681$).

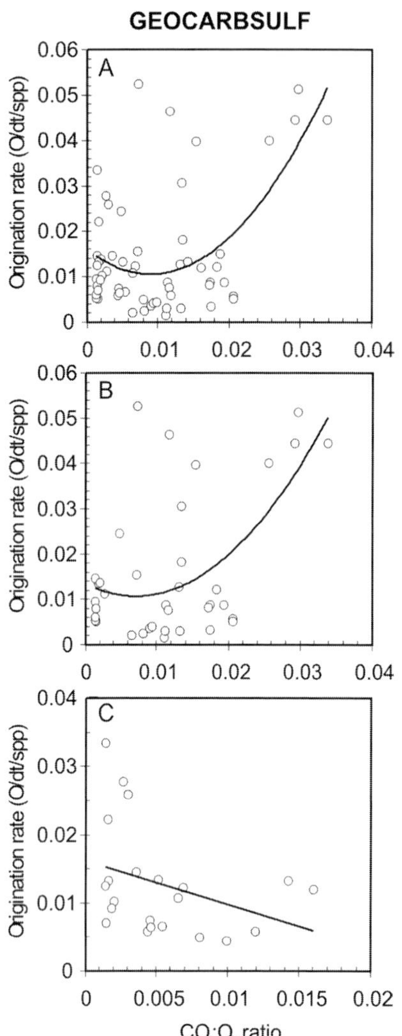

Figure 5.5 Relationship between fossil plant species speciation rate and GEOCARBSULF modelled atmospheric $CO_2:O_2$ for the last 400 million years for all major plant grades. Speciation rate is based on dS/dt/standing species diversity (Niklas et al., 1983; Niklas, 1997). Atmospheric $CO_2:O_2$ is from GEOCARBSULF (Berner, 2006). (A) Pteridophytes, gymnosperms and angiosperms ($y = 66.934x^2 - 1.212x + 0.016$, $r^2 = 0.3008$). (B) Pteridophytes and gymnosperms only, for the period 400–140 mya ($y = 55.203x^2 - 0.7888x + 0.0135$, $r^2 = 0.3179$). (C) Predominantly angiosperms, for the last 140 million years ($y = -0.6384x + 0.0161$, $r^2 = 0.1192$).

CO_2 enrichment – FACE – Curtis and Wang, 1998), or in competition experiments (Poorter and Navas, 2003) the fertilisation effect of CO_2 is minimised as essential nutrients become limiting. Thus we suggest, in accordance with experimental

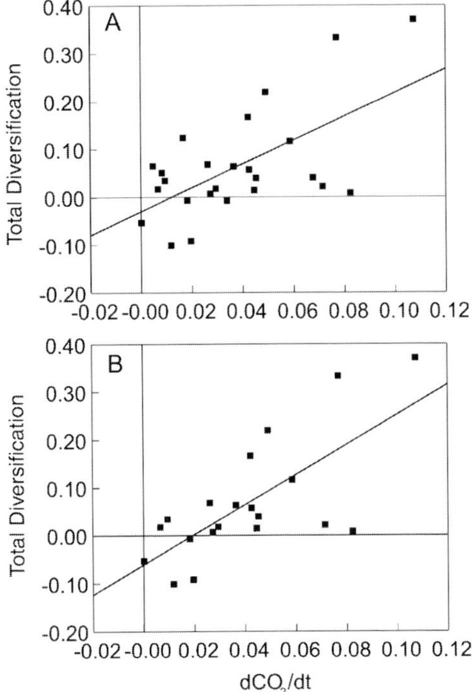

Figure 5.6 Relationship between total plant diversification (speciation rate minus extinction rate) and rate of CO_2 change during times of increasing palaeo-CO_2 concentration. CO_2 calculated from Berner (1991) and Tajika (1999): (A) for the entire Palaeozoic, Mesozoic and Cenozoic; (B) pre-Cretaceous record only. Simple linear regressions have been fitted to both data sets ($y = a0 + a1x$). r^2 values of (A) 0.375 and (B) 0.50388, r^2 (adj) of (A) 0.348 and (B) 0.476 were determined at the 95% confidence level. F-stats are equal to (A) 13.837 and (B) 18.28.

findings (Bazzaz et al., 1995), that episodes of rapid CO_2 rise in the ancient past may have enhanced speciation rates (Fig 5.6) as a consequence of increased selection for species more capable of efficiently using resources (e.g. nitrogen) that limit the 'fertilisation effect' of CO_2. Additional support for this suggestion comes from a recent study on the responses of plant-community composition and diversity to a rise in atmospheric CO_2 at the Triassic–Jurassic boundary 200 mya (McElwain et al., 2009). The slope of 'best fit models' used by McElwain et al. (2009) to characterise the shape of the relative abundance distribution of fossil plant taxa was found to increase with increasing CO_2, indicating that palaeo-plant communities were experiencing increasing ecological competition as CO_2 and global temperatures rose. Although it is difficult to tease apart the direct effects of CO_2 from the indirect effects mediated through greenhouse warming, this study provides a multimillion-year view of the effects of global change on plant evolutionary processes.

Another example from the plant fossil record that supports our thesis comes from a 65-million-year-old record of tropical plant speciation rates based on fossil pollen (Jaramillo et al., 2006). This record shows that highest tropical plant speciation rates occurred in the Eocene coincident with the highest prolonged global temperatures for the past 65 million years. Although the authors argue that the patterns in this speciation record were likely driven by a species–area effect, with high speciation coinciding with geographical expansion of tropical everwet climate and vice versa, an equally parsimonious explanation is that high global temperatures in the Eocene were forced by elevated CO_2, which directly promoted speciation rates through increased substrate availability and enhanced interplant competition.

Elevated CO_2 experiments on multiple plant generations have indicated that the potential for CO_2 to act as a selective agent may be very high (Ward et al., 2000). Ward and Kelly (2004) have argued that under certain conditions photosynthetic acclimation to elevated CO_2 may not be adaptive, based on the availability of other resources (e.g. high nitrogen levels). This idea is supported by palaeoecological data spanning the Late Palaeocene thermal maximum (LPTM), a rapid and presumed CO_2-driven global warming event approximately 55 million years ago, where there is evidence for a massive influx of nitrogen-fixing legumes replacing native vegetation in fossil plant localities in Wyoming (Wing et al., 2005). This pattern of vegetation change over tens of thousands of years supports the idea that elevated CO_2 may enhance competition for resources, such as nitrogen, that become increasingly depleted. Leguminosae were the ecological 'winners' during the LPTM because they were able to outcompete non-nitrogen-fixing taxa. If nitrogen depletion under elevated CO_2 conditions in turn leads to an enhancement of plant–plant competition, we see yet another potential mechanistic explanation for the positive relationship between atmospheric CO_2 content and fossil plant speciation rates over the past 400 million years. An additional factor which may contribute to lower plant nitrogen content in higher CO_2 conditions has been identified by Rachmilevitch et al. (2004), who proposed that nitrate assimilation in plants is highly dependent on photorespiration, which is depressed under conditions of elevated CO_2. Our analysis provides indirect support for this suggestion in that we observe increased speciation rate at high $CO_2 : O_2$ ratios when photorespiration would have been minimised (Fig 5.6).

We suggest that the global effects of elevated CO_2 also increased plant speciation rates by decreasing generation times, in turn by temporally advancing reproduction and concomitantly increasing the probability of fixing potentially beneficial genetic mutations in rapidly diversifying populations. Indeed, it is reasonable to infer that elevated CO_2 levels drove down plant body size. Since reproductive rates and body size are inversely proportional, and since body size allometrically scales as the –1/4 power of generation time (Niklas and Enquist, 2001), speciation rates would have increased during episodes of elevated atmospheric CO_2. Incremental increases in species richness (with the successive evolution and radiation of reproductive innovations) would, in turn, increase species packing and thus enhance the probability of establishing beneficial genetic recombinations. In this respect, evidence from the plant fossil record indicates that rapidly radiating clades tend to be composed of herbaceous species, with arborescent species evolving later in each clade (Niklas, 1997). Since herbaceous species are small, their rates of reproduction tend to be high, thereby further fostering additional speciation within clades.

5.6 High temperature and plant macroevolution

The direct effects of higher temperatures due to CO_2-induced greenhouse warming may also have played an important role in promoting plant speciation. It has been suggested that elevated CO_2 and global warming increased fossil plant species-level turnover at the Triassic–Jurassic boundary due to the detrimental effects of high temperature on leaf energy budgets (McElwain et al., 1999). Additionally, heat shock proteins, which protect organisms from high thermal injury and generally high-stress environments (by maintaining the three-dimensional configuration of key developmental proteins), are also thought to mask high levels of genotypic variability (Rutherford and Lindquist, 1998). When heat shock proteins are activated due to heat stress, Rutherford and Lindquist (1998) have argued that this can result in significant genotypic variability being unmasked, enabling rapid rates of phenotypic change during times of environmental stress. Over the long term, could high global temperatures drive plant speciation rates via such a mechanism? It will not be possible to address this question and tease apart the relative roles of atmospheric CO_2 and global temperature on plant speciation rate until detailed and independent proxy records for both are available for significantly long time intervals over the past 500 million years. The significant positive influence of the rate of CO_2 change on fossil plant diversification during episodes of increasing CO_2 ($r^2 = 0.375–0.504$; Fig 5.6) points to the role of rapid atmospheric change in promoting speciation over extinction in the plant fossil record. However, the observation that rapid rates of rising CO_2 had a more pronounced effect on pre-Cretaceous plant speciation rates ($r^2 = 0.504$) than for the entire Phanacrozoic ($r^2 = 0.375$) suggests that factors other than CO_2, such as climate or biotic factors, were more important, especially in the case of angiosperms (Barrett and Willis, 2001; Crepet and Niklas, 2009).

5.7 Conclusions

In summary, we propose that the direct effects of CO_2 on interplant competition, nitrogen status, genotypic diversity, reproductive rates and generation time, in combination with the indirect effects of global climatic warming, were major factors influencing the tempo of land-plant macroevolution. The fact that changes in atmospheric CO_2 are global further supports this hypothesis, as opportunities for plant migration would be minimised during episodes of rapid atmospheric CO_2 rise, thereby increasing selection pressures on plants. The major drivers of floral and faunal evolution may therefore be distinct, as we propose 'CO_2 change' rather than 'mass extinction' *sensu* Gould (1985) as an uppermost factor driving the rate of plant macroevolution.

Although extremely reduced CO_2 concentrations below critical thresholds (< 500 ppm) may also play an important role in plant evolution, such as the evolution of the megaphyllous leaf (Beerling et al., 2001) and the C_4 photosynthetic pathway (Ehleringer et al., 1991; Cowling, 2001), these evolutionary events represent vegetative and biochemical innovations respectively. We propose that episodes of rapidly rising atmospheric CO_2 were a critical driving force in larger-scale patterns in plant evolution (i.e. those at the level of reproductive grade), contributing to the timing of major reproductive innovations in earth's history by greatly enhancing plant speciation.

If we are correct in proposing that atmospheric CO_2 and plant speciation rate are positively correlated, this by no means suggests that we should have a rosy outlook regarding earth's high-CO_2 future. Future atmospheric CO_2 levels are predicted to reach a maximum of *c.* 900 ppm by the year 2100, which is below the threshold CO_2 concentration of 1000 ppm. The rate of current CO_2 change is unprecedented in earth's history. We have reported data for times in the geological past when CO_2 levels were up to 10 times higher than today and when CO_2 was rising at a rapid rate; however, the pace of past CO_2 change was at least an order of magnitude slower than the earth is experiencing today. Furthermore, the future diversity and functioning of terrestrial ecosystems is dependent on plant speciation as well as extinction rates and the changing ecological role of existing and new species. The fossil record provides examples of turnover events (DiMichele and Phillips, 1996; Dimichele et al., 2009), extinction events (Looy et al., 2001; Vajda et al., 2001; Wing, 2004; McElwain and Punyasena, 2007) and ecological replacement events (Wing et al., 2005), many of which coincide with rapid rises in atmospheric CO_2 and/or global temperature. We therefore conclude that plant evolutionary responses to elevated CO_2 are likely to reinforce the short-term acclimation responses observed in high-CO_2 experiments, but that the key to predicting the impact on future ecosystems is through an understanding of plant ecological interactions and species competition.

Acknowledgements

We thank Scott Lidgard of the Field Museum, Chicago, for helpful discussions on an earlier version of the manuscript and Eiichi Tajika and Robert Berner for use of their CO_2 data. Funding from the Science Foundation Ireland (08/RFP/EOB1131) and a Marie Curie Excellence Grant (MEXT-CT-2006-042531) are gratefully acknowledged by the first author.

References

Algeo, T. J. and Scheckler, S. E. (1998). Terrestrial-marine teleconnections in the Devonian: links between the evolution of land plants, weathering processes, and marine anoxic events. *Philosophical Transactions of the Royal Society of London B*, **353**, 113–128.

Barrett, P. M. and Willis, K. J. (2001). Did dinosaurs invent flowers? Dinosaur-angiosperm coevolution revisited. *Biological Reviews*, **76**, 411–447.

Bazzaz, F. A., Jasienski, M., Thomas, S. C. and Wayne, P. (1995). Microevolutionary responses in experimental populations of plants to CO_2-enriched environments: parallel results from 2 model systems. *Proceedings of the National Academy of Sciences of the USA*, **92**, 8161–8165.

Beerling, D. J., Osborne, C. P. and Chaloner, W. G. (2001). Evolution of leaf-form in land plants linked to atmospheric CO_2 decline in the Late Palaeozoic era. *Nature*, **410**, 352–354.

Berner, R. A. (1991). A model for atmospheric CO_2 over Phanerozoic time. *American Journal of Science*, **291**, 339–375.

Berner, R. A. (1997). The rise of plants and their effect on weathering and atmospheric CO_2. *Science*, **276**, 544–546.

Berner, R. A. (2006). GEOCARBSULF: a combined model for Phanerozoic atmospheric O_2 and CO_2. *Geochimica et Cosmochimica Acta*, **70**, 5653–5664.

Berner, R. A. and Canfield, D. E. (1989). A new model for atmospheric oxygen over Phanerozoic time. *American Journal of Science*, **289**, 333–361.

Berner, R. A. and Kothavala, Z. (2001). GEOCARB III: a revised model of atmospheric CO_2 over Phanerozoic time. *American Journal of Science*, **301**, 182–204.

Boyce, C. K., Brodribb, T. J., Feild, T. S. and Zwieniecki, M. A. (2009). Angiosperm leaf vein evolution was physiologically and environmentally transformative. *Proceedings of the Royal Society of London B*, **276**, 1771–1776.

Cerling, T. E. (1991). Carbon dioxide in the atmosphere: evidence from Cenozoic and Mesozoic paleosols. *American Journal of Science*, **291**, 377–400.

Christin, P. A., Besnard, G., Samaritani, E. et al. (2008). Oligocene CO_2 decline promoted C-4 photosynthesis in grasses. *Current Biology*, **18**, 37–43.

Cornette, J. L., Lieberman, B. S. and Goldstein, R. H. (2002). Documenting a significant relationship between macroevolutionary origination rates and Phanerozoic pCO_2 levels. *Proceedings of the National Academy of Sciences of the USA*, **99**, 7832–7835.

Cowling, S. A. (2001). Plant carbon balance, evolutionary innovation and extinction in land plants. *Global Change Biology*, **7**, 231–239.

Crepet, W. L. and Niklas, K. J. (2009). Darwin's second abominable mystery: why are there so many angiosperms? *American Journal of Botany*, **96**, 366–381.

Curtis, P. S. and Wang, X. (1998). A meta analysis of elevated CO_2 effects on woody plant mass, form and physiology. *Oecologia*, **113**, 299–313.

DeLucia, E. H., Hamilton, J. G., Naidu, S. L. et al. (1999). Net primary production of a forest ecosystem with experimental CO_2 enrichment. *Science*, **284**, 1177–1179.

DiMichele, W. A. and Phillips, T. L. (1996). Climate change, plant extinctions, and vegetation recovery during the middle-late Pennsylvanian transition: the case of tropical peat-forming environments in North America. In *Biotic Recovery from Mass Extinction*, ed. M. L. Hart. London: Geological Society of London, pp. 201–221.

Dimichele, W. A., Montanez, I. P., Poulsen, C. J. and Tabor, N. J. (2009). Climate and vegetational regime shifts in the late Paleozoic ice age earth. *Geobiology*, **7**, 200–226.

Ehleringer, J. R., Sage, R. F., Flanagan, L. B. and Pearcy, R. W. (1991). Climate change and the evolution of C_4 photosynthesis. *Trends in Ecology and Evolution*, **6**, 95–99.

Fletcher, B. J., Beerling, D. J., Brentnall, S. J. and Royer, D. L. (2005). Fossil bryophyte as recorders of ancient CO_2 levels: experimental evidence and a Cretaceous case study. *Global Biogeochemical Cycles*, **19**.

Fletcher, B. J., Brentnall, S. J., Quick, W. P. and Beerling, D. J. (2006). BRYOCARB: a process-based model of thallose liverwort carbon isotope fractionation in response to CO_2, O_2, light and temperature. *Geochimica et Cosmochimica Acta*, **70**, 5676–5691.

Franks, P. J. and Beerling, D. J. (2009). CO_2-forced evolution of plant gas exchange capacity and water-use efficiency over the Phanerozoic. *Geobiology*, **7**, 227–236.

Gould, S. J. (1985). The paradox of the first tier: an agenda for paleobiology. *Paleobiology*, **11**, 2–12.

Huh, Y. and Edmond, J. M. (1999). The fluvial geochemistry of the rivers of Eastern Siberia: III. Tributaries of the Lena and Anabar draining the basement terrain of the Siberian Craton and the Trans-Baikal Highlands. *Geochimica et Cosmochimica Acta*, **63**, 967–987.

Hussain, M., Kubiske, M. E. and Connor, K. F. (2001). Germination of CO_2-enriched *Pinus taeda* L. seeds and subsequent seedling growth responses to CO_2 enrichment. *Journal of Functional Ecology*, **15**, 344–350.

Jablonski, D. (2001). Micro and macroevolution: scale and hierarchy in evolutionary biology and paleobiology. *Palaeobiology*, **26**, 15–53.

Jaramillo, C., Rueda, M. J. and Mora, G. (2006). Cenozoic plant diversity in the Neotropics. *Science*, **311**, 1893–1896.

Knoll, A. H. (1985). Patterns of evolution in the Archean and Proterozoic eons. *Palaeobiology*, **11**, 53–64.

Korner, C. (2006). Plant CO_2 responses: an issue of definition, time and resource supply. *New Phytologist*, **172**, 393–411.

Lau, J. A., Peiffer, J., Reich, P. B. and Tiffin, P. (2008). Transgenerational effects of global environmental change: long-term CO_2 and nitrogen treatments influence offspring growth response to elevated CO_2. *Oecologia*, **158**, 141–150.

Lidgard, S. and Crane, P. R. (1988). Quantitative analyses of the early angiosperm radiation. *Nature*, **331**, 344–346.

Lidgard, S. and Crane, P. R. (1990). Angiosperm diversification and Cretaceous florisitic trends: a comparison of palynofloras and leaf macrofloras. *Paleobiology*, **16**, 77–93.

Looy, C. V., Twitchett, R. J., Dilcher, D. L., van Konijnenburg-van Cittert, J. H. A. and Visscher, H. (2001). Life in the end-Permian dead zone. *Proceedings of the National Academy of Sciences of the USA*, **98**, 7879–7883.

Lupia, R., Lidgard, S. and Crane, P. R. (1999). Comparing palynological abundance and diversity: implications for biotic replacement during the Cretaceous angiosperm radiation. *Paleobiology*, **25**, 305–340.

McElwain, J. C. and Chaloner, W. G. (1995). Stomatal density and index of fossil plants track atmospheric carbon-dioxide in the Paleozoic. *Annals of Botany*, **76**, 389–395.

McElwain, J. C. and Punyasena, S. (2007). Mass extinction events and the plant fossil record. *Trends in Ecology and Evolution*, **22**, 548–557.

McElwain, J. C., Beerling, D. J. and Woodward, F. I. (1999). Fossil plants and global warming at the Triassic–Jurassic boundary. *Science*, **285**, 1386–1390.

McElwain, J. C., Willis, K. J. and Lupia, R. (2005). Cretaceous CO_2 decline and the radiation and diversification of angiosperms. In *A History of Atmospheric CO_2 and its Effects on Plants, Animals and Ecosystems*, ed. J. R. Ehleringer, T. E. Cerling and M. D. Dearind. New York, NY: Springer, pp. 133–166.

McElwain, J. C., Wagner, P. J. and Hesselbo, S. P. (2009). Fossil plant relative abundances indicate sudden loss of Late Triassic biodiversity in East Greenland. *Science*, **324**, 1554–1556.

Moulton, K. L. and Berner, R. A. (1998). Quantification of the effect of plants on weathering: studies in Iceland. *Geology*, **26**, 895–898.

Moulton, K. L., West, J. and Berner, R. A. (2000). Solute flux and mineral mass balance approaches to the quantification of plant effects on silicate weathering. *American Journal of Science*, **300**, 539–570.

Niklas, K. J. (1997). *The Evolutionary Biology of Plants*. Chicago, IL: University of Chicago Press.

Niklas, K. J. and Enquist, B. J. (2001). Invariant scaling relationships for interspecific plant biomass production rates and body size. *Proceedings of the National Academy of Sciences of the USA*, **98**, 2922–2927.

Niklas, K. J., Tiffney, B. H. and Knoll, A. H. (1983). Patterns in vascular land plant diversification. *Nature*, **303**, 614–616.

Niklas, K. J., Tiffney, B. H. and Knoll, A. H. (1985). Patterns in vascular land plant diversification: an analysis at the species level. In *Phanerozoic Diversity Patterns: Profiles in Macroevolution*, ed. J. W. Valentine. Princeton, NJ: Princeton, University Press, pp. 97–128.

Pagani, M., Arthur, M. A. and Freeman, K. H. (1999). Miocene evolution of atmospheric carbon dioxide. *Paleoceanography*, **14**, 273–292.

Pearson, P. N. and Palmer, M. R. (2000). Atmospheric carbon dioxide concentrations over the past 60 million years. *Nature*, **406**, 695–699.

Phillips, O. L. and Gentry, A. H. (1994). Increasing turnover through time in tropical forests. *Science*, **263**, 954–958.

Poorter, H. and Navas, M. L. (2003). Plant growth and competition at elevated CO_2: on winners, losers and functional groups. *New Phytologist*, **157**, 175–198.

Rachmilevitch, S., Cousins, A. B. and Bloom, A. J. (2004). Nitrate assimilation in plant shoots depends on photorespiration. *Proceedings of the National Academy of Sciences of the USA*, **101**, 11506–11510.

Robinson, J. M. (1994). Speculations on carbon-dioxide starvation, Late Tertiary evolution of stomatal regulation and floristic modernization. *Plant Cell and Environment*, **17**, 345–354.

Royer, D. L. (2001). Stomatal density and stomatal index as indicators of paleoatmospheric CO_2 concentration. *Review of Palaeobotany and Palynology*, **114**, 1–28.

Royer, D. L., Berner, R. A. and Beerling, D. J. (2001). Phanerozoic atmospheric CO_2 change: evaluating geochemical and paleobiological approaches. *Earth-Science Reviews*, **54**, 349–392.

Royer, D. L., Berner, R. A., Montanez, I. P., Tabor, N. J. and Beerling, D. J. (2004). CO_2 as a primary driver of Phanerozoic climate. *GSA Today*, **14**, 4–10.

Rutherford, S. L. and Lindquist, S. (1998). Hsp90 as a capacitor for morphological evolution. *Nature*, **396**, 336–342.

Sage, R. F. (2004). The evolution of C-4 photosynthesis. *New Phytologist*, **161**, 341–370.

Tajika, E. (1999). Carbon cycle and climate change during the Cretaceous inferred from a biogeochemical carbon cycle model. *The Island Arc*, **8**, 293–303.

Taylor, L. L., Leake, J. R., Quirk, J. et al. (2009). Biological weathering and the long-term carbon cycle: integrating mycorrhizal evolution and function into the current paradigm. *Geobiology*, **7**, 171–191.

Traverse, A. (1988). Plant evolution dances to a different beat: plant and evolutionary mechanisms compared. *Historical Biology*, **1**, 277–356.

Vajda, V., Raine, J. I. and Hollis, C. J. (2001). Indication of global deforestation at the Createaceous–Tertiary boundary by New Zealand fern spike. *Science*, **294**, 1700–1702.

Ward, J. K. and Kelly, J. K. (2004). Scaling up evolutionary responses to elevated CO_2: lessons from *Arabidopsis*. *Ecology Letters*, **7**, 427–440.

Ward, J. K., Antonovics, J., Thomas, R. B. and Strain, B. R. (2000). Is atmospheric CO_2 a selective agent on model C3 annuals? *Oecologia*, **123**, 330–341.

Willis, K. J. and Bennett, K. D. (1995). Mass extinction, punctuated equilibrium and the fossil plant record. *Trends in Ecology and Evolution*, **10**, 308–309.

Willis, K. J. and McElwain, J. C. (2002). *The Evolution of Plants*. Oxford: Oxford University Press.

Willis, K. J., Bennett, K. D. and Birks, H. J. B. (2009). Variability in thermal and UV-B energy fluxes through time and their influence on plant diversity and speciation. *Journal of Biogeography*, **36**, 1630–1644.

Wing, S. L. (2004). Mass extinctions in plant evolution. In *Extinctions in the History of Life*, ed. P. D. Taylor. Cambridge: Cambridge University Press, pp. 61–97.

Wing, S. L., Harrington, G. J., Smith, F. A. et al. (2005). Transient floral change and rapid global warming at the Paleocene-Eocene boundary. *Science*, **310**, 993–996.

Woodward, F. I. (1987). Stomatal numbers are sensitive to increase in CO_2 from pre-industrial levels. *Nature*, **327**, 617–618.

6

Wood anatomy and climate change

P. Baas

National Herbarium of the Netherlands, Leiden, the Netherlands

E. A. Wheeler

Department of Wood and Paper Science, North Carolina State University, NC, USA

Abstract

This chapter reviews the potential of comparative wood anatomy for climate reconstruction and for assessing the possible risks of global warming to extant woody plants. There is growing evidence that wood evolution has been driven by functional adaptations to climate change in vessel-bearing woody angiosperms, giving rise to multiple parallelisms and reversals in vessel, fibre, parenchyma and ray modifications. Despite this homoplasy, wood anatomical character complexes are phylogenetically constrained, often allowing different clades at various levels of the taxonomic hierarchy (families, genera and groups of closely related species) to be reliably identified by wood anatomical attributes alone. Examples are presented of how wood anatomical characters can be used as climate proxies, especially for mean annual temperature (MAT), and its covariables latitude and altitude. One of the great challenges of modern wood research is to model the

Climate Change, Ecology and Systematics, ed. Trevor R. Hodkinson, Michael B. Jones, Stephen Waldren and John A. N. Parnell. Published by Cambridge University Press.

relationships between climate and wood anatomical diversity patterns of extinct and extant plant communities in such a way that the impact of current and future climate change can be predicted reliably.

6.1 Introduction

Secondary xylem is a multifunctional, complex plant tissue that provides an archive of the external signals that modified its functional attributes at different timescales, from the lifespan of a single tree to millions of years of biological evolution (Baas, 1986; Wheeler and Baas, 1991, 1993; Carlquist, 2001; Sperry, 2003; Baas et al., 2004; Poole and van den Bergen, 2006; Wheeler et al., 2007). In this chapter we review the major structure–function relationships of wood and our current understanding of xylem evolution in woody angiosperms. From a summary of general ecological trends that are evident in both extant plants and the fossil record, we argue that xylem evolution has to a large extent been driven by functional adaptation to changing environmental conditions, especially climate change. Many wood anatomical parameters may thus be used as proxies for MAT, seasonality, latitude and altitude of provenance, and/or as indicators of xeric or mesic environmental conditions. We will only discuss the potential of vessel-bearing dicot woods as a source of climate proxies. We are aware that the realisation of the full potential of wood anatomy as a proxy for environmental conditions will require multidisciplinary cooperation among wood anatomists, climatologists, ecologists and mathematicians, and the establishment of even more, multilayered, detailed data sets. One extremely informative wood anatomical proxy for multiple climate variables, the growth ring and its various anatomical attributes, has been successfully developed in this respect, and has played a crucial role in our awareness of recent global warming and climate changes over the last 10 000 years (e.g. Briffa, 2000). However, in this chapter we do not deal with this highly specialised field of wood anatomy, namely dendrochronology and dendroclimatology. Much of the strength of dendroclimatology lies in the use of large data sets for single species, whereas the links between systematics and climate change research are more evident in analyses of the wood anatomical features of all woody plants within diverse species assemblages (florulas) or within species-rich clades.

6.2 The functions of wood and their relationships with climate

Wood in the living tree carries out four crucial functions: (1) hydraulic, or long-distance sap transport from roots to transpiring leaves in the dead cells called vessel elements and tracheids; (2) mechanical support, especially by fibres and/or tracheids, but with significant contributions from parenchyma and vessel cell walls; (3) contributions to metabolism (storage and mobilisation of carbohydrates, for instance); and (4) defence mechanisms to control microorganisms and insects by wound responses of the living axial and ray parenchyma of the sapwood and the production of secondary metabolites during heartwood formation.

Vessel diameter and density, type of perforation plate (scalariform or simple) and pit membrane porosity determine the hydraulic conductivity (Sperry, 2003). They also play a role in the vulnerability of water columns to cavitation and the spread of embolisms under strong negative pressure, and probably also in the spread of air bubbles after freeze–thaw cycles (Zimmermann, 1978; Tyree and Zimmermann, 2002; Sperry, 2003; Choat et al., 2008). Hydraulic demands are different in different climatic conditions along gradients of temperature, moisture availability and seasonality, which explains why there are significant ecological trends along these three gradients for most hydraulic wood anatomical attributes.

Ring porosity, the presence of relatively wide vessels at the beginning of a growth ring, is a hydraulic strategy of adaptive value in seasonal climates. The wide earlywood vessels offer little resistance to flow, but become embolised fairly early in the growing season, when sap transport continues via the much more numerous but more transport-resistant narrow latewood vessels. Ring-porous woods are restricted to deciduous species of seasonal temperate climates (mostly northern hemisphere) and some highly monsoonal forest species (e.g. teak, *Tectona grandis* L.f.) in the tropics (Boura and De Franceschi, 2007), but by no means are all deciduous woody plants ring-porous. In the northern hemisphere, from the Cretaceous to the present day there has been an increase in the incidence of species with ring-porous woods and semi-ring-porous woods (Fig 6.1), reflecting the increasing seasonality and the increased incidence of deciduous species that occurred (e.g. Wolfe, 1987; Graham, 1999).

The hydraulic, mechanical, metabolic and defence functions and their associated structural attributes show important trade-offs (cf. Ewers' trade-off triangle in Baas et al., 2004), and each extant species has presumably found an appropriate balance for its natural distribution area. For instance, wide vessels have high conductive efficiency but they expose the tree to high risks of cavitation, and if they are too numerous per unit of volume they may reduce the mechanical strength of the wood (as in many lianas); narrower vessels reduce cavitation risks but are conductively less efficient. If a woody species produces an overabundance of narrow vessels to compensate for low conductivity, fibre volume and mechanical strength would decrease. However, a slight increase in relative fibre wall thickness can compensate for high tissue proportions of non-mechanical cell types such as vessel elements and parenchyma. Thus, fibre walls are usually thicker in drought-adapted species than in mesic species, so that overall there is a positive correlation between hydraulic safety, wood density and mechanical strength (Hacke et al., 2001, 2006).

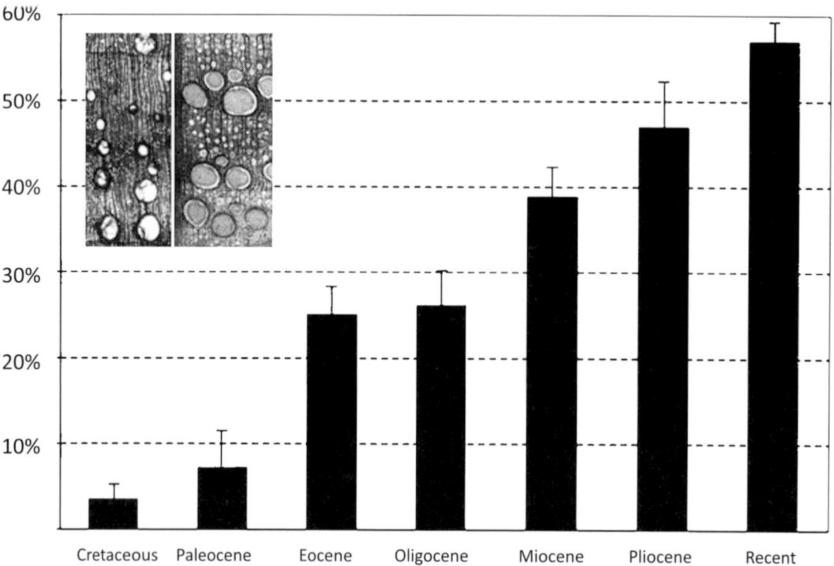

Figure 6.1 Incidence of ring-porosity and semi-ring-porosity in woods from Cretaceous to Recent in the northern hemisphere. Fine line represents estimated standard deviation. Insets: left to right, semi-ring-porous and ring-porous fossil *Quercus*. For background and statistics see Wheeler and Baas (1991).

6.3 The evolution of wood anatomical diversity

Bailey and Tupper (1918) proposed several major trends in xylem evolution of angiosperms, postulating a vessel-less ancestry with very long tracheids, evolving into wood with long tracheid-like vessel elements with scalariform perforations and pitting, to very short vessel elements with simple perforations and alternate pits (Fig 6.2). Such vessel specialisation went hand in hand with the evolution of fibres with minute pits, which function chiefly as mechanical supports. Many other wood anatomical features have been correlated with these major trends, so that there are many well-defined hypotheses on the main directions of parenchyma and ray evolution (Carlquist, 1975, 2001; Baas, 1986). There is good experimental evidence that scalariform perforations, especially if their bars are closely spaced, substantially add to the flow resistance in vessels (Sperry et al., 2007), putting selective pressure on their elimination in environments demanding high conductive efficiency, such as the lowland tropics or drought-stressed environments (Baas, 1976; Jansen et al., 2004). On the other hand, scalariform perforations have been hypothesised to be functional in trapping gas bubbles in thawing xylem sap at frost-prone latitudes and altitudes (Zimmermann, 1978), possibly adding a selective bonus for their retention in boreal, alpine and arctic regions.

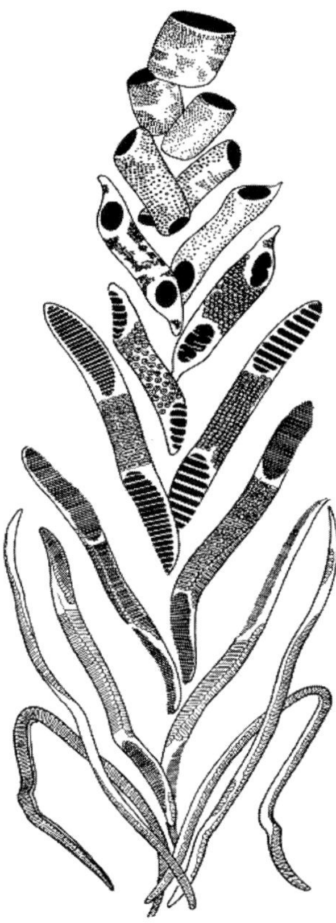

Figure 6.2 Cartoon of Baileyan trends in the evolution of tracheary elements, from long tracheids to short vessel elements. Cartoon by B. G. L. Swamy, reproduced with permission from Bailey (1954).

The fossil record, despite its incompleteness, largely supports the Baileyan trends (Wheeler and Baas, 1991, 1993), as does phylogenetic DNA analysis, including the APG II phylogenetic tree for all extant angiosperm families (Angiosperm Phylogeny Group, 2003). *Amborella*, the phylogenetically most outlying angiosperm, is vessel-less; scalariform perforations are more common in the eumagnoliids and basal eudicot clades than in the more derived rosids and asterids (Baas et al., 2003; Wheeler and Baas, unpublished results). However, despite supporting the Baileyan trends, the APG II tree, as well as many published robust DNA phylogenies of woody plant orders and families (e.g. Baas et al., 2000; Lens et al., 2007), have revealed homoplasy in most wood anatomical features. This homoplasy had earlier been hypothesised on the basis of morphological phylogenies (Loconte and Stevenson, 1991; Baas and Wheeler, 1996), and provides the source of variation that can be analysed for its functional significance and climatic and other environmental signals. The presence of strong ecological trends through adaptive evolution, resulting in rampant homoplasy, raises the question whether wood anatomical features are of reduced or even negligible diagnostic and phylogenetic significance. The answer is no, because it is still possible to identify woody families, genera or infrageneric clades (groups of closely related species) on the basis of these highly functional and adaptive wood characters (cf. online wood identification application of InsideWood – http://insidewood.lib.ncsu.edu). Presence or absence of two hydraulically important features, scalariform perforation plates and vestured pits, constitutes a good example. Both features are highly diagnostic, often above the genus level. Perforation type often characterises families; vestured pits are in some instances diagnostic at the level of order (Gentianales and Myrtales) and superfamily (Fabaceae *s.l.*) – with very few exceptions. Yet Jansen et al. (2004) have shown that both features show very strong but contrasting ecological trends (Fig 6.3), consistent with their presumed roles in hydraulic safety in frost-prone habitats (scalariform perforations – Zimmermann, 1978) or in environments subject to strong negative xylem pressures and high cavitation risks such as deserts, Mediterranean climates, lowland tropical monsoon forests (vestured pits – Jansen et al., 2003). The much enlarged APG order of the Ericales constitutes another example of combined phylogenetic and functional/ecological signal in its

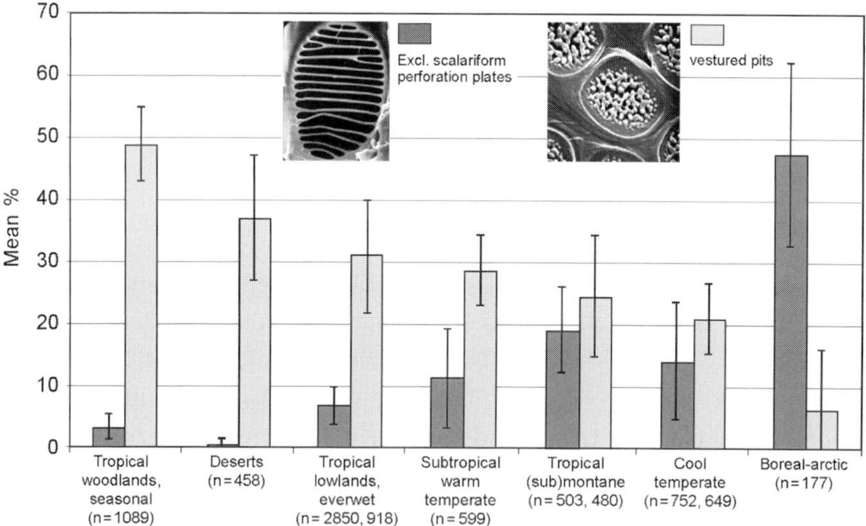

Figure 6.3 The incidence of scalariform perforations and vestured pits in different floristic zones showing opposite ecological trends. Scalariform perforations and vestured pits are illustrated by scanning electron microscopy pictures. Reproduced from Jansen et al. (2004) with permission of the National Academy of Sciences of the USA (© National Academy of Sciences of the USA, 2004).

wood anatomical diversity. The 'total evidence' DNA and wood anatomical phylogeny shows a major dichotomy: (1) a mainly tropical lowland clade with simple vessel perforations; and (2) a temperate and tropical montane clade with scalariform perforations. These are hydraulic attributes that are thought to be functional in the respective climate zones occupied by the two clades (Lens et al., 2007).

This phenomenon of combined phylogenetic/diagnostic value at fairly high taxonomic levels and functionally adaptive sorting along ecological gradients of the same wood anatomical attributes indicates that many wood anatomical adaptations are deep-seated and have been conserved in many clades. Therefore, we believe that patterns of wood anatomical diversity in extant and extinct dicot trees and shrubs can best, if not fully, be understood as a result of adaptive changes over the last 100 million years to climate and other environmental factors affecting the three main functions of wood (Baas et al., 2003).

6.4 Comparative wood anatomy, latitude, altitude and temperature

The notion that wood from tropical regions differs from that in temperate zones is as old as wood anatomy itself. Leeuwenhoek in the seventeenth century commented on the lack of growth rings in wood from Ceylon, a feature arising because trees could grow continuously in the tropics (Baas, 1982). In the twentieth century, latitude and altitude have been shown to be significantly correlated with a whole range of wood anatomical parameters, especially vessel diameter, vessel frequency, vessel element length, incidence of scalariform perforations and helical vessel wall thickenings, in a large number of studies (see reviews by Baas, 1982, 1986; Baas et al., 2003; Wheeler et al., 2007). Table 6.1 summarises the nature of the most significant correlations, which apply both when comparing species assemblages (local floras) from different regions, and species within species-rich and widespread genera. Some latitudinal trends, especially those on the incidence of scalariform perforations and ring porosity, have also been evident in the fossil record from the early Tertiary onwards (Wheeler and Baas, 1991, 1993). We still do not understand why these trends generally do not apply at the species level (van der Graaff and Baas, 1974; Noshiro and Baas, 2000; Liu and Noshiro, 2003).

Since latitude and altitude of provenance can serve as reasonably good proxies for temperature, these aforementioned correlations provide a tool for reconstructing MAT from the incidence of certain wood anatomical attributes in a local flora. Wiemann et al. (1998, 1999, 2001) have developed and discussed a number of regression equations to estimate MAT from wood anatomical spectra of a number of local floras throughout the Americas. The ability of these equations to estimate

Table 6.1 Selected wood anatomical features and the nature of their correlations with major climate gradients (cold to warm, mesic to xeric). Data from multiple sources, especially Wheeler et al. (2007).

Anatomical feature	Cold to warm	Mesic to xeric
Vessel frequency	–	+
Vessel diameter	+	–
Scalariform perforations	–	–
Vessel element length	+	–
Vestured pits	+	+
Helical thickenings	–	+
Ring porosity	–	?
Storied structure	+	?
Parenchyma rare or absent	–	?
Paratracheal parenchyma	+	?
Marginal parenchyma	–	?
Septate fibres	+	?

+, positive correlation; –, negative correlation; ?, correlation not yet robustly analysed or ambiguous.

MAT was tested for a range of additional present-day sites with meteorological data (Wiemann et al., 1999). Two equations (given below) appeared to perform best. Both are based on arcsine transformations of the relative proportions of species having the anatomical features storied rays (*stor*), marginal parenchyma (*marg*), axial parenchyma rare to absent (*abs*) and septate fibres (*sept*):

$$\text{MAT °C} = 24.78 + 36.57\,(stor) - 15.61\,(marg) - 16.41\,(abs)$$

$$\text{MAT °C} = 17.07 + 25.23\,(stor) - 23.17\,(abs) + 13.79\,(sept)$$

Perhaps surprisingly, on an individual basis, these particular features are not those that correlate best with MAT. As older studies predict (e.g. Carlquist, 1975; Baas, 1986), vessel diameter and incidence of spiral thickenings are more highly correlated with MAT.

Wheeler and Manchester (2002) and Wheeler and Dillhoff (2009) have used these equations to predict MAT for the wood assemblages of the mid-Eocene Clarno Nut Beds and mid-Miocene Vantage Fossil Forest, both in the northwestern USA. There is agreement between their results and early independent temperature

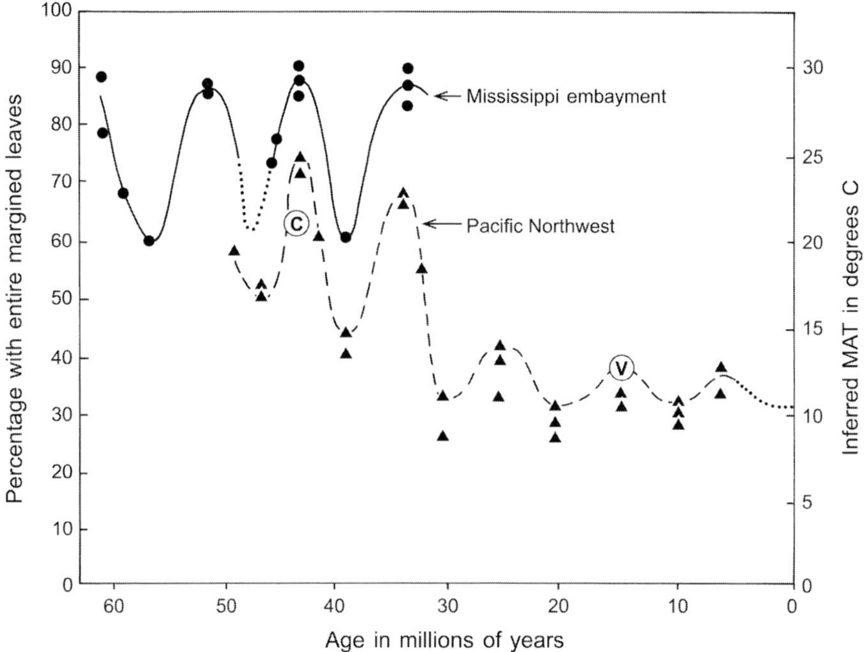

Figure 6.4 Mean annual temperature (MAT, °C) estimates for the Clarno (C) and Vantage (V) fossil wood assemblages plotted onto Wolfe's (1978) graph of MAT °C. Estimates for the Tertiary of the Pacific Northwest based on the incidence of leaf margin types in fossil floras.

reconstructions by Wolfe (1978) based on the incidence of leaf margin types in fossil floras and oxygen isotope data (Fig 6.4). The estimated MATs also agree with the composition of the assemblages, as the nearest living relatives of species of these assemblages occur in warmer and more equitable climates than currently prevail in eastern and central Oregon and Washington State. Sakala (2007) found Wiemann's equations also applicable to rich Miocene wood assemblages in southern Germany, but he noted their limitations for less species-rich Tertiary assemblages in central Europe. Wiemann et al. (2001) were very much aware of the limitations of their equations. Their MAT calculations sometimes produced values that were over 5 °C off the real value in modern test sites, and they found that different physiognomies of temperate and tropical vegetations caused by different types of seasonality could seriously affect the results.

Another confounding factor in MAT reconstruction must be that wood anatomical diversity patterns are also very much influenced by rainfall and water retention capacity of the soil. Thus many of the wood anatomical parameters correlated with temperature and latitude also show strong correlations with moisture availability, as shown in Table 6.1 (see also Baas and Carlquist, 1985; Carlquist and

Hoekman, 1985; Baas and Schweingruber, 1987). It is self-evident that this complicates climate reconstruction from these wood anatomical parameters.

From the brief summary of our current understanding of the complex functions of wood and of prevailing ecological trends in wood anatomy, it follows that wood should be considered a key life-history trait, on equal par with leaves and reproductive organs, when it comes to niche partitioning and explaining the evolutionary success of vascular plant species (Chave et al., 2009). To tease out the temperature component from all the other physical and biological influencing factors in the evolution of xylem diversity requires much more research than simple or multiple correlation analyses. Currently, interesting progress is being made in our understanding of wood density, a parameter that integrates a number of wood anatomical variables such as tissue proportions, cell wall thickness and relative cell lumen areas. Hacke et al. (2001, 2006) found wood density to be a good predictor for hydraulic safety. Various ecological studies emphasise the role of wood density as an important life-history trait, showing much wider variation in tropical lowland communities than in communities at higher latitudes and altitudes (e.g. Chudnoff, 1976), while also being rather highly conserved phylogenetically (Swenson and Enquist, 2007).

6.5 Current and future effects of rising temperature and CO_2 levels

So far we have discussed wood anatomical proxies for climate that can be related to climate gradients and climate change over very large spatial and evolutionary timescales. The resulting patterns suggest that with ongoing global warming, present-day species assemblages will not be optimally adapted hydraulically and mechanically to these changing conditions and may shift their latitudinal and altitudinal ranges. This is in fact what has already been reported from careful floristic monitoring in Europe. Lenoir et al. (2008) noted that over the last century there has been an upward altitudinal shift of 29 m per decade in vascular plants in western Europe. Tamis et al. (2005) found that in the last 25 years the vascular plant flora of the Netherlands had been 'enriched' by numerous species that before had their northern latitudinal limits further south in Europe. Huge latitudinal and altitudinal shifts are well known from relatively recent glacial and interglacial periods (e.g. Davis and Shaw, 2001). The existence of continuous ecological networks to allow species to migrate is normally a precondition for these range shifts to occur, and in our current ecologically fragmented world that condition is no longer always met. Furthermore, shrubs adapted to arctic or alpine conditions have 'nowhere to go' under global warming, and Gorschuch et al. (2001) hypothesised on the basis of comparative wood anatomy that *Vaccinium uliginosum* L. in Alaska may become hydraulically overchallenged with rising temperatures by developing the wrong vessel-diameter distribution profile to survive freeze–thaw cycles.

Understanding the immediate influence of increasing atmospheric temperatures and CO_2 levels on wood attributes and hydraulic fitness of species requires experimental evidence. Following in the tradition of research on the effects of acid rain and air pollution from the 1980s onwards, there have been numerous publications on large-scale and long-term experiments, in open-top chambers or in altered field conditions. The results from these do not show a clear pattern. For instance, Overdieck et al. (2007) found that increasing temperature resulted in increased growth of beech (*Fagus sylvatica* L.) saplings and that changes in CO_2 levels resulted in quantitative wood anatomical changes. Wood density is usually positively related with ambient temperature (Roderick and Berry, 2001; Thomas et al., 2007). Thomas et al. (2007) also found a decrease in vessel lumen area in *Eucalyptus* with higher temperatures. An important aspect of global warming is its inevitable effect on the length of the growing season and changes in the onset and termination dates of cambial activity (Gricar, 2007).

We concur with several authors who have suggested that, under global warming, woody plants that are already in stressful environments are most at risk, such as arctic and alpine shrubs that are highly tuned to hydraulic safety with their multitude of very narrow vessels. Increasing temperatures will mean higher demands on conductive efficiency, which might make their xylem too vulnerable to irregular freeze–thaw cycles during the growing season that will certainly continue to exist (Gorschuch et al., 2001). Similarly, xeric trees and shrubs may be at risk if their growth conditions become even drier and hotter, exceeding their already extremely high cavitation resistance. Increased irregularities in local climates and a dramatic increase in tornadoes and hurricanes would be a mechanical threat to multiple species in hurricane-prone regions.

6.6 Conclusions

In order to make full use of the great potential of comparative wood anatomy for research on the effects of climate change on woody plants in the past, present and future, we believe that the Wiemann approach (Wiemann et al., 1998, 1999, 2001) deserves to be revisited on a much more refined scale, using currently available tools for taking into account evolutionary backgrounds with comprehensive phylogenetic supertrees (cf. Swenson and Enquist, 2007). Apart from wood anatomical character states, recorded for the purposes of wood anatomical identification in databases such as InsideWood, critical features that are more functional, such as vessel diameter distribution and vessel frequency spectra, should be included in local, regional and global profiles. Such profiles could then be subjected to climate change scenarios for predicting which species will survive, escape or go extinct. We are not claiming that this is the most efficient way to record effects of global change on present and future ecosystems, but it would certainly improve and refine our understanding of the biology of trees and forest ecosystems.

Acknowledgements

We are grateful to the organisers of the Trinity College Dublin conference on climate change and systematics, who challenged us to think about the potential of our favourite research subject, systematic wood anatomy, for the field of climate research.

References

Angiosperm Phylogeny Group (2003). An update of the Angiosperm Phylogeny Group classification for the orders and families of flowering plants: APG II. *Botanical Journal of the Linnean Society of London*, **141**, 399–436.

Baas, P. (1976). Some functional and adaptive aspects of vessel member morphology. In *Wood Structure in Biological and Technological Research*, ed. P. Baas, A. J. Bolton and D. M. Catling. Leiden Botanical Series No. 3, Leiden: Leiden University Press, pp. 157–181.

Baas, P. (1982). Systematic, phylogenetic, and ecological wood anatomy – history and perspectives. In *New Perspectives in Wood Anatomy*, ed. P. Baas. The Hague: Nijhoff and Junk, pp. 23–58.

Baas, P. (1986). Ecological patterns in xylem anatomy. In *On the Economy of Plant Form and Function*, ed. J. Givnish. Cambridge, NY: Cambridge University Press, pp. 327–352.

Baas, P. and Carlquist, S. (1985). A comparison of the ecological wood anatomy of the floras of southern California and Israel. *International Association of Wood Anatomists Bulletin New Series*, **6**, 349–353.

Baas, P. and Schweingruber, F. H. (1987). Ecological trends in the wood anatomy of trees, shrubs and climbers from Europe. *International Association of Wood Anatomists Bulletin New Series*, **8**, 245–274.

Baas, P. and Wheeler, E. A. (1996). Parallelism and reversibility of xylem evolution – a review. *International Association of Wood Anatomists Journal*, **17**, 351–364.

Baas, P., Wheeler, E. A. and Chase, M. W. (2000). Dicotyledonous wood anatomy and the APG system of angiosperm classification. *Botanical Journal of the Linnean Society of London*, **134**, 3–17.

Baas P., Jansen, S. and Wheeler, E. A. (2003). Ecological adaptations and deep phylogenetic splits: evidence from the secondary xylem. In *Deep Morphology: Toward a Renaissance of Morphology in Plant Systematics*, ed. T. F. Stuessy, V. Mayer and E. Hörandl. *Regnum Vegetabile*, **141**, 221–239.

Baas, P., Ewers, F. W., Davies, S. D. and Wheeler, E. A. (2004). The evolution of xylem physiology. In *Evolution of Plant Physiology from Whole Plants to Ecosystems*, ed. A. R. Hemsley and I. Poole. Linnean Society Symposium Series No. 21. London: Elsevier Academic Press, pp. 273–296.

Bailey, I. W. (1954). Contributions to plant anatomy. *Chronica Botanica*, **15**, xxvi and 262.

Bailey, I. W. and Tupper, W. W. (1918). Size variation in tracheary cells I: a comparison between the secondary xylems of vascular cryptogams, gymnosperms and angiosperms. *Proceedings of the American Academy of Arts and Science*, **54**, 149–204.

Boura, A. and De Franceschi, D. (2007). Is porous wood structure exclusive of deciduous trees? *Comptes Rendus Palevol*, **6**, 385–391.

Briffa, K. R. (2000). Annual climate variability in the Holocene: interpreting the message of ancient trees. *Quaternary Science Reviews*, **19**, 87–105.

Carlquist, S. (1975). *Ecological Strategies of Xylem Evolution*. Berkeley, CA: University of California Press.

Carlquist, S. (2001). *Comparative Wood Anatomy: Systematic, Ecological, and Evolutionary Aspects of Dicotyledon Wood*. Berlin, NY: Springer Verlag.

Carlquist, S. and Hoekman, D. A. (1985). Ecological wood anatomy of the woody southern California flora. *International Association of Wood Anatomists Journal*, **6**, 319–347.

Chave, J., Coomes, D., Jansen, S. et al. (2009). Towards a worldwide wood economics spectrum. *Ecology Letters*, **12**, 351–366.

Choat, B., Cobb, A. R. and Jansen, S. (2008). Structure and function of bordered pits: new discoveries and impacts on whole-plant hydraulic function. *New Phytologist*, **177**, 608–626.

Chudnoff, M. (1976). Density of tropical timbers as influenced by climatic life zones. *Commonwealth Forestry Review*, **55**, 203–217.

Davis, M. B. and Shaw, R. G. (2001). Range shifts and adaptive responses to Quaternary climate change. *Science*, **292**, 673–679.

Gorschuch, D. M., Oberbauer, S. F. and Fisher, J. B. (2001). Comparative vessel anatomy of arctic deciduous and evergeen dicots. *American Journal of Botany*, **88**, 1643–1649.

Graham, A. (1999). *Late Cretaceous and Cenozoic History of North American Vegetation*. Oxford: Oxford University Press.

Gricar, J. (2007). Xylo- and phloemogenesis in silver fir (*Abies alba* Mill.) and Norway spruce (*Picea abies* (L.) Karts.). *Studia Forestalia Slovenica*, **132**, 1–106.

Hacke, U. G., Sperry, J. S., Pockman, W. T., Davies, S. D. and McCulloh, K. A. (2001). Trends in wood density and structure are linked to prevention of xylem implosion by negative pressure. *Oecologia*, **126**, 457–461.

Hacke, U. G., Sperry, J. S., Wheeler, J. K. and Castro, L. (2006). Scaling of angiosperm xylem structure with safety and efficiency. *Tree Physiology*, **26**, 689–701.

Jansen, S., Baas, P. and Smets, E. (2003). Vestured pits: do they promote safer water transport? *International Journal of Plant Science*, **164**, 405–413.

Jansen, S., Baas, P. Gasson, P., Lens, F. and Smets, E. (2004). Variation in xylem structure from tropics to tundra: evidence from vestured pits.

Proceedings of the National Academy of Sciences of the USA, **101**, 8833–8837.

Lenoir, J., Gégout, J. C., Marquet, P. A., de Ruffray, P. and Brisse, H. (2008). A significant upward shift in plant species optimum elevation during the 20th century. *Science*, **320**, 1768–1771.

Lens, F., Schöneberger, J., Baas, P., Jansen, S. and Smets, E. (2007). The role of wood anatomy in phylogeny reconstruction of Ericales. *Cladistics*, **23**, 229–254.

Liu, J. and Noshiro, S. (2003). Lack of latitudinal trends in wood anatomy of *Dodonaea viscosa* (Sapindaceae), a species with a worldwide distribution. *American Journal of Botany*, **90**, 532–539.

Loconte, H. and Stevenson, D. W. (1991). Cladistics of the Magnoliidae. *Cladistics*, **7**, 267–296.

Noshiro, S. and Baas, P. (2000). Latitudinal trends in wood anatomy within species and genera: a case study in *Cornus s.l.* (Cornaceae). *American Journal of Botany*, **87**,1495–1506.

Overdieck, D., Ziche, D. and Böttcher-Jungclaus, K. (2007). Temperature responses of growth and wood anatomy in European beech saplings grown in different carbon dioxide concentrations. *Tree Physiology*, **27**, 261–268.

Poole, I. and van den Bergen, P. F. (2006). Physiognomic and chemical characters in wood as paleoclimatic proxies. *Plant Ecology*, **182**, 175–195.

Roderick, M. L. and Berry, S. L. (2001). Linking wood density with tree growth and environment: a theoretical analysis based on the motion of water. *New Phytologist*, **149**, 473–485.

Sakala, J. (2007). The potential of fossil angiosperm wood to reconstruct the palaeoclimate in the Tertiary of central Europe (Czech Republic, Germany). *Acta Palaeobotanica*, **47**, 127–133.

Sperry, J.S. (2003). Evolution of water transport and xylem structure. *International Journal of Plant Science*, **164**, S115-S127.

Sperry, J. S., Hacke, J. G., Field, T. S ., Sano, Y. and Sikkema, E. H. (2007). Hydraulic consequences of vessel evolution in angiosperms. *International Journal of Plant Science*, **168**, 1127–1139.

Swenson, N. G. and Enquist, B. J. (2007). Ecological and evolutionary determinants of a key plant functional trait: wood density and its community-wide variation across latitude and elevation. *American Journal of Botany*, **94**, 451–459.

Tamis, W., van' t Zelfde, M., van der Meijden, R. and Udo de Haes, H. A. (2005). Changes in vascular plant biodiversity in the Netherlands in the 20th century explained by their climatic and other environmental characteristics. *Climate Change*, **72**, 37–56.

Thomas, D. S., Montagu, K. D. and Conroy, J. P. (2007). Temperature effects on wood anatomy, wood density, photosynthesis and biomass partitioning of *Eucalyptus grandis* seedlings. *Tree Physiology*, **27**, 251–260.

Tyree, M. T. and Zimmermann, M. H. (2002). *Xylem Structure and the Ascent of Sap*, 2nd edn. Berlin: Springer Verlag.

van der Graaff, N. A. and Baas, P. (1974). Wood anatomical variation in relation to latitude and altitude. *Blumea*, **22**, 101–121.

Wheeler, E. A. and Baas, P. (1991). A survey of the fossil record from dicotyledonous wood and its significance for evolutionary and ecological wood anatomy. *International Association of Wood Anatomists Bulletin New Series*, **12**, 275–332.

Wheeler, E. A. and Baas, P. (1993). The potentials and limitations of dicotyledonous wood anatomy for climatic reconstructions. *Paleobiology*, **14**, 486–497.

Wheeler, E. A. and Dillhoff, T. (2009). The Middle Miocene wood flora from Vantage, Washington, USA. *International Association of Wood Anatomists Journal*, Suppl. 7.

Wheeler, E. A. and Manchester, S. R. (2002). Woods of the Eocene Nuts Beds flora, Clarno Formation, Oregon, USA. *International Association of Wood Anatomists Journal*, Suppl. 3.

Wheeler, E. A., Baas, P. and Rodgers, S. A. (2007). Variations in dicot wood anatomy: a global analysis based on the InsideWood database. *International Association of Wood Anatomists Journal*, **28**, 229–258.

Wiemann, M. C., Wheeler, E. A., Manchester, S. R. and Portier, K. M. (1998). Dicotyledonous wood anatomical characters as predictors of climate. *Palaeography, Palaeoclimatology, Palaeoecology*, **139**, 83–100.

Wiemann, M. C., Manchester, S. R. and Wheeler, E. A. (1999). Paleotemperature estimation from dicotyledonous wood anatomical characters. *Palaios*, **14**, 459–474.

Wiemann, M. C., Dilcher, D. L. and Manchester, S. R. (2001). Estimation of mean annual temperature from leaf and wood physiognomy. *Forest Science*, **47**, 141–149.

Wolfe, J. A. (1978). A paleobotanical interpretation of Tertiary climates in the northern hemisphere. *American Scientist*, **66**, 694–703.

Wolfe, J. A. (1987). Late Cretaceous-Cenozoic history of deciduousness and the terminal Cretaceous event. *Paleobiology*, **13**, 215–226.

Zimmermann, M. H. (1978). Structural requirements for optimal water conduction in tree stems. In *Tropical Trees as Living Systems*, ed. P. B. Tomlinson and M. H. Zimmermann. Cambridge: Cambridge University Press, pp. 517–532.

7

Savanna biome evolution, climate change and the ecological expansion of C_4 grasses

Y. Bouchenak-Khelladi

Department of Botany, University of Cape Town, South Africa

T. R. Hodkinson

School of Natural Sciences, Trinity College Dublin, Ireland

Abstract

This chapter discusses the abiotic (climatic, atmospheric, fire) and biotic (herbivory, competition) factors driving the origin of savanna biomes and the evolution of their dominant plant group, the grasses. C_4 photosynthesis is a key innovation in grass evolution, and we outline how phylogenetic approaches have helped us understand the multiple origins of this trait and the factors driving its evolution, such as atmospheric CO_2 concentrations, drought, heat and fire. C_4 grasses have interacted closely with other organisms throughout their evolution, and we describe evidence for their coevolution with ungulate herbivores in savanna habitats (an evolutionary arms-race scenario). We also describe phylogenetic approaches for reconstructing ancestral niches and geographical ranges of

Climate Change, Ecology and Systematics, ed. Trevor R. Hodkinson, Michael B. Jones, Stephen Waldren and John A. N. Parnell. Published by Cambridge University Press.

grasses over evolutionary time. These studies reveal a close link between climate and savanna evolution, with the first C_4 grasses evolving in open habitats of Africa. By reviewing the findings of several major studies, we hope to provide predictions about the fate of savannas under future global change scenarios.

7.1 Introduction

Biomes are natural communities of wide geographical extent, characterised by distinctive, climatically controlled groups of organisms (Raven et al., 2005). Savannas are among the most charismatic of such biomes because they are extremely species-rich, because their predominantly C_4 grasses coevolved with a large diversity of mammalian grazers, and because their history is intimately linked with the opening up of tropical forests that occurred during the Cenozoic (in the last 65 million years), due to climate change. C_4 grasses exhibit the Hatch–Slack photosynthetic pathway (Slack and Hatch, 1967). This chapter describes the close link between climate change and the evolution of savanna habitats and their C_4 grasses. Although not discussed here, it should also be noted that photosynthesis affects climates and atmosphere (Cerling et al., 1997) through carbon fixation, because photosynthetic plants are so abundant on earth.

The definition of savanna is not always clear or consistently applied, and hence various terms have been used to describe them. Tropical grasslands by definition lie between the tropics of Cancer and Capricorn, but for many applications it does not make ecological sense to divide grasslands into groups based on their latitude (Shaw, 2000). It is often more meaningful to classify grassland biomes that lie in both the tropics and subtropics, frequently sandwiched latitudinally between equatorial rainforest on one side and deserts/semi-deserts on the other (Huntley and Walker, 1982). We can also divide tropical and subtropical grasslands into two groups depending on whether or not they have an overstorey of trees and shrubs. If they have an overstorey they are sometimes classified as savanna (in the strict sense) in contrast to other types of tropical and subtropical grassland. However, it is not always appropriate to classify them in this way, because in reality a gradation exists from one to another (Fig 7.1).

The term savanna is frequently used to mean both types of tropical and subtropical grassland, with trees (woodland savanna) and without trees (grassland savanna). It can therefore be used to include regions that stretch outside of the tropics such as South Africa and parts of Australia and India. We adopt this broad definition of savanna, which is also taken to be equivalent with 'natural C_4 grasslands' (Long and Jones, 1991). Major savannas occur on four continents: Africa holds the largest biome surface (including velds), followed by Australia, South America (campo, pampa, llano and cerrado) and southern India (Fig 7.2). They

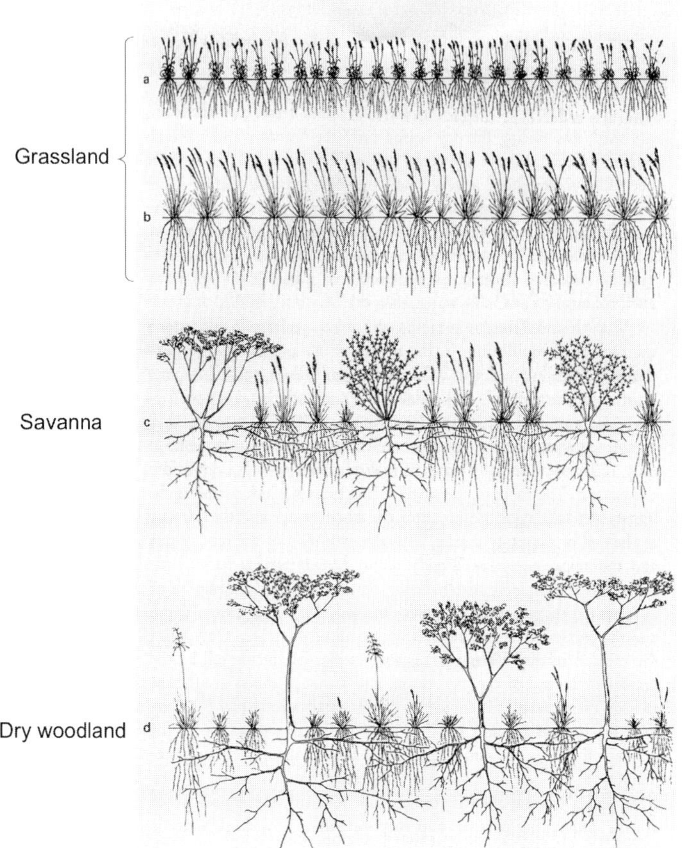

Figure 7.1 Transition of grassland to woodland, via savanna. Reproduced with permission from Walter (1979).

occur in areas with winterless climates, with no month having a mean temperature lower than 18 °C, and in areas with pronounced dry seasons. Precipitation is a key factor in the balance between grassland and forest (Walter, 1979). If precipitation is even and high throughout the year it supports tropical rainforest vegetation. However, if there is a pronounced dry season of several months, then grasslands develop (Walter, 1979). Fire also maintains sharp boundaries between forest and grasslands (Shaw, 2000; Bond et al., 2003).

Savannas are species-rich, and many modern mammal lineages (including hominids and ungulates) evolved in these regions (MacFadden and Cerling, 1994; Janis et al., 2002). Therefore, studying patterns of savanna biome evolution throughout geological time can help us understand evolutionary processes and mechanisms that have led to present-day biodiversity in those areas.

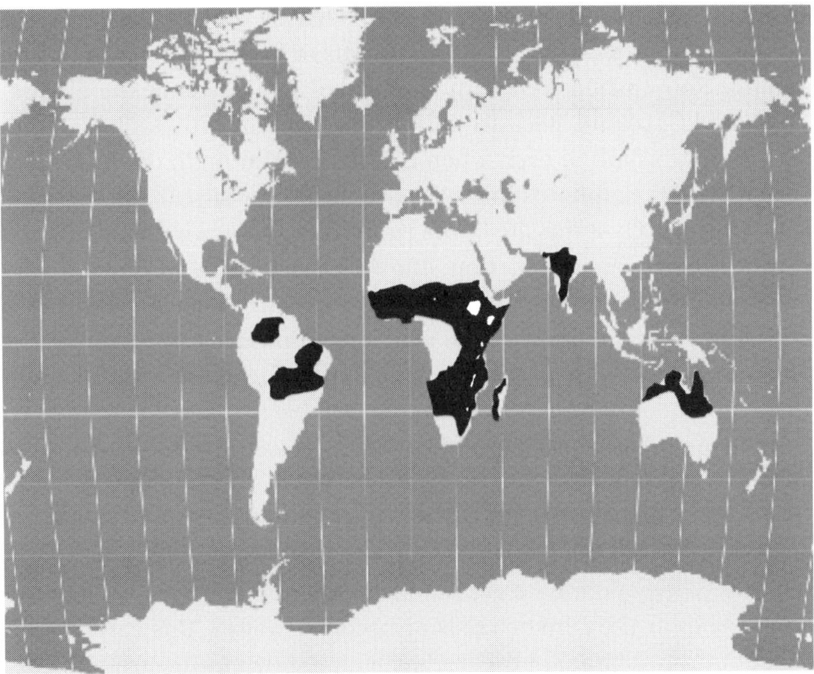

Figure 7.2 Major tropical/subtropical grasslands of the world. The main areas lie between equatorial rainforests and the deserts/semi-deserts near the tropics of Cancer and Capricorn, but they also extend into subtropical areas such as South Africa, parts of India and Australia. (Source: Encyclopaedia Britannica, www.britannica.com/EBchecked/topic-art/648562/46/Worldwide-distribution-of-savannas).

The C_4 photosynthetic pathway is advantageous to grasses growing in warmer climates and at high light intensities (Sage, 2004), because they attain higher photosynthetic light-use efficiency than C_3 plants (Ehleringer and Monson, 1993). C_3 and C_4 plants have different ways of fixing CO_2 via the Calvin cycle (Sage, 2004). C_3 plants use CO_2 directly from the atmosphere, and the first organic compound produced is a three-carbon molecule called 3-phosphoglyceric acid (3-PGA). In hot and dry weather, C_3 plants partially close their stomata to reduce water loss, but this also reduces CO_2 uptake. Furthermore, the first enzyme of the Calvin cycle, known as rubisco, incorporates O_2 instead of CO_2, leading to a wasteful process known as photorespiration (Salisbury and Ross, 1992; Campbell et al., 2008). In photorespiration the Calvin cycle produces a two-carbon compound instead of its usual three-carbon product and this is broken down into CO_2 and H_2O (instead of the desired three-carbon sugar). C_4 plants have evolved an innovation to both save water and prevent wasteful photorespiration (Fig 7.3). In hot conditions, C_4 plants also partially close their stomata to save water, but can continue making sugars via photosynthesis, because they have an enzyme (phosphoenolpyruvate carboxylase – PEP-C) that has a high affinity for CO_2 and fixes carbon into a four-carbon compound instead of 3-PGA; hence the distinction between C_3 and C_4 plants. Because PEP-C in C_4 photosynthesis has a high affinity for CO_2 it can fix carbon even when CO_2 concentration in the leaf is low (Campbell et al., 2008). C_4 photosynthesis also avoids wasteful photorespiration by concentrating CO_2 at the site of the rubisco enzyme, which favours the carboxylase over the oxygenase reaction. The PEP-C fixes CO_2 in one cell and other molecules shuttle CO_2 to the Calvin cycle in a nearby cell that continues to photosynthesise, even though the CO_2 available is reduced (because of the closing of the stomata). Because of the special partitioning of photosynthesis in different cells, C_4 plants often display a unique and easily identifiable 'Kranz' anatomy (German for wreath, because of the wreath-like arrangement of bundle sheaths – Salisbury and Ross, 1992).

Amongst C_4 plants, C_4 grasses are globally the most important group because of their dominance in many biomes (Sage and Monson, 1999; Sage, 2004; Bouchenak-Khelladi et al., 2009). They dominate grasslands in hot and dry climates, but are replaced by C_3 species as the climate cools or becomes wetter. This can be seen clearly with mountain vegetation in the tropics, which shows a cline from C_4 species composition and dominance at sea level to C_3 species in cooler and wetter mountain top conditions (Fig 7.3 – Rundel, 1980). It is therefore essential to understand the evolutionary/ecological history of C_4 grasses in order to comprehend the origin and evolutionary processes that have affected these biomes.

Reconstructing the evolution of C_4 grasses requires phylogenetic inferences (i.e. hypotheses of evolutionary relationships of biological taxa), which allow us to consider the rates of diversification and speciation within a clade and the correlation of such rates with morphological or ecological traits. The rise of phylogenetics, linked to the rise in available DNA sequence data, has had a great impact on our understanding of how lineages and their inherent characteristics have evolved over time. The integration of phylogenetics, historical biogeography and ecology offers great potential for studying past and future dynamics of organisms and their ecosystems, especially with respect to climate change (Edwards et al., 2007; Culham and Yesson, Chapter 10; Rodríguez-Sánchez and Arroyo, Chapter 13). The use of phylogenetic trees to reconstruct ancestral ecologies can help to inform the processes and factors driving and maintaining species diversity in communities (Slingsby and Verboom, 2006), in biomes (Hardy and Linder, 2005), or even at a global scale (Bouchenak-Khelladi et al., 2009).

Recent phylogenetic studies with good taxon sampling have helped resolve the main taxonomic groups in the grass family (GPWG, 2001; Sánchez-Ken et al., 2007; Bouchenak-Khelladi et al., 2008). Such robust phylogenetic trees of the grasses, enriched by palaeontological and ecological data, can be used to identify

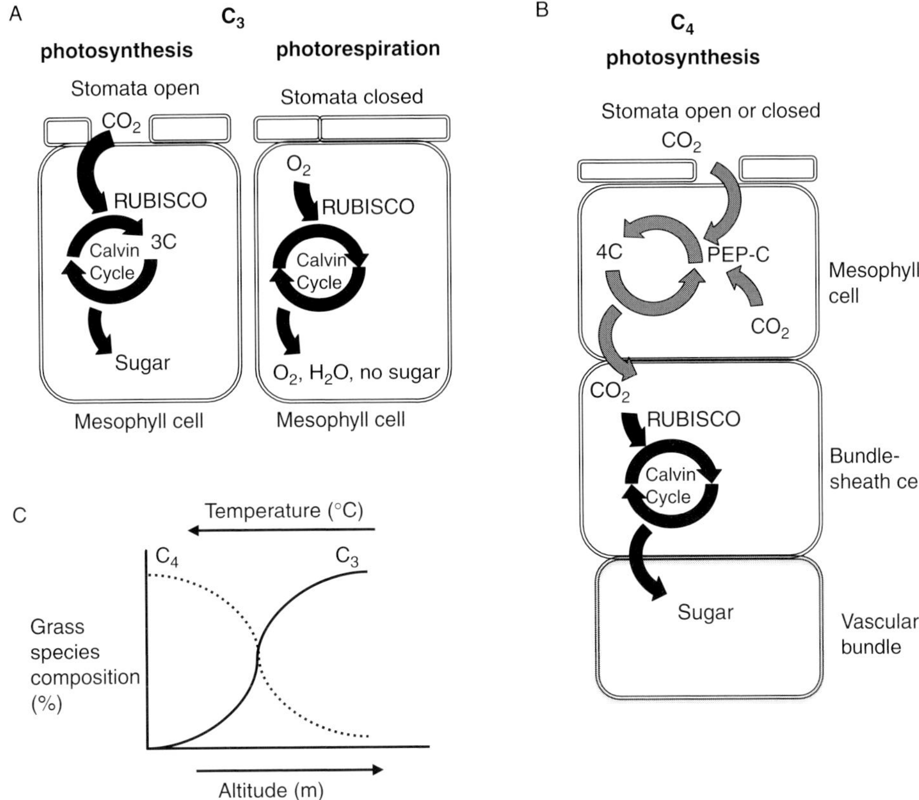

Figure 7.3 C_3 and C_4 photosynthetic pathways and the response of C_3 and C_4 grass communities to altitude and temperature. (A) C_3 pathway; (B) C_4 pathway; (C) change in C_3 and C_4 plant species composition and dominance on a hypothetical mountain in the tropics.

environmental and/or biological factors that may have led to ecological adaptations of grasses in their savanna biomes. Several abiotic factors are thought to have driven tropical grassland evolution, such as CO_2 levels (Sage, 2001), fire regimes (Bond et al., 2003, 2005) and rainfall seasonality (Keeley and Rundel, 2005; Osborne, 2008). However, biotic factors (such as the coevolution of ungulate grazers in the African savanna regions) are also likely to have been important.

Here, we firstly outline the main abiotic and biotic factors that have driven the evolution and expansion of savannas and their constituent C_4 grasses. We then discuss the role of phylogenetic studies in understanding the origins and diversification of C_4 photosynthetic grasses in geographical space and geological time, and finally outline future approaches that may be used to study the evolution of these important ecosystems, especially in relation to predicting the possible impacts of higher CO_2 concentrations and climate change.

7.2 Abiotic and biotic factors driving the evolution of C_4 savanna grasses

There have been several recent advances in our knowledge of grass family evolution thanks to fossils (MacFadden and Cerling, 1994; Cerling et al., 1998; Jacobs et al., 1999; Janis et al., 2002), isotopes (Latorre et al., 1997; Zazzo et al., 2000; Cerling, 2009), grass phytoliths (Prasad et al., 2005; Strömberg, 2005) and molecular data (GWPG, 2001; Bremer, 2002; Bouchenak-Khelladi et al., 2009). However, the patterns and processes of C_4 grass evolution are largely unknown (Strömberg, 2005). The environmental factors that have most likely affected the evolution of C_4 grasses and their diversification are fluctuating CO_2 levels, increased aridity, greater rainfall seasonality, fire and coevolution, especially with herbivores.

Atmospheric CO_2 levels declined during the late Oligocene (Tipple and Pagani, 2007), and this may have triggered the appearance of C_4 photosynthesis in grasses (Sage, 2001; Vicentini et al., 2008). However, the decline does not coincide with the ecological rise to dominance of C_4 grasses that occurred between 5 and 10 million years ago (mya) (Cerling et al., 1997; Pagani et al., 2005). The C_4 pathway is advantageous to plants growing under conditions of CO_2 limitation (Sage, 2004), but this does not necessarily mean that CO_2 limitation was the sole driver of C_4 evolution in grasses (Bouchenak-Khelladi et al., 2009), or of the expansion of savannas dominated by C_4 grasses.

Increasing aridity coupled with greater rainfall seasonality is thought to have favoured the ecological expansion of C_4 grasses into warmer and lower-latitude regions of the world (Edwards and Still, 2008; Osborne, 2008). Studies have also shown, however, that C_4 variants (namely NAD-ME, NADP-ME and PCK) respond differently to different precipitation regimes (Hattersley and Watson, 1992; Taub and Lerdau, 2000; Osborne, 2008). Therefore, aridity and rainfall may have played a role in C_4 grass evolution, but certainly not uniformly among all C_4 grass lineages and across all continents (Osborne, 2008). For instance, Ortiz-Jaureguizar and Cladera (2006), using data from the literature on plate tectonics, volcanism, sea-level changes and land mammal fossil records, found that from the early Palaeocene to the late Pleistocene (*c.* 55–4 mya), southern South American biomes changed from tropical forest to steppe, across a sequence of woodland savanna and grassland savanna. This suggests that climatic change through the Cenozoic, from warm, wet and non-seasonal to colder, drier and seasonal, allowed a decrease in forest cover and the evolution and expansion of steppe and savanna in South America.

Fire has also played a significant role in savanna expansion (Bond et al., 2003; Osborne, 2008). It accelerates forest loss and C_4 grassland expansion through multiple positive feedback loops that each promote drought and more fire. An increase in climate seasonality (with a pronounced dry season) may have allowed the instalment of fire climates (Keeley and Rundel, 2005). C_4 grasses are thought to be well adapted to fire because of their high below-ground biomass allocation, fast primary growth, low decomposition rates and above-ground 'biofuel' accumulation (Bond et al., 2003). Palaeoecological charcoal sediment profiles indicate a higher intensity of fire in the Miocene, which may be the result of increased combustible fuel loads due to an elevated productivity of C_4 grasses (Keeley and Rundel, 2005).

Herbivory may be one potential factor promoting the expansion of C_4 grasslands through the Miocene (Coughenour, 1985; MacFadden and Cerling, 1994; Cerling et al., 1997; Janis et al., 2002; Bouchenak-Khelladi et al., 2009), even though there is a substantial temporal mismatch between savanna expansion and the radiation of ungulates during this epoch. The shifts from low-crowned to high-crowned teeth in ungulates seem to be related to a shift in their diets: the appearance of more fibrous vegetation (Jernvall et al., 1996) and a change from browsing to grazing (Janis et al., 2002). African savannas, dominated by C_4 grasses, hold the largest ungulate diversity on earth. This tremendous diversity of grazers co-occurring with browser ungulates makes the ecological and evolutionary aspects of the relationship between grasses and megaherbivores central to understanding the evolution and functioning of savanna ecosystems.

It appears, therefore, that a range of factors have played a role in driving the evolution of C_4 grasses and their savanna biomes, but that none is compelling by itself. It is more likely that a combination of these biotic and abiotic factors, through multiple positive feedbacks, promoted the expansion of savanna biomes and their C_4 grasses throughout the globe. Recent and detailed studies suggest that late Miocene C_4 expansion was regionally heterogeneous rather than globally synchronous, indicating that local or regional environmental factors also played an important role in driving C_4 plant expansion, or at least that these interacted significantly with global climatic factors (Tipple and Pagani, 2007; Edwards and Still, 2008). For instance, Beerling and Osborne (2006) proposed that the 'CO_2 starvation hypothesis' (i.e. atmospheric CO_2 concentrations falling below a critical threshold during the Miocene – Cerling et al., 1997), which favoured C_4 species, is an essential primer for the origination of C_4 savanna ecosystems but that coevolution of herbivores may disrupt or enhance the network. Explaining the rapid appearance and subsequent persistence of C_4 savanna ecosystems therefore remains a major scientific challenge, although phylogenetic studies are offering new insights and approaches to their study.

7.3 Phylogenetic approaches for studying savanna evolution through geological time

7.3.1 Major clades of C_4 grasses

The grass family (Poaceae) is particularly rich in C_4 species (*c.* 4500 species, representing 60% of all C_4 plants – Sage, 2004). Recent advances in grass phylogenetics have resolved the major relationships at subfamilial and tribal levels (GPWG, 2001; Sánchez-Ken et al., 2007; Bouchenak-Khelladi et al., 2008; Sungkaew et al., 2008), including the C_4 clades. Two major clades have been resolved, namely BEP (Bambusoideae, Ehrhartoideae and Pooideae) and PACCMAD (Panicoideae, Arundinoideae, Chloridoideae, Centothecoideae *s.l.*, Micrairoideae, Aristidoideae and Danthonioideae) (Fig 7.4). BEP contains exclusively C_3 species, while PACCMAD includes some C_3 species, as well as all known C_4 grasses (Fig 7.4). C_4 taxa are mainly distributed within Aristidoideae, Chloridoideae and Panicoideae, and include representatives of NAPD-ME, NAD-ME and PCK C_4 subtypes, respectively (Hattersley and Watson, 1992; GPWG, 2001; Hodkinson et al., 2007; Sánchez-Ken et al., 2007; Bouchenak-Khelladi et al., 2008).

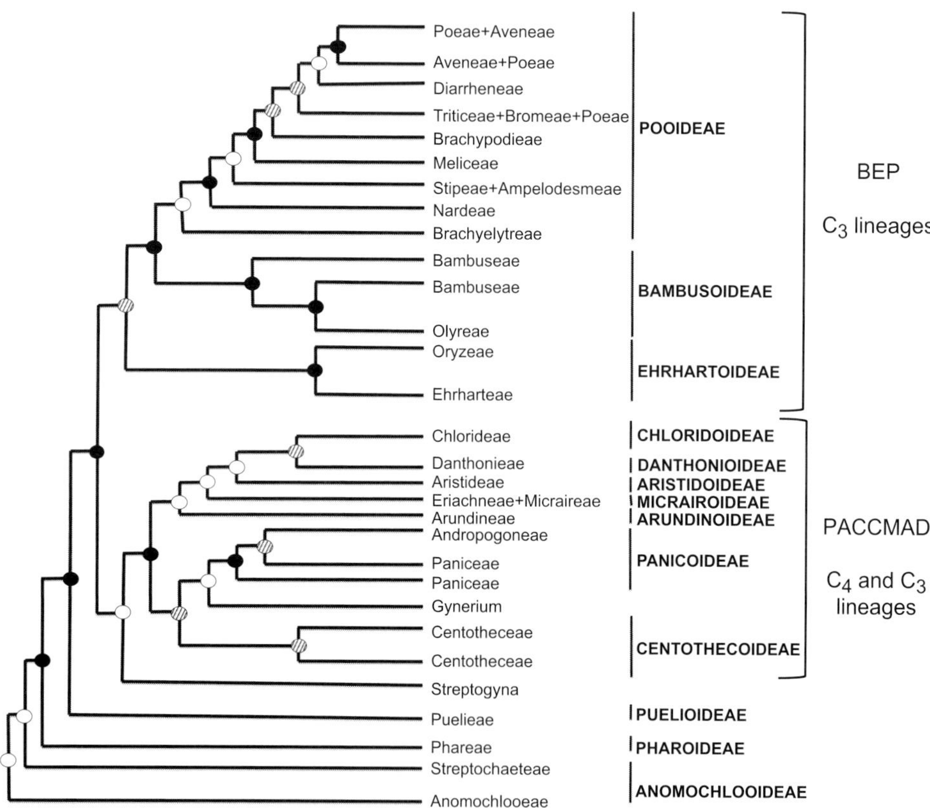

Figure 7.4 Summary tree showing subfamilial and intertribal relationships of grasses (Poaceae). ○, weak support (55–75 bootstrap percentage – BP); ◍, moderate support (75–85 BP); ●, strong support (85–100 BP). Modified from Bouchenak-Khelladi et al. (2008).

7.3.2 Phylogenetic origins of C_4 photosynthesis

Phylogenetic trees can be used to infer the timing of origin of the C_4 photosynthetic pathway among grass lineages and hence reveal the evolutionary period at which C_4 grass lineages diversified. Such dates can be used as proxies for estimating the origin and expansion of savanna biomes worldwide. Some recent phylogenetic studies have attempted to estimate the number of times lineages of C_4 grasses have evolved and the ages of these clades (Kellogg, 2001; Christin et al., 2008; Vicentini et al., 2008; Bouchenak-Khelladi et al., 2009). Kellogg (2001) estimated five origins by mapping the C_4 trait onto the phylogenetic trees of the GPWG (2001), which only included 62 grass genera amongst the 13 currently recognised subfamilies (GPWG, 2001; Sánchez-Ken et al., 2007). Christin et al. (2008) sampled 187 species and showed that C_4 photosynthesis might have evolved 17 or 18 times, apparently being promoted by the decline of CO_2 in the Oligocene. Vicentini et al. (2008) sampled 97 species, with an emphasis on Panicoideae, and found four to five origins for C_4 photosynthesis in grasses also occurring in the Oligocene. Bouchenak-Khelladi et al. (2009), using a comprehensive generic-level phylogenetic tree of the grasses encompassing all 800 grass genera, estimated 12 shifts from C_3 to C_4 photosynthesis between the late Oligocene and the late Quaternary. The differences in the estimations of number of C_4 origins among the three studies are due to the differences in taxon sample size and strategy. Vicentini et al. (2008) placed emphasis on the subfamily Pancioideae, whereas Christin et al. (2008) sampled representatives from all C_4 grass subfamilies (i.e. Aristidoideae, Chloridoideae and Panicoideae) but included only 65 out of 355 C_4 genera. Although Bouchenak-Khelladi et al. (2009) sampled all grass genera, they missed the origin of the pathway within genera. Indeed, several grass genera include both C_3 and C_4 species (*Alloteropsis*, *Neurachne* and *Panicum* amongst others), which would lead to an increased detection of the number of C_4 origins at species level. However, all three studies agreed that C_4 photosynthesis first arose in grasses during the Oligocene, suggesting that the CO_2 decline in the Oligocene (Tipple and Pagani, 2007) was correlated with, and possibly a triggering or contributing factor for, the origin of C_4 photosynthesis. However, as explained by Sage (2001, 2004) and discussed earlier, the importance of low CO_2 levels at the Oligocene transition is not necessarily the sole driver for C_4 evolution, but rather a precondition. Studies on additional ecological factors are needed to understand the origins of this biological innovation (Roalson, 2008) and the expansion of C_4 biomes.

7.3.3 Expansion of C_4 grasslands and phylogenetic diversification of C_4 grasses

Identifying when C_4 lineages of grasses diversified (major periods of speciation and divergence into what we today label as genera, tribes and higher taxonomic ranks) over time can help estimate when C_4 grass savannas expanded globally. Using a comprehensive generic-level phylogenetic tree, it is possible to assess diversification patterns across grass lineages by comparing the topological distribution of species diversity on the phylogeny (Chan and Moore, 2005; Bouchenak-Khelladi et al., 2009). Such an approach looks for imbalance on the tree in terms of species richness and identifies which clades are significantly more species-rich than would be expected by purely stochastic processes. Bouchenak-Khelladi et al. (2009) showed that phylogenetic diversifications detected within the PACCMAD clade (Figs 7.4, 7.6) occurred between the mid-Miocene and the late Quaternary, which suggests that C_4 grass lineages did not dominate ecosystems until the Miocene. Based on plant carbon isotopes, soil carbonates and herbivore teeth, Cerling et al. (1997) dated the origin of the savanna biome to the late Miocene (*c.* 8 mya), with an expansion around 1 mya. Using pollen and carbon isotopes from western and eastern Africa, Jacobs (2004), however, showed that the grass-dominated savannas began to expand by the mid-Miocene (16 mya) and became widespread only in the late Miocene (around 8 mya). These data suggest that C_4 grass lineages may have diversified as early as the Miocene but only became ecologically dominant by the late Miocene through to the Pleistocene.

7.3.4 Phylogenetic evidence for coevolutionary processes in savannas

Other factors may have promoted C_4 species to expand globally, including herbivory (Coughenour, 1985; Macfadden and Cerling, 1994; Sage, 2001; Beerling and Osborne, 2006). The spread of C_4 grasslands may have been associated with increased grazing rates in the Miocene (Chapman, 1996). The evolution of anti-herbivore defence mechanisms may have occurred in response to the diversification of megaherbivore fauna (Coughenour, 1985 – an 'evolutionary arms race' scenario). For instance, silica bodies are thought to reduce palatability and digestibility of grasses (Coughenour, 1985; Massey and Hartley, 2006; Piperno, 2006). Although the evolution of C_4 grasses led to the formation of new habitats in which we would expect grazing mammals, such as ungulates, to evolve and diversify, few studies have tried to investigate a possible coevolutionary scenario between grasses and grazers. A recent attempt used silica density of grass epidermal cells as a proxy for leaf palatability (Bouchenak-Khelladi et al., 2009). Silica density indexes were measured for 90 grass species, and ancestral values were reconstructed onto the phylogeny. Using a correlation analysis, increases in silica densities were significantly correlated with the appearance of C_4 grass lineages. Using molecular dating techniques, the results showed that there was an increase in C_4 grass silica densities in the late Miocene (Bouchenak-Khelladi et al., 2009), coincident with an increase in the abundance of hypsodont (i.e. high-crowned teeth, which enable herbivores to deal with more fibrous vegetation) fossils (Jernvall and Fortelius, 2002). It is plausible that C_4 grasses

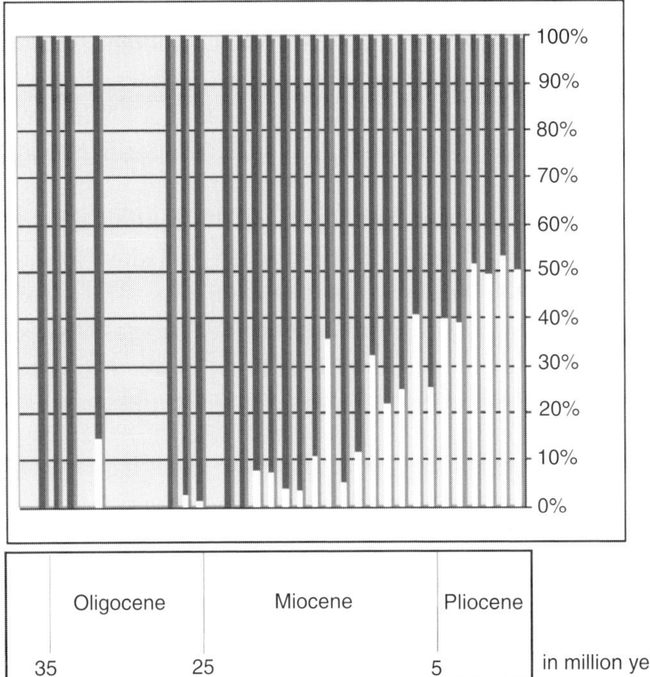

Figure 7.5 Proportion of brachydont/mesodont (dark), and hypsodont (light) teeth recorded in three-million-year intervals (throughout the Neogene in the Old World), showing that the occurrence of hypsodonty increased through the Miocene. Reproduced with permission from Jernvall et al. (1996).

responded to increased selective pressure imposed by grazers from the late Miocene, as illustrated in Fig 7.5. However, no strong body of evidence is available to date to confirm this hypothesis. Clearly, grasses have an extraordinary ability to respond to grazing, a feature that is best explained by adaptive coevolutionary processes. It confers selective advantage over other plants via increased survival and competitiveness. Such traits were critical for the expansion of savanna ecosystems.

7.3.5 Ancestral niche and geographical range reconstruction of C_4 grasses

Reconstruction of the evolution of geographical ranges and ancestral ecological habitats of C_4 grasses can also be undertaken with phylogenetic approaches. It is possible to reconstruct the evolutionary history of the C_4 clade in order to locate, in space and time, the most likely distribution of C_4 grass lineage ancestors (Hardy and Linder, 2005; Donoghue, 2008). Such ancestral reconstructions involve the use of phylogenetic trees for extant species and the character states (i.e. ecological traits or geographical distributions) observed in these species to make inferences about the ancestral states of ancestral species (Hardy, 2006). Such inferences result from the use of optimisation techniques (algorithms), which carry character states in extant species back to ancestral species (i.e. internal nodes of a phylogeny), according to an optimality criterion such as parsimony or maximum likelihood (Hardy, 2006).

Using these approaches, Bouchenak-Khelladi et al. (2010) applied a comprehensive generic-level phylogenetic tree of the PACCMAD clade (Fig 7.6), geographical and habitat data for C_4 grass genera, and several ancestral reconstruction methods, to infer the ancestral niche and geographical ranges of major grass lineages. The origin of the PACCMAD clade (crown node) was estimated

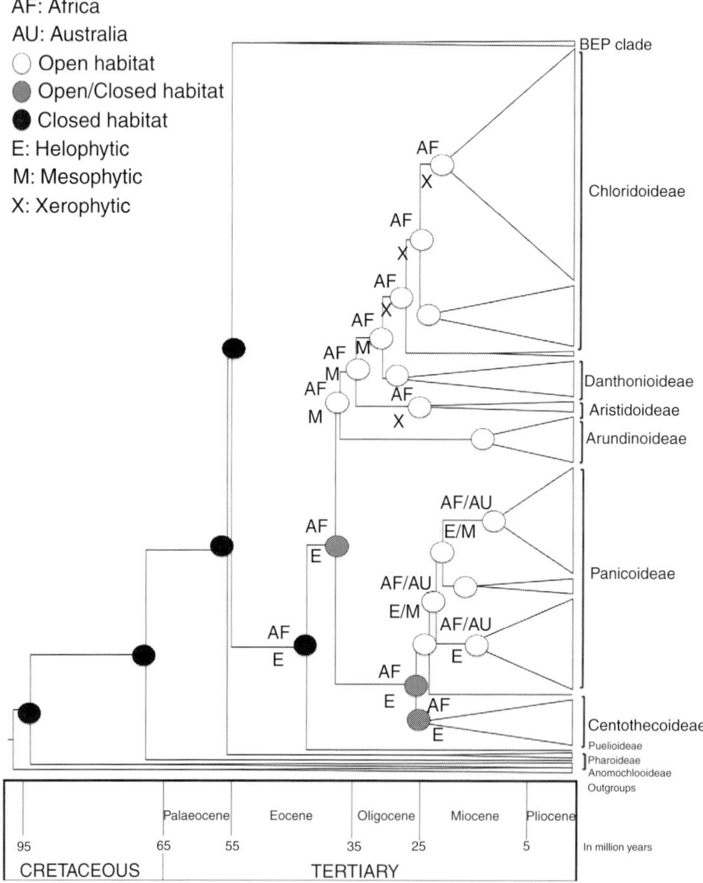

Figure 7.6 Biogeography and ecological niche reconstructions for the PACCMAD clade, using maximum likelihood. Adapted from Bouchenak-Khelladi et al. (2010).

at about 45 mya in the mid-Eocene (Fig 7.6). The ancestors of this clade were inferred to be adapted to forested habitats. Occupation of open habitats in Africa occurred in the late Eocene to early Oligocene among the ancestors of Chloridoideae. The Oligocene period was considerably drier than the rest of the Tertiary, and this might have had an effect on the decrease of forest cover and expansion of open habitats (Janis, 1993). This date for the earliest major shift of grasses from closed to open habitat types is consistent with the date suggested by Strömberg (2005) based on phytolith data, thus adding considerable weight to the hypothesis that the late Eocene marked the early taxonomic diversification of open-habitat grasses in Africa. Strömberg (2005) found that despite this earlier taxonomic diversification of open-habitat types, typically PACCMAD grasses, a greater spread at the expense of closed-habitat types occurred during the late Oligocene and the early Miocene. It is plausible that the occupation of open habitats was triggered in Africa well before they became ecologically dominant in other parts of the world, such as in North America (Strömberg, 2005) or southern Europe (Strömberg et al., 2007).

The adaptive transition from a wetter and shaded environment to a drier and more open habitat occurred for Chloridoideae and Aristidoideae between 35 and 25 mya, most likely in Africa (Fig 7.6). Indeed, the environmental history of Africa since the Eocene involves the expansion of arid-adapted vegetation (Bobe, 2006). Aristidoideae and Chloridoideae include exclusively C_4 grasses, and the dates of appearance are in agreement with recent studies (Christin et al., 2008; Vicentini et al., 2008; Bouchenak-Khelladi et al., 2009, 2010), even though the molecular dating techniques differed. Based on the evidence presented in Bouchenak-Khelladi et al. (2010), C_4 photosynthesis appears to have first originated in Africa, at least for Chloridoideae, with C_4 ancestors then dispersing to other continents. Panicoideae are inferred to have originated about 26 mya in the early Miocene in mixed open/closed habitats (Fig 7.6). Even though there is considerable uncertainty in ancestral reconstructions for this clade, it seems that Paniceae and Andropogoneae dispersed from Africa to Australia from the early Miocene (Fig 7.6) and that their ancestor was probably adapted to a relatively moist habitat in contrast with Aristidoideae and Chloridoideae. This is a finding consistent with Osborne and Freckleton (2009). Therefore, adaptations to open and arid habitats did not occur simultaneously among C_4 grass lineages, suggesting that climatic changes did not affect C_4 grass evolution uniformly. For instance, it has been shown that the abundance of NADP-ME types (mainly members of Panicoideae) is positively correlated with increasing precipitation, whereas the abundance of NAD-ME and PCK types (mainly Aristidoideae and Chloridoideae) is negatively correlated (Hattersley and Watson, 1992; Taub and Lerdau, 2000; Bond et al., 2003; Osborne, 2008; Osborne and Freckleton, 2009).

7.4 Conclusions, and the fate of savanna grassland ecosystems

The phylogenetic studies highlighted above have helped reveal processes that may have driven C_4 grass evolution and resultant savanna biomes (Bouchenak-Khelladi et al., 2009, 2010; Osborne and Freckleton, 2009). However, limitations inherent to phylogenetic methods arose. Firstly, no comprehensive species-level phylogenies of the PACCMAD clade have been produced. A comprehensive sampling would allow a more robust evolutionary reconstruction because missing species introduce bias by removing the most recent speciation events (Barraclough and Nee, 2001). Secondly, the C_4 syndrome may be an oversimplification of the actual evolution of the different pathways in grasses; further phylogenetic studies should focus on mapping the evolution of the three C_4 variants among grass subfamilies within the PACCMAD clade. Studies of C_4 occurrence are also required in other plant groups such as Cyperaceae (Simpson et al., Chapter 19). Finally, the identification and quantification of C_4 grass functional traits need to be performed on a broad range of taxa in order to test for an evolutionary conservatism of particular traits that would be related to grass leaf palatability, adaptation to fire and aridity, and dispersal potential.

Given our knowledge of savanna evolution in relation to climatic change and geological time, it would be useful if we could develop models to better understand how savannas and their C_4 grasses might respond to future human-induced climate change. To date, studies have shown that the evolution of C_4 grasses was correlated with, and may have been triggered by, a drop in CO_2 level in the Oligocene, but that a suite of other environmental and ecological factors drove the expansion of the C_4 biomes. It seems that there was a significant delay in the expansion of savanna biomes after the origins of C_4 species. The drop in CO_2 levels may have contributed to the origin of C_4 photosynthesis in the Oligocene, which in turn may have had complex biological implications for more than 15 million years, before the rise and ecological expansion of savannas in the late Miocene. It is very likely that there were highly complex abiotic and biotic interactions at the ecosystem and organismal levels. For instance, one can speculate that the Oligocene drop in CO_2 level allowed C_4 photosynthetic grasses to evolve and dominate ecosystems at the expense of trees. This could have caused increased ecosystem flammability due to accumulation of biofuel by C_4 grasses, and allowed fire regimes to intensify, thereby permitting the opening of forest cover until the late Miocene. Such large open habitats promoted the diversification of grasses with fire and aridity adaptations. Grazers could have radiated in these biomes, leading to the extreme diversity of ungulates we see today in African savannas.

Studying evolutionary responses to historical climate shifts can give us insights into current and potential future changes (Yesson and Culham, 2006; Culham and Yesson, Chapter 10). Because biological organisms are very likely to respond to changing environments through evolutionary time, it is essential to use information from their evolutionary history and inferred palaeoclimates to better understand the impacts of climatic change on biodiversity. Several studies have used present-day bioclimatic models and examined them from an evolutionary perspective (Hugall et al., 2002; Graham et al., 2004; Yesson and Culham, 2006). By examining climate-preference characteristics of sampled species and optimising them across the phylogeny, it is possible to reconstruct ancestral bioclimatic models (Yesson and Culham, 2006). In terms of grasses, for instance, existing data on fire regime (Bond et al., 2005) and global climate (BRIDGE project – www.ggy.bris.ac.uk/research/bridge) could be used to reconstruct ancestral fire regimes and palaeoclimates to evaluate how C_4 grass lineages responded to past climates. Then, based on these results, it would be possible to predict how human-induced climate change that influences rainfall seasonality, and thus fire regimes, may have an impact on C_4 grasses and savanna biomes.

Acknowledgements

The authors wish to thank the following for their assistance: W. Bond, G. A. Verboom, the South African National Parks Scientific Services and the National Research Foundation. They also thank Enterprise Ireland, the University of Cape Town and the Smuts Memorial Botanical Postdoctoral Fellowship for funding.

References

Barraclough, T. G. and Nee, S. (2001). Phylogenetics and speciation. *Trends in Ecology and Evolution*, **16**, 391–399.

Beerling, D. J. and Osborne, C. P. (2006). The origin of the savanna biome. *Global Change Biology*, **12**, 2023–2031.

Bobe, R. (2006). The evolution of arid ecosystems in eastern Africa. *Journal of Arid Environments*, **66**, 564–584.

Bond, W. J., Midgley, G. F. and Woodward, F. I. (2003). What controls South African vegetation: climate or fire? *South African Journal of Botany*, **69**, 79–91.

Bond, W. J., Woodward, F. I. and Midgley, G. F. (2005). The global distribution of ecosystems in a world without fire. *New Phytologist*, **165**, 525–538.

Bouchenak-Khelladi, Y., Salamin, N., Savolainen, V. et al. (2008). Large multi-gene phylogenetic trees of the grasses (Poaceae): progress towards complete tribal and generic level sampling. *Molecular Phylogenetics and Evolution*, **47**, 488–505.

Bouchenak-Khelladi, Y., Verboom, G. A., Hodkinson, T. R. et al. (2009).

The origins and diversification of C4 grasses and savanna-adapted ungulates. *Global Change Biology*, **15**, 2397–2417.

Bouchenak-Khelladi, Y., Verboom, G. A., Savolainen, V. and Hodkinson, T. R. (2010). Biogeography of the grasses (Poaceae): a phylogenetic approach to reveal evolutionary history in geographical space and geological time. *Botanical Journal of the Linnean Society*, **162**, 543–557.

Bremer, K. (2002). Gondwanan evolution of the grass alliance of families (Poales). *Evolution*, **56**, 1374–1387.

Campbell, N. A., Reece, J. B., Urry, L. A. et al. (2008). *Biology*, 8th edn. London: Pearson Benjamin Cummings.

Cerling, T. E. (2009). Paleorecords of C_4 plants and ecosystems. In *C_4 Plant Biology*, ed. R. F. Sage and R. K. Monson. San Diego, CA: Academic Press, pp. 445–469.

Cerling, T. E., Harris, J. M., MacFadden, B. J. et al. (1997). Global vegetation change through the Miocene/Pliocene boundary. *Nature*, **389**, 153–158.

Cerling, T. E., Ehleringer, J. R. and Harris, J. M. (1998). Carbon dioxide starvation, the development of C_4 ecosystems, and mammalian evolution. *Phylosophical Transactions of the Royal Society of London B*, **353**, 159–171.

Chan, K. M. A. and Moore, B. R. (2005). SYMMETREE: whole-tree analysis of differential diversification rates. *Bioinformatics*, **21**, 1709–1710.

Chapman, G. P. (1996). From extinct to present-day grasses. In *The Biology of Grasses*, ed. G. P. Chapman. Oxford: CAB International, pp. 112–123.

Christin, P. A., Besnard, G., Samaritani, E. et al. (2008). Oligocene CO_2 decline promoted C4 photosynthesis in grasses. *Current Biology*, **18**, 1–7.

Coughenour, M. B. (1985). Graminoid responses to grazing by large herbivores: adaptations, exaptations, and interacting processes. *Annals of the Missouri Botanical Gardens*, **72**, 852–863.

Donoghue, M. J. (2008). A phylogenetic perspective on the distribution of plant diversity. *Proceedings of the National Academy of Sciences of the USA*, **105**, 11549–11555.

Edwards, E. J. and Still, C. J. (2008). Climate, phylogeny and the ecological distribution of C_4 grasses. *Ecology Letters*, **11**, 266–276.

Edwards, E. J., Still, C. J. and Donoghue, M. J. (2007). The relevance of phylogeny to studies of global climate change. *Trends in Ecology and Evolution*, **22**, 243–249.

Ehleringer, J. R. and Monson, R. K. (1993). Evolutionary and ecological aspects of phtotosynthetic pathway variation. *Annual Review of Ecology and Systematics*, **24**, 411–439.

Graham, C. H., Ron, S. H., Santos, J. C., Schneider, C. J. and Moritz, C. (2004). Integrating phylogenetics and environmental niche models to explore speciation mechanisms in dendrobatid frogs. *Evolution*, **58**, 1781–1793.

Grass Phylogeny Working Group (GPWG) (2001). Phylogeny and subfamilial classification of the grasses (Poaceae). *Annals of the Missouri Botanical Gardens*, **88**, 373–457.

Hardy, C. R. (2006). Reconstructing ancestral ecologies: challenges and possible solutions. *Diversity and Distributions*, **12**, 7–19.

Hardy, C. R. and Linder, H. P. (2005). Intraspecific variability and timing in ancestral ecology reconstructions: a test case from the Cape flora. *Systematic Biology*, **54**, 299–316.

Hattersley, P. W. and Watson, L. (1992). Diversification of photosynthesis. In *Grass Evolution and Domestication*, ed. G. P. Chapman. Cambridge: Cambridge University Press.

Hodkinson, T. R., Savolainen, V., Jacobs, S. W. et al. (2007). Supersizing: progress in documenting and understanding grass richness. In *Towards the Tree of Life: Systematics of Species Rich Groups*, ed. T. R. Hodkinson and J. A. N. Parnell. Boca Raton, FL: Taylor and Francis CRC Press, pp. 279–298.

Hugall, A., Moritz, C., Moussalli, A. and Stanisic, J. (2002). Reconciling paleodistribution models and comparative phylogeography in the wet tropics rainforest land snail *Gnarosophia bellendenkerensis* (Brazier 1875). *Proceedings of the National Academy of Sciences of the USA*, **99**, 6112–6117.

Huntley, B. J. and Walker, B. H. (1982). *Ecology of Tropical Savannas*. Berlin: Springer.

Jacobs, B. F. (2004). Paleobotanical studies from tropical Africa: relevance to the evolution of forest, woodland, and savanna biomes. *Philosophical Transactions of the Royal Society of London B*, **359**, 1573–1583.

Jacobs, B. F., Kingston, J. D. and Jacobs, L. L. (1999). The origin of grass-dominated ecosystems. *Annals of the Missouri Botanical Gardens*, **86**, 590–643.

Janis, C. M. (1993). Tertiary mammal evolution in the context of changing climates, vegetation, and tectonic events. *Annual Review of Ecology and Systematics* **24**, 467–500.

Janis, C. M., Damuth, J. and Theodor, J. M. (2002). The origins and evolution of North American grassland biome: the story from the hoofed mammals. *Paleogeography, Paleoclimatology, Paleoecology*, **117**, 183–198.

Jernvall, J. and Fortelius, N. (2002). Common mammals drive the evolutionary increase of hypsodonty in the Neogene. *Nature*, **417**, 538–540.

Jernvall, J., Hunter, J. P. and Fortelius, M. (1996). Molar tooth diversity, disparity, and ecology in Cenozoic ungulate radiations. *Science*, **274**, 1489–1492.

Keeley, J. E. and Rundel, P. W. (2005). Fire and the Miocene expansion of C_4 grasslands. *Ecology Letters*, **8**, 683–690.

Kellogg, E. A. (2001). Evolutionary history of the grasses. *Plant Physiology*, **125**, 1198–1205.

Latorre, C., Quade, J. and McIntosh, W. C. (1997). The expansion of C_4 grasses and global change in the late Miocene: stable isotope evidence from the Americas. *Earth and Planetary Science Letters*, **146**, 83–96.

Long, S. P. and Jones, M. B. (1991). Introduction, aims, goals and general methods. In *Primary Productivity of Grass Ecosystems of the Tropics and Subtropics*, ed. S. P. Long, M. B. Jones and M. J. Roberts. London: Chapman and Hall, pp. 1–24.

MacFadden, B. J. and Cerling, T. E. (1994). Fossil horses, carbon isotopes and global change. *Trends in Ecology and Evolution*, **9**, 481–486.

Massey, F. P. and Hartley, S. E. (2006). Experimental demonstration of the anti-herbivore effects of silica in grasses: impacts on foliage digestibility and vole growth rates. *Proceedings of the Royal Society of London B*, **273**, 2299–2304.

Ortiz-Jaureguizar, E. and Cladera, G. A. (2006). Paleoenvironmental evolution of southern South America during the Cenozoic. *Journal of Arid Environments*, **66**, 385–388.

Osborne, C. P. (2008). Atmosphere, ecology and evolution: what drove the Miocene expansion of C4 grasslands. *Journal of Ecology*, **96**, 35–45.

Osborne, C. P. and Freckleton, R. P. (2009). Ecological selection pressures for C_4 photosynthesis. *Proceedings of the Royal Society of London B*, **276**, 1753–1760.

Pagani, M., Zachos, J., Freeman, K. H., Bohaty, S. and Tipple, B. (2005). Marked change in atmospheric carbon dioxide concentrations during the Oligocene. *Science*, **309**, 600–603.

Piperno, D. R. (2006). *Phytoliths: a Comprehensive Guide for Archeaologists and Paleo-ecologists*. Lanham, MD: Rowman Altamira.

Prasad, V., Strömberg, C. A. E., Alimohammadian, H. and Sahni, A. (2005). Dinosaur coprolites and the early evolution of grasses and grazers. *Science*, **310**, 1177–1180.

Raven, P. H., Evert, R. F. and Eichhorn, S. E. (2005). *Biology of Plants*, 7th edn. New York, NY: Freeman and Worth.

Roalson, E. H. (2008). C_4 photosynthesis: differentiating causation and coincidence. *Current Biology*, **18**, 167–168.

Rundel, P. W. (1980). The ecological distribution of C_4 and C_3 grasses in the Hawaiian islands. *Oecologia*, **45**, 354–359.

Sage, R. F. (2001). Environmental and evolutionary preconditions for the origin and diversification of the C_4 photosynthetic syndrome. *Plant Biology*, **3**, 202–213.

Sage, R. F. (2004). The evolution of C_4 photosynthesis. *New Phytologist*, **161**, 341–370.

Sage, R. F. and Monson, R. K. (1999). *C_4 Plant Biology*. San Diego, CA: Academic Press.

Salisbury, F. B. and Ross, C. W. (1992). *Plant Physiology*. Belmont, CA: Wadsworth.

Sánchez-Ken, J. G., Clark, L. G., Kellogg, E. A. and Kay, E. E. (2007). Reinstatement and emendation of subfamily Micrairoideae (Poaceae). *Systematic Botany*, **32**, 71–80.

Shaw, R. B. (2000). Tropical grasslands and savannas. In *Grasses, Systematics and Evolution*, ed. S. Jacobs and J. Everett. Melbourne: CSIRO, pp. 351–355.

Slack, C. R. and Hatch, M. D. (1967). Comparative studies on the activity of carboxylases and other enzymes in relation to the new pathway of photosynthetic carbon dioxide fixation in tropical grasses. *Biochemical Journal*, **103**, 660–665.

Slingsby, J. A. and Verboom, G. A. (2006). Phylogenetic relatedness limits coexistence at fine spatial scales: evidence from the schoenoid sedges (Cyperaceae: Schoeneae) of the Cape Floristic Region, South Africa. *American Naturalist*, **168**, 14–27.

Strömberg, C. A. E. (2005). Decoupled taxonomic radiation and ecological expansion of open-habitat grasses in the Cenozoic of North America. *Proceedings of the National Academy of Sciences of the USA*, **102**, 11980–11984.

Strömberg, C. A. E., Werdelin, L., Friis, E. M. and Sarac, G. (2007). The spread of grass-dominated habitats in Turkey and surrounding areas during the Cenozoic: phytolith evidence. *Paleogeology, Paleoclimatology, Paleoecology*, **250**, 18–49.

Sungkaew, S., Stapleton, C. M. A., Salamin, N. and Hodkinson, T. R. (2008). Non-monophyly of the woody bamboos (Bambuseae; Poaceae): a multi-gene region phylogenetic analysis of Bambuseae s.s. *Journal of Plant Research*, **122**, 95–108.

Taub, D. R. and Lerdau, M. T. (2000). Relationship between leaf nitrogen and photosynthetic rate for three NAD-ME and three NADP-ME C_4 grasses. *American Journal of Botany*, **87**, 412–417.

Tipple, B. J. and Pagani, M. (2007). The early origins of terrestrial C_4 photosynthesis. *Annual Review of Earth and Planetary Sciences*, **35**, 435–461.

Vicentini, A., Barber, J. C., Aliscioni, S. S., Giussani, L. M. and Kellogg, E. A. (2008). The age of grasses and clusters of C_4 photosynthesis. *Global Change Biology*, **14**, 1–15.

Walter, H. (1979). *Vegetation of the Earth and Ecological Systems of the Geobiosphere*. Heidelberg: Springer Verlag, pp. 66–71.

Yesson, C. and Culham, A. (2006). Phyloclimatic modeling: combining phylogenetics and bioclimatic modeling. *Systematic Biology*, **55**, 785–802.

Zazzo, A., Bocherens, H., Brunet, M. et al. (2000). Herbivore paleo-diet and paleo-environmental changes in Chad during the Pliocene using stable isotope ratios of tooth enamel carbonate. *Paleobiology*, **26**, 294–309.

Climate warming results in phenotypic and evolutionary changes in spring events: a mini-review

A. Donnelly

School of Natural Sciences, Trinity College Dublin, Ireland

A. Caffarra

Department of Environmental Sciences, Fondazione E. Mach, Istituto Agrario San Michele all'Adige, Italy

E. Diskin

School of Natural Sciences, Trinity College Dublin, Ireland

C. T. Kelleher

National Botanic Gardens, Glasnevin, Dublin, Ireland

A. Pletsers

School of Natural Sciences, Trinity College Dublin, Ireland

Climate Change, Ecology and Systematics, ed. Trevor R. Hodkinson, Michael B. Jones, Stephen Waldren and John A. N. Parnell. Published by Cambridge University Press.

H. Proctor

School of Natural Sciences, Trinity College Dublin, Ireland

R. Stirnemann

School of Natural Sciences, Trinity College Dublin, Ireland

M. B. Jones

School of Natural Sciences, Trinity College Dublin, Ireland

J. O'Halloran

Biological, Earth and Environmental Sciences,
University College Cork, Ireland

B. F. O'Neill

School of Natural Sciences, Trinity College Dublin, Ireland

J. Peñuelas

Center for Ecological Research and Forestry Applications (CSIC),
Campus Universitat Autònoma de Barcelona, Spain

T. Sparks

Fachgebiet für Ökoklimatologie, Technische Universität München, Germany and *Institute of Zoology, Poznań University of Life Sciences, Poland* and *Department of Zoology, University of Cambridge, UK*

Abstract

The impact of climate change, in particular increasing spring temperatures, on life-cycle events of plants and animals has gained scientific attention in recent years. Leafing of trees, appearance and abundance of insects, and migration of birds, across a range of species and countries, have been cited as phenotrends that are advancing in response to warmer spring temperatures. The ability of organisms to acclimate to variations in environmental conditions is known as phenotypic plasticity. Plasticity allows organisms to time developmental stages to coincide with optimum availability of environmental resources. There may, however, come a time when the limit of this plasticity is reached and the species needs to adapt genetically to survive. Here we discuss evidence of the impact of climate warming on plant, insect and bird phenology through examination of: (1) phenotypic plasticity in (a) bud burst in trees, (b) appearance of insects and (c) migration of birds; and (2) genetic adaptation in (a) gene expression during bud burst in trees, (b) the timing of occurrence of phenological events in insects and (c) arrival and breeding times of migratory birds. Finally, we summarise the potential consequences of future climatic changes for plant, insect and bird phenology.

8.1 Introduction

The recent resurgence of interest in phenology (the timing of recurring life-cycle events in plants and animals) has stemmed from research on the impact of climate change, in particular, global warming. As many life-cycle events are influenced by temperature, trends in the timing of phenophases, in both plants and animals, reflect the impact of warming on the environment. For example, when spring temperatures are relatively high, leaves on trees tend to emerge earlier than usual (Peñuelas and Filella, 2001; Donnelly et al., 2006; Menzel et al., 2006), insects appear earlier (Roy and Sparks, 2000; Stefanescu et al., 2003; Gordo and Sanz, 2006) and migratory birds arrive earlier (Lehikoinen et al., 2004; Sparks et al., 2005; Donnelly et al., 2009). These trends have been observed in many countries around the world (Peñuelas and Filella, 2001; Menzel et al., 2006). Long-term historic records of the timing of spring phenology provide us with indicators of climate change. The inclusion of pan-European phenological data (Menzel et al., 2006) in the 2007 Intergovernmental Panel on Climate Change *Fourth Assessment Report on Impacts, Adaptation and Vulnerability* (IPCC, 2007) illustrates the strength of these data in convincing policy makers that climate change is having a direct impact on the environment.

Traditionally, climate change and phenology studies have focused on examining historical trends in the timing of key phenophases and relating these to climate variables (Roy and Sparks, 2000; Menzel et al., 2006; Donnelly et al., 2009). While these trends provide some excellent environmental indicators of climate warming, phenological research is now moving towards an examination of genetic adaptation occurring through natural selection as a result of environmental pressure (Bearhop et al., 2005; Jonzén et al., 2006).

Phenotypic plasticity is a mechanism through which an organism can adjust the timing of development in response to environmental pressures (Bradshaw, 1965). It occurs over the short term and ensures the continued survival of organisms in a changing environment. It enables plants and animals to acclimate to seasonal changes in, for example, temperature. It also allows birds to adjust their behaviour according to real-time environmental cues, which they may experience at wintering grounds or during migration (Hüppop and Hüppop, 2003; Vähätalo et al., 2004). However, when an organism is no longer able to adjust to these environmental cues by way of phenotypic plasticity, selective pressure can result in adaptation at a population level. If there is sufficient selective pressure, climate change has the potential to result in genetic adaptation and eventually the evolution of new species. For example, recent evidence suggests that rapid climate change has been implicated in the evolution of poorly adapted wild bird populations in the UK (Charmantier et al., 2008).

Here we discuss evidence of the impact of climate warming on plant, insect and bird phenology through examination of (1) phenotypic plasticity in (a) bud burst in trees, (b) appearance of insects and (c) migration of birds; and (2) genetic adaptation in (a) gene expression during bud burst in trees, (b) the timing of occurrence of phenological events in insects and (c) arrival and breeding times of migratory birds. Finally, we summarise the potential consequences of future climatic changes for plant, insect and bird phenology.

Given the diversity of research methods used to study these discrete categories of organism, it was a challenge to find some common ground upon which to evaluate the effects of global warming on spring phenology. For example, we found in the literature many more research publications related to phenological traits of plant genes than either insect or bird genes, which hindered direct comparisons. However, we have chosen to adopt an interdisciplinary approach to address the issue. We believe that this is the first review that attempts to integrate the impact of rising temperature on three distinct categories of organism.

8.2 Plant phenology

With a rapidly changing climate that brings about rising average global temperatures, increasing frequency of extreme weather events, and new pests and diseases, trees may no longer be able to adapt to their changing environment. The only solutions for the survival of trees are to adapt rapidly to the changing conditions or to migrate to ecological niches that are more suitable, or alternatively to become extinct (Aitken et al., 2008). Shifts in plant phenology and species' ranges in response to changing temperature have been widely reported (Chmielewski and Rötzer, 2001; Parmesan, 2006; Cleland et al., 2007), while evidence of rapid evolutionary change in response to climate warming has also been found (Jump et al., 2006). As with other organisms, certain questions remain unanswered. How far can phenotypic plasticity stretch? What possibilities exist for genetic adaptation of populations? How drastic are the consequences of climate change in terms of migration and extinction?

8.2.1 Evidence of evolutionary responses to climate warming

Evolutionary responses to climate are apparent from biogeographic patterns. Species' distributions are generally defined by their climate envelope (Pearson and Dawson, 2003). Over time species have adapted and evolved in response to climatic pressures. Here, we focus on evidence from studies involving populations of tree species, as these often have wide distributions across multiple climatic zones. Many of these studies involve common garden experiments, where plants from different regions are grown at one site with common environmental conditions. These studies can be used to distinguish between environmental and genetic influences to determine, for example, if phenological events are genetically controlled. We report on current adaptations of populations and assess the potential for future changes in response to warming. We examine evidence of phenotypic plasticity as a strategy used by trees to respond to climate change and assess the underlying genetic mechanisms responsible for adaptation.

Variation across climatic gradients

Populations of different tree species across Eurasia can be used to demonstrate the evolutionary response of plants to climatic gradients. For example, species such as *Populus tremula* L. (aspen) are distributed from the Mediterranean to the subarctic, and across a broad longitudinal range from the European Atlantic coast to inland regions as far as continental Russia. These species are adapted to the local growing conditions as a result of an interaction between physical limitations and competition (Savolainen et al., 2007). Local adaptation to climate has been well documented in numerous tree species through provenance testing (Aitken et al., 2008). Plant species may be capable of surviving outside their normal range, but are prevented from doing so by competition with other species. This is clearly demonstrated by the fact that a species from a discrete climatic zone is able to grow, when placed in a series of botanic gardens, across a diverse range of environmental conditions in a number of continents. In the case of *P. tremula*, a common garden study of Swedish populations showed that trees from lower latitudes flushed earlier than ones from higher latitudes (Hall et al., 2007). A similar study on bud set of *Pinus* (pine) across a latitudinal gradient from Spain to Finland showed that the timing of bud set was determined by the origin of the plant rather than the conditions of growth (García-Gil et al., 2003).

Other studies confirm that phenology-related traits such as bud set and bud burst are highly heritable rather than plastic responses to changes in conditions (Yakovlev et al., 2006). Bradshaw and Stettler (1995) calculated that up to 98% of the total phenotypic variance of bud burst in *Populus* hybrids could be explained by heritability. Gene flow is a process that helps homogenise populations, but adaptive differentiation of populations in temperate and boreal forests persists despite substantial gene flow taking place (Savolainen et al., 2007). Although they are well adapted to their environment, forest trees can survive and grow outside their natural range (Savolainen et al., 2007), but different co-occurring tree species can also show considerable interspecific variation in their response to the environment they are adapted to (Lechowics, 1984; Ogaya and Peñuelas, 2007). Thus, while populations of trees do show phenotypic variation along climatic clines, questions remain as to the potential of an evolutionary response to rapid climate change within tree species. In particular, separating phenotypic plasticity from underlying genetic variation will be an important step in understanding phenological changes.

Phenotypic plasticity in phenological traits in plants

Recent phenotypic changes are well documented in tree phenology studies across Europe (Menzel et al., 2006). Tree species are responding to warmer spring temperatures and are adjusting bud burst to coincide with an earlier spring. Thus, trees are capitalising on earlier spring warming and benefiting from an extended growing season. However, whether these changes are due predominantly to phenotypic plasticity or to genetic variation is still unknown, although some studies are showing evidence of an inadequate capacity of trees to adapt swiftly enough to climate changes (Aitken et al., 2008).

Response to recent climate change: adaptation and genetic variation

Large-scale biogeographic patterns are the result of thousands of years of migration, adaptation and competition. However, recent climate change relates to a timescale spanning only decades. It is important to know whether or not plant species can respond to this rapid change in climatic conditions and, if so, how fast they can do this. Unlike insects and birds, plants cannot readily migrate when environmental conditions change rapidly. Potential outcomes of climate change on plant populations depend on their ability to change, which in turn, depends on phenotypic plasticity, underlying genetic variation, dispersal ability and establishment rates (Savolainen et al., 2007). The ability of plants to acclimate to a changing environment is evident from historical records of the timing of phenological events in the recent past in which tree species have shown earlier bud burst and flowering times across Europe in response to increases in spring temperatures (Menzel et al., 2006). As trees are long-living sedentary organisms they must withstand considerable variation in environmental conditions over their lifespan. In addition, as a species can be spread over a large range, each population must acclimate to local conditions to ensure survival. Therefore, there must be a balance between local adaptation (genotype) and phenotypic plasticity.

Some studies have assessed genetic variation related to climate and have shown population differentiation based on differing climatic conditions (e.g.

Jump and Peñuelas, 2005). A small number of studies have assessed levels of variation in phenology-related genes in wild plant populations across broad geographic ranges (García-Gil et al., 2003; Ingvarsson et al., 2006; Savolainen et al., 2007; Savolainen and Pyhäjärvi, 2007). These studies aim to identify selection pressure on genes in wild populations and to provide associations between gene variants and phenotypes. Association studies are a relatively new method in plant genetics, but it is expected that they will become useful in determining the mechanisms of complex traits and provide an understanding of the interaction between genotype and phenotype (Neale and Savolainen, 2004). A study carried out on *Populus tremula* showed significant variation in a gene involved in the control of light response in plants (phytochrome B2 – *phyB2*) along a latitudinal gradient (Ingvarsson et al., 2006). However, this was not the case in a homologue of *phyB2* in *Pinus sylvestris* L. populations across Europe (García-Gil et al., 2003). The difference in response between species highlights the need for further investigation in this area. These analyses are vital to understanding the genetic variation underlying phenological responses. Most studies have used a small sample size or a restricted geographical distribution, and therefore extrapolation of the results to different species or to larger areas is not feasible. Further studies will be key to identifying natural selection pressures resulting from current and future climate change, firstly to establish the potential inherent in species, and secondly to assess whether climate change results in selection of favoured variants.

8.2.2 Dormancy

In climate studies, the most commonly reported plant spring phenophase is leaf unfolding (Donnelly et al., 2006; Menzel et al., 2006). In temperate climates deciduous trees lose their leaves in autumn, remain dormant during winter months and resume growth in spring when environmental conditions become favourable. Dormancy is a complex process that allows trees to survive adverse conditions (Lang et al., 1987). Dormancy release is triggered by environmental cues and precedes leaf unfolding; thus, it is an important determinant of the timing of spring phenology in temperate trees. Bud dormancy is a mechanism induced by short day lengths and colder temperatures. It allows the vulnerable meristems to be protected and to cease growth while conditions are unfavourable (Lang et al., 1987). Bud burst has been shown to be controlled by the cumulative sum of temperatures to which buds are exposed after a requisite cold period. This relationship with temperature shows how dependent plants are on their environment and how they need to be finely tuned, or adapted, to the conditions in which they grow. Dormancy and bud burst may also be indirectly affected by climate change through, for example, changes in the availability of resources, such as carbon, soil nutrients and water (Rathcke and Lacey, 1985), susceptibility to embolism in early- versus late-leafing species, heterogeneity in leafing dates to avoid herbivory, and natural selection on interdependent traits. These factors can act as constraints on phenological patterns (i.e. limit the ability to evolve). It is therefore important to understand the mechanisms underlying dormancy, to enable predictions of how subsequent phenophases, such as bud burst and flowering, may be influenced by future climatic changes. In the following sections, we examine the various phases of dormancy and explore current knowledge on gene expression involved in its regulation.

Regulation of dormancy: molecular aspects and gene expression

Despite the role of dormancy in determining the time course of phenophases such as bud burst, flowering and bud set, our current knowledge of the molecular mechanisms involved in dormancy are still limited. The slow progress in understanding the biology of dormancy is perhaps due to the theory, prevalent in the last century, that the transition from dormancy to growth was the result of the balance between promoting and inhibiting hormones (Arora et al., 2003). However, this view has been questioned recently by a number of different studies showing that dormant status could not be unequivocally linked to hormone concentrations (Arora et al., 2003) and suggesting a multi-tiered control of dormancy in plants (Crabbe, 1994).

Current research on the impacts of climate change on plant phenology has focused on the climatic control of dormancy (Menzel and Fabian, 1999; IPCC, 2007). The advent of functional genomics and the availability of new technologies have resulted in a series of studies that have shed new light on possible pathways involved in dormancy induction and release (Yanovsky and Kay, 2002; Böhlenius et al., 2006; Rohde and Bhalerao, 2007; Ruttink et al., 2007). Many of these studies have used *Populus* as test material, as its genome has been sequenced and it is considered the model species for tree biology (Jansson and Douglas, 2007) in the same way as *Arabidopsis* is for most other plants.

Growth cessation and dormancy induction

The signal for growth cessation is triggered by detection of short days (SDs – i.e. days below a critical photoperiod) by phytochrome (*phy*) in plant leaves. Recent findings show that in *Populus tremula* the response to SDs is mediated by orthologues to the gene *Constans* (*CO*) and the floral integrator gene *Flowering Locus T* (*FT*) in *Arabidopsis* (Böhlenius et al., 2006). Transcript profiles from *P. tremula* grown with different critical photoperiods showed different diurnal oscillation patterns in the expression of *PtCO* (the *Populus CO* homologue). The importance of this finding was supported by the fact that the diurnal phase of *CO* expression affects flowering time and the expression of *FT* in *Arabidopsis*. In addition, SD-insensitive transgenic trees did not show any repression in *PtFT* expression. The findings of Böhlenius et al. (2006) support the view that the transition between long days (LDs) and SDs was perceived by the plant as the moment when the expression of

PtCO peaked in darkness. This, in turn, triggered a downregulation of *PtFT*, and induced growth cessation (Böhlenius et al., 2006). Regulation of growth and floral transition by *FT* has also been found in potato (Rodriguez-Falcon et al., 2006) and poplar (Hsu et al., 2006).

Phytochrome A (*phyA*), a photoreceptor that is sensitive to red light, also plays a role in growth cessation, through the regulation of *FT* and *CO* transcription (Yanovsky and Kay, 2002). Transgenic *P. tremula* overexpressing the oat *phyA* gene initiated growth cessation only in response to very short photoperiods of six hours, compared to the usual 14–16 hours (Olsen et al., 1997). The amount of *phyA* expressed by the plant might thus affect photoperiodic responses in trees (Olsen et al., 1997). Ruttink et al. (2007) applied metabolite and transcript profiling to bud samples taken at weekly intervals during the time course of the study from SD-induced dormancy induction to endodormancy in *P. tremulus* × *P. alba*. Their results showed that light, ethylene and abscissic acid transduction pathways consecutively controlled the transition from bud formation to acclimation and to dormancy. However, while considerable molecular and biochemical changes occurred in the first few weeks of SDs, changes in gene expression were not significant during the time of transition to endodormancy, suggesting that other factors may be involved (Ruttink et al., 2007).

Little is known about the changes occurring during the establishment of endodormancy. This particular phase is the final step in a series of transformations that are difficult to separate. Ruttink et al. (2007) showed that photoperiod was responsible for a signalling cascade, which probably triggered subsequent molecular events and activated different pathways. In an extensive survey, Rohde and Bhalerao (2007) examined gene expression during the induction, maintenance and release from dormancy in *P. tremula* × *P. alba*, and found a global change in expression patterns after 24 SDs, in accordance with previous findings (Ruttink et al., 2007). This termination of dormancy occurred concurrently with bud set and was thus related to changes in apical bud morphology (Rohde and Bhalerao, 2007). The transition to endodormancy was not marked by dramatic changes in gene expression, but a cluster of novel candidate genes was proposed for functions during chilling requirement. One of these is a DNA binding protein with linker-histone domains that has a potential regulatory role in dormancy release (Rohde and Bhalerao, 2007).

Once endodormancy is established, chilling temperatures are required to release it and restore growing ability. However, the control of this process is still a matter of speculation. The similarities between chilling fulfilment and vernalisation might suggest a role of genes that are orthologous to *Flowering Locus C* (*FLC*), whose repression in *Arabidopsis* occurs after exposure to cold temperatures (Sung and Amasino, 2005). Indeed, *FLC*-like genes have been found to be differentially expressed during dormancy release in poplar (Coleman and Chen, 2008).

Similar to endodormancy fixation, dormancy release is also characterised by the expression of DNA binding proteins (Yakovlev et al., 2006). This, and additional evidence for DNA methylation observed in potato buds (Law and Suttle, 2003), suggests that epigenetic changes (changes that can alter gene expression that do not depend on the DNA sequence but involve other types of chemical modifications) might play an important role in dormancy (Horvath et al., 2003).

8.2.3 Plant phenology conclusions

There is no doubt that plants have the ability to respond to climate warming through a plastic response in the timing of bud burst. However, there is no clear understanding of the limits of this phenotypic plasticity or of the underlying potential variation in the genes controlling it. Many studies have shown genetic variation associated with climate and phenology, but evidence for an evolutionary response to climate change remains scarce (Jump and Peñuelas, 2005; Jump et al., 2006, 2008). Tree species are known to be highly variable, but the extent of their ability to adapt genetically to rapid change is currently unknown. Although the importance of evolutionary adaptation in response to short-term climate change has been downplayed (Huntley, 2007), it is likely to be a key element in the long term. Short-term changes are likely to be dominated by plasticity, migration and selection of competitive species over others. In the longer term, evolution of tree species could be affected through climate change pressures selecting favoured variants.

The impact of future increases in temperature on dormancy is complex, as it has the potential to impact on both dormancy induction and release. Our understanding of the mechanisms underlying dormancy is improving rapidly, despite the fact that many of its components are still unclear. The information that molecular biology is offering for model plant systems like poplar is of critical value in unravelling the different processes and signals involved in dormancy induction and release. The integration of this knowledge into a coherent framework and the testing of functional hypotheses will be crucial to the development of a holistic model of dormancy that will help us to understand and mitigate the impacts of climate change on plants.

8.3 Insect phenology

Recent global warming has been cited as the driving force behind the advancement (i.e. earlier occurrence) of phenological events in insects, including the first appearance of cabbage root fly (Collier et al., 1991), butterflies (Roy and Sparks, 2000; Stefanescu et al., 2003) and bees (Gordo and Sanz, 2006). As is the case with plants and birds, there is a strong association in insects between development and temperature. When environmental change is so great that the limits of phenotypic plasticity are reached, genotypic change within a population becomes necessary to ensure survival (Tauber and Tauber, 1976; Bale et al., 2002; van Asch and Visser, 2007).

The following sections discuss how increasing spring temperature associated with climate change has been shown to influence phenological events in insects, thereby acting as the driving force behind evolution through natural selection. In addition, further impacts of climate change on insects are presented by examining evolutionary adaptation with reference to research that considers several of the intratrophic interactions exhibited by insects.

8.3.1 Evidence of evolutionary responses to climate warming

Much of the empirical evidence for the effects of climate change on natural selection in insects comes from fruit, pomace or vinegar flies (*Drosophila* spp.). Indeed, several studies on *Drosophila* have demonstrated genetic change in response to warming. In Australia (Umina et al., 2008) and the USA (Levitan and Etges, 2005), latitudinal variation in the chromosomal arrangements and allele frequencies, respectively, have been reported in *Drosophila* populations. In both of these studies, microevolution has occurred, as demonstrated by an increased selection for, and resultant increase in, the relative abundance of types that display greater ability to survive in warmer climates. Similar selection within warm-preferring populations has also been noted in Spain (Rodriguez-Trelles and Rodriguez, 1998). For such genetic change to occur, van Asch et al. (2007) suggested two prerequisites: (1) that significant genetic variation existed, and (2) that severe fitness consequences existed. However, none of the studies described above made specific reference to pressures on phenological events. Indeed, van Asch et al. (2007) reported that, with respect to phenology, there existed only a few examples of genetic change in response to climate change. It has also been suggested that the extent of genetic variation required for adaptation is largely unknown (Davis and Shaw, 2001). In their study of the pitcher plant mosquito (*Wyeomyia smithii* Coquillett, 1901), Bradshaw and Holzapfel (2001) reported genetic change as a response to an increase in the length of the growing season, suggesting that the change was evidence of a genetic response to climate warming.

Although the populations studied to date exhibited an ability to respond to increased temperatures, it is worthwhile considering the implications of further increases in temperature. It has been suggested that the large-scale viability of populations will be affected where natural selection cannot keep up with climate change (van Asch and Visser, 2007). Consideration of fitness consequences becomes particularly important when considering insects' interactions with other trophic levels on which their fitness is reliant. The potential for mismatch in interdependent phenophases exists when responses to climate change are asynchronous, meaning that the potential for negative fitness consequences is increased.

8.3.2 Insect–plant interactions

Insects often rely on mutualisms with a host plant (e.g. plant–pollinator), which in some cases are very specific, having coevolved over long periods of time (Herre et al., 1999; Pellmyr and Leebens-Mack, 1999). It is important to consider evolutionary processes in insects within the context of their interactions with plants. As a result of different responses to climate change between trophic levels, the synchrony of phenological events can be disrupted, resulting in mismatches either spatially (e.g. range shifts) or temporally (timing of phenological events) (Harrington et al., 1999; Stenseth et al., 2002; Edwards and Richardson, 2004; Both et al., 2006). Here we focus on temporal mismatches due to climate change.

Mismatches have been investigated for both plant–herbivore (insect) interactions (van Asch et al., 2007; Forkner et al., 2008) and plant–pollinator interactions (Memmott et al., 2007; Hegland et al., 2008). Van Asch and Visser (2007) considered the fitness consequences of hatching time in forest caterpillars, for which an optimum time exists, to ensure maximum feeding potential for the population. Asynchrony can disrupt this optimum, and negative fitness consequences may result. Hegland et al. (2008) suggested that such asynchrony can be projected by considering the 'potential for adaptation'. Where asynchronous changes are found, selection will occur for those individuals best able to match the changing environmental conditions at the same rate as the host plant (van Asch et al., 2007). The extent of mismatch, resulting through asynchrony and fitness consequences, is likely to be amplified by the degree of specificity of the relationship (specialist or generalist – Gilchrist, 1995; van Asch and Visser, 2007). Hegland et al. (2008) further described future synchrony through the use of models of interactions, suggesting that responses are not only driven by the organism's ability to respond to climate change itself, but are also compounded by the insect's ability to adapt to the changing nature of the interaction with its host plant.

8.3.3 Insect phenology conclusions

Hegland et al. (2008) warned that most models produced to project potential future plant–pollinator mismatches are only approximations, given the limited amount of research used to produce the models. This suggests the need for further research on the insects themselves, as well as the species with which they interact, in order to produce models that will be able to predict future interactions under future climate change conditions. Our understanding of insects' potential to evolve as a result of climate change is important in predicting future population trends both of insects and of interdependent species. It is therefore critical that future research should consider both genotypic and phenotypic flexibility, and be carried out over the long term to ensure that models are both accurate and reliable (Bale et al., 2002; van Asch et al., 2007).

8.4 Bird phenology

Migratory birds are particularly vulnerable to global climatic change, as their complex annual life cycle involves breeding, moult and two migration events (Pulido et al., 2001), all of which are influenced by temperature. In general, migratory birds respond to increasing spring temperature by arriving earlier at their breeding grounds (Hüppop and Hüppop, 2003; Lehikoinen et al., 2004; Sparks et al., 2005; Donnelly et al., 2009) and by laying their eggs earlier (Both and Visser, 2001; Both et al., 2006), thus increasing their potential for breeding success. However, some long-distance migrants, such as the willow warbler (*Phylloscopus trochilus* (Linnaeus, 1758)), have been shown to arrive later in response to increasing temperature at their breeding ground (Barrett, 2002).

In the following sections we consider evidence of climate-driven impacts on phenotypic plasticity in bird migration and on mismatches between interdependent phenophases. In addition, we present examples of evolutionary responses to warming through assortative mating and genetic selection for earlier breeding. Finally, we examine the consequences of earlier arrival for populations.

8.4.1 Phenotypic plasticity in phenological traits in birds

As is the case for plants and insects, phenotypic plasticity enables birds (both resident and migratory) to adjust the timing of their development in response to changing environmental conditions. Short-distance migrants are able to respond relatively quickly to environmental changes at the breeding grounds, but long-distance migrants may be constrained in their plastic responses by endogenous rhythms that control migration, as migration onset is unlikely to be directly linked to climate at the breeding ground (Visser et al., 1998; Cotton, 2003; Lehikoinen et al., 2004; Jonzén et al., 2006; Pulido, 2007). However, recent research proposed that trans-Saharan migrant birds may be able to gauge climatic conditions in the breeding grounds if meteorological conditions in Europe (during the breeding season) co-vary with those in Africa (during late winter) and phenotypically adjust their migration to optimise arrival time (Saino and Ambrosini, 2008). At present, it is unclear whether the observed changes in migratory behaviour that have been attributed to climatic variability are due to phenotypic plasticity, or whether they are a consequence of adaptive evolution (Pulido et al., 2001; Jonzén et al., 2006).

Recently, Both and te Marvelde (2007) compared geographical variation in egg-laying dates of a short-distance migrant (European starling – *Sturnus vulgaris* Linnaeus, 1758) and a long-distance migrant (pied flycatcher – *Ficedula hypoleuca*

(Pallas, 1764)) over a 25-year period in Europe. The authors reported that spatial and temporal heterogeneity in annual median egg-laying dates across Europe resulted from climate warming being stronger in some regions than in others. Møller et al. (2008) also reported that the impact of climate change on the timing of spring migration may have increased in recent years. Therefore, conditions along the migration routes appear to vary over space and time, and egg-laying date varies accordingly. It is thus evident that egg-laying date is strongly linked to temperature, which varies over space and time. As a result, within a species, phenological change may differ at different locations.

8.4.2 Mismatch between interdependent phenophases

As stated earlier, mismatches between interdependent phenophases can result in negative consequences for population survival. A study in the UK by Charmantier et al. (2008) has shown that, over a 47-year period, the egg-laying dates of a resident great tit (*Parus major* Linnaeus, 1758) population have been brought forward by two weeks. In addition, an advance in peak larval biomass of the winter moth (*Operopthera brumata* Linnaeus, 1758), a key food source for great tit offspring, has also been observed over the same time period. The authors have reported a strong correlation between these events and increasing spring temperature. Therefore, synchronisation between the egg-laying date and the maximal food source was found to be maintained over the study period. They concluded that changing environmental conditions resulted in phenotypic change alone, with no evidence of adaptive evolution being reported.

However, this was not found to be the case for other populations of passerine birds, where a mismatch in phenophases has been reported. For example, Visser et al. (1998) reported no advance in the egg-laying date of resident great tits in the Netherlands over a 23-year period, even though spring temperature increased, and both leaf and caterpillar emergence occurred earlier. The authors concluded that this mismatch caused increased natural selection for early-egg-laying birds, thus indicating a potential evolutionary change due to climate change. In addition, Both et al. (2005) reported a mismatch between arrival time in the Netherlands of the long-distance pied flycatcher and its main food source. These results suggest that warming temperatures are causing a mismatch between both resident and migratory birds and their main food sources.

Lyon et al. (2008) suggested a possible reason for the different responses in the timing of egg laying between the Dutch and UK tit populations. They proposed that the birds may be using different environmental cues to trigger egg laying. For example, birds that showed little plasticity in response to temperature (many of the Dutch population) may be using photoperiod as their sole cue. Accordingly, it is important to ensure that the cue under examination is the cue that is driving the phenological response.

It is evident that migratory birds depend on a plentiful food supply being available when they arrive at their breeding grounds. However, not all organisms respond to increasing spring temperature to the same degree, as response has been shown to be species-specific (Sparks and Tryjanowski, 2007). Therefore, as in the case of insects, any asynchrony between interdependent phenophases of different species (i.e. between arrival time of migratory birds and peak in abundance of their caterpillar food source) could have negative consequences on the reproductive success of birds and, thus, their population size.

Given the fact that insects are appearing earlier in warmer springs, it may also be the case that resident birds competing for the same food resource as their migrant counterparts have a competitive advantage, as they are already in situ when the insect food supply emerges (Ahola et al., 2007). This could, again, have negative implications for migratory birds, given that their food resource may be even further depleted by the time they arrive at their breeding grounds.

8.4.3 Evidence of evolutionary responses to climate warming

Assortative mating

Assortative mating is the non-random selection of mating partners with respect to one or more traits; it is positive when like phenotypes mate more frequently than would be expected by chance and is negative when the reverse occurs (Hartl and Jones, 2009). A recent study found evidence of assortative mating in populations of blackcaps (*Sylvia atricapilla* (Linnaeus, 1758)) in Europe (Bearhop et al., 2005). In the 1960s, blackcaps that spent their summers in Germany/Austria wintered in Iberia and northern Africa. However, since then, more and more of these birds have begun to overwinter in Britain and Ireland. Thus a change in migration pattern emerged. This resulted in the birds that spent the winter in Britain and Ireland arriving at their breeding grounds earlier, because critical photoperiods that trigger migration were found to be 10 days earlier than in more southern latitudes. In addition, because of the shorter migratory distance, these birds were possibly in better condition on arrival. The birds that arrived early tended to mate together and choose the best breeding territories, all of which resulted in greater reproductive success. The birds that arrived later also mated together, and therefore these two populations paired assortatively. According to Bearhop et al. (2005), this temporal separation can promote speciation. Consequently, it may be that changes in environmental conditions that result in new migration routes may lead to the evolution of new species. It is therefore likely that, for some birds, future climate warming that results in earlier arrival times at breeding grounds has the potential to lead to speciation, especially if coupled with other factors such as geographical allopatry.

Jonzén et al. (2006) suggest that climate-driven evolutionary change in migration is evident in long-distance migrant birds from Africa as their arrival time at breeding grounds in Europe has advanced. They argue that, because these birds reproduce from the age of one year and migration timing is heritable, there is potential for a rapid evolutionary response to climate change to occur. However, this claim is disputed by Both (2007). He agrees that evolutionary change is expected but, given the absence of evidence that early arrival and breeding are selected for (Jonzén et al., 2006), he considers phenotypic responses more likely than an evolutionary response in driving these processes.

It has been suggested by Bradshaw and Holzapfel (2006) that rapid climate change results in genetic change related to altered seasonal events rather than to higher temperature alone. The authors reported that genetic changes in a range of animals, including birds, are an adaptation to changes in seasonal events resulting in earlier reproduction. In addition, Sparks and Tryjanowski (2007) suggested that the earlier arrival time of the sand martin (*Riparia riparia* (Linnaeus, 1758)) in Britain may be an adaptive response to changes in food supply. Therefore, indirect effects of climate change should be borne in mind when considering observed phenological changes in bird migration.

Genetic selection for earlier breeding in migrant birds

The extent to which birds can, and are, tracking changing climatic conditions by altering the timing of reproduction has been explained largely by phenotypic plasticity (Wingfield et al., 1992). However, changing climatic conditions may also be selecting for changes in the frequency of genes that regulate the timing of reproduction in populations. This may allow species to adapt and move past the limits imposed by phenotypic plasticity (Nussey et al., 2005). However, in order for natural selection to occur and allow species to adapt to climatic warming over time, there must be a genetic foundation with sufficient genetic variability between individuals for directional selection of particular traits to take place (Kellermann et al., 2006). Therefore, phenotypic plasticity does not appear adequate to allow birds to match changes at lower trophic levels.

The degree of plasticity in the timing of reproduction has been shown to be a heritable trait (Nussey et al., 2005; Reed et al., 2008). Selection of this heritable component could allow some individuals to track climatic changes better than others. Selection of these individuals may enable the population to track food resources and reduce phenological mismatches beyond points imposed by current plastic limits (Stenseth and Mysterud, 2002). However, not all species may be able to select for these more plastic individuals. In some species, very little variation in plasticity occurs between individuals in a population (Nussey et al., 2005; Charmantier et al., 2008). This might not be indicative of a lack of plasticity, since populations may be highly plastic in response to a large-scale environmental cue (Reed et al., 2006). For instance, although the timing of breeding in the common guillemot (*Uria aalge* (Pontoppidan, 1763)) is highly plastic at the population level in response to the North Atlantic oscillation, there is very little variation between individuals (Reed et al., 2006). Instead, this lack of an individual response may be because social cues dominate the regulation of the timing of reproduction (Reed et al., 2006). In these species the dominance of sociality may limit an individual's potential response and the possibility for directional selection of a trait (Reed et al., 2006). This will affect the ability of these species to genetically adapt to changing climatic conditions, and they will only be able to rely on the degree of phenotypic plasticity already within the population.

There appears to be little direct genetic evidence of evolution in the timing of reproduction in birds. However, in wild bird populations the genetic components of variance in reproduction dates have been calculated using crossbreeding experiments and models that estimate genetic parameters (Kruuk, 2004; Nussey et al., 2005). Using this method, Nussey et al. (2005) concluded that significant genetic variation for laying-date plasticity existed in the Dutch Hoge Veluwe great tit population, and that laying-date plasticity was significantly heritable. The models used for quantitative genetics enabled an analysis of genetic (co)variances in populations in the wild (Kruuk, 2004). This technique relies on parents resembling offspring more closely than randomly sampled individuals from the population. However, closely related individuals are also more likely to experience similar environmental conditions. Therefore variation in the timing of reproduction may appear to be genetically based but may instead be due to environmental conditions. Despite these studies being very convincing, evidence at the DNA level would be useful to confirm that the observed trends in the timing of bird breeding do indeed have a genetic basis and that observed trends are not due to parental effects (Kruuk, 2004).

Even if some populations or species are tracking climatic changes at the genetic level, this is not feasible for all populations. For instance, no heritable variation has been shown in the plastic responses of reproduction in collared flycatchers (*Ficedula albicollis* (Temminck, 1815)) (Brommer et al., 2005). If traits for earlier reproduction are not heritable, or if only a small proportion of individuals carry the genetic traits required for natural selection, selective processes may not be possible or may be too slow to allow species to track climatic changes (Nussey et al., 2005). Species that rely on cues that do not reliably indicate changing seasons or physiological requirements are particularly at risk, as evolutionary processes may have little or nothing to select upon. For example, migratory bird species which rely on large-scale climatic patterns may be relying on cues at the wintering ground that no longer match seasonal conditions at the breeding grounds. The capacity for evolutionary change in phenological events may enable some species or populations to reduce mismatches and ultimately increase chances of population viability (Stenseth and Mysterud, 2002; Walther et al., 2002). However, it remains to be seen if evolutionary change can occur fast enough to keep up with the rate of change observed in environmental conditions.

8.4.4 Consequences of early arrival for populations

The timing of arrival at breeding grounds has implications for population success. The earlier arrivals have a better chance of finding a mate and securing the best territory, which, as we have seen, can lead to assortative mating. In addition, if annual and in particular overwinter survival is higher because of warmer temperatures, resident birds can begin to breed earlier, which may result in a depletion of resources by the time migrants arrive. Mismatches in the timing of interdependent phenophases can result in reduced reproductive output, leading to a decline in population size and an ultimate risk of extinction. It has been shown that, of 100 European migratory bird species examined, those that did not demonstrate a phenological response to climate change over the period 1990–2000 showed a declining population trend (Møller et al., 2008). Population sizes that had either stable or increasing trends showed a phenological response in the form of an advance in mean arrival date. However, Møller et al. (2008) also reported that mean arrival date was not a good predictor of population trend for an earlier period (1970–90) and concluded that the impact of climate change on spring migration may have increased in recent years.

8.4.5 Bird phenology conclusions

It appears that, in different migrant bird species, there is strong evidence indicating that phenotypic responses have occurred as a consequence of climate warming. However, the rate and magnitude of this change is both species- and population-specific. Pulido (2007) suggested that selection for early breeding and arrival was likely to increase if the trend of increasing temperature persisted. The most likely evolutionary processes leading to a change in migration timing are adaptive changes in migration distance and changes in phenotypic plasticity of departure date in response to day length at the wintering grounds. In addition, over recent decades, climate change has led to a number of heritable genetic changes in bird populations as a result of both direct and indirect impacts.

8.5 Consequences of future climate change for plant and animal phenology

It is evident from this review that climate change is having a detectable impact on spring phenophases of both plants and animals. The effects reported result from either direct effects of increasing temperature causing an advance in the timing of spring events, or indirect effects on interdependent organisms causing a mismatch between dependent phenophases, thus having negative consequences on population size of insects and birds. As temperature is predicted to rise over the coming decades, it is assumed that these changes in phenology will continue.

The degree to which organisms can respond will be dependent on how far their phenotypic plasticity can stretch. This characteristic is, of course, species-specific. In addition, we expect to see climate change exerting selective pressure on organisms that adapt to the changing environmental conditions through heritable genetic changes. This may, in turn, under some circumstances lead to speciation. If organisms cannot keep pace with the changes in their environmental conditions, by means of either phenotypic plasticity or genetic adaptation, the consequences will be severe and extinction is a clear possibility.

Acknowledgements

The authors would like to thank Dr Trevor Hodkinson, Professor John Parnell and Dr Sinéad Boyce for their comments, which greatly enhanced this manuscript. In addition we would like to express our gratitude to the Irish Environmental Protection Agency for providing financial assistance for this work, under the STRIVE programme Climate Change Impacts on Phenology: Implications for Terrestrial Ecosystems (project number 2007-CCRP-2.4).

References

Ahola, M. P., Laaksonen, T., Eeva, T. and Lehikoinen, E. (2007). Climate change can alter competitive relationships between resident and migratory birds. *Journal of Animal Ecology*, **76**, 1045–1052.

Aitken, S. N., Yeaman, S., Holliday, J. A., Wang, T. and Curtis-McLane, S. (2008). Adaptation, migration or extirpation: climate change outcomes for tree populations. *Evolutionary Applications*, **1**, 95–111.

Arora, R., Rowland, L. J. and Tanino, K. (2003). Induction and release of bud dormancy in woody perennials, a science comes of age. *Hortscience*, **38**, 911–921.

Bale, J. S., Masters, G. J., Hodkinson, I. D. et al. (2002). Herbivory in global climate change research: direct effects of rising temperature on insect herbivores. *Global Change Biology*, **8**, 1–16.

Barrett, R. T. (2002). The phenology of spring bird migration to north Norway. *Bird Study*, **49**, 270–277.

Bearhop, S., Fiedler, W., Furness, R. W. et al. (2005). Assortative mating as a mechanism for rapid evolution of a migratory divide. *Science*, **310**, 502–504.

Böhlenius, H., Huang, T., Charbonnel-Campaa, L. et al. (2006). The conserved CO/FT regulatory module controls timing of flowering and seasonal growth cessation in trees. *Science*, **312**, 1040–1043.

Both, C. (2007). Comment on 'rapid advance of spring arrival dates in long-distance migratory birds'. *Science*, **315**, 1959–1961.

Both, C. and te Marvelde, L. (2007). Climate change and timing of avian breeding and migration through Europe. *Climate Research*, **35**, 93–105.

Both, C. and Visser, M. (2001). Adjustment to climate change is constrained by arrival date in a long-distance migrant bird. *Nature*, **411**, 296–298.

Both, C., Bijlsma, R. G. and Visser, M. (2005). Climatic effects on timing of spring migration and breeding in a long-distance migrant, the pied flycatcher *Ficedula hypoleuca*. *Journal of Avian Biology*, **36**, 368–373.

Both, C., Bouwhuis, S., Lessells, C. M. and Visser, M. E. (2006). Climate change and population declines in a long-distance migratory bird. *Nature*, **441**, 81–83.

Bradshaw, A. D. (1965). Evolutionary significance of phenotypic plasticity in plants. *Advances in Genetics*, **13**, 115–155.

Bradshaw, H. D. and Stettler, R. F. (1995). Molecular genetics of growth and development in *Populus*. IV. Mapping QTLs with large effects on growth, form, and phenology traits in a forest tree. *Genetics*, **139**, 963–973.

Bradshaw, W. E. and Holzapfel, C. M. (2001). Genetic shift in photoperiodic response correlated with global warming. *Proceedings of the National Academy of Sciences of the USA*, **98**, 14509–14511.

Bradshaw, W. E. and Holzapfel, C. M. (2006). Evolutionary response to rapid climate change. *Science*, **312**, 1477–1478.

Brommer, J. E., Merilä, J., Sheldon, B. C. and Gustafsson, L. (2005). Natural selection and genetic variation for reproductive reaction norms in a wild bird population. *Evolution*, **59**, 1362–1371.

Charmantier, A., McCleery, R. H., Cole, L. R. et al. (2008). Adaptive phenotypic plasticity in response to climate change in a wild bird population. *Science*, **322**, 800–803.

Chmielewski, F. M. and Rö tzer, T. (2001). Responses of tree phenology to climatic changes across Europe. *Agricultural and Forest Meteorology*, **108**, 101–112.

Cleland, E. E., Chuine, I., Menzel, A., Mooney, H. A. and Schwartz, M. D. (2007). Shifting plant phenology in response to global change. *Trends in Ecology and Evolution*, **22**, 357–365.

Coleman, G. D. and Chen, K. Y. (2008). Regulation of FLC-like genes and bud dormancy in poplar. Plant and Animal Genome Conference XVI, San Diego, CA. www.intl-pag.org/16/abstracts/PAG16_W10_80.html.

Collier, R. H., Finch, S., Phelps, K. and Thompson, A. R. (1991). Possible impact of global warming on cabbage root fly (*Delia radicum*) activity in the UK. *Annals of Applied Biology*, **118**, 261–271.

Cotton, P. A. (2003). Avian migration phenology and global climate change. *Proceedings of the National Academy of Sciences of the USA*, **100**, 12219–12222.

Crabbe, J. (1994). Dormancy. *Encyclopedia of Agricultural Sciences. Volume I*. New York, NY: Academic Press.

Davis, M. B. and Shaw, R. G. (2001). Range shifts and adaptive responses to Quaternary climate change. *Science*, **292**, 673–679.

Donnelly, A., Salamin, N. and Jones, M. B. (2006). Changes in tree phenology: an indicator of spring warming in Ireland? *Biology and Environment: Proceedings of the Royal Irish Academy*, **106**, 47–55.

Donnelly, A., Cooney, T., Jennings, E., Buscardo, E. and Jones, M. B. (2009). Response of birds to climatic variability; evidence from the western fringe of Europe. *International Journal of Biometeorology*, **53**, 211–220.

Edwards, M. and Richardson, A. J. (2004). Impact of climate change on marine pelagic phenology and trophic mismatch. *Nature*, **430**, 881–884.

Forkner, R. E., Marquis, R. J., Lill, J. T. and Le Corff, J. (2008). Timing is everything? Phenological synchrony and population variability in leaf-chewing herbivores of *Quercus*. *Ecological Entomology*, **33**, 276–285.

Garcia-Gil, M. R., Mikkonen, M. and Savolainen, O. (2003). Nucleotide diversity at two phytochrome loci along a latitudinal cline in *Pinus sylvestris*. *Molecular Ecology*, **12**, 1195–1206.

Gilchrist, G. W. (1995). Specialists and generalists in changing environments. I. Fitness landscapes of thermal sensitivity. *American Naturalist*, **146**, 252–270.

Gordo, O. and Sanz, J. J. (2006). Temporal trends in phenology of the honey bee *Apis mellifera* (L.) and the small white *Pieris rapae* (L.) in the Iberian Peninsula (1952–2004). *Ecological Entomology*, **31**, 261–268.

Hall, D., Luquez, V., Garcia, V. M. et al. (2007). Adaptive population differentiation in phenology across a latitudinal gradient in European aspen (*Populus tremula*, L.): a comparison of neutral markers, candidate genes and phenotypic traits. *Evolution*, **61**, 2849–2860.

Harrington, R., Woiwod, I. and Sparks, T. (1999). Climate change and trophic interactions. *Trends in Ecology and Evolution*, **14**, 146–150.

Hartl, D. L. and Jones, E. W. (2009). *Genetics: Analysis of Genes and Genomes*, 7th edn. Boston, MA: Jones and Bartlett.

Hegland, S. J., Nielsen, A., Lázaro, A., Bjerknes, A. and Totland, Ø. (2008). How does climate warming affect plant–pollinator interactions? *Ecology Letters*, **12**, 184–195.

Herre, E. A., Knowlton, N., Mueller, U. G. and Rehner, S. A. (1999). The evolution of mutualisms: exploring the paths between conflict and cooperation. *Trends in Ecology and Evolution*, **14**, 49–53.

Horvath, D. P., Anderson, J. V., Chao, W. S. and Foley, M. F. (2003). Knowing when to grow: signals regulating bud dormancy. *Trends in Plant Science*, **8**, 534–540.

Huntley, B. (2007). Evolutionary response to climate change? *Heredity*, **98**, 247–248.

Hüppop, O. and Hüppop, K. (2003). North Atlantic Oscillation and timing of spring migration in birds. *Proceedings of the Royal Society of London B*, **270**, 233–240.

Hsu, C. Y., Liu, Y., Luthe, D. S. and Yuceer, C. (2006). Poplar FT2 shortens the juvenile phase and promotes seasonal flowering. *Plant and Cell Physiology*, **49**, 291–300.

Ingvarsson, P. K., García, M. V., Hall, D., Luquez, V. and Jansson, S. (2006). Clinal variation in *phyB2*, a candidate gene for day-length-induced growth cessation and bud set, across a latitudinal gradient in European aspen (*Populus tremula*). *Genetics*, **172**, 1845–1853.

Intergovernmental Panel on Climate Change (IPCC) (2007). *Climate Change 2007: Impacts, Adaptation and Vulnerability. Contribution of Working Group II to the Fourth Assessment Report of the Intergovernmental Panel on Climate Change*, ed. M. Parry, O. Canziani, J. Palutikof, P. van der Linden and C. Hanson. Cambridge: Cambridge University Press.

Jansson, S. and Douglas, C. J. (2007). *Populus*: a model system for plant biology. *Annual Review of Plant Biology*, **58**, 435–458.

Jonzén, N., Lindén, A., Ergon, T. et al. (2006). Rapid advance of spring arrival dates in long-distance migratory birds. *Science*, **312**, 1959–1961.

Jump, A. S. and Peñuelas, J. (2005). Running to stand still: adaptation and the response of plants to rapid climate change. *Ecology Letters*, **8**, 1010–1020.

Jump, A. S., Hunt, J. M., Martinez-Izquierdo, J. A. and Peñuelas, J. (2006). Natural selection and climate change: temperature-linked spatial and temporal trends in gene frequency in *Fagus sylvatica*. *Molecular Ecology*, **15**, 3469–3480.

Jump, A. S., Peñuelas, J., Rico, L. et al. (2008). Simulated climate change provokes rapid genetic change in the Mediterranean shrub *Fumana thymifolia*. *Global Change Biology*, **14**, 637–643.

Kellermann, V. M., van Heerwaarden, B., Hoffmann, A. A. and Sgrò, C. M. (2006). Very low additive genetic variance and evolutionary potential in multiple populations of two rainforest *Drosophila* species. *Evolution*, **60**, 1104–1108.

Kruuk, L. E. B. (2004). Estimating genetic parameters in natural populations using the 'animal model'. *Philosophical Transactions of the Royal Society of London B*, **359**, 873–890.

Lang, G. A., Early, J. D., Martin, G. C. and Darnell, R. L. (1987). Endodormancy, paradormancy, and ecodormancy – physiological terminology and classification for dormancy research. *Hortscience*, **22**, 371–377.

Law, R. D. and Suttle, J. C. (2003). Transient decreases in methylation at 5'-cCGG-3'sequences in potato (*Solanum tuberosum*) meristem DNA during progression of tubers through dormancy precede the resumption of sprout growth. *Plant Molecular Biology*, **51**, 437–447.

Lechowics, M. J. (1984). Why do temperate deciduous trees leaf out at different times? Adaptation and ecology of forest communities. *American Naturalist*, **124**, 821–842.

Lehikoinen, E., Sparks, T. H. and Zalakevicius, M. (2004). Arrival and departure dates. In *Birds and Climate Change. Advances in Ecological Research*, Vol. 35, ed. A. P. Møller, W. Fiedler and P. Berthold. London, UK: Academic Press, pp. 1–31.

Levitan, M. and Etges, W. J. (2005). Climate change and recent genetic flux in populations of *Drosophila robusta*. *BMC Evolutionary Biology*, **5**, 4.

Lyon, B. E., Chaine, A. S. and Winkler, D. W. (2008). A matter of timing. *Science*, **321**, 1051–1052.

Memmott, J., Craze, P. G., Waser, N. M. and Price, M. V. (2007). Global warming and the disruption of plant–pollinator interactions. *Ecology Letters*, **10**, 710–717.

Menzel, A. and Fabian, P. (1999). Growing season extended in Europe. *Nature*, **397**, 659.

Menzel, A., Sparks, T. H., Estrella, N. et al. (2006). European phenological response to climate change matches the warming pattern. *Global Change Biology*, **12**, 1–8.

Møller, A. P., Rubolini, D. and Lehikoinen, E. (2008). Populations of migratory bird species that did not show a phenological response to climate change are declining. *Proceedings of the National Academy of Sciences of the USA*, **105**, 16195–16200.

Neale, D. B. and Savolainen, O. (2004). Association genetics of complex traits in conifers. *Trends in Plant Science*, **9**, 325–330.

Nussey, D. H., Postma, E., Gienapp, P. and Visser, M. E. (2005). Selection on heritable phenotypic plasticity in a wild bird population. *Science*, **310**, 304–306.

Ogaya, R. and Peñuelas, J. (2007). Tree growth, mortality, and above-ground biomass accumulation in a holm oak forest under a five-year experimental field drought. *Plant Ecology*, **189**, 291–299.

Olsen, J. E., Junttila, O., Nilsen, J. et al. (1997). Ectopic expression of oat phytochrome A in hybrid aspen changes critical daylength for growth and prevents cold acclimatization. *The Plant Journal*, **12**, 1339–1350.

Parmesan, C. (2006). Ecological and evolutionary responses to recent climate change. *Annual Review of Ecology, Evolution, and Systematics*, **37**, 637–669.

Pearson, R. G. and Dawson, T. P. (2003). Predicting the impacts of climate change on the distribution of species: are bioclimate envelope models useful? *Global Ecology and Biogeography*, **12**, 361–371.

Pellmyr, O. and Leebens-Mack, J. (1999). Forty million years of mutualism: evidence for Eocene origin of the yucca–yucca moth association. *Proceedings of the National Academy of Sciences of the USA*, **96**, 9178–9183.

Peñuelas, J. and Filella, I. (2001). Responses to a warming world. *Science*, **294**, 793–795.

Pulido, F. (2007). Phenotypic changes in spring arrival: evolution, phenotypic plasticity, effects of weather and condition. *Climate Research*, **35**, 5–23.

Pulido, F., Berthold, P., Mohr, G. and Querner, U. (2001). Heritability of the timing of autumn migration in a natural bird population. *Proceedings of the Royal Society of London B*, **268**, 953–959.

Rathcke, B. and Lacey, E. P. (1985). Phenological patterns of terrestrial plants. *Annual Review of Ecology and Systematics*, **16**, 179–214.

Reed, T. E., Wanless, S., Harris, M. P. et al. (2006). Responding to environmental change: plastic responses vary little in a synchronous breeder. *Proceedings of the Royal Society of London B*, **273**, 2713–2719.

Reed, T. E., Kruuk, L. E., Wanless, S. et al. (2008). Reproductive senescence in a long-lived seabird: rates of decline in late-life performance are associated with varying costs of early reproduction. *American Naturalist*, **171**, 89–101.

Rodriguez-Falcon, M., Bou, J. and Prat, S. (2006). Seasonal control of tuberization in potato: conserved elements with the flowering response. *Annual Review of Plant Biology*, **57**, 151–180.

Rodriguez-Trelles, F. and Rodriguez, M. A. (1998). Rapid micro-evolution and loss of chromosomal diversity in *Drosophila* in response to climate warming. *Evolutionary Ecology*, **12**, 829–838.

Rohde, A. and Bhalerao, R. P. (2007). Plant dormancy in the perennial context. *Trends in Plant Science*, **12**, 217–223.

Roy, D. B. and Sparks, T. H. (2000). Phenology of British butterflies and climate change. *Global Change Biology*, **6**, 407–416.

Ruttink, T., Arend, M., Morreel, K. et al. (2007). A molecular timetable for apical bud formation and dormancy induction in poplar. *The Plant Cell*, **19**, 2370–2390.

Saino, N. and Ambrosini, R. (2008). Climatic connectivity between Africa and Europe may serve as a basis for phenotypic adjustment of migration schedules of trans-Saharan migratory birds. *Global Change Biology*, **14**, 250–263.

Savolainen, O. and Pyhäjärvi, T. (2007). Genomic diversity in forest trees. *Current Opinion in Plant Biology*, **10**, 162–167.

Savolainen, O., Pyhäjärvi, T. and Knürr, T. (2007). Gene flow and local adaptation in trees. *Annual Review of Ecology, Evolution, and Systematics*, **38**, 595–619.

Sparks, T. H. and Tryjanowski, P. (2007). Patterns of spring arrival dates differ in two hirundines. *Climate Research*, **35**, 159–164.

Sparks, T. H., Bairlein, F., Bojarinova, J. G. et al. (2005). Examining the total arrival distribution of migratory birds. *Global Change Biology*, **11**, 22–30.

Stefanescu, C., Peñuelas, J. and Filella, I. (2003). Effects of climatic change on the phenology of butterflies in the northwest Mediterranean Basin. *Global Change Biology*, **9**, 1494–1506.

Stenseth, N. C. and Mysterud, A. (2002). Climate, changing phenology, and other life history traits: non-linearity and match-mismatch to the environment. *Proceedings of the National Academy of Sciences of the USA*, **99**, 13379–13381.

Stenseth, N. C., Mysterud, A., Ottersen, G. et al. (2002). Ecological effects of climate fluctuations. *Science*, **23**, 1292–1296.

Sung, S. and Amasino, R. M. (2005). Remembering winter: toward a molecular understanding of vernalization. *Annual Review of Plant Biology*, **56**, 491–508.

Tauber, M. J. and Tauber, C. A. (1976). Insect seasonality: diapause maintenance, termination, and postdiapause development. *Annual Review of Entomology*, **21**, 81–107.

Umina, P. A., Weeks, A. R., Kearney, M. R., McKechnie, S. W. and Hoffmann, A. A. (2008). A rapid shift in classical clinal pattern in *Drosophila* reflecting climate change. *Science*, **308**, 691–693.

Vähätalo, A. V., Rainio, K., Lehikoinen, A. and Lehikoinen, E. (2004). Spring arrival of birds depends on the North Atlantic Oscillation. *Journal of Avian Biology*, **35**, 210–216.

van Asch, M. and Visser, M. E. (2007). Phenology of forest caterpillars and their host trees: the importance of synchrony. *Annual Review of Entomology*, **52**, 37–55.

van Asch, M., van Tienderen, P. H., Holleman, L. J. M. and Visser, M. E. (2007). Predicting adaptation of phenology in response to climate change, an insect herbivore example. *Global Change Biology*, **13**, 1596–1604.

Visser, M. E., van Noordwijk, A. J., Tinbergen, J. M. and Lessells, C. M. (1998). Warmer springs lead to mistimed reproduction in great tits (*Parus major*). *Proceedings of the Royal Society of London B*, **265**, 1867–1870.

Walther, G. R., Post, E., Convey, P. et al. (2002). Ecological responses to recent climate change. *Nature*, **416**, 389–395.

Wingfield, J. C., Hahn, T. P., Levin, R. and Honey, P. (1992). Environmental predictability and control of gonadal cycles in birds. *Journal of Experimental Zoology*, **261**, 214–231.

Yakovlev, I. A., Fossdal, C. G., Johnsen, O., Junttila, O. and Skrøppa, T. (2006). Analysis of gene expression during bud burst initiation in Norway spruce via ESTs from subtracted cDNA libraries. *Tree Genetics and Genomes*, **2**, 1614–2950.

Yanovsky, M. J. and Kay, S. A. (2002). Molecular basis of seasonal time measurement in *Arabidopsis*. *Nature*, **419**, 308–312.

9

Terrestrial green algae: systematics, biogeography and expected responses to climate change

F. Rindi

Dipartimento di Scienze del Mare,
Università Politecnica delle Marche, Ancona, Italy

Abstract

Terrestrial green microalgae are among the most widespread and evolutionarily diverse organisms inhabiting terrestrial environments. In the last 30 years, ultrastructural and molecular data have led to important insights into the evolution of these organisms. It has become clear that terrestrial green algae are a highly polyphyletic group originating from the colonisation of terrestrial environments by many separate lineages of aquatic algae, both freshwater and marine. Such diversity implies great differences in physiological and biochemical attributes, with the consequence that different taxa are expected to exhibit different responses to climatic changes. Elevated carbon dioxide (CO_2), variations in rainfall and humidity and increased photosynthetically active radiation (PAR) and ultraviolet (UV) radiation are the aspects of global change that will most likely affect terrestrial green algae. The published information on impacts of global change is largely based

Climate Change, Ecology and Systematics, ed. Trevor R. Hodkinson, Michael B. Jones, Stephen Waldren and John A. N. Parnell. Published by Cambridge University Press.

on short-term studies, which have examined the immediate response of algae to experimental manipulation of climatic parameters. However, recent experimental long-term studies have shown that green microalgae evolve in response to climatic change, and the physiological responses of algal strains in present-day conditions might not reflect the responses of the same strains in future climate scenarios.

9.1 Introduction

As generally defined, the term algae includes all photosynthetic eukaryotes with the exception of land plants (Brodie and Zuccarello, 2007; Delwiche, 2007). Members of this highly diverse, non-monophyletic set of organisms occur in any habitat in which sufficient photon irradiance for photosynthesis is available, and they contribute to global primary production to an extent which may reach 50% (Beardall and Raven, 2004).

Due to their ecological importance, algae have received great consideration in studies concerning climate change. Most studies have focused on phytoplankton communities, which, due to their short generation times and fast turnover, respond quickly to environmental changes. Variations in phytoplankton communities linked to climate change have been documented (Falkowski et al., 2004; Tozzi et al., 2004; Servais et al., 2008) and predictions on the effects of climate change on their composition have been proposed (Falkowski and Oliver, 2007). Many algal groups do not produce any fossils because they possess cell walls made of soft materials. However, some groups with resistant cell walls, such as diatoms, dinoflagellates and coccolithophorids, have left a well-documented fossil record (Falkowski et al., 2004). Some recent studies based on fossil records have suggested that climatic changes have played a major role in phytoplankton evolution (Miller et al., 2005). In particular, climatically induced changes in oceanic mixing have driven macroevolutionary changes in size and other morphological characters of several microalgal groups (Finkel et al., 2005, 2007).

Terrestrial algae represent a heterogeneous assemblage of microscopic organisms belonging primarily to two groups: the green algae (Chlorophyta and Streptophyta) and the diatoms (Bacillariophyceae, Ochrophyta). Compared to living in the aquatic environment, life for algae in terrestrial habitats involves exposure to harsher conditions, such as near complete desiccation, wider and faster variations of temperature, stronger insolation and exposure to higher levels of UV radiation (Reisser and Houben, 2001; Gray et al., 2007; Karsten et al., 2007a). Thus, terrestrial algae are arguably affected more directly than aquatic algae by climatic changes and can be expected to respond in a more immediate way. It is therefore surprising that terrestrial algae have been largely neglected in studies related to climate change. Here, the systematics and biogeography of terrestrial green algae are reviewed and considered in relation to matters of climatic change.

9.2 Systematics and evolution of terrestrial green algae

Green algae are photosynthetic eukaryotes bearing double membrane-bound plastids containing chlorophyll *a* and *b*, accessory pigments found in embryophytes (betacarotene and xanthophylls) and a unique stellate structure linking nine pairs of microtubules in the flagellar base (Lewis and McCourt, 2004). Molecular dating places the most recent common ancestor of all extant green algae at 1100 million to 1200 million years ago (Yoon et al., 2004). In general, reconstruction of the past evolution of these organisms is problematic. The fossil record is very scanty and almost entirely circumscribed to a few groups with calcified cell walls (such as the marine orders Dasycladales and Bryopsidales – Verbruggen et al., 2009). This prevents the possibility of reliable phylogenetic calibrations and reconstruction of ancestral character states.

Microscopic green algae are distributed in marine, freshwater and terrestrial habitats. In terrestrial environments, they represent a major component of the microbial flora occurring on exposed aerial surfaces and occur in almost every habitat, including the most extreme, such as surfaces of urban buildings (Rindi, 2007), biotic crusts in hot deserts (Lewis and Flechtner, 2002; Flechtner, 2007), Antarctic rocks covered by ice for most of the time (Broady, 1996) and air at an altitude of 2000 m (Sharma et al., 2007). Species of green microalgae occurring on aerial surfaces are among the earliest-described organisms. Some species were described by Linnaeus (1753, 1759); others were described from the first half of the nineteenth century (e.g. Agardh, 1824; Kützing, 1843; Nägeli, 1849). For a long time the systematics and taxonomy of green microalgae were based entirely on morphological features (e.g. Smith, 1950; Ettl and Gärtner, 1995). Taxonomic studies of these organisms received a boost in the mid twentieth century, when the development of culture techniques allowed easy observation of growth and reproduction in controlled conditions. Classification schemes based on morphology were surrounded by controversy and the systematic significance of several morphological characters was highly debated. These difficulties were due to some typical traits of the biology of green microalgae. Most terrestrial algae have a small size (typically 5–50 μm) and very few species produce thalli observable by the unaided eye (*Prasiola crispa* (Lightfoot) Kützing, a leafy alga widespread in Antarctica and other cold regions, represents the best-known example). Their morphology is simple and usually referable to three habits: (1) single cells, such as *Chlorella*, *Chlorococcum*, *Dictyochloropsis*, *Stichococcus* and *Trebouxia* (Fig 9.1); (2) sarcinoid habit, i.e. packet-like colonies formed by a limited number of cells, such as *Apatococcus*, *Chlorokybus*, *Chlorosarcina*, *Chlorosarcinopsis* and *Desmococcus* (Fig 9.2); and (3) uniseriate filaments, either branched or not, such as *Klebsormidium*, *Printzina*, *Rosenvingiella* and *Trentepohlia* (Fig 9.3). These morphologies offer very

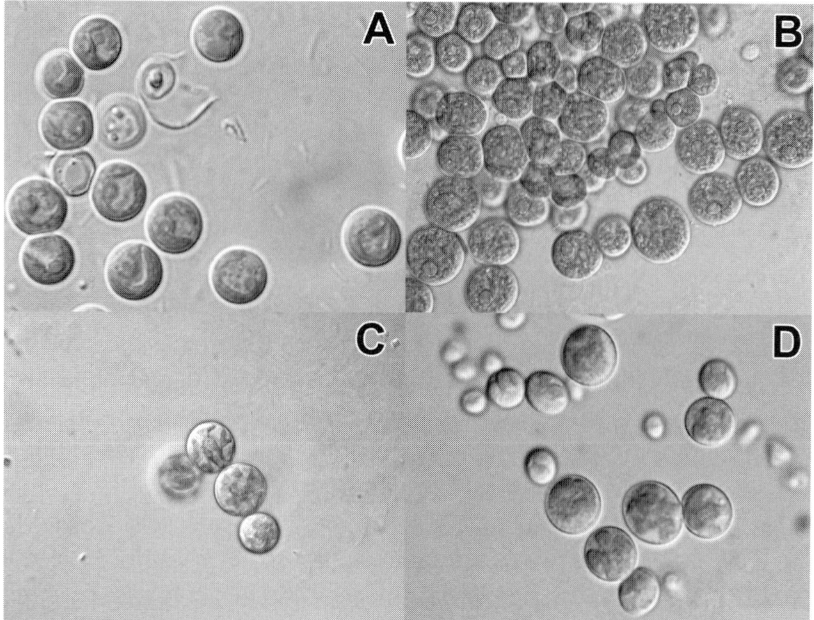

Figure 9.1 Examples of terrestrial green algae with unicellular morphology: (A) *Chlorella vulgaris* Beijerinck (cells 7–9 μm in diameter); (B) *Chlorococcum infusionum* (Schrank) Meneghini (cells 6–30 μm in diameter); (C) *Asterochloris* sp. (cells 5–8 μm in diameter); (D) *Trebouxia arboricola* Puymaly (cells 8–16 μm in diameter). All images courtesy of Pavel Škaloud, http://botany.natur.cuni.cz/skaloud/index_Borec.htm.

few characters useful for taxonomic and systematic purposes and their value is often disputed. It has been demonstrated that several characters are very plastic and may vary dramatically in relation to environmental conditions (Rindi and Guiry, 2002; Luo et al., 2006).

The development of electron microscopy in the 1970s facilitated a revolution in the systematics of green algae, unravelling a wealth of ultrastructural features (in particular the arrangement of the flagellar apparatus) that were used as a basis for a new classification (Mattox and Stewart, 1984). Most importantly, ultrastructural data suggested that several morphological vegetative features have evolved numerous times and are generally unreliable as characters marking monophyletic groups (Lewis and McCourt, 2004). This has become dramatically clear in the last 20 years, with the advent of molecular systematics. DNA sequence data have generally confirmed conclusions based on ultrastructure (Lewis et al., 1992; Friedl and Zeltner, 1994) and have shown that green microalgae groups based on general morphological similarity do not reflect phylogenetic patterns. At present, the amount of molecular data available for terrestrial algae is much more limited than for aquatic algae and vascular plants. The necessity of isolation in unialgal cultures

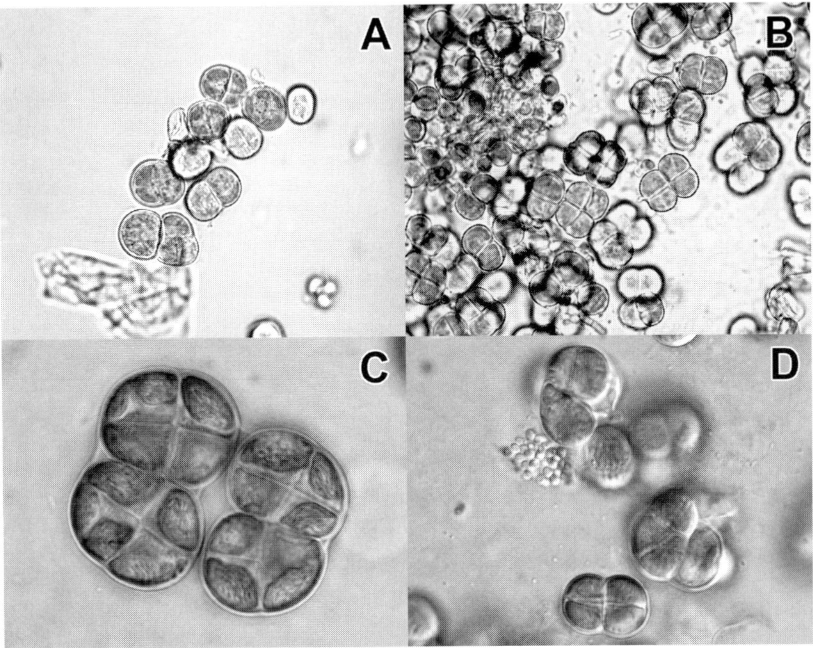

Figure 9.2 Examples of terrestrial green algae with sarcinoid morphology: (A) *Apatococcus lobatus* (Chodat) J. B. Petersen (cells 6–9 μm wide); (B) *Desmococcus olivaceus* (Peerson ex Acharius) Laundon (cells 3–6 μm wide); (C) *Chlorokybus atmophyticus* Geitler (cells up to 16 μm wide); (D) *Interfilum massjukiae* Mikhailyuk et al. (cells 11–15 μm wide). Images C and D courtesy of Pavel Škaloud, http://botany.natur.cuni.cz/skaloud/index_Borec.htm.

and the comparatively low number of systematists working on these organisms has slowed the production of high-quality molecular data sets. For example, by January 2009 no sequences for *Apatococcus* and a mere five for *Desmococcus* were deposited in GenBank, despite these being two of the most widespread genera. For other common taxa, such as the order Klebsormidiales (Novis, 2006; Mikhailyuk et al., 2008; Rindi et al., 2008a; Sluiman et al., 2008), the genus *Chlorosarcinopsis* and related taxa (Watanabe et al., 2006), the order Prasiolales (Rindi et al., 2007) and the order Trentepohliales (Rindi et al., 2009), large molecular data sets have become available only recently. Furthermore, the number of molecular markers sequenced is still limited (18S rRNA gene sequences represent the vast majority of the data available). At present, exhaustive phylogenetic analyses are available for relatively few groups. Further still, new species continue to be described in steady fashion, especially from tropical regions (Neustupa, 2005; Novis, 2006; Rindi et al., 2006; Rindi and López-Bautista, 2007; Eliáš et al., 2008; Mikhailyuk et al., 2008; Zhang et al., 2008), which shows that our knowledge of this group is still largely

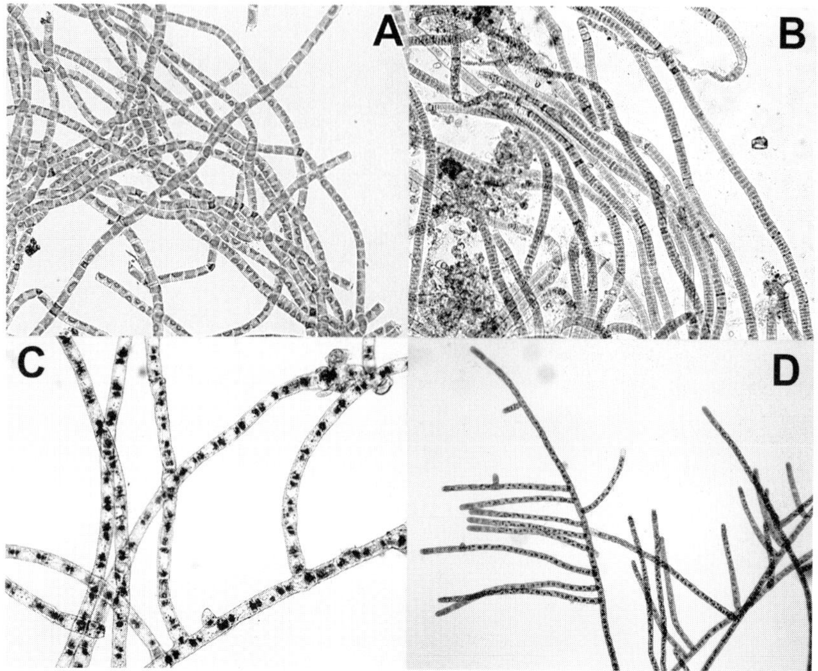

Figure 9.3 Examples of terrestrial green algae with uniseriate filamentous morphology: (A) *Klebsormidium* cfr. *flaccidum* (Kützing) P. C. Silva, Mattox and Blackwell (filaments 6–8 μm wide); (B) *Rosenvingiella radicans* (Kützing) Rindi, McIvor and Guiry (filaments 6–10 μm wide); (C) *Trentepohlia arborum* (C. Agardh) Hariot (filaments 12–15 μm wide); (D) *Printzina bosseae* (De Wildeman) Thompson and Wujek (filaments 10–13 μm wide).

incomplete. It is clear that in the general scenario of reconstructing the tree of life (Hodkinson and Parnell, 2007) terrestrial green algae are one of the groups for which most effort will be required in the future.

However, the molecular data produced in the last decade are sufficient to highlight some important traits in the evolution of these organisms. It is now established that green algae belong to a well-supported monophyletic group, the Viridiplantae, and have evolved in two major lineages, the Chlorophyta and the Streptophyta (Lewis and McCourt, 2004; Pröschold and Leliaert, 2007). The Chlorophyta consist of three well-supported groups (Fig 9.4), the Chlorophyceae, Trebouxiophyceae and Ulvophyceae, that are taxonomically separated at the class level. Outlying these clades is a grade comprising the Prasinophytes, a non-monophyletic agglomeration of unicellular algae whose classification is still in need of rearrangement (Fig 9.4). They are viewed as the form of cell most similar to the ancestral green alga (believed to be a flagellate aquatic unicellular alga – Lewis and McCourt, 2004). The Streptophyta contain the Embryophytes (or land plants) and several

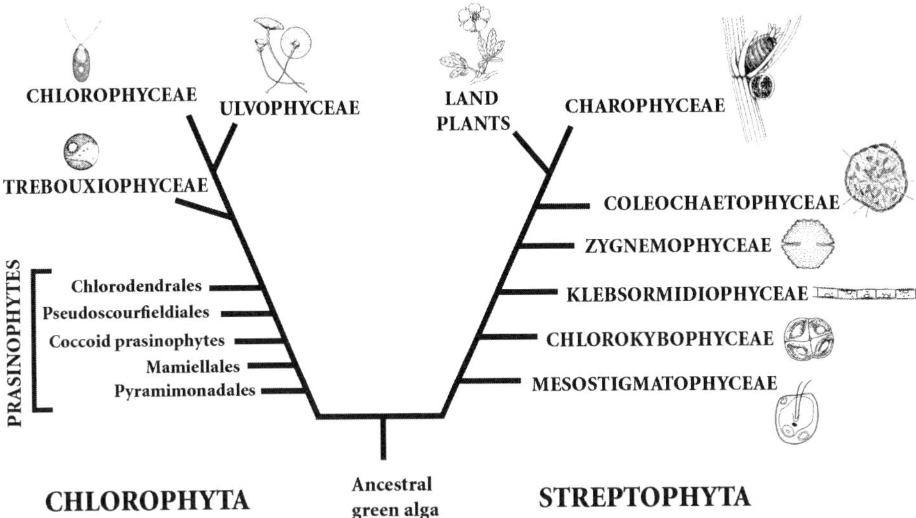

Figure 9.4 Summary of the phylogenetic relationships among the major lineages of green algae, based on DNA sequence data.

groups of green algae (Charophyceae, Chlorokybophyceae, Coleochaetophyceae, Klebsormidiophyceae, Mesostigmatophyceae and Zygnemophyceae) that form well-supported clades, but whose interrelationships are not well resolved (Fig 9.4).

Molecular data indicate that in the green algae the terrestrial lifestyle has been acquired many times independently. Terrestrial green algae represent a polyphyletic group derived from many independent transitions from aquatic habitats (both marine and freshwater) to the land (Lewis and Lewis, 2005; Lewis, 2007; López-Bautista et al., 2007; Cardon et al., 2008). Recent studies suggest that after the passage to land, green algae evolved actively even in environments considered too hostile to support the life of these organisms, such as hot deserts (Lewis and Lewis, 2005; Lewis, 2007). At present it is not clear how many transitions to land have taken place, but it has been shown that terrestrial members occur in at least six groups, namely Trebouxiophyceae, Chlorophyceae, Ulvophyceae, Chlorokybophyceae, Klebsormidiophyceae and Zygnemophyceae (Fig 9.4). The class Trebouxiophyceae, which was described on the basis of 18S rRNA sequence data (Friedl, 1995), includes most of the genera and species currently known (in particular several common genera, such as *Chlorella, Prasiola, Stichococcus, Trebouxia*). The Chlorophyceae, Klebsormidiophyceae, Chlorokybophyceae and Zygnemophyceae are mainly freshwater algae and include a more limited number of terrestrial members; some, however, such as *Klebsormidium* (Klebsormidiophyceae), and *Bracteacoccus* and *Chlorococcum* (Chlorophyceae), are among the most common terrestrial algae. The class Ulvophyceae is formed primarily by the green seaweeds, but it includes also one of the most diverse groups of terrestrial algae, the order Trentepohliales.

Molecular data have demonstrated that the morphology of subaerial green algae is highly homoplasious. The evolution of these organisms has been characterised by an extreme morphological convergence that restricted their morphology to a narrow range, not indicative of their great genetic diversity. Many taxa described on a morphological basis have been shown to be polyphyletic complexes of separate species. The best-known example is the widespread genus *Chlorella*, which includes algae belonging to two different classes, the Chlorophyceae and the Trebouxiophyceae (Huss et al., 1999; Krienitz et al., 2004; Zhang et al., 2008). Other common genera that have been revealed as non-monophyletic include *Chlorococcum* (Buchheim et al., 2002), *Klebsormidium* (Mikhailyuk et al., 2008), *Planophila* (Friedl and O'Kelly, 2002), *Printzina* (Rindi et al., 2009) and *Trentepohlia* (Rindi et al., 2009). It is now accepted that the taxonomy and classification of many taxa are in urgent need of a rearrangement based on phylogenetic principles (López-Bautista et al., 2007; Pröschold and Leliaert, 2007). Such a rearrangement, and the redefinition of species concepts associated with it, will be fundamental to understanding how these organisms will be affected by climate change in terms of species distributions and community compositions.

9.3 Biogeography of terrestrial green algae and climate change

The dispersal of small-sized terrestrial organisms is primarily operated by wind and is affected by atmospheric humidity and rainfall (Sharma et al., 2007). Consequently, climate changes involving variations in these parameters may affect considerably the range of dispersal, and therefore the biogeography, of many terrestrial microorganisms.

Two different viewpoints on the biogeography of small-sized eukaryotes have been established in recent years. On one side, a view based on ecological grounds assumes that prokaryotes, unicellular eukaryotes and small multicellular organisms have a cosmopolitan distribution because of their minute sizes and their ability to form dormant stages (cysts, eggs, spores), which facilitates dispersal by air, dust and migrating animals (Finlay, 2002; Fenchel and Finlay, 2003). Conversely, a school of thought based on a taxonomic viewpoint believes that most microorganisms have a more or less restricted distribution, and that there is a mistaken impression of ubiquity due to undersampling, misidentifications and reliance on morphological species concepts (Foissner, 2006, 2008). For green terrestrial algae, information on geographical distribution is generally vague and so far has been based entirely on morphological species concepts. Several species (often early described taxa) are considered to have a cosmopolitan or very wide distribution; typical examples are represented by *Apatococcus lobatus* (Chodat) J. B. Petersen, *Desmococcus olivaceus* (Peerson ex Acharius) Laundon, *Chlorella vulgaris* Beijerinck, *Klebsormidium flaccidum* (Kützing) P. C. Silva, Mattox and Blackwell, *Printzina lagenifera* (Hildebrand) Thompson and Wujek, *Stichococcus bacillaris* Nägeli, *Trentepohlia aurea* (Linnaeus) Martius and *Trentepohlia arborum* (C. Agardh) Hariot. In treatments reporting geographical information on these species, expressions such as 'probably cosmopolitan', 'very common both in the old and the new world' or 'very common in the tropics' are typically used (Printz, 1939; Smith, 1950; Ettl and Gärtner, 1995; John, 2002). Conversely, many other species are known only from the type locality or from a restricted number of locations (e.g. Thompson and Wujek, 1992, 1997; Neustupa and Šejnohová, 2003; Novis, 2006; Rindi et al., 2006). Unfortunately, the small size of most terrestrial green algae and their homoplasious morphology leads to high risks of misidentification. At the same time, many species occur in nature in very limited amounts and cannot be observed by the unaided eye; their presence can be revealed only with isolation in culture (John, 1988; Broady, 1996).

These problems considerably limit a correct characterisation of the biogeography of green microalgae. In general, reliable conclusions can be drawn only for a few 'flagship' taxa with characteristic morphology, easily recognisable in the field. For example, species of the order Trentepohliales are easily identified due to their red, orange or yellow colour produced by the accumulation of carotenoids. It is well documented that humid areas in the tropics represent the centre of distribution for this group (Wee and Lee, 1980; John, 1988; Rindi and López-Bautista, 2008; Rindi et al., 2008a, 2008b). Its highest species diversity has been recorded in tropical regions with consistently humid climate and high rainfall, such as Queensland, Java, French Guiana and central Panama (Rindi and López-Bautista, 2008; Rindi et al., 2008b). *Prasiola crispa* is one of the few species of terrestrial green algae with relatively large size; it forms irregularly curled blades that grow in dense dark-green layers on bare ground or rock (Fig 9.5 – Rindi et al., 1999). Its distinctive morphology facilitates accurate biogeographic distribution mapping. This alga is distributed in polar and cold temperate regions of both hemispheres and is the most common terrestrial alga in Antarctica, where it forms large populations in areas where penguin colonies deposit large amounts of guano (Hoyer et al., 2001). A similar association with polar and cold temperate zones is also typical of the other terrestrial members of the order Prasiolales (Rindi et al., 2007). However, for other species of this order the detailed distribution is more difficult to assess than for *P. crispa*, since they are impossible to identify without microscopical inspection and they are more prone to misidentification.

Apart for a few cases like these, the detailed geographical distribution of terrestrial green algae is very difficult to establish. Molecular data are reshaping

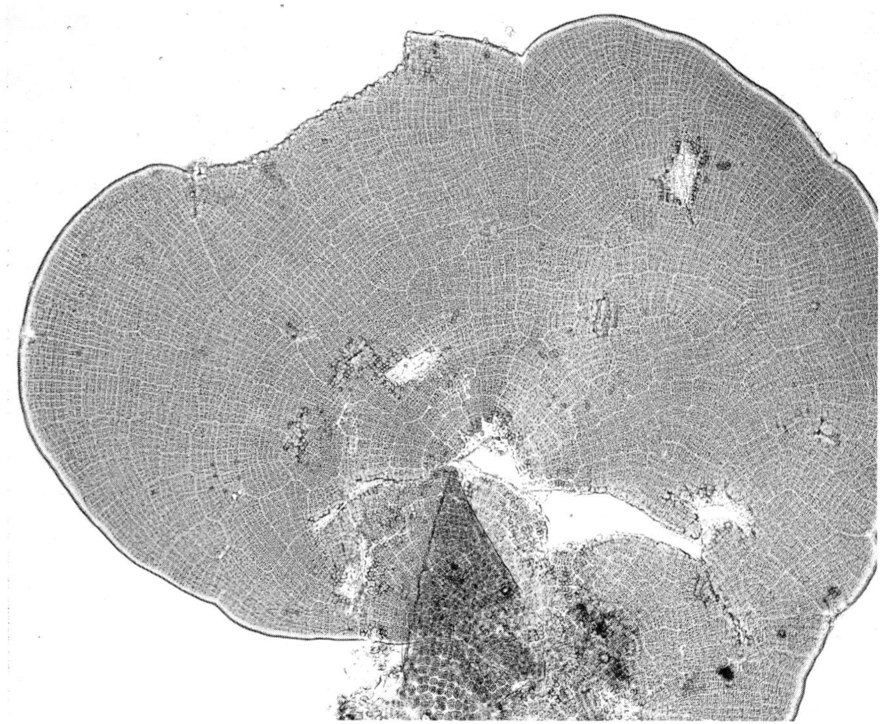

Figure 9.5 *Prasiola crispa* (Lightfoot) Kützing. Habit of a specimen (about 1 mm wide).

species concepts and are revealing that records based only on morphology are inadequate to assess geographical distributions. For example, recent molecular analyses (Rindi et al., 2009) have shown that some species of Trentepohliales with distribution considered pantropical (such as *Printzina lagenifera* and *Trentepohlia arborum*) represent complexes of cryptic species with convergent morphology; individual species probably have a much more restricted geographical distribution. It is therefore clear that molecular data have become a mandatory requirement for distribution assessments of terrestrial green algae – and therefore for associated climate change research. Accurate assessments will only be possible after species' delimitations are well defined and taxonomy is reassessed for most groups of terrestrial green algae.

9.4 Terrestrial green algae and climate change

So far, little consideration has been given to terrestrial green algae in relation to matters of climatic change. Effects of climate change on algal communities involve the interaction of several environmental parameters and are not easy to predict, particularly in the case of microalgal communities (Bell and Collins, 2008). The vast majority of the investigations that have attempted to predict the effects of climate change on microalgae are ecophysiological studies, in which the physiological responses of algae to simulated climatic changes have been examined in the field or in the laboratory. These studies can tell a great deal about how these organisms respond to immediate, short-term changes in environmental parameters. However, caution is necessary when their results are used to predict the effects of long-term climate change. Most microalgae have large population sizes, fast growth and short generation times. Long-term experiments on *Chlamydomonas reinhardtii* Dangeard simulating increased CO_2 have shown that it may drive selection of new phenotypes, even on a time span as short as 1000 generations (which in the laboratory can be obtained in a few years). Lineages selected in high CO_2 showed substantial physiological differences from lineages maintained at ambient CO_2 (Collins and Bell, 2004; Collins et al., 2006a, 2006b). Although no similar experimental evidence is available for other environmental variables, it is conceivable that modifications of other parameters may also produce similar forms of selection. In other words, green microalgae evolve in response to climatic changes, and the physiological responses of algal strains in present-day conditions might not reflect the responses of the same strains in a future, changed climate. Keeping this in mind, the environmental parameters most directly related to climatic changes (CO_2 concentration, rainfall and humidity, temperature and irradiation) are considered here, and predictions about their possible effects on terrestrial algae are based on the information currently available.

9.4.1 Elevated CO_2

Due to human activities, atmospheric CO_2 concentration is currently rising at an unprecedented rate, and is expected to show a two- to threefold increase over pre-industrial levels within this century (Beardall and Raven, 2004; Bell and Collins, 2008). Elevated CO_2 is therefore the most immediate atmospheric change affecting algae, and many studies have considered its possible effects, especially in phytoplankton communities (Beardall and Giordano, 2002; Beardall and Raven, 2004; Schippers et al., 2004; Falkowski and Oliver, 2007), but limited information is available for terrestrial green algae.

In general, the short-term response of green microalgae to a moderate increase in CO_2 concentration is an increase in photosynthetic rate and growth, whereas extremely high CO_2 concentrations may be inhibitory for photosynthesis (Miyachi et al., 2003). There is, however, substantial variation from taxon to taxon. Some strains of *Chlorococcum* and *Chlorella* have the capacity to adapt well to strong increases in CO_2, and when transferred from low CO_2 to very high CO_2, after an initial lag period, they grow rapidly. Conversely, some strains of *Stichococcus bacillaris* are intolerant of extremely high CO_2 conditions and do not show any significant state transition after being transferred to elevated concentrations (Miyachi et al., 2003). A drop of the pH value in the chloroplast stroma, which affects the activity of rubisco and other key enzymes, is considered to be the main factor causing inhibition of photosynthesis at extremely high CO_2 concentrations (Miyachi et al., 2003).

Information about long-term responses is based mostly on selection experiments performed on *Chlamydomonas reinhardtii* (Collins and Bell, 2004; Collins et al., 2006a, 2006b). From one original individual strain, five cell lines were grown at ambient CO_2 concentrations and five other lines at elevated CO_2 concentrations for 1000 generations (Collins and Bell, 2004). In these experiments, the algal lines grown in high CO_2 concentrations did not develop a metabolism suited for these conditions (i.e. with faster growth in elevated CO_2). By the end of the selection experiment, the growth of the lines in high CO_2 was no greater than, and perhaps even less than, the growth of the ambient lines. A particularly important result of this experiment was that two of the lines grown at high CO_2 evolved a syndrome involving high rates of photosynthesis and respiration, combined with higher chlorophyll content and reduced cell size. The growth of these lines was markedly impaired at ambient CO_2 concentrations. This outcome was interpreted as a consequence of accumulation of conditionally neutral mutations in genes controlling the carbon-concentrating mechanisms (CCMs). In a subsequent study, Collins and Bell (2006) also found maladaptive responses in strains of *Chloranomala, Chlorella, Chlorococcum* and *Tetracystis* isolated from the soil of natural CO_2 springs in Italy and Slovenia. This suggests that some loss of function in carbon uptake or metabolism is a general feature of long-term exposure to elevated CO_2 in populations of green microalgae. However, the nature and extent of this loss are not constant and appear to vary case by case. Even in different lines grown from the same strain, evolution under high CO_2 conditions may lead to several possible outcomes, introducing a source of uncertainty into predictions of how microalgal populations will change as CO_2 rises. The studies cited above suggest the possibility that many strains of green microalgae may become metabolically less efficient (in particular will evolve less efficient CCMs); however, these changes will differ in extent and intensity in different species, as well as in different populations of the same species. Therefore, the composition of microalgal communities can be expected to change, but at present it is impossible to predict how. Such predictions would require information based on assemblages that have had the opportunity to undergo both ecological interactions and evolutionary change (Bell and Collins, 2008), which is not available presently.

9.4.2 Rainfall and humidity

Surviving dehydration was the main challenge that green algae had to face in order to colonise terrestrial environments. This passage was completed in many separate lineages by developing adaptations aimed at limiting water loss and preventing damage caused by desiccation. At a cellular level, desiccation causes several types of biochemical and ultrastructural damage that algal cells must be able to prevent or alleviate (Gray et al., 2007). Very little information is available about the mechanisms adopted by terrestrial green algae to prevent such damage. In other groups of photosynthetic eukaryotes, these mechanisms include osmotic adjustments to the cytoplasm that reduce damage to membranes and organelles, synthesis of proteins that protect cellular components during the desiccation process, and production of antioxidants and scavenging enzymes that neutralise reactive oxygen species during drying (Gray et al., 2007). It is likely that such mechanisms also take place in terrestrial green algae. Characteristics of the algal cells such as surface/volume ratio, thickness of the cell walls and presence of mucilaginous layers affect their capacity to retain cellular water (Callaghan et al., 2004; Häubner et al., 2006; Karsten et al., 2007b). Mucoid substances excreted by some species may hinder evaporation and hence support at least some basic photosynthetic activity under desiccation (Häubner et al., 2006). Some organic compounds such as sugar alcohols (Karsten et al., 2007b; Oren, 2007) and sporopollenins (Good and Chapman, 1978; Xiong et al., 1997; Reisser and Houben, 2001), whose presence has been reported in some species of terrestrial green algae, have multiple metabolic functions and are believed to enhance resistance to desiccation.

Regardless of their desiccation tolerance, all terrestrial green algae require liquid water or an atmosphere saturated with humidity to be metabolically active. These organisms can recover photosynthetic functionality quickly after rehydration, even after very long periods of desiccation (Trainor and Gladych, 1995; Chen and Lai, 1996; Häubner et al., 2006; Gray et al., 2007; Cardon et al., 2008). Häubner et al. (2006) found that a biofilm of green algae (*Stichococcus* sp. and *Chlorella luteoviridis* Chodat) in Germany photosynthesised mostly in the early hours of the morning, due to the presence of condensation water on the facade of the buildings on which the biofilm occurred; later in the day, when the water film evaporated, the microalgae became desiccated and photosynthesis was inhibited. The same phenomenon was reported by Ong et al. (1992) for *Trentepohlia odorata* (Wiggers) Wittrock growing on buildings in Singapore. Interestingly, this alga seems to have developed an internal rhythm adapted to this situation; depression of photosynthesis occurred also on rainy days, when favourable conditions of humidity and light intensity persisted later in the day. In *Apatococcus lobatus*, active uptake of CO_2 for photosynthesis can take place at a relative air humidity as low as 68% (Bertsch, 1966). Besides being required for the internal metabolism of the cell, liquid water is also important as an external vector of nutrients. It is known that the atmosphere carries airborne nutrients in the form of aerosols, dust particles and gaseous emissions (Wright et al., 2001; Sharma et al., 2007). Several species of green algae colonise walls, roofs and other artificial surfaces where availability of nutrients is extremely limited (Rindi and Guiry, 2002; Häubner et al., 2006; Rindi,

2007). Nutrient acquisition in terrestrial algae is not well understood, but it is likely that airborne nutrients carried by rainwater may be an important resource for organisms living in these habitats (Karsten et al., 2007b).

For these reasons, changes in rainfall and atmospheric humidity are aspects of climate change that may be expected to have a strong impact on terrestrial algae. Unfortunately, few investigations have examined the responses of individual taxa to moisture variation in the field, and the data available on this topic are limited (Bertsch, 1966; Häubner et al., 2006). The situation is not helped by their small size and the taxonomic problems of species identity, hampering attempts to design models or draw major generalisations about the effect of this factor on large-scale distributions. Therefore, predictions about the effect of changes in humidity and rainfall on terrestrial algae have to be based largely on observational evidence.

In general, increases in rainfall and atmospheric humidity can be expected to benefit terrestrial green algae and increase their abundance. All terrestrial algae proliferate in conditions of high humidity, and it is common knowledge that well-developed populations of these organisms occur at sites with local conditions of high moisture (John, 1988; Nienow, 1996; Rindi and Guiry, 2004). Decreased humidity and availability of liquid water would be a limitation for the growth of these organisms, but would not necessarily affect the diversity of their assemblages, as shown for example by green algae of deserts. Green algae living in biotic crusts in hot deserts represent a surprisingly diverse group that includes organisms belonging to at least three different classes (Chlorophyceae, Trebouxiophyceae and Klebsormidiophyceae) (Lewis and Flechtner, 2002; Flechtner, 2007). They are the green algae most tolerant of desiccation, and some can survive up to 35 years without water (Trainor and Gladych, 1995). Molecular evidence shows that these organisms are not a transient presence, but have actively evolved in desert habitats (Lewis and Lewis, 2005; Lewis, 2007). This indicates that green algae are versatile organisms, in which many separate lineages can develop tolerance to extreme desiccation. Thus, decreased humidity may potentially be an environmental constraint stimulating adaptation and evolution of new taxa.

However, algal groups associated with high atmospheric humidity would be affected negatively in the short term by drier climate conditions. Species of the ulvophycean order Trentepohliales represent a good example of this. These algae are most diverse in humid tropical regions, where they represent the dominant green algal component of the terrestrial microbial vegetation. Many tropical Trentepohliales are associated with damp rainforest habitats, characterised by deep shade and humidity (Thompson and Wujek, 1997; Neustupa, 2005; Rindi and López-Bautista, 2007, 2008). In the tropics they are also well known for their profuse development (Wee and Lee, 1980; John, 1988; Rindi and López-Bautista, 2008; Rindi et al., 2008b), which may cause major practical nuisances such as disfiguration of buildings (Wee and Lee, 1980) and infection of plants of commercial interest (Chapman and Waters, 2001). Although in temperate regions the diversity of this group is considerably lower, profuse growth of Trentepohliales can still occur in areas with high rainfall, such as western Ireland (Rindi and Guiry, 2002). Given such a strict association with high humidity, it can be expected that a decrease in abundance and diversity would be the first response of this group to drier climatic conditions. Climatic changes and other forms of anthropogenic disturbance causing alteration of these conditions are likely to cause reduction or disappearance of these algae, which would be highly significant because several studies suggest that their diversity may potentially be immense (Thompson and Wujek, 1997; López-Bautista et al., 2007; Neustupa and Škaloud, 2008).

The belts or patches of green algae commonly occurring at the base of walls in older European cities represent an interesting case in which atmospheric humidity, in conjunction with other climatic factors, may affect taxonomic composition (Rindi, 2007). Studies conducted in recent years have shown that there are strong differences in the composition of these communities in different parts of Europe (Rindi et al., 1999, 2008a; Rindi and Guiry, 2004). In most of the continent, they are formed by species of the genus *Klebsormidium*. However, in moist Atlantic regions affected by the North Atlantic Drift of the Gulf Stream, such as Ireland and northwestern Spain, they are formed by algae of the order Prasiolales (*Rosenvingiella radicans* (Kützing) Rindi, McIvor and Guiry, *Prasiola calophylla* (Carmichael ex Greville) Kützing and filamentous forms of *P. crispa*). These regions are characterised by a wet and rainy climate, with higher mean rainfall than in continental or Mediterranean areas, and cooler summer temperatures. It is likely that in the short term, drier and warmer conditions would promote a replacement of the Prasiolales with *Klebsormidium* in this type of habitat. It is clear that further distributional data from other Atlantic regions are necessary to map the relative distribution of these assemblages with better resolution, and that other environmental variables will need to be considered (e.g. not only the absolute amount of rainfall but also its distribution in the annual cycle). Predictions should also be based on physiological experiments documenting desiccation and temperature tolerances of the species involved. Once these become available, it will be possible to design distribution models that would allow the use of the composition of wall-base assemblages as a simple climatic indicator.

9.4.3 Temperature

Different microalgal species have different ranges of tolerance and physiological responses to changes in temperature. Elevated temperatures can be expected to affect different species in different ways, causing increased metabolic activity and growth in some and pushing others beyond their temperature optima, thus changing species composition (Beardall and Raven, 2004). In terrestrial environments, temperature variations are much wider and take place faster than in aquatic environments. Being adapted to this situation, terrestrial algae can be expected to be less immediately affected by increased temperature than aquatic algae. It is in fact generally reported that most terrestrial algae are able to survive in a wide temperature range (Reisser, 2007). It is difficult to predict how the composition of green algal communities will be affected, since data concerning thermal tolerances and optima of individual species are limited. Furthermore, rather than temperature per se, the effect of temperature in conjunction with other climatic factors (especially solar radiation, to which temperature is directly correlated) can be expected to have the greatest effect on these algae. Strains of *Chlorella* and *Stichococcus* isolated from building facades in Germany grew with relatively high rates between 1 and 30 °C, with optimum rates at 21–24 °C (Häubner et al., 2006). Since the Trentepohliales are most abundant in the tropics and other regions with warm climates, it is likely that elevated temperatures, if associated with an increase in humidity, would favour an expansion of these algae. Aptroot and van Herk (2007) remarked that this might already be taking place in northwestern Europe. They observed that in recent years species of lichens with southern affinities have expanded their distribution; they believe that this may be due to the effect of global warming on their *Trentepohlia* phycobionts. Conversely, for terrestrial members of the Prasiolales, which occur only in polar or cold temperate regions, increased temperatures may be expected to cause a restriction of the geographical range towards higher latitudes.

9.4.4 PAR and UV radiation

Solar radiation affects photosynthetic organisms in two biologically important ways, as photosynthetically active radiation (PAR) and ultraviolet (UV) radiation. PAR (which includes the spectral range 400–700 nm) is the portion of solar radiation used for photosynthesis, and therefore represents an essential resource for any photosynthetic organism. Published data indicate that terrestrial algae can photosynthesise under high photon flux densities (Karsten et al., 2007b). However, excessively high levels of PAR can cause inhibition and several types of damage to their photosynthetic apparatus. Cases of photoinhibition or other forms of photodamage have even been reported in algae living in habitats exposed to full sunlight, which would be expected to be tolerant of very high irradiances (Ong et al., 1992; Hughes, 2006; Gray et al., 2007). Gray et al. (2007) documented a case in unicellular green algae isolated from desert crusts. Despite their ability to withstand desiccation, these algae were susceptible to photodamage at relatively low light intensity (130 μmol m^{-2} s^{-1}) when dry. This was unexpected, because such light intensity is considerably lower than the full sun intensity (2000 μmol m^{-2} s^{-1}) to which the crusts are exposed. They suggested that in the field these algae may occupy microenvironments within the crust where they are protected from damaging light levels. The differential susceptibility to illumination expressed by the taxa examined suggests that there may be a complex spatial arrangement of green algal species in the crust, structured in response to the attenuation of light with depth through the crust profile. Hughes (2006) showed an inhibitory effect of ambient solar radiation in *Stichococcus bacillaris* from Antarctica and reported that this alga is unlikely to inhabit soil surfaces, but colonises shaded crevices beneath soil surface particles. These examples suggest that an enhanced PAR, due to a decreased cloud coverage or an increase in the number of sunny days per year, might represent a form of stress for many terrestrial green algae and favour the species most adapted to high irradiation.

The production of photoprotective pigments is considered the main mechanism by which terrestrial green algae prevent inhibition or damage of their photosynthetic machinery. Pigments involved in the light-dependent conversion of violaxanthin to zeaxanthin, the so-called xanthophyll cycle, have been recorded in several genera of terrestrial green algae (Schubert et al., 1994; Chen and Lai, 1996). In vascular plants they serve as a short-term light acclimation mechanism, promoting thermal dissipation of surplus excitation energy. It can therefore be expected that these pigments play a similar role in terrestrial green algae. Masojidek et al. (2004), however, considered that in algae the function of xanthophyll-cycle pigments is ambiguous, since their contribution to energy dissipation can vary significantly among species. Carotenoid pigments are produced by a large number of green algae, including terrestrial species (Abe et al., 1999; Del Campo et al., 2000). Their production is generally regarded as a form of protection against strong light irradiances (Ong et al., 1992; Callaghan et al., 2004). The red, orange or yellow colour of the Trentepohliales is due to an accumulation of carotenoid pigments, in particular β-carotene and haematochrome (López-Bautista et al., 2002). Typically, specimens of Trentepohliales growing in shaded rainforest habitats or other low-irradiation sites produce fewer carotenoids and look green; conversely, at exposed sites these algae accumulate large amounts of carotenoids, which give them their typical red, orange or yellow pigmentation.

UV radiation is arbitrarily divided into UVA (315–400 nm), UVB (280–315 nm) and UVC (200–280 nm). In recent decades, the depletion of stratospheric ozone caused by gaseous emissions of human origin has enhanced UV radiation, particularly in polar zones (Robinson et al., 2003; Callaghan et al., 2004), and UV levels are expected to increase in regions with decreased cloud cover. UVA and UVB have multiple harmful effects, which have been well documented (e.g. Beardall and Raven, 2004; Karsten et al., 2007a). Terrestrial algae are naturally adapted to higher levels of UV than aquatic algae. It is therefore unsurprising that studies comparing the resistance of aquatic and terrestrial taxa showed a higher tolerance in terrestrial taxa (Karsten et al., 2007a). Terrestrial green algae have developed several mechanisms to prevent and/or counteract the damage caused by UV radiation. Their type and extent seem to differ from species to species, as indicated by the fact that there is not a uniform physiological response to artificially increased UVB levels (Xiong et al., 1999; Reisser and Houben, 2001; Karsten et al., 2007a; Wong et al., 2007). The best-known mechanism is the production of mycosporine-like amino acids (MAAs).

MAAs are water-soluble, low-molecular-weight molecules with maximum absorption bands between 310 and 360 nm (Cockell and Knowland, 1999). These compounds occur in organisms from a wide taxonomic range, both aquatic and terrestrial (Cockell and Knowland, 1999). In terrestrial green algae the presence of MAAs has been reported in 13 genera: *Apatococcus, Bracteacoccus, Chlorella, Elliptochloris, Klebsormidium, Myrmecia, Pabia, Prasiola, Prasiolopsis, Pseudochlorella, Pseudococcomyxa, Stichococcus* and *Trichophilus* (Xiong et al., 1999; Reisser and Houben, 2001; Karsten et al., 2005, 2007a). The fact that these genera belong to three different classes (Chlorophyceae, Klebsormidiophyceae and Trebouxiophyceae) indicates that the production of MAAs is an effective strategy, with wide taxonomic distribution among green algae. The chemical nature of UV-absorbing MAAs appears to be related to phylogenetic patterns, suggesting that these compounds may be good chemosystematic markers. For example, an MAA with an absorption maximum at 324 nm was described by Hoyer et al. (2001) in *Prasiola crispa* ssp. *antarctica* (Kützing) Knebel. Subsequently Karsten et al. (2005, 2007a) identified this MAA in other members of the Trebouxiophyceae, but they could not find it in the Chlorophyceae and Ulvophyceae. They therefore advocated its chemosystematic value as a marker for the class Trebouxiophyceae. Experimental evidence shows that in many taxa, MAAs can be induced or stimulated by UV exposure. After four hours' exposure of algae to UVB on four consecutive days, Reisser and Houben (2001) obtained production of MAAs in strains of *Chlorella, Klebsormidium* and *Stichococcus* in which these compounds were not detectable before exposure. Similarly, Karsten et al. (2007a) were able to induce production of MAAs in *Chlorella luteoviridis, Myrmecia incisa* and *Stichococcus* sp. by UV exposure. Conversely, Lud et al. (2001) could not induce formation of MAAs or other UV-absorbing substances in *Prasiola crispa* from Antarctica, and concluded that this alga counteracts the effect of UV radiation primarily by a fast and efficient elimination of cyclobutyl pyrimidine dimers (CPD). Karsten et al. (2007a) also found evidence, besides UV, for an effect of nitrogen availability on biosynthesis of MAAs. They showed an accumulation of a 324 nm MAA under increasing potassium nitrate (KNO_3) concentrations and the simultaneous presence of UVA or UVA/B, concluding that a lack of sufficient nitrogen in a high-radiation environment may depress MAA biosynthesis. In consideration of such a relationship, they concluded that an adequate supply of nitrogen might be a critical factor for the good physiological performance of terrestrial algae in high-UV regimes. The presence of the previously discussed sporopollenins in cell walls, which have been found in several genera, is also considered to play a role in UV protection. Xiong et al. (1997) suggested that sporopollenin provides a constant protection, while MAAs are induced by radiation stress and occur with some delay.

Thus, mechanisms of protection and responses to high irradiation vary greatly from species to species. In the short term, increases in PAR and/or UV levels will select communities of terrestrial green algae in which the species with best protection will be favoured. Species of the predominantly terrestrial Trebouxiophyceae appear to have a larger set of photoprotective pigments and to be better equipped to withstand strong irradiation. These algae can therefore be expected to perform better than members of other classes. In the long run, increased radiation might act as an evolutionary force promoting the development of better protection mechanisms in lineages with weak or moderate resistance.

9.5 Conclusions and directions for future work

In comparison to other groups of algae and plants, our general knowledge of the diversity and biology of terrestrial algae is limited. This represents a major impediment to predict how these organisms will respond to climatic changes, both as individual species and as communities. The concept of species in green microalgae is in a phase of redefinition. Molecular genetic data are revolutionising our views of the diversity of these organisms and are revealing that terrestrial green algae are a much more heterogeneous group than previously understood. It is clear that species concepts based only on morphological grounds are inadequate. Species should be defined using a polyphasic approach combining as many different types of data as possible (morphological, molecular, ultrastructural, biochemical, physiological and ecological). This is of critical importance, because until individual species are clearly and unambiguously defined it will be extremely difficult to state how species will respond to climate changes. A high genetic diversity implies a high biochemical and physiological diversity and therefore, indirectly, a high diversity in responses to climatic changes. At present, we do not know how exhaustive our taxonomic knowledge of terrestrial green algae is, but there is reason to believe that a very large number of species is still undescribed. In this regard, special efforts should be dedicated to habitats and regions of the world that have been poorly explored and that are likely to host many new species. This is especially true for tropical regions, tropical rainforests in particular. The high humidity and the great variety of habitats typical of these environments are ideal to promote an immense diversity of terrestrial algae. The limited information available supports this idea; unfortunately studies on terrestrial algae of rainforests are rare and molecular data obtained from these organisms are almost non-existent.

Once species concepts and taxonomy of the main groups of terrestrial algae are better understood, the production of high-quality biogeographical records and new physiological and biochemical data, at least for the most common and widespread species, should be the next logical step in order to predict the responses of individual species to climatic changes. Assessments of geographical distribution should be made only after the identity of a species has been fully clarified, using records based on both morphological and molecular data. Records based on herbarium specimens or other public collections should not be used without a critical re-evaluation. Physiological analyses concerning desiccation tolerance, survival ranges and optimal temperatures, and mechanisms of protection against and recovery from photodamage, should be combined with distribution records. Production of this type of data is a top priority at least for common species that produce large populations in nature. The combination of distribution records and physiological information would allow the design of ecological niche models incorporating the effects of climatic parameters, which would be very useful to predict shifts in distributions. At present, such models are non-existent for terrestrial algae.

It should be appreciated that responses in present-day conditions might not reflect the responses in a future, changed climate. The studies of Collins and Bell (2004, 2006) and Collins et al. (2006a, 2006b) on *Chlamydomonas* showed that climatic changes may select new phenotypes and drive the evolution of green microalgae. It is clear that the main practical limitation to understanding microalgal evolutionary responses to global change is simply the lack of published studies (Bell and Collins, 2008). It is highly desirable that similar studies become available for a range of terrestrial algal species.

Terrestrial green algae have the potential to provide a great contribution to our understanding of how climate change will affect terrestrial microbial communities. Our knowledge of these organisms is still generally limited, and any type of new information on their diversity and biology will be of great value.

Acknowledgements

Work on the systematics and ecology of terrestrial algae carried out in the past 10 years has greatly contributed to shaping the ideas presented in this chapter. This work has been funded by the US National Science Foundation (Systematics Program DEB-0542924) and the Higher Education Authority of Ireland (Cycles 2 and 3 of the Programme for Research in Third Level Institutions). I am very grateful to the numerous colleagues who have supported my work in various forms, in particular to Professor Michael Guiry and Dr Juan López-Bautista for their enthusiastic support. I express sincere thanks to Dr Pavel Škaloud and Professor Jiří Neustupa for the use of their images.

References

Abe, K., Nishimura, N. and Hirano, M. (1999). Simultaneous production of beta-carotene, vitamin E and vitamin C by the aerial microalga *Trentepohlia aurea*. *Journal of Applied Phycology*, **11**, 331–336.

Agardh, C. A. (1824). *Systema algarum*. Lund: Berling.

Aptroot, A. and van Herk, C. M. (2007). Further evidence of the effects of global warming on lichens, particularly those with *Trentepohlia* phycobionts. *Environmental Pollution*, **146**, 293–298.

Beardall, J. and Giordano, M. (2002). Ecological implications of microalgal and cyanobacterial CO_2 concentrating mechanisms, and their regulation. *Functional Plant Biology*, **29**, 335–347.

Beardall, J. and Raven, J. A. (2004). The potential effects of global climate change on microalgal photosynthesis, growth and ecology. *Phycologia*, **43**, 26–40.

Bell, G. and Collins, S. (2008). Adaptation, extinction and global change. *Evolutionary Applications*, **1**, 3–16.

Bertsch, A. (1966). CO_2 Gaswechsel der Grünalge *Apatococcus lobatus*. *Planta*, **70**, 46–72.

Broady, P. A. (1996). Diversity, distribution and dispersal of Antarctic terrestrial algae. *Biodiversity and Conservation*, **5**, 1307–1335.

Brodie, J. and Zuccarello, G. C. (2007). Systematics of the species rich algae: red algal classification, phylogeny and speciation. In *Reconstructing the Tree of Life: Taxonomy and Systematics of Species Rich Taxa*, ed. T. R. Hodkinson and J. A. N. Parnell. Systematics Association Special Volume 72. Boca Raton, FL: CRC Press, pp. 317–330.

Buchheim, M. A., Buchheim, J. A., Carlson, T. and Kugrens, P. (2002). Phylogeny of Lobocharacium (Chlorophyceae) and allies: a study of 18S and 26S rDNA data. *Journal of Phycology*, **38**, 376–383.

Callaghan, T. V., Björn, L. O., Chernov, Y. et al. (2004). Biodiversity, distributions and adaptations of Arctic species in the context of environmental change. *Ambio*, **33**, 404–417.

Cardon, Z. G., Gray, D. W. and Lewis, L. A. (2008). The green algal underground: evolutionary secrets of desert cells. *Bioscience*, **58**, 114–122.

Chapman, R. L. and Waters, D. A. (2001). Lichenization of the Trentepohliales: complex algae and odd relationships. In *Symbiosis*, ed. J. Seckbach. Dordrecht: Kluwer, pp. 361–371.

Chen, P. C. and Lai, C. L. (1996). Physiological adaptation during cell dehydration and rewetting of a newly-isolated *Chlorella* species. *Physiologia Plantarum*, **96**, 453–457.

Cockell, C. S. and Knowland, J. (1999). Ultraviolet radiation screening compounds. *Biological Reviews*, **74**, 311–345.

Collins, S. and Bell, G. (2004). Phenotypic consequences of 1,000 generations of selection at elevated CO_2 in a green alga. *Nature*, **431**, 566–569.

Collins, S. and Bell, G. (2006). Evolution of natural algal populations at elevated CO_2. *Ecology Letters*, **9**, 129–135.

Collins, S., Sültemeyer, D. and Bell, G. (2006a). Changes in C uptake in populations of *Chlamydomonas reinhardtii* selected at high CO_2. *Plant, Cell and Environment*, **29**, 1812–1819.

Collins, S., Sültemeyer, D. and Bell, G. (2006b). Rewinding the tape: selection of algae adapted to high CO_2 at current and Pleistocene levels of CO_2. *Evolution*, **60**, 1392–1401.

Del Campo, J. A., Moreno, J., Rodriguez, H. et al. (2000). Carotenoid content of chlorophycean microalgae: factors determining lutein accumulation in *Muriellopsis* sp. Chlorophyta. *Journal of Biotechnology*, **76**, 51–59.

Delwiche, C. F. (2007). Algae in the warp and weave of life: bound by plastids. In *Unravelling the Algae: the Past, Present and Future of Algal Systematics*, ed. J. Brodie and J. Lewis. Systematics Association Special Volume 75. Boca Raton, FL: CRC Press, pp. 7–20.

Eliáš, M., Neustupa, J. and Škaloud, P. (2008). *Elliptochloris bilobata* var. *corticola* var. nov. (Trebouxiophyceae, Chlorophyta), a novel subaerial coccal green alga. *Biologia*, **63**, 791–798.

Ettl, H. and Gärtner, G. (1995). *Syllabus der Boden-, Luft- und Flechtenalgen*. Stuttgart: Gustav Fischer Verlag.

Falkowski, P. G. and Oliver, M. J. (2007). Mix and match: how climate selects phytoplankton. *Nature Reviews Microbiology*, **5**, 813–819.

Falkowski, P. G., Katz, M. E., Knowll, A. H. et al. (2004). The evolution of modern eukaryotic phytoplankton. *Science*, **305**, 354–360.

Fenchel, T. and Finlay, B. J. (2003). Is microbial diversity fundamentally different from biodiversity of larger animals and plants? *European Journal of Protistology*, **39**, 486–490.

Finkel, Z. V., Katz, M. E., Wright, J. D., Schofield, O. M. E. and Falkowski, P. G. (2005). Climatically driven macroevolutionary patterns in the size of marine diatoms over the Cenozoic. *Proceedings of the National Academy of Sciences of the USA*, **102**, 8927–8932.

Finkel, Z. V., Sebbo, J., Feist-Burkhardt, S. et al. (2007). A universal driver of macroevolutionary change in the size of marine phytoplankton over the Cenozoic. *Proceedings of the National Academy of Sciences of the USA*, **104**, 20416–20420.

Finlay, B. J. (2002). Global dispersal of free-living microbial eukaryote species. *Science*, **296**, 1061–1063.

Flechtner, V. R. (2007). North American microbiotic soil crust communities: diversity despite challenge. In *Algae and Cyanobacteria in Extreme Environments*, ed. J. Seckbach. Dordrecht: Springer, pp. 539–551.

Foissner, W. (2006). Biogeography and dispersal of micro-organisms: a review emphasizing protests. *Acta Protozoologica*, **45**, 111–136.

Foissner, W. (2008). Protist diversity and distribution: some basic considerations. *Biodiversity and Conservation*, **17**, 235–242.

Friedl, T. (1995). Inferring taxonomic positions and testing genus level assignments in coccoid green lichen algae: a phylogenetic analysis of 18S ribosomal RNA sequences from *Dictyochloropsis reticulata* and from members of the genus *Myrmecia* (Chlorophyta, Trebouxiophyceae cl. nov.). *Journal of Phycology*, **31**, 632–639.

Friedl, T. and O' Kelly, C. J. (2002). Phylogenetic relationships of green algae assigned to the genus *Planophila* (Chlorophyta): evidence from 18S rDNA sequence data and ultrastructure. *European Journal of Phycology*, **37**, 373–384.

Friedl, T. and Zeltner, C. (1994). Assessing the relationships of some coccoid green lichen algae and the Microthamniales (Chlorophyta) with 18S gene sequence comparisons. *Journal of Phycology*, **30**, 500–506.

Good, B. H. and Chapman, R. L. (1978). The ultrastructure of *Phycopeltis* (Chroolepidaceae: Chlorophyta). 1. Sporopollenin in the cell walls. *American Journal of Botany*, **65**, 27–33.

Gray, D. W., Lewis, L. A. and Cardon, Z. G. (2007). Photosynthetic recovery following desiccation of desert green algae (Chlorophyta) and their aquatic relatives. *Plant, Cell and Environment*, **30**, 1240–1255.

Häubner, N., Schumann, R. and Karsten, U. (2006). Aeroterrestrial algae growing in biofilms on facades: response to temperature and water stress. *Microbial Ecology*, **51**, 285–293.

Hodkinson, T. R. and Parnell, J. A. N. (2007). Introduction to the systematics of species rich groups. In *Reconstructing the Tree of Life: Taxonomy and Systematics of Species Rich Taxa*, ed. T. R. Hodkinson and J. A. N. Parnell. Systematics Association Special Volume 72. Boca Raton, FL: CRC Press, pp. 3–20.

Hoyer, K., Karsten, U., Sawall, T. and Wiencke, C. (2001). Photoprotective substances in Antarctic macroalgae and their variation with respect to depth, distribution, different tissues and developmental stages. *Marine Ecology Progress Series*, **211**, 117–129.

Hughes, K. A. (2006). Solar UV-B radiation, associated with ozone depletion, inhibits the Antarctic terrestrial microalga *Stichococcus bacillaris*. *Polar Biology*, **29**, 327–336.

Huss, V. A. R., Frank, C., Hartmann, E. C. et al. (1999). Biochemical taxonomy and molecular phylogeny of the genus *Chlorella* sensu lato (Chlorophyta). *Journal of Phycology*, **35**, 587–598.

John, D. M. (1988). Algal growths on buildings: a general review and methods of treatment. *Biodeterioration Abstracts*, **2**, 81–102.

John, D. M. (2002). Orders Chaetophorales, Klebsormidiales, Microsporales, Ulotrichales. In *The Freshwater Algal Flora of the British Isles*, ed. D. M. John, B. A. Whitton and A. J. Brook. Cambridge: Cambridge University Press, pp. 433–468.

Karsten, U., Friedl, T., Schumann, R., Hoyer, K. and Lembcke, S. (2005). Mycosporine-like amino acids and phylogenies in green algae: *Prasiola* and its relatives from the Trebouxiophyceae (Chlorophyta). *Journal of Phycology*, **41**, 557–566.

Karsten, U., Lembcke, S. and Schumann, R. (2007a). The effects of ultraviolet radiation on photosynthetic performance, growth, and sunscreen compounds in aeroterrestrial biofilm algae isolated from building facades. *Planta*, **225**, 991–1000.

Karsten, U., Schumann, R. and Mostaert, A. S. (2007b). Aeroterrestrial algae growing on man-made surfaces: what are the secrets of their ecological success? In *Algae and Cyanobacteria in Extreme Environments*, ed. J. Seckbach. Dordrecht: Springer, pp. 583–597.

Krienitz, L., Hegewald, E. H., Hepperle, D. et al. (2004). Phylogenetic relationship of *Chlorella* and *Parachlorella* gen. nov. (Chlorophyta, Trebouxiophyceae). *Phycologia*, **43**, 529–542.

Kützing, F. T. (1843). *Phycologia generalis.* Leipzig: F. A. Brockhaus.

Lewis, L. A. (2007). Chlorophyta on land: independent lineages of green eukaryotes from arid lands. In *Algae and Cyanobacteria in Extreme Environments*, ed. J. Seckbach. Dordrecht: Springer, pp. 571–582.

Lewis, L. A. and Flechtner, V. R. (2002). Green algae (Chlorophyta) of desert microbiotic crusts: diversity of North American taxa. *Taxon*, **51**, 443–451.

Lewis, L. A. and Lewis, P. O. (2005). Unearthing the molecular phylodiversity of desert soil green algae (Chlorophyta). *Systematic Biology*, **54**, 936–947.

Lewis, L. A. and McCourt, R. M. (2004). Green algae and the origin of land plants. *American Journal of Botany*, **91**, 1535–1556.

Lewis, L. A., Wilcox, L. W., Fuerst, P. A. and Floyd, G. L. (1992). Concordance of molecular and ultrastructural data in the study of zoosporic green algae. *Journal of Phycology*, **28**, 375–380.

Linnaeus, C. (1753). *Species plantarum, Vol II.* Stockholm.

Linnaeus, C. (1759). *Systema naturae per regna tria naturae, Vol II.* Stockholm.

López-Bautista, J. M., Waters, D. A. and Chapman, R. L. (2002). The Trentepohliales revisited. *Constancea*, **83**. ucjeps.berkeley.edu/constancea/83/lopez_etal/trentepohliales.html.

López-Bautista, J. M., Rindi, F. and Casamatta, D. (2007). The systematics of subaerial algae. In *Algae and Cyanobacteria in Extreme Environments*, ed. J. Seckbach. Dordrecht: Springer, pp. 601–617.

Lud, D., Buma, A. G. J., van de Poll, W., Moerdijk, T. C. W. and Huiskes, H. L. (2001). DNA damage and photosynthetic performance in the Antarctic terrestrial alga *Prasiola crispa* ssp. *antarctica* (Chlorophyta) under manipulated UV-radiation. *Journal of Phycology*, **37**, 459–467.

Luo, W., Pflugmacher, S., Pröschold, T., Walz, N. and Krienitz, L. (2006). Genotype versus phenotype variability in *Chlorella* and *Micractinium* (Chlorophyta, Trebouxiophyceae). *Protist*, **157**, 315–333.

Masojidek, J., Kopecky, J., Koblizek, M. and Torzillo, G. (2004). The xanthophyll cycle in green algae (Chlorophyta): its role in the photosynthetic apparatus. *Plant Biology*, **6**, 342–349.

Mattox, K. R. and Stewart, K. D. (1984). Classification of the green algae: a concept based on comparative cytology. In *The Systematics of the Green Algae*, ed. D. E. G. Irvine and D. M. John. London: Academic Press, pp. 29–72.

Mikhailyuk, T. I., Sluiman, H. J., Massalski, A. et al. (2008). New streptophyte green algae from terrestrial habitats and an assessment of the genus *Interfilum* (Klebsormidiophyceae, Streptophyta). *Journal of Phycology*, **44**, 1586–1603.

Miller, K. G., Kominz, M. A., Browning, J. V. et al. (2005). The Phanerozoic record of global sea-level change. *Science*, **310**, 1293–1298.

Miyachi, S., Iwasaki, I. and Shiraiwa, Y. (2003). Historical perspective on microalgal and cyanobacterial acclimation to low- and extremely high-CO_2 conditions. *Photosynthesis Research*, **77**, 139–153.

Nägeli, C. (1849). Gattungen einzelliger Algen, physiologisch und systematisch bearbeitet. *Neue Denkschriften der Allgemeine Schweizerischen Gesellschaft für die Gesammten Naturwissenschaften*, **10**, 1–139.

Neustupa, J. (2005). Investigations on the genus *Phycopeltis* (Trentepohliaceae, Chlorophyta) from South-East Asia, including the description of two new species. *Cryptogamie Algologie*, **26**, 229–242.

Neustupa, J. and Šejnohová, L. (2003). *Marvania aerophytica* sp. nov., a new subaerial tropical green alga. *Biologia*, **58**, 503–507.

Neustupa, J. and Škaloud, P. (2008). Diversity of subaerial algae and cyanobacteria on tree bark in tropical mountain habitats. *Biologia*, **63**, 806–812.

Nienow, J. A. (1996). Ecology of subaerial algae. *Nova Hedwigia Beiheft*, **112**, 537–552.

Novis, P. M. (2006). Taxonomy of *Klebsormidium* (Klebsormidiales, Charophyceae) in New Zealand streams and the significance of low-pH habitats. *Phycologia*, **45**, 293–301.

Ong, B. L., Lim, M. and Wee, Y. C. (1992). Effects of desiccation and illumination on photosynthesis and pigmentation of an edaphic population of *Trentepohlia odorata* (Chlorophyta). *Journal of Phycology*, **28**, 768–772.

Oren, A. (2007). Diversity of organic osmotic compounds and osmotic adaptation in cyanobacteria and algae. In *Algae and Cyanobacteria in Extreme Environments*, ed. J. Seckbach. Dordrecht: Springer, pp. 641–655.

Printz, H. (1939). Vorarbeiten zu einer Monographie der Trentepohliaceen. *Nytt Magasin for Naturvbidenskapene*, **80**, 137–210.

Pröschold, T. and Leliaert, F. (2007). Systematics of the green algae: conflict of classic and modern approaches. In *Unravelling the Algae: the Past, Present and Future of Algal Systematics*, ed. J. Brodie and J. Lewis. Systematics Association Special Volume 75. Boca Raton, FL: CRC Press, pp. 123–153.

Reisser, W. (2007). The hidden life of algae underground. In *Algae and Cyanobacteria in Extreme Environments*, ed. J. Seckbach. Dordrecht: Springer, pp. 47–58.

Reisser, W. and Houben, P. (2001). Different strategies of aeroterrestrial algae in reacting to increased levels of UV-B and ozone. *Nova Hedwigia Beiheft*, **123**, 291–296.

Rindi, F. (2007). Diversity, distribution and ecology of green algae and cyanobacteria in urban habitats. In *Algae and Cyanobacteria in Extreme Environments*, ed. J. Seckbach. Dordrecht: Springer, pp. 571–582.

Rindi, F. and Guiry, M. D. (2002). Diversity, life history and ecology of *Trentepohlia* and *Printzina* (Trentepohliales, Chlorophyta) in urban habitats in western Ireland. *Journal of Phycology*, **38**, 39–54.

Rindi, F. and Guiry, M. D. (2004). Composition and spatial variability of terrestrial algal assemblages occurring at the bases of urban walls in Europe. *Phycologia*, **43**, 225–235.

Rindi, F. and López-Bautista, J. M. (2007). New and interesting records of *Trentepohlia* (Trentepohliales, Chlorophyta) from French Guiana, including the description of two new species. *Phycologia*, **46**, 698–708.

Rindi, F. and López-Bautista, J. M. (2008). Diversity and ecology of Trentepohliales (Ulvophyceae, Chlorophyta) in French Guiana. *Cryptogamie Algologie*, **29**, 13–43.

Rindi, F., Guiry, M. D., Barbiero, R. P. and Cinelli, F. (1999). The marine and terrestrial Prasiolales (Chlorophyta) of Galway City, Ireland: a morphological and ecological study. *Journal of Phycology*, **35**, 469–482.

Rindi, F., López-Bautista, J. M., Sherwood, A. R. and Guiry, M. D. (2006). Morphology and phylogenetic position of *Spongiochrysis hawaiiensis* gen. et sp. nov., the first known terrestrial member of the order Cladophorales (Ulvophyceae, Chlorophyta). *International Journal of Systematic and Evolutionary Microbiology*, **56**, 913–922.

Rindi, F., McIvor, L., Sherwood, A. R. et al. (2007). Molecular phylogeny of the green algal order Prasiolales (Trebouxiophyceae, Chlorophyta). *Journal of Phycology*, **43**, 811–822.

Rindi, F., Guiry, M. D. and López-Bautista, J. M. (2008a). Distribution, morphology and phylogeny of *Klebsormidium* (Klebsormidiales, Charophyceae) in urban environments in Europe. *Journal of Phycology*, **44**, 1529–1540.

Rindi, F., Lam, D. W. and López-Bautista, J. M. (2008b). Trentepohliales (Ulvophyceae, Chlorophyta) from Panama. *Nova Hedwigia*, **87**, 421–444.

Rindi, F., Lam, D. W. and López-Bautista, J. M. (2009). Phylogenetic relationships and species circumscription in *Trentepohlia* and *Printzina* (Trentepohliales, Chlorophyta). *Molecular Phylogenetics and Evolution*, **52**, 329–339.

Robinson, S., Wasley, J. and Tobin, A. K. (2003). Living on the edge: plants and global change in continental and maritime Antarctica. *Global Change Biology*, **9**, 1681–1717.

Schippers, P., Lurling, M. and Scheffer, M. (2004). Increase of atmospheric CO_2 promotes phytoplankton productivity. *Ecology Letters*, **6**, 446–451.

Schubert, H., Kroom, B. M. A. and Matthijs, H. C. P. (1994). In-vivo manipulation of the xanthophyll cycle and the role of zeaxanthin in the protection against photodamage in the green alga *Chlorella pyrenoidosa*. *Journal of Biological Chemistry*, **269**, 7267–7272.

Servais, T., Lehnert, O., Li, J., et al. (2008). The Ordovician biodiversification: revolution in the oceanic trophic chain. *Lethaia*, **41**, 99–109.

Sharma, N. K., Rai, A. K., Singh, S., and Brown, R. M. (2007). Airborne algae: their present status and relevance. *Journal of Phycology*, **43**, 615–627.

Sluiman, H. J., Guihal, C. and Mudimu, O. (2008). Assessing phylogenetic affinities and species delimitations in Klebsormidiales (Streptophyta): nuclear-encoded rDNA phylogeny and ITS secondary structure models in *Klebsormidium*, *Hormidiella* and *Entransia*. *Journal of Phycology*, **44**, 183–195.

Smith, G. M. (1950). *The Freshwater Algae of the United States*. New York, NY: McGraw-Hill.

Thompson, R. H. and Wujek, D. E. (1992). *Printzina* gen. nov. (Trentepohliaceae), including a description of a new species. *Journal of Phycology*, **28**, 232–237.

Thompson, R. H. and Wujek, D. E. (1997). *Trentepohliales: Cephaleuros, Phycopeltis and Stomatochroon. Morphology, Taxonomy and Ecology.* Enfield, NH: Science Publishers.

Tozzi, S., Schofield, O. and Falkowski, P. (2004). Historical climate change and ocean turbulence as selective agents for two key phytoplankton functional groups. *Marine Ecology Progress Series*, **274**, 123–132.

Trainor, F. R. and Gladych, R. (1995). Survival of algae in desiccated soil: a 35-year study. *Phycologia*, **34**, 191–192.

Verbruggen, H., Ashworth, M., LoDuca, S. T. et al. (2009). A multi-locus time calibrated phylogeny of the siphonous green algae. *Molecular Phylogenetics and Evolution*, **50**, 642–653.

Watanabe, S., Mitsui, K., Nakayama, T. and Inouye, I. (2006). Phylogenetic relationships and taxonomy of sarcinoid green algae: *Chlorosarcinopsis*, *Desmotetra*, *Sarcinochlamys*, gen. nov., *Neochlorosarcina* and *Chlorosphaeropsis* (Chlorophyceae, Chlorophyta). *Journal of Phycology*, **42**, 679–695.

Wee, Y. C. and Lee, K. B. (1980). Proliferation of algae on surfaces of buildings in Singapore. *International Biodeteration Bulletin*, **16**, 113–117.

Wong, C. Y., Chu, W. L., Marchant, H. and Phang, S. M. (2007). Comparing the responses of Antarctic, tropical and temperate microalgae to ultraviolet radiation stress. *Journal of Applied Phycology*, **19**, 689–699.

Wright, R. F., Alewell, C., Cullen, J. M. et al. (2001). Trends in nitrogen deposition and leaching in acid-sensitive streams. *Hydrological and Earth Systems Science*, **5**, 299–310.

Xiong, F., Komenda, J., Kopecky, J. and Nedbal, L. (1997). Strategies of ultraviolet-B protection in microscopic algae. *Physiologia Plantarum*, **100**, 378–388.

Xiong, F., Kopecky, J. and Nedbal, L. (1999). The occurrence of UV-B adsorbing mycosporine-like amino acids in freshwater and terrestrial microalgae (Chlorophyta). *Aquatic Botany*, **63**, 37–49.

Yoon, H. S., Hackett, J. D., Ciniglia, C., Pinto, G. and Bhattacharya, D. (2004). A molecular timeline for the origin of photosynthetic eukaryotes. *Molecular Biology and Evolution*, **21**, 809–818.

Zhang, J. M., Huss, V. A. R., Sun, X. P., Chang, K. J. and Pang, D. B. (2008). Morphology and phylogenetic position of a trebouxiophycean green alga (Chlorophyta) growing on the rubber tree, *Hevea brasiliensis*, with the description of a new genus and species. *European Journal of Phycology*, **43**, 185–193.

SECTION 3

Biogeography, migration and ecological niche modelling

10

Biodiversity informatics for climate change studies

A. Culham

School of Biological Sciences and *The Walker Institute for Climate Change, University of Reading, UK*

C. Yesson

Institute of Zoology, Zoological Society of London, UK

Abstract

Modelling the impacts of climate change on biodiversity in a phylogenetic context combines the disparate disciplines of phylogenetics, geographic information systems, niche ecology and climate change research. Each subjcct has its own approach, literature and data. The strength of an integrative research, known as 'phyloclimatic modelling', is that it provides novel insights into the possible interactions of life and climate over millions of years. However, the risk is that problems associated with each subject area might be compounded if analyses are not conducted with care. The continuous development of analytical approaches and the steady increase in data availability have offered new opportunities for data combination. Modelling techniques and output for climate, ecological niche modelling,

Climate Change, Ecology and Systematics, ed. Trevor R. Hodkinson, Michael B. Jones, Stephen Waldren and John A. N. Parnell. Published by Cambridge University Press.

phylogeny reconstruction and temporal calibration are becoming stronger, and the reliability of results is quantifiable. In contrast, there is still a desperate lack of fundamental data on organismal distribution and on fossil history of lineages. When theories of taxonomic delimitation change, there are subsequent changes in organismal names. This creates difficulty for name-based data retrieval, but techniques are being developed to reduce this problem. Improvements in theory, associated tools and data availability will broaden the applicability of phyloclimatic modelling.

10.1 Background

Modelling the impact of climate change on the world's biota is an aspirational goal dependent on the availability of both large amounts of data and substantial computing resources. These models can be used to help us understand evolutionary relationships and ecological requirements of species, and to estimate their past, present and future distributions. The impacts of climate change on plant life are of major concern to humans because plants, apart from their intrinsic interest, play a vital role in ecosystem function and in food production and security (Heywood, 2009). The data required for modelling include species' occurrence locations, climatic variables, edaphic information and characters for phylogenetic reconstructions, while computing resources are required to build climatic and niche models and to analyse the data. The integration of these wide-ranging variables is known as 'phyloclimatic modelling' (Yesson and Culham, 2006a). There are now vast repositories of data available through distributed systems that offer the potential to allow modelling of biotic distribution patterns, phylogenies, ecological niches and the impacts of climate change without researchers having to leave their desks. However, these data should not be approached naively. Caution, and awareness of their weaknesses as well as their strengths, is needed. Before modelling can take place it is essential to consider what is being modelled. The relationships between an organism and its environment are complex, encompassing biotic and abiotic factors, functioning from microscales of a few millimetres through to macroscales of continental expanses.

One popular approach, which can be used to help understand the ecological requirements of species and estimate their distribution, is ecological niche modelling (Rödder et al., Chapter 11). Current niche modelling techniques necessarily focus on abiotic factors that show continuous variation, such as climatic conditions, and are usually referred to as models of the 'fundamental niche' (Hutchinson, 1957). Soberón (2007) reviewed niche definitions in this context, referring to these macroscale models as the 'Grinnellian niche' and explicitly excluding biotic interactions from such models. This definition is appropriate to much of the current niche modelling activity (Elith et al., 2006).

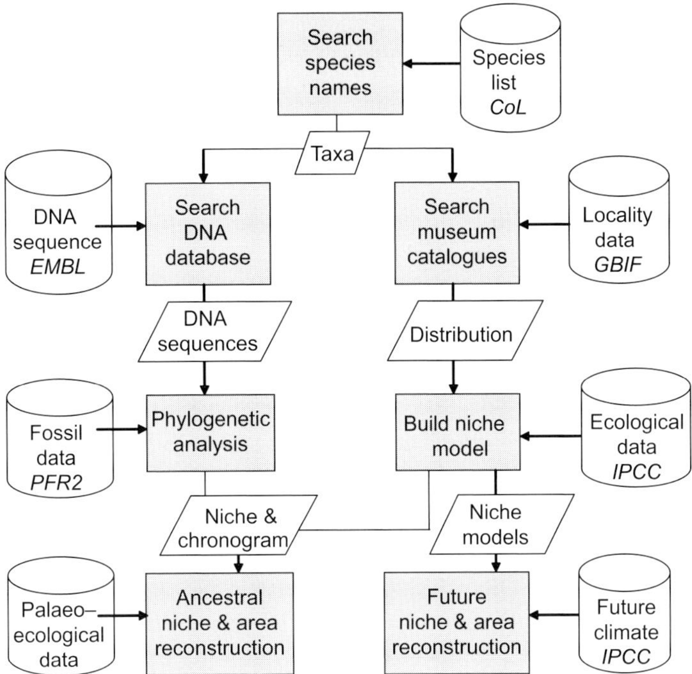

Figure 10.1 A simplified flowchart indicating data and processes for a phyloclimatic modelling workflow. Data sources are represented as cylinders, with an example database in italics. Processes are in grey boxes, outputs are in rhomboids. CoL, Catalogue of Life; GBIF, Global Biodiversity Information Facility; IPCC, Intergovernmental Panel on Climate Change; PFR2, Plant Fossil Record 2; EMBL, European Molecular Biology Laboratory.

However, in order to understand fully the interactions of species with climate, it is desirable to combine knowledge of present distribution and climate with evolutionary history (Yesson and Culham, 2006a), and hence to identify patterns for phylogenetic lineages as well as extant individual species. Such phyloclimatic modelling work requires access to substantial amounts of distributed data (Graham et al., 2004; Peterson, 2006; Yesson et al., 2007; Guralnick and Hill, 2009) and combination of these using appropriate analytical techniques (Pahwa et al., 2006). Figure 10.1 shows an example workflow for such an approach. Data on current climatic, distributional and edaphic factors are brought together to model the current bioclimatic niche of a series of species (or populations, or higher taxa). The physical data defining the niches are coded as characters on a chronogram for the same group of species (ideally based on DNA sequence data calibrated against fossil data). Reconstructions of character states at internal nodes parameterise ancestral niche models for those hypothetical taxa. The ancestral niche models are then fitted to palaeoclimate models to establish areas of potential palaeodistribution of species. Part of the process can also be worked forward in time to estimate the possible impact of climate change on monophyla in the future. This chapter reviews some of the available data, some services that draw on those data, and quality-control issues with distributed data sources. It also highlights challenges for the future.

10.2 Biodiversity informatic data sources

10.2.1 Climate model data

Without doubt the single most focused research investment in this field is in the production of future climate models. Source climate data on which these models are based are gathered from weather stations and atmospheric probes around the world. A range of models are used (Caballero and Lynch, Chapter 2) but the predominant ones for use in predictions of global climate change are coupled atmosphere–ocean models such as HadCM3 (resolution 2.5 × 3.75 degrees latitude × longitude and 19 levels of atmosphere) and GFDL CM2.X (resolution 2.5 × 2 degrees and 25 levels of atmosphere) as adopted by the Intergovernmental Panel on Climate Change (IPCC, 2007). Such models are extremely complex and demand high-performance computing resources (Slingo et al., 2009; Washington et al., 2009) that are now at the petaflop level (a thousand trillion floating point operations per second). A new generation of massive parallel computers is allowing a bridge between climate and weather models (Slingo et al., 2009). In contrast, there is relatively little work on palaeoclimate modelling (Sellwood and Valdes, 2006; Williams et al., 2007), but this is essential if biotic evolution is to be understood in relation to climate change over evolutionary time (Yesson and Culham, 2006a, 2006b).

10.2.2 Distributional data

Perhaps the best place to look for distributional data is the Global Biodiversity Information Facility (GBIF - www.gbif.org), the largest data portal to herbarium, museum and other specimen data; in April 2009 it included *c.* 8000 data sets from *c.* 300 data providers comprising *c.* 175 000 000 occurrence records. Large figures such as this look impressive and offer a mean of nearly 100 records per named species for the roughly 1.8 million species currently recognised (Bisby et al., 2009). If we assume a minimum data requirement of between 5 and 50 independent observations (Hernandez et al., 2006; Wisz et al., 2008), and if these distributional records were spread evenly for both taxonomy and geography, then we already have the geographic data needed to attempt ecological niche modelling for almost all of the world's biota. The reality is that coverage is uneven for both taxonomic and geographic reporting (Graham et al., 2004; Yesson et al., 2007; Collen et al., 2008). Some major taxonomic groups, and many species, are completely lacking data

Table 10.1 Locality data available via the Global Biodiversity Information Facility (GBIF) in April 2009.

Kingdom	Approximate number of recorded species[a]	GBIF species[b]	GBIF occurrences[c]
Animalia	5 500 000	> 250 000	90 062 361
Archaea	n/a	320	1
Bacteria	1 000 000	11 304	34 550
Chromista	200 000	6 782	386 212
Fungi	1 500 000	103 670	1 429 696
Plantae	440 000	> 250 000	32 858 998
Protozoa	260 000	11 124	1 270 532
Viruses	400 000	277	0

[a] Species numbers from www.environment.gov.au/biodiversity/abrs/publications/other/species-numbers.

[b] GBIF species downloaded using the 'species from results' link from the kingdom occurrence summary pages. Note that this download is capped at 250 000 – this limit was reached for Animalia and Plantae.

[c] GBIF occurrences from the kingdom overview page.

(Table 10.1). For example, no georeferenced data are available for the viruses, and the entire kingdom Archaea is represented by one data point, while others, such as the class Aves (birds) represent almost half the total georeferenced data (60 261 221 records in April 2009) for only about 10 000 species (http://avibase.bsc-eoc.org/avibase.jsp), giving an impressive average in excess of 6000 records per species!

Geographic coverage shows similar patchiness. A glance at the data density maps for Plantae (plants) and Animalia (animals) suggests that Europe and North America are areas of highest biodiversity, while the rainforests of Brazil are shown as biodiversity-poor areas (Fig 10.2). It should be noted that many Brazilian data are available through the species link portal (http://splink.cria.org.br), which includes many data not currently accessible via the GBIF portal. A detailed investigation using the Fabaceae (= Leguminosae; pea and bean family) as an exemplar group shows this pattern to scale through all levels of geography and taxonomy within GBIF data (Yesson et al., 2007). Patchy coverage is combined with inconsistent data quality, something that is not surprising considering that data sources range from modern global positioning system (GPS) data accurate to a few millimetres through to museum specimens collected long before most of the world was mapped accurately and for which data have had to be interpreted and digitised manually.

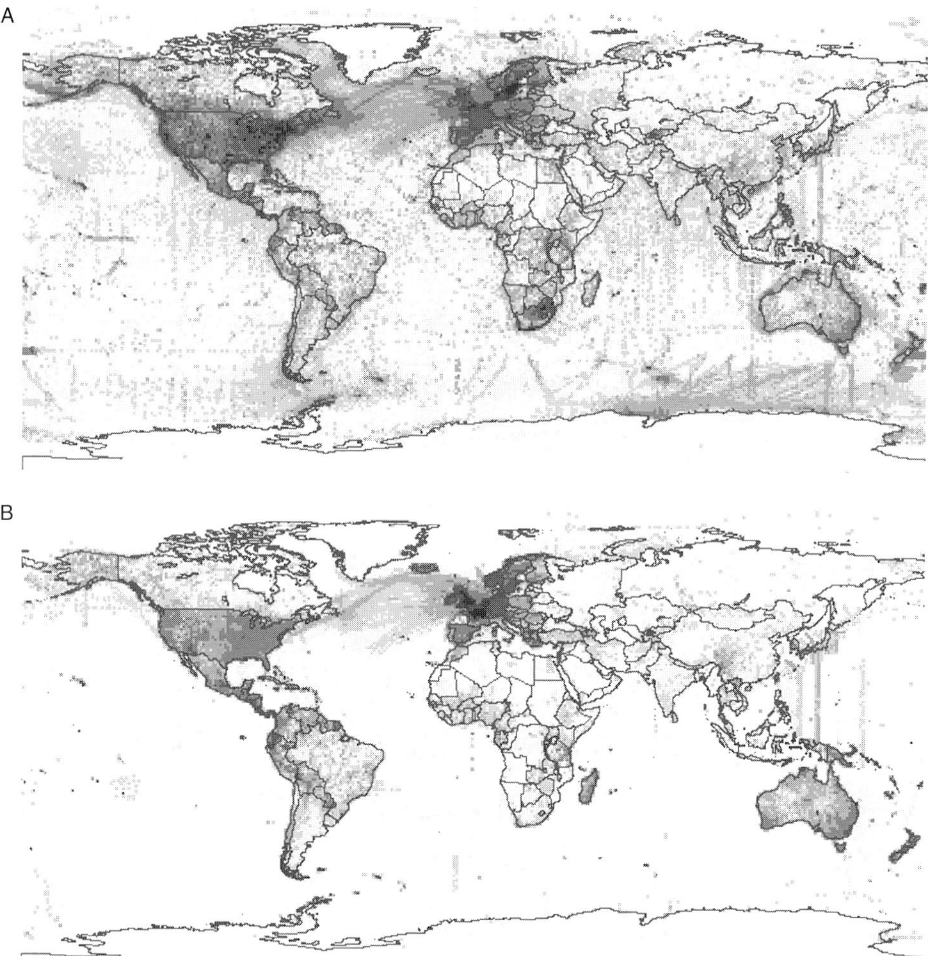

Figure 10.2 Distribution of locality data available through the Global Biodiversity Information Facility (GBIF) in April 2009 for (A) Animalia and (B) Plantae. Darker areas indicate higher frequency.

10.2.3 Taxonomic data

On top of issues with distribution data are taxonomic errors, caused by problems such as ambiguous synonyms (Page, 2005). Although organisations such as GBIF are integrating taxonomic lists into their data, there are still problems. For example, of the 21 000 data points for the tropical tree family Ebenaceae some 11 000 are in the north Atlantic Ocean, not because of problems with georeferencing, but because the family includes in its synonymy the genus *Paralia*, a name also used for a genus of phytoplankton! Blind use of such data in this case would lead to more than 53% of the distribution data being attributed wrongly. Other cases may be less obvious. However, new developments such as the Life Science

Identifier (LSID) (Clark et al., 2004) and taxonomically intelligent network services (Patterson et al., 2006) may help to reduce these difficulties.

10.2.4 DNA sequence data

Phylogenetic studies are allowing patterns of change in lineages, rather than just species, to be investigated. Such research is based on phylogenetic trees predominantly built using DNA sequence data. Not only is DNA sequencing commonly used to establish the relationships among morphological species, it is also now widely used to assess species- and populational-level boundaries that might be invisible using purely morphological data (e.g. Hartmann et al., 2006; Leliaert et al., 2009; Bateman, Chapter 3; Bernardo, Chapter 18). DNA sequence data are accessed via the three data portals for molecular data: Genbank (www.ncbi.nlm.nih.gov/genbank), EMBL-BANK (www.ebi.ac.uk/embl) and the DNA Data Bank of Japan (DDBJ) (www.ddbj.nig.ac.jp). These data portals share data, so each of the three underlying databases is largely similar in content. These databases were established during the early days of DNA sequencing and were in place for subsequent large-scale sequencing work. Many scientific journals adopted a requirement for authors to deposit data in these databases before publication of their results, and that obligation has ensured a comprehensive record of DNA sequencing activity. Quality control of data is primarily dependent on the data provider, although submissions are reviewed by specialist teams for the receiving database to ensure appropriate attempts at annotation of sequence data. There remains the problem that explanation of the sequencing method, data reliability and species/sample authentication is reported in the source publication rather than recorded in the database. This century, at least, the opportunity to cite voucher specimens that allow independent authentication of identification has become more common. Data quality, assessed through DNA sequencing electronic trace files, is an ongoing issue because of the large storage size of trace files versus text-based DNA sequence files. Trace files show the quality of the DNA read and not just the sequence of bases in the recorded DNA sequence. Trace files are now being stored for several specialist projects, such as whole genome studies and expressed sequence tag (EST) studies (e.g. www.ncbi.nlm.nih.gov/Traces and http://trace.ensembl.org), but the number of records is still trivial in comparison with the number of text-based sequence depositions.

10.2.5 Fossil data

The least complete source of data for phyloclimatic modelling studies, by far, is that for the fossil record. There are two main online databases for macrofossil data: the Paleobiology Database (http://paleodb.org), which offers a form-based search, and the Fossil Record (www.fossilrecord.net), which offers a series of downloadable files organised by taxonomic groups. In addition, there are other specialist databases such as the Fossil Pollen Database (http://pollen.cerege.fr/fpd-epd), the

Palaeoflora database (www.palaeoflora.de), which specifically includes climate preference data for the closest living relatives of fossil species, and Chronos (www.chronos.org), a data portal for data sets covering geological timescales. However, data are still scattered over a range of websites and in a broad variety of formats.

10.3 Tools

10.3.1 Niche modelling

There is a range of competing algorithms for ecological niche modelling. One of the earliest developed and most straightforward is BIOCLIM (Busby, 1991), but others have developed the approach using genetic algorithms (Stockwell and Peters, 1999), maximum entropy (Phillips et al., 2006) and many others (Elith et al., 2006). Some of these approaches have dedicated software such as DesktopGarp (www.nhm.ku.edu/desktopgarp) and Maxent (www.cs.princeton.edu/~schapire/maxent), but there are also modelling packages that offer a range of algorithms such as OpenModeller (http://openmodeller.sourceforge.net) within one desktop interface. A strength of these packages is that they are offered free of charge for research use, and in the case of OpenModeller as an open source project with a team of contributors around the world.

10.3.2 Online systems

Several online facilities are available that give an idea of the potential for future web-based biodiversity services. There have been two contrasting approaches to providing the computing power for niche modelling using distributed data. One is the provision of an application on a dedicated server via a web interface: for example, the model used for WhyWhere (http://landshape.org/enm/whywhere-20-server). The other is the use of distributed computing via a web interface and through a managing server that distributes jobs to desktop PCs: for example, the system for Lifemapper (www.lifemapper.org). Both systems allow use of GBIF distribution data, but both are limited in their niche modelling approaches when compared with desktop software such as OpenModeller. These systems begin to show the opportunities given by large distributed data sets. However, they continue to be reliant on trust in the quality and consistency of those data and still require substantial human input for large modelling projects.

10.4 Conclusions: present uses and future needs

Phyloclimatic modelling approaches have already been used to investigate climate-related evolution and distribution in several genera. In the Mediterranean basin, speciation in *Anthemis* has been linked to aridification 9 million years ago (mya) and to climatic oscillation in the past 3.5 million years (Lo Presti and Oberprieler, 2009), while *Cyclamen* appears to have been influenced more by geographic separation caused by fluctuating sea level over a similar period (Yesson et al., 2009). The Pacific Northwest mesic forest organisms of North America have been studied through palaeo-niche modelling to better understand current biogeography in the light of putative palaeogeographical distributions during cycles of glaciation (Carstens and Richards, 2007). Niche evolution over phylogenetic time has been applied to a range of terrestrial species and genera including Icteridae (American blackbirds – Eaton et al., 2008), *Oenothera* (evening primroses – Evans et al., 2009), *Drosera* (sundews – Yesson and Culham, 2006a) and Poaceae (grasses – Jakob et al., 2009) as well as marine algae (Verbruggen et al., 2009). These papers highlight the potential of a phyloclimatic approach to gain insights into, and deeper understanding of, biogeography and evolutionary history. They provide examples of the high explanatory power of past distributions on present ones over long (Yesson and Culham, 2006a) and intermediate (Carstens and Richards, 2007) timescales. Yesson and Culham (Chapter 12) outline how phyloclimatic modelling approaches have been applied to study genera in the Mediterranean-type climatic zones of Australia and the Mediterranean basin. Knowledge of the past informs plans for the future.

For the future, several developments are under way. For example, integrated data pipelines that allow experimental modelling systems to be automated over large numbers of species are in development. The Kepler project (https://kepler-project.org/users/projects-using-kepler) is an example using workflow management tools. The success of developments in such integrative science bringing together taxonomy, ecology, climatology and computer science will ultimately rest on the security of research funding in this area and on the development of open source tools that can be built collaboratively on an international scale.

Acknowledgements

We wish to thank the Biotechnology and Biological Sciences Research Council for funding the BioDiversity World project, the University of Reading for funding the second author's PhD, and numerous colleagues for feedback and discussion of our ideas over the past five years.

References

Bisby, F., Roskov, Y., Orrell, T. et al. (2009). Species 2000 and ITIS Catalogue of Life: 2009 Annual Checklist. Reading: Species 2000. www.Catalogueoflife.Org/annual-checklist/2009.

Busby, J. R. (1991). BIOCLIM: a bioclimatic analysis and prediction system. In

Nature Conservation: Cost Effective Biological Surveys and Data Analysis, ed. C. R. Margules and M. P. Austin. Melbourne: CSIRO, pp. 64–68.

Carstens, B. C. and Richards, C. L. (2007). Integrating coalescent and ecological niche modeling in comparative phylogeography. *Evolution*, **61**, 1439–1454.

Clark, T., Martin, S. and Liefeld, T. (2004). Globally distributed object identification for biological knowledgebases. *Briefings in Bioinformatics*, **5**, 59–70.

Collen, B., Ram, M., Zamin, T. and McRae, L. (2008). The tropical biodiversity data gap: addressing disparity in global monitoring. *Tropical Conservation Science*, **1**, 75–88.

Eaton, M. D., Soberon, J. and Peterson, T. (2008). Phylogenetic perspective on ecological niche evolution in American blackbirds (family Icteridae). *Biological Journal of the Linnean Society*, **94**, 869–878.

Elith, J., Graham, C. H., Anderson, R. P. et al. (2006). Novel methods improve prediction of species' distributions from occurrence data. *Ecography*, **29**, 129–151.

Evans, M. E. K., Smith, S. A., Flynn, R. S. and Donoghue, M. J. (2009). Climate, niche evolution, and diversification of the 'Bird-cage' Evening primroses (*Oenothera*, sections *Anogra* and *Kleinia*). *American Naturalist*, **173**, 225–240.

Graham, C. H., Ferrier, S., Huettman, F., Moritz, C. and Peterson, A. T. (2004). New developments in museum-based informatics and applications in biodiversity analysis. *Trends in Ecology and Evolution*, **19**, 497–503.

Guralnick, R. and Hill, A. (2009). Biodiversity informatics: automated approaches for documenting global biodiversity patterns and processes. *Bioinformatics*, **25**, 421–428.

Hartmann, F. A., Wilson, R., Gradstein, S. R., Schneider, H. and Heinrichs, J. (2006). Testing hypotheses on species delimitations and disjunctions in the liverwort *Bryopteris* (Jungermanniopsida: Lejeuneaceae). *International Journal of Plant Sciences*, **167**, 1205–1214

Hernandez, P. A., Graham, C. H., Master, L. L. and Albert, D. L. (2006). The effect of sample size and species characteristics on performance of different species distribution modeling methods. *Ecography*, **29**, 773–785.

Heywood, V. H. (2009). The impacts of climate change on plant species in Europe, with contributions by A. Culham. Strasbourg: Convention on the Conservation of European Wildlife and Natural Habitats Standing Committee.

Hutchinson, G. E. (1957). Concluding remarks. Cold Spring Harbor Symposium. *Quantitative Biology*, **22**, 415–427.

Intergovernmental Panel on Climate Change (IPCC) (2007). *Climate Change 2007: Synthesis Report. Contribution of Working Groups I, II and III to the Fourth Assessment Report of the Intergovernmental Panel on Climate Change*, ed. R. K. Pachauri and A. Reisinger. Geneva: IPCC.

Jakob, S. S., Martinez-Meyer, E. and Blattner, F. R. (2009). Phylogeographic analyses and paleodistribution modeling indicate Pleistocene in situ survival of *Hordeum* species (Poaceae) in southern Patagonia without genetic or spatial restriction. *Molecular Biology and Evolution*, **26**, 907–923.

Leliaert, F., Verbruggen, H., Wysor, B. and De Clerck, O. (2009). DNA taxonomy in morphologically plastic taxa: algorithmic species delimitation in the *Boodlea* complex (Chlorophyta: Cladophorales). *Molecular Phylogenetics and Evolution*, **53**, 122–133.

Lo Presti, R. M. and Oberprieler, C. (2009). Evolutionary history, biogeography and eco-climatological differentiation of the genus *Anthemis* L. (Compositae, Anthemideae) in the circum-Mediterranean area. *Journal of Biogeography*, **36**, 1313–1332.

Page, R. D. M. (2005). A taxonomic search engine: federating taxonomic databases using web services. *BMC Bioinformatics*, **6**, 48.

Pahwa, J. S., Jones, A. C., White, R. J. et al. (2006). Supporting the construction of workflows for biodiversity problem-solving accessing secure, distributed resources. *Scientific Programming*, **14**, 195–208.

Patterson, D. J., Remsen, D., Marino, W. A. and Norton, C. (2006). Taxonomic indexing: extending the role of taxonomy. *Systematic Biology*, **55**, 367–373.

Peterson, A. T. (2006). Uses and requirements of ecological niche models and related distributional models. *Biodiversity Informatics*, **3**, 59–72.

Phillips, S. J., Anderson, R. P. and Schapire, R. E. (2006). Maximum entropy modeling of species geographic distributions. *Ecological Modelling*, **190**, 231–259.

Sellwood, B. W. and Valdes, P. J. (2006). Mesozoic climates: general circulation models and the rock record. *Sedimentary Geology*, **190**, 269–287.

Slingo, J., Bates, K., Nikiforakis, N. et al. (2009). Developing the next-generation climate system models: challenges and achievements. *Philosophical Transactions of the Royal Society of London A*, **367**, 815–831.

Soberón, J. (2007). Grinnellian and Eltonian niches and geographic distributions of species. *Ecological Letters*, **10**, 1115–1123.

Stockwell, D. R. B. and Peters, D. P. (1999). The GARP modelling system: problems and solutions to automated spatial prediction. *International Journal of Geographic Information Systems*, **13**, 143–158.

Verbruggen, H., Tyberghein, L., Pauly, K. et al. (2009). Macroecology meets macroevolution: evolutionary niche dynamics in the seaweed *Halimeda*. *Global Ecology and Biogeography*, **18**, 393–405.

Washington, W. M., Buja, L. and Craig, A. (2009). The computational future for climate and earth system models: on the path to petaflop and beyond. *Philosophical Transactions of the Royal Society of London A*, **367**, 833–846.

Williams, M., Haywood, A. M., Gregory, J. and Schmidt, D. N. (2007). *Deep-time Perspectives on Climate Change: Marrying the Signal From Computer Models and Biological Proxies*. London: Geological Society of London on behalf of the Micropalaeontological Society.

Wisz, M. S., Hijmans, R. J., Li, J. et al. (2008). Effects of sample size on the performance of species distribution models. *Diversity and Distributions*, **14**, 763–773.

Yesson, C. and Culham, A. (2006a). Phyloclimatic modelling: combining phylogenetics and bioclimatic modelling. *Systematic Biology*, **55**, 785–802.

Yesson, C. and Culham, A. (2006b). A phyloclimatic study of *Cyclamen*. *BMC Evolutionary Biology*, **6**, 72.

Yesson, C., Brewer, P. W., Sutton, T. et al. (2007). How global is the Global Biodiversity Information Facility? *PLoS ONE*, **2**, e1124.

Yesson, C., Toomey, N. H. and Culham, A. (2009). *Cyclamen*: time, sea and speciation biogeography using a temporally calibrated phylogeny. *Journal of Biogeography*, **36**, 1234–1252.

11

Climate envelope models in systematics and evolutionary research: theory and practice

D. RÖDDER

Herpetology Department, Zoologisches Forschungsmuseum Alexander Koenig, Bonn and *Biogeography Department, Trier University, Germany*

S. SCHMIDTLEIN

Vegetation Geography Department, Bonn University, Germany

S. SCHICK

Biogeography Department, Trier University, Germany

S. LÖTTERS

Biogeography Department, Trier University, Germany

Abstract

Climatic information from distribution data of a species can be used to compute its climate envelope. Climate envelope models (CEMs) are employed to predict

---

Climate Change, Ecology and Systematics, ed. Trevor R. Hodkinson, Michael B. Jones, Stephen Waldren and John A. N. Parnell. Published by Cambridge University Press.

potential geographic ranges of species as a function of climate by comparing the climate envelope with climatic conditions at locations of unknown occurrence. CEMs find their way into applied sciences such as conservation management and risk assessment, but they also perform well in systematics and evolutionary research, often supplementary to other methods. Although the application of CEM approaches is developing rapidly, there is a considerable lack of theoretical background. We summarise theoretical assumptions behind CEMs, describe how they work and discuss possible pitfalls when interpreting results. In addition, we provide examples from our ongoing research on the Afrotropical reed frogs, genus *Hyperolius* (Hyperoliidae). We delimit the potential distribution of a recently recognised taxon within the *Hyperolius cinnamomeoventris* species complex and propose possible speciation scenarios for *H. mitchelli* and *H. puncticulatus*.

11.1 Introduction

11.1.1 Climate and the geographic distribution of species

It is known that climate elements and factors have an important influence on the distribution of plant and animal species; likewise, the ecological niche concept has been well discussed (Grinnell, 1917; James et al., 1984). In recent years, there has been a remarkable increase in availability of information on climatic parameters in geographic space, including remote regions. There has also been improved recording of species distribution data. Accompanied by improved computation capacities, these have allowed for an increase in large-scale assessments of the relationship between observed species distributions and explanatory environmental (climatic) parameters. Such studies can be approached by modelling climate niches of species and then projecting them into geographic space (Guisan and Zimmermann, 2000).

In combination with climate scenarios, niche models are frequently used for an assessment of future climate-change impacts on species distributions (Thuiller et al., 2008). In the disciplines of biological systematics and evolution, climate niche models can provide helpful information (Hirzel and Le Lay, 2008; Kozak et al., 2008). Possible applications include the assessment of species' geographic ranges and delimitations for taxonomic purposes, or testing of evolutionary scenarios such as prehistoric distributions or niche evolution (Graham et al., 2004; Raxworthy et al., 2007; Kozak et al., 2008; Rödder and Dambach, 2010).

11.1.2 Ecological niche concepts

Following the concept of Hutchinson (1957, 1978), a species' climate niche or climate envelope is part of its fundamental niche (Fig 11.1). This is defined as the complete set of abiotic environmental conditions under which a species can persist. The realised niche of a species is understood to represent a subset of the fundamental niche (abiotic) considering biotic interaction such as food availability, competition, predation or interaction with pathogens (Fig 11.1).

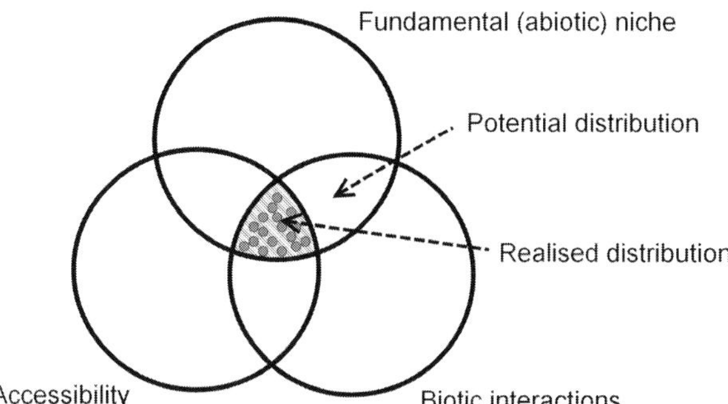

Figure 11.1 Relationships between fundamental niche, biotic interaction and accessibility, according to Hutchinson (1957) as modified by Soberón and Peterson (2005). The potential distribution is a subset of the fundamental niche considering biotic interactions, whereas the realised distribution is a subset of the potential distribution considering accessibility. Dots represent species records.

The variables determining such niches can be subdivided into specific classes depending on the spatial extent in which they operate and whether or not competition plays a role. The Grinnellian niche is defined by 'external' drivers such as climate, which are not subject to species–species interactions (Grinnell, 1917). In contrast, the Eltonian niche focuses on biotic interaction and resource–consumer dynamics (Elton, 1927). The former usually operates on coarser scales whereas the latter is related more closely to the local scales (Soberón, 2007). As a result, climate niche models, by definition, address the Grinnellian niche only. It is important to note that Grinnellian (and Eltonian) niches are not always completely reflected in the actual range of a species because of historical factors. Remnant populations, source–sink dynamics and incomplete 'filling' of a niche due to dispersal limitations or other factors may lead to deviant patterns (Puliam, 2000; Guisan and Thuiller, 2005). This implies that the range of environmental conditions observed at the sites of species occurrences may not necessarily cover the entire niche spectrum suitable for the species, or may go well beyond the range of conditions suitable for long-term persistence (Tilman et al., 1994). Species may be in disequilibrium with climate conditions (Araújo and Pearson, 2007; Rödder et al., 2008).

11.2 How to assess niches in geographic space

11.2.1 CEMs

Mapped climate data offer remarkable opportunities to approach variation in environmental factors belonging to the Grinnellian niche within the ranges of species, especially when combined with spatial modelling and geographical information systems (GIS) techniques (Kozak et al., 2008). Such approaches are known as climate envelope models (CEMs – Fig 11.2). For some time now, they have been

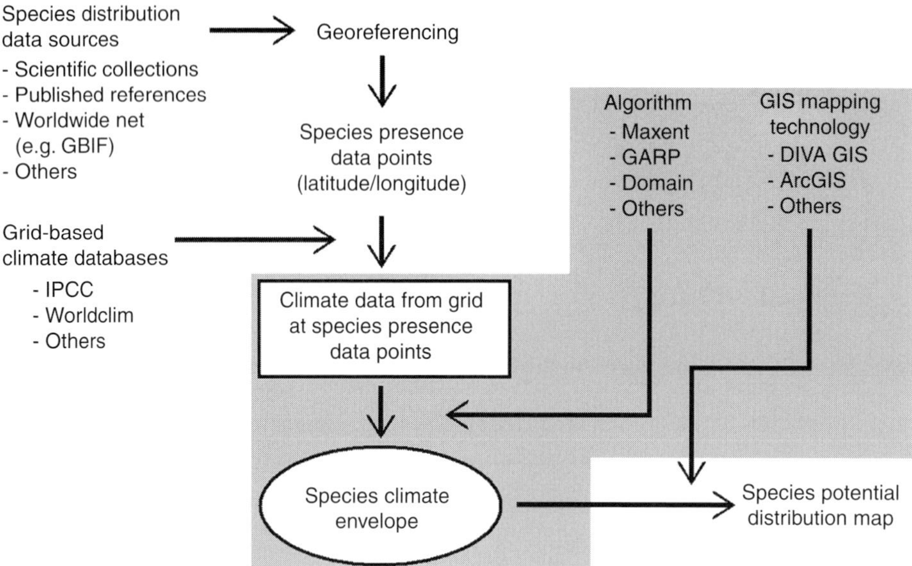

Figure 11.2 Flowchart illustrating the main steps for building a CEM.

applied to predict species' potential distributions under current, past and future climate scenarios (e.g. Araújo et al., 2004; Araújo and Guisan, 2006; Heikkinen et al., 2006; Hijmans and Graham, 2006; Waltari et al., 2007; Pearman et al., 2008; Waltari and Guralnick, 2009). They play an important role in applied science, for instance in: (1) invasive species biology and pest control/risk assessments (Peterson and Vieglais, 2001; Peterson, 2003; Rödder et al., 2008; Rödder, 2009); (2) the determination of the potential ranges of crops (Hijmans et al., 2005a); (3) conservation priority setting (Araújo et al., 2004; Kremen et al., 2008; Lötters et al., 2008; Rödder et al., 2010); and (4) several aspects of ecology and evolutionary biology such as speciation studies (Peterson et al., 1999; Graham et al., 2004; Kozak et al., 2008).

11.2.2 What is needed for CEM building?

Species records

Prior to constructing a CEM, it is necessary to compile a list of georeferenced distribution records of the target species (i.e. with information on latitude/longitude) (Fig 11.2). Fortunately, the amount of available data has increased significantly during the last decades as, for instance, the Global Biodiversity Information Facility (GBIF) makes available > 10^8 species records for free (www.gbif.org). If geographical coordinates are not available, gazetteers such as the Alexandria Digital Library Gazetteer of the University of California (www.alexandria.ucsb.edu) are helpful. Despite being able to georeference records one by one, batch processing can facilitate the procedure (Soberón and Peterson, 2004) by using, for example, the BioGeomancer software (http://biogeomancer.org). The spatial accuracy of the geographical coordinates necessary for robust model building depends on the spatial resolution of the environmental layers used and the algorithm applied (Graham et al., 2008). A minimum number of at least 10–30 distribution records for the species (or occasionally subspecies, superspecies etc.) under study is necessary depending on the algorithm later applied (e.g. Elith et al., 2006; Pearson et al., 2007; Wisz et al., 2008).

When compiling species records it is important to evaluate possible bias (Soberón and Peterson, 2004), which can comprise spatial sample selection bias (Jiménez-Valverde and Lobo, 2006), historical factors (Hortal et al., 2008) or taxonomic components (Soberón and Peterson, 2004). Another concern is spatial autocorrelation, which is always present in spatial data sets and which may bear heavily on models (Dormann et al., 2007). Spatial autocorrelation is basically a lack of independence between observations due to the fact that vicinity in space alters the chance of occurrence. An analogous phenomenon is observed in time if multitemporal data sets are used. Methods dealing with spatial autocorrelation, such as simultaneous autoregressive (SAR) models or principal coordinate analysis of neighbour matrices (PCNM) approaches, are treated in detail by Dormann et al. (2007).

Most algorithms build models based on species' presence records, but there are also a variety of applications that can, in addition, deal with species' absence data. Presence-only methods may be preferable since true absence records of species remain difficult to prove (Gu and Swihart, 2004), especially for rare or highly mobile species. It is also often unclear whether a species is absent from a given locality because the site is outside its climate envelope or because of other factors such as biotic interaction, disturbance or dispersal limits. This can lead to misinterpretations; e.g. if the climate at a locality treated as an absence locality is within the target species' climate envelope, the model algorithm misinterprets the climate at this site as unsuitable. Constructing models for migratory species is a special challenge, since temperospatial patterns need to be considered when compiling species records and predictors (Martinez-Meyer et al., 2004; Hirzel and Le Lay, 2008).

Climate layers

Climate information can be incorporated into CEMs from various sources, and the selection of the most suitable data set depends on the spatial extent of the target area and the goals of the study. Data on current climate from all over the world can be obtained from the WorldClim database, which is based on weather conditions recorded at roughly 50 000 locations for precipitation and 25 000 locations for temperature between 1950 and 2000 (www.worldclim.org – Hijmans et al., 2005b). This grid-based database (resolution 30 arcseconds) was created by interpolation using a thin plate smoothing spline of observed climate at weather stations, with latitude, longitude and elevation as independent variables using the ANUCLIM software (Hutchinson, 1995, 2004). Regional climate data, which, if available, are usually preferred for non-global studies, are provided by national weather agencies and other local sources.

When projecting species' CEMs into past or future climate scenarios it is important to acknowledge that different scenarios will reveal different results and no single 'best' model exists. Strengths and weaknesses of different climate models should be considered (Beaumont et al., 2008). Evaluation of a variety of scenarios may help to assess variations in outputs. For example, climate change projections based on the CCCma, CSIRO and HadCM3 models (Flato et al., 2000; Gordon et al., 2000) and the emission scenarios reported in the Special Report on Emissions Scenarios (SRES) by the Intergovernmental Panel on Climate Change (IPCC) (www.grida.no/climate/ipcc/emission) for the years 2020, 2050, 2080 can be obtained via the WorldClim homepage. Future projections using other IPCC scenarios are also available from the Climate Research Unit (CRU) of the University of East Anglia in the UK (www.cru.uea.ac.uk/cru/data/hrg.htm – New et al., 1999, 2000). Furthermore, upcoming regional models are providing more spatially detailed information and take account of regional scale topographic variability from the very beginning.

A set of different families of emission scenarios was formulated based on future production of greenhouse gases and aerosol precursor emissions. The SRES scenarios of A2a and B2a each describe one possible demographic, politico-economic, social and technological future. Scenario B2a emphasises more environmentally conscious and more regionalised solutions to economic, social and environmental sustainability. Scenario A2a also emphasises regionalised solutions to economic and social development, but it is less environmentally conscious.

For the palaeoclimate during the Last Glacial Maximum of *c.* 21 000 years before present (bp), general circulation model (GCM) simulations from the Community Climate System Model (CCSM) are available (www.ccsm.ucar.edu – Kiehl and Gent, 2004). The Model for Interdisciplinary Research on Climate (MIROC version 3.2 – www.ccsr.u-tokyo.ac.jp/~hasumi/MIROC) can also be used.

Ecological niche modelling commonly uses results from a climate model, including a broad range of variables, typically including minimum and maximum temperatures and the mean precipitation per month (= 36 climate parameters). Based on these monthly layers, 19 bioclimatic parameters can be generated, e.g. with DIVA-GIS (www.diva-gis.org – Hijmans et al., 2002, 2005a). These are often used in CEMs, and represent annual seasonality and extreme or limiting climate factors (Table 11.1). Bioclimatic parameters are more useful than 'raw' monthly values since they are independent of latitudinal variation. This becomes obvious when considering that the maximum temperature of the warmest month is more informative with respect to a species' biology than the

Table 11.1 Bioclimatic parameters and their abbreviations representing annual trends, seasonality and extreme or limiting climate factors. From www.worldclim.org/bioclim.

Abbreviation	Parameter
BIO1	Annual mean temperature
BIO2	Mean diurnal range (mean of monthly (max temp – min temp))
BIO3	Isothermality (BIO2/BIO7) (× 100)
BIO4	Temperature seasonality (standard deviation × 100)
BIO5	Max temperature of warmest month
BIO6	Min temperature of coldest month
BIO7	Temperature annual range (BIO5 – BIO6)
BIO8	Mean temperature of wettest quarter
BIO9	Mean temperature of driest quarter
BIO10	Mean temperature of warmest quarter
BIO11	Mean temperature of coldest quarter
BIO12	Annual precipitation
BIO13	Precipitation of wettest month
BIO14	Precipitation of driest month
BIO15	Precipitation seasonality (coefficient of variation)
BIO16	Precipitation of wettest quarter
BIO17	Precipitation of driest quarter
BIO18	Precipitation of warmest quarter
BIO19	Precipitation of coldest quarter

maximum temperature of a specific month, because the latter varies with latitude (Nix, 1986; Busby, 1991).

Multicollinearity among predictor variables may hamper the analysis of species–environment relationships because its ecologically more causal variables may be excluded from models if other intercorrelated variables explain the variation in response variables better in statistical terms (Heikkinen et al., 2006). For example, if two variables are similarly distributed in space, both are similarly represented in a species model. Independent variation of the two variables may lead to false predictions when one of them is causally linked to a species' distribution and one is not. Therefore, variable selection should be guided by a thorough assessment of the target species' ecology, and rather a minimalistic set of predictors should be preferred depending on the focal species. For example, BIO1, 10, 11, 12, 16 and 17 from the WorldClim data set reflect the availability and range of thermal energy and humidity and may be suitable for CEM projections between different climate scenarios, according to different authors (Carnaval and Moritz, 2008). Specific adjustments of variables according to specific ecological needs of the target species may improve the model output (Beaumont et al., 2005; Rödder et al., 2009). It needs to be noted that negative effects of multicollinearity may vary among algorithms.

Model algorithms

Once species records and predictor variables are compiled, a subsequent step involves the development of a multidimensional view of the climate envelope of a species. This is a considerable challenge, given the complex nature of a species' ecological niche (Peterson and Vargas, 1993). In other words, in CEMs, species occurrences are mapped into the climate space. The part of the climate space occupied by a species can be mapped into geographic reality using spatial climate information. The selection of a suitable algorithm for the computation of the CEM depends on the quantity of distribution records available, their quality, and the specific goal of the study.

Among the earlier applied algorithms used for presence-only data were BIOCLIM (Nix, 1986; Busby, 1991) and DOMAIN (Carpenter et al., 1993), as implemented in DIVA-GIS. Whereas BIOCLIM measures the distance to the midpoint of the training sites in suitable climate space as suggested by conditions at training records, DOMAIN measures the environmental similarity of each grid cell to the most similar training site (Nix, 1986; Carpenter et al., 1993) (Fig 11.3). More sophisticated algorithms are Genetic Algorithm for Rule-set Production (GARP – Stockwell and Noble, 1992; Stockwell and Peters, 1999) and Maxent (Phillips et al., 2004, 2006). These more recently developed methods derive predictions by developing sets of rules or by machine learning approaches, and they are considered superior to most other methods (for a comparison of performance, see Elith et al., 2006; Heikkinen et al., 2006; Wisz et al., 2008). If absence records are available, or even records of abundance, algorithms such as artificial neuronal networks, classification and regression trees, generalised additive models or generalised dissimilarity models can be applied. These algorithms are, for example, implemented in the BIOMOD tool (Thuiller, 2003).

11.2.3 Evaluation of CEM results

Assessing the accuracy of a model's prediction is a vital step in model development, since applications of it will have little merit if the accuracy of its prediction is unknown (Pearce and Ferrier, 2000). One of the most commonly used approaches

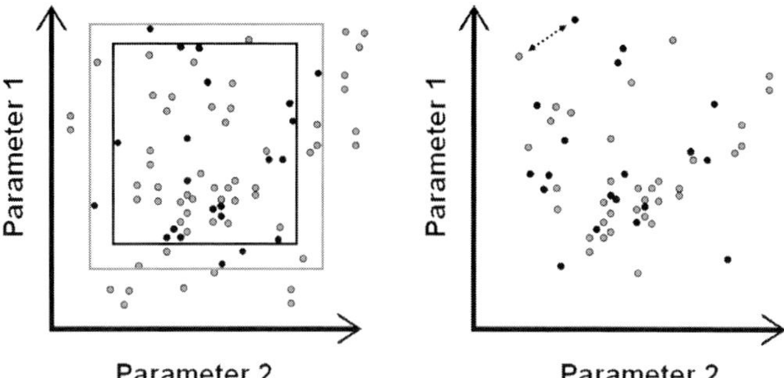

Figure 11.3 Examples of two different approaches to the analysis of environmental niches. Assumptions and concepts may vary between different modelling algorithms, whereby conditions at species records (black dots) and conditions at locations in question (grey dots) are indicated. In BIOCLIM (left) the environmental niche is defined as a boxcar environmental envelope in climate space, whereby 'core' (black box, enclosing 90% of all species records) and 'marginal' areas (grey box, enclosing 100% of the records) are defined. All grey dots enclosed by the boxcar envelope are suggested to be suitable for the target species. In DOMAIN (right) the relative distance between conditions as observed at species records (black dots) and at locations to be assessed (grey dots) is measured in climate space.

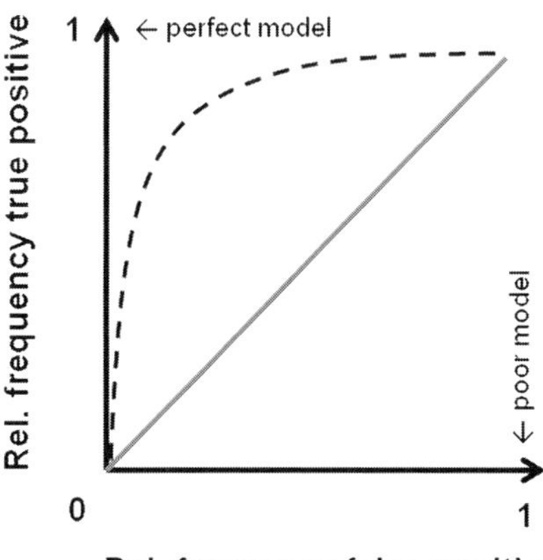

Figure 11.4 The receiver operating characteristic (ROC) curve. The ROC is formed by plotting values of the relative frequency of true positive records predicted by a given model against the values of the relative frequency of false positive records. The solid 1 : 1 line signifies random predictive ability, where there is no ability to distinguish occupied and unoccupied sites. The dashed line may be characteristic for a model with good predictive abilities.

to test a model is to use area under the curve (AUC) statistics, referring to the receiver operating characteristic (ROC) curve (Fig 11.4). Between a quarter and a third of the species records are commonly set aside from modelling to act as test points (Hanley and McNeil, 1982; Phillips et al., 2006). This method is recommended for ecological applications because it is non-parametric (Pearce and Ferrier, 2000). Values of AUC range from 0.5 (i.e. random) for models with no predictive ability to 1.0 for models giving perfect predictions. According to the classification of Swets (1988), AUC values > 0.9 describe 'very good', > 0.8 'good' and > 0.7 'useful' discrimination ability. Lobo et al. (2008) recently criticised AUC values as potentially misleading. However, for presence-only models, no alternative testing measure exists and it has therefore been frequently used in subsequent studies (e.g. Graham et al., 2008; Wisz et al., 2008).

Another way to test model output is Cohen's kappa statistic of similarity (κ), applicable when presence and absence data are available (Fielding and Bell, 1997; Pearce and Ferrier, 2000). This measure is adjusted to account for chance agreement between predicted and observed values by taking the proportion of correct predictions expected by chance into account. Cohen's kappa yields values ranging from 0.0 (no predictive ability) to 1.0 (perfect predictive ability), with κ values above 0.7 being described as having 'very good' discrimination ability (Monserud and Leemans, 1992).

It needs to be pointed out that test statistics such as AUC or Cohen's kappa are measures of model fit compared to the observed distribution of a species and must not be interpreted as a measure of ecological sense, i.e. characterising a model best meeting the biology of the target species. A profound knowledge of a species' natural history is necessary to evaluate whether a model meets the ecological requirements of the target species, and this is best found through expert opinions.

11.3 CEMs in practice: Afrotropical reed frogs

Reed frogs (Hyperoliidae: *Hyperolius*) are a monophyletic group of small nocturnal and arborical amphibians which are known from savannas and forests in sub-Saharan Africa (Schiøtz, 1999; Veith et al., 2009). More than 130 species have been recognised (Frost, 2010). Due to limited interspecific and remarkable intraspecific morphological variation, the taxonomy of numerous *Hyperolius* taxa is poorly understood. Certain nominal species may actually represent complexes of distinct taxa. Bioacoustics and DNA barcoding have provided useful tools for species discrimination, but the availability of samples is still sparse (Köhler et al., 2005; Veith et al., 2009). We have studied reed frog systematics for around a decade and provide here some examples of CEM and how it can supplement other methods used to study the systematics and evolution of these amphibians.

11.3.1 *Hyperolius cinnamomeoventris* species complex

Hyperolius cinnamomeoventris Bocage, 1866 is a reed frog from the Congo Basin and vicinities. Results from Lötters et al. (2004), Veith et al. (2009) and the authors'

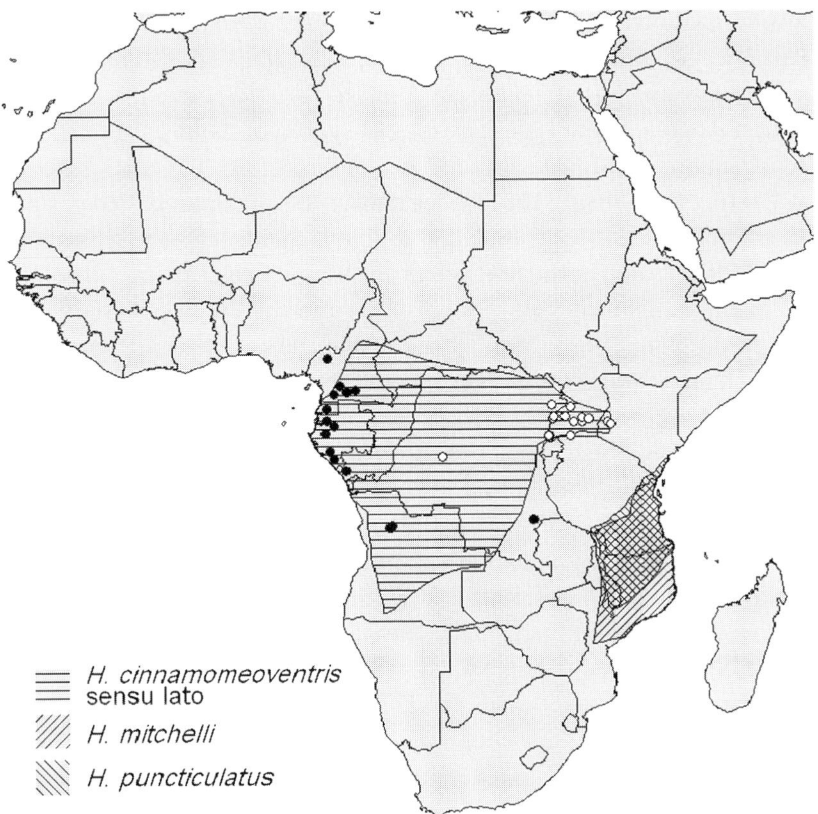

Figure 11.5 Distribution of three reed frog species, genus *Hyperolius*, according to the International Union for the Conservation of Nature (IUCN) Red List (www.iucnredlist.org). In the case of the *H. cinnamomeoventris* species complex, DNA barcoding (*c.* 550 bp of the 16S mitochondrial rRNA gene) of samples from different localities (all circles, some of which lay outside the geographic range according to the IUCN Red List) revealed that different sibling species are involved. White circles indicate samples which belong to an eastern taxon within this species complex (authors' unpublished data).

unpublished molecular data suggest that several cryptic species have been included in this name. Samples studied from part of the geographic range encompassed by this reed frog complex in the eastern part of the Democratic Republic of Congo (DRC), Uganda and western Kenya are genetically distinct from those from elsewhere within its entire geographic range (Fig 11.5, white circles). We tentatively conclude that they represent an 'eastern taxon' within the *H. cinnamomeoventris* complex. A problem is that the relatively low number of genetic samples does not allow for an appreciation of spatial delimitation of the eastern taxon or any other species behind the name *H. cinnamomeoventris*. It also remains unclear how to apply the nomenclature, as there are several different scientific names currently treated as junior synonyms of *H. cinnamomeoventris* (Frost, 2010), namely: *H. fimbriolatus* Bocage, 1866; *H. ituriensis* Bocage, 1866; *H. olivaceus* Peters, 1876; *H. tristis* Bocage, 1866; and *H. wittei* Laurent, 1957. *Hyperolius cinnamomeoventris* is described from Angola, while type localities of synonym names are from this country, DRC or Gabon (Fig 11.6A).

A Maxent CEM (Fig 11.6A) using 17 presence data points of the suggested eastern taxon and based on BIO1, 10, 11, 12, 16 and 17 (Table 11.1) advocates that this reed frog is potentially distributed in the northern Lake Victoria catchment, part of the northern Congo Basin, the Eastern Arc Mountains and the Ethiopian Highlands (AUC = 0.992). Eastern Arc and Ethiopia are outside the realised distribution of any *Hyperolius* referable to *H. cinnamomeoventris* (Fig 11.5), and can be ignored when analysing the geographic range of the hypothetical eastern taxon. Today, the type locality of *H. cinnamomeoventris* and those of all its junior synonym names are outside the potential distribution of the suggested eastern taxon, except that of *H. ituriensis* (Fig 11.6A), suggesting that this name may be applicable. In addition, it may be possible that type localities of other synonym names represent Pleistocene relicts. In order to assess this hypothesis we projected the Maxent CEM as shown in Fig 11.6A onto palaeoclimate data derived from GCM simulations from the CCSM, as explained above. Figure 11.6B indicates how the eastern taxon was potentially distributed during the Last Glacial Maximum. During this cooler and drier period, it might have been more widely distributed in the Congo Basin than today. Regarding the type localities of the different names available, they are all situated outside the potential distribution of the suggested eastern taxon under Last Glacial Maximum conditions, again with exception of *H. ituriensis* (Fig 11.6A,B). This underlines that *H. ituriensis* may be the best applicable name for the suggested eastern taxon.

We lack distribution data to generate CEMs for other species previously included under the name *H. cinnamomeoventris* and identified through molecular markers. As a result, it cannot be tested if any of these occurs in sympatry with the eastern taxon, especially if any of them potentially occurs in the area of the type locality of *H. ituriensis*. It may be speculated that this is not the case, as sister species commonly exhibit allopatric distributions (Fisher et al., 2001; Graham et al., 2004). This is well explained, as geographic separation likely causes some degree of shift in climate envelopes as a response to local environmental conditions (Graham et al., 2004).

11.3.2 *Hyperolius mitchelli* and *H. puncticulatus*

Although it is often observed and readily explained that sister species are allopatric and differ in their climate envelopes, this is not always corroborated. The East African reed frog species *H. mitchelli* Loveridge, 1953 and *H. puncticulatus* (Pfeffer, 1893) can be distinguished clearly on the basis of vocalisations and DNA

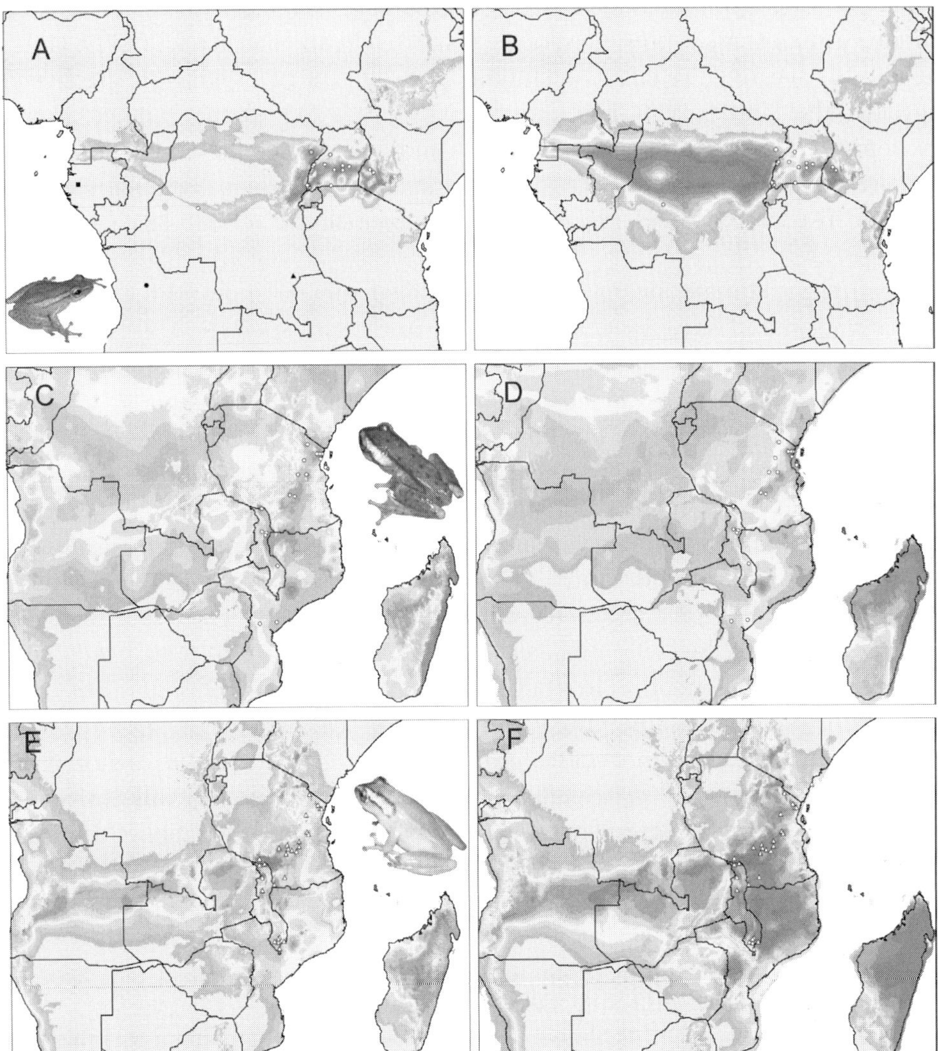

Figure 11.6 CEMs for three African reed frog species (*Hyperolius*). See Fig 11.5 for names and known distributions. CEMs computed with Maxent for different African reed frog species under current climate conditions (A, C, E) and palaeoclimate conditions suggested for the Last Glacial Maximum, *c.* 21 000 years bp (B, D, F, considering past sea-level fluctuation). White circles indicate presence data for species used for modelling; other symbols used in (A) represent type localities of nominal species currently treated as junior synonyms of *H. cinnamomeoventris* (filled square, *H. fimbriolatus* and *H. olivaceus*; filled circle, *H. cinnamomeoventris* sensu stricto and *H. tristis*; filled triangle, *H. wittei*; grey circle, *H. ituriensis*). *See colour plate section.*

sequences (Schiøtz, 1999; Rödder and Böhme, 2009; authors' unpublished data). Their known geographic ranges largely overlap (Fig 11.5). Furthermore, a Maxent model (using 18 and 27 presence points, respectively, and based on BIO1, 10, 11, 12, 16 and 17 – see Table 11.1; AUC = 0.903 and 0.952 for the two species, respectively) revealed that similarity in their climate envelopes is so high that the two cannot be separated using CEMs (Fig 11.6C,E). As perhaps expected, even when projecting the Maxent models onto CCSM palaeoclimate simulations, the potential geographic distributions of *H. mitchelli* and *H. puncticulatus* remain largely similar (Fig 11.6D,F).

Apparently, climate niches in these two sister species show a high degree of conservatism and have not changed significantly with speciation. This gives an interesting insight into their evolution and poses some questions. We would like to know if *H. mitchelli* and *H. puncticulatus* have speciated in sympatry, or if they have speciated in allopatry but have retained similar climate envelopes. Both patterns are uncommon in amphibians. To the best of our knowledge, sympatric speciation has never been demonstrated in amphibian species. For us, the most likely explanation is that speciation has taken place in isolated refuges during a warm phase allowing isolation. Climatic conditions at these isolated refuges (likely mountains) were apparently similar, thus explaining the lack of niche divergence during speciation. The example of *H. mitchelli* and *H. puncticulatus* also illustrates the effects of limited accessibility and availability of microhabitats. Both species may find climatically suitable regions outside their realised distribution in great parts of central Africa and Madagascar. While range expansions to Madagascar are restricted by the sea, the Albertine Rift valley, with its numerous lakes, also prevents westward range extensions. In addition, this lowland region is climatically relatively unsuitable for both species (Fig 11.6C,E, indicated in green).

11.4 Potential pitfalls of CEMs

CEMs are derived from a subset of environmental conditions and species records (i.e. variables selected), and therefore capture only a general estimate of the niche. They are unable to capture the niche completely, and output maps show regions with similar conditions as observed at the training records according to the selected predictor variables, rather than 'niches' (Fig 11.7).

The question of whether climate niches are conservative or not over phylogenetic history is important when applying CEMs. 'Niche conservatism' of closely related species is a phenomenon that has been observed in several different taxonomic groups (Peterson et al., 1999; Wiens and Graham, 2005). On the other hand, niche shifts have also been documented (Graham et al., 2004; Broennimann et al.,

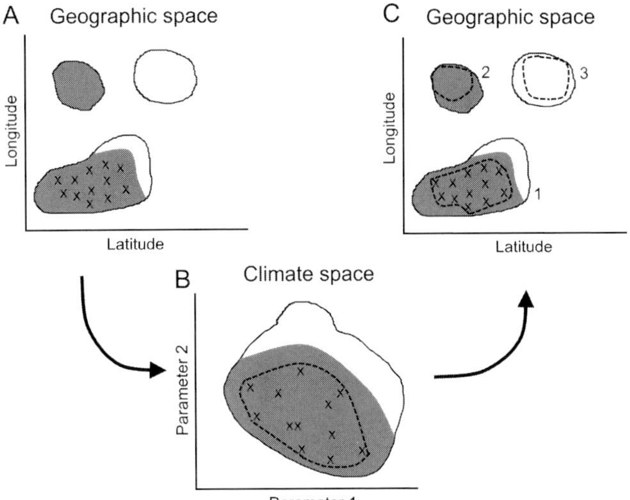

Figure 11.7 Potential error sources and uncertainties in CEMs. Species records (x) available for model training commonly reflect the species' entire realised distribution (A, C: grey), its potential distribution (solid lines), or the complete suitable niche space (B: grey). Since CEMs are fitted based on a subset of the suitable niche space (C: dashed line), the model may not identify the entire realised or potential distribution. Subsequent projection of the model in geographic space may identify three different distribution types: projected area 1 identifies the known distribution of the training records, area 2 identifies a part of the realised distribution from which no species records were available, and area 3 identifies a potential distribution that is actually not inhabited, e.g. due to limited accessibility or some other controlling factors. After Pearson (2007).

2007; Fitzpatrick et al., 2007; Rödder and Lötters, 2009), so the extent of either niche shifts or niche conservatism is unknown and still debated (Wiens and Graham, 2005; Pearman et al., 2007). In a recent review, it has been shown that the fundamental niche can remain stable for tens of thousands of years, independent of the taxonomic group, or it can shift substantially within only a few generations (Pearman et al., 2007). In general, there is a considerable lack of knowledge regarding the processes triggering shifts in climate niches, as well as a shortage of suitable methods for analysing them.

Sax et al. (2008) pointed out that alien invasive species can offer sources of unexpected experiments providing valuable insights into ecological and evolutionary processes. Indeed, some recent studies have addressed the question of rapid fundamental niche shifts during invasion processes. Using CEMs, Broennimann et al. (2007) found that in the spotted knapweed (*Centaurea maculosa* Lam.) the climate envelopes in its native range (western North

America) differed from its invasive range in Europe. Similarly, Fitzpatrick et al. (2007) demonstrated, using a CEM approach, that fire ants (*Solenopsis invicta* Buren, 1972) can be ascribed to climate envelopes in their invaded range (North America) from which they are absent in their native South American range. On the other hand, Rödder (2009) and Rödder and Weinsheimer (2009) were even able to successfully predict the invasive range of island frog species (*Eleutherodactylus coqui* Thomas, 1966; *Osteopilus septentrionalis* (Duméril and Bibron, 1841)) with data derived from climatic conditions within their native range. Their results suggest a high degree of niche conservatism even when climate is not the range-limiting factor.

A first attempt towards quantitative analyses of niche conservatism was presented by Warren et al. (2008). The authors proposed two niche overlap indices for quantitative comparisons between potential distributions as well as null models for significance tests. Further studies are needed to assess the functional link of predictors and distribution and the factors triggering niche divergence or conservatism (Rödder and Lötters, 2009).

11.5 Conclusions

Predictions of species distributions derived from correlative models can help to understand the spatial patterns of biodiversity and speciation processes. The amount of available data and software and the number of studies applying niche model techniques are increasing steadily. However, there is an overwhelming discrepancy between increasingly complex, often large-scale studies involving hundreds or thousands of species, and the understanding of underlying processes, derivation of valid assumptions and development of conceptual backgrounds (Jiménez-Valverde et al., 2008). One problem may be that models predicting potential distributions of species are easily computed and test statistics sometimes misleading (Lobo et al., 2008). This, together with the lack of a conceptual background, facilitates misinterpretations. CEMs can be powerful tools, providing ideas and helping to generate hypotheses, but they must be critically evaluated during each step from model building to final interpretation.

Acknowledgements

We are grateful to Trevor Hodkinson and an anonymous reviewer for valuable suggestions. The work of the first author was funded by the Graduiertenförderung des Landes Nordrhein-Westfalen.

References

Araújo, M. B. and Guisan, A. (2006). Five (or so) challenges for species distribution modelling. *Journal of Biogeography,* **33**, 1677–1688.

Araújo, M .B. and Pearson, R. G. (2007). Equilibrium of species' distributions with climate. *Ecography,* **28**, 693–695.

Araújo, M. B., Cabeza, M., Thullier, W., Hannah, L. and Williams, P. H. (2004). Would climate change drive species out of reserves? An assessment of existing reserve-selection methods. *Global Change Biology,* **10**, 1618–1626.

Beaumont, L. J., Hughes, L. and Poulsen, M. (2005). Predicting species distributions: use of climatic parameters in BIOCLIM and its impact on predictions of species' current and future distributions. *Ecological Modelling,* **186**, 250–269.

Beaumont, L. J., Hughes, L. and Pitman, A. J. (2008). Why is the choice of future climate scenarios for species distribution modelling important? *Ecology Letters,* **11**, 1135–1146.

Broennimann, O., Treier, U. A., Müller-Schärer, H. et al. (2007). Evidence of climatic niche shift during biological invasion. *Ecology Letters,* **10**, 701–709.

Busby, J. R. (1991). BIOCLIM: a bioclimatic analysis and prediction system. In *Nature Conservation: Cost Effective Biological Surveys and Data Analysis,* ed. C. R. Margules and M. P. Austin. Melbourne: CSIRO, pp. 64–68.

Carnaval, A. C. and Moritz, C. (2008). Historical climate modelling predicts patterns of current biodiversity in the Brazilian Atlantic forest. *Journal of Biogeography,* **25**, 1187–1201.

Carpenter, G., Gillison, A. N. and Winter, J. (1993). DOMAIN: a flexible modeling procedure for mapping potential distributions of plants and animals. *Biodiversity and Conservation,* **2**, 667–680.

Dormann, C. F., McPherson, J., Araújo, M. B. et al. (2007). Methods to account for spatial autocorrelation in the analysis of species distributional data: a review. *Ecography,* **30**, 609–628.

Elith, J., Graham, C. H., Anderson, R. P. et al. (2006). Novel methods improve prediction of species' distributions from occurrence data. *Ecography,* **29**, 129–151.

Elton, C. (1927). *Animal Ecology.* London: Sedgwick and Jackson.

Fielding, A. H. and Bell, J. F. (1997). A review of methods for the assessment of prediction errors in conservation presence/absence models. *Environmental Conservation,* **24**, 38–49.

Fisher, J., Lindenmayer, D. B., Nix, H. A., Stein, J. L. and Stein, J. A. (2001). Climate and animal distribution: a climatic analysis of the Australian marsupial *Trichosurus caninus. Journal of Biogeography,* **28**, 293–304.

Fitzpatrick, M. C., Weltzin, J. F., Sanders, N. J. and Dunn, R. R. (2007). The biogeography of prediction error: why does the introduced range of the fire ant over-predict its native range? *Global Ecology and Biogeography,* **16**, 24–33.

Flato, G. M., Boer, G. J., Lee, W. G. et al. (2000). The Canadian Centre for Climate Modelling and Analysis global coupled model and its climate. *Climate Dynamics,* **16**, 451–467.

Frost, D. R. (2010). Amphibian species of the world: an online reference. Version

5.4. New York, NY: American Museum of Natural History. http://research.amnh.org/vz/herpetology/amphibia.

Gordon, C., Cooper, C., Senior, C. A. et al. (2000). The simulation of SST, sea ice extents and ocean heat transports in a version of the Hadley Centre coupled model without flux adjustments. *Climate Dynamics,* **16**, 147–168.

Graham, C. H., Ron, S. R., Santos, J. C., Schneider, C. J. and Moritz, C. (2004). Integrating phylogenetics and environmental niche models to explore speciation mechanisms in dendrobatid frogs. *Evolution,* **58**, 1781–1793.

Graham, C. H., Elith, J., Hijmans, R. J. et al. (2008). The influence of spatial errors in species' occurrence data used in distribution models. *Journal of Applied Ecology,* **45**, 239–247.

Grinnell, J. (1917). The niche-relationships of the California Thrasher. *Auk,* **34**, 427–433.

Gu, W. and Swihart, R. K. (2004). Absent or undetected? Effects of non-detection of species occurrence on wildlife-habitat models. *Biological Conservation,* **116**, 195–203.

Guisan, A. and Thuiller, W. (2005). Predicting species distributions: offering more than simple habitat models. *Ecology Letters,* **8**, 993–1009.

Guisan, A. and Zimmermann, N. (2000). Predictive habitat distribution models in ecology. *Ecological Modelling,* **135**, 147–186.

Hanley, J. and McNeil, B. (1982). The meaning of the use of the area under a receiver operating characteristic (ROC) curve. *Radiology,* **143**, 29–36.

Heikkinen, R. K., Luoto, M., Araújo, M. B. et al. (2006). Methods and uncertainties in bioclimatic envelope modeling under climate change. *Progress in Physical Geography,* **30**, 751–777.

Hijmans, R. J. and Graham, C. H. (2006). The ability of climate envelope models to predict the effect of climate change on species' distributions. *Global Change Biology,* **12**, 2272–2281.

Hijmans, R. J., Guarino, L. and Rojas, E. (2002). *DIVA-GIS: A Geographical Information System for the Analysis of Biodiversity Data. Manual.* Lima: International Potato Center.

Hijmans, R. J., Guarino, L., Jarvis, A. et al. (2005a). DIVA-GIS, version 5.2. Manual. www.diva-gis.org.

Hijmans, R. J., Cameron, S. E., Parra, J. L., Jones, P. G. and Jarvis, A. (2005b). Very high resolution interpolated climate surfaces for global land areas. *International Journal of Climatology,* **25**, 1965–1978.

Hirzel, A. H. and Le Lay, G. (2008). Habitat suitability modelling and niche theory. *Journal of Applied Ecology,* **45**, 1372–1381.

Hortal, J., Jiménez-Valverde, A., Gómez, J. F., Lobo, J. M. and Baselga, A. (2008). Historical bias in biodiversity inventories affects the observed environmental niche of the species. *OIKOS,* **117**, 847–858.

Hutchinson, G. E. (1957). Concluding remarks. *Cold Spring Harbor Symposia on Quantitative Biology,* **22**, 415–427.

Hutchinson, G. E. (1978). *An Introduction to Population Ecology.* New Haven, CT: Yale University Press.

Hutchinson, M. F. (1995). Interpolating mean rainfall using thin plate smoothing splines. *International Journal of Geographical Information Systems,* **9**, 385–403.

Hutchinson, M. F. (2004). *Anusplin version 4.3*. Canberra: Centre for Resource and Environment Studies, Australian National University.

James, F. C., Johnston, R. F., Warner, N. O., Niemi, G. and Boecklen, W. (1984). The Grinnellian niche of the Wood Thrush. *American Naturalist*, **124**, 17–47.

Jiménez-Valverde, A. and Lobo, J. M. (2006). The ghost of unbalanced species distribution data in geographical model predictions. *Diversity and Distributions*, **12**, 512–524.

Jiménez-Valverde, A., Lobo, J. M. and Hortal, J. (2008). Not as good as they seem: the importance of concepts in species distribution modeling. *Diversity and Distributions*, **14**, 885–890.

Kiehl, J. T. and Gent, P. R. (2004). The Community Climate System Model, version 2. *Journal of Climate*, **17**, 1666–1669.

Köhler, J., Scheelke, K., Schick, S., Veith, M. and Lötters, S. (2005). Contribution to the taxonomy of hyperoliid frogs (Amphibia, Anura, Hyperoliidae): advertisement calls of twelve species from East and Central Africa. *African Zoology*, **40**, 127–142.

Kozak, K. H., Graham, C. H. and Wiens, J. J. (2008). Integrating GIS-based environmental data into evolutionary biology. *Trends in Ecology and Evolution*, **23**, 141–148.

Kremen, C., Cameron, A., Moilanen, A. et al. (2008). Aligning conservation priorities across taxa in Madagascar with high-resolution planning tools. *Science*, **320**, 222–226.

Lobo, J. M., Jiménez-Valverde, A. and Real, R. (2008). AUC: a misleading measure of the performance of predictive distribution models. *Global Ecology and Biogeography*, **17**, 145–151.

Lötters, S., Rotich, D., Scheelke, K. et al. (2004). Bio-sketches and partitioning of syntopic reed frogs, genus *Hyperolius* (Amphibia: Hyperoliidae), in two humid tropical African forest regions. *Journal of Natural History*, **38**, 1969–1997.

Lötters, S., Rödder, D., Bielby, J. et al. (2008). Meeting the challenge of conserving Madagascar's megadiverse amphibians: addition of a risk assessment for the chytrid fungus. *PLoS Biology*, **6**.

Martinez-Meyer, E., Peterson, A. T. and Navarro-Siguenza, A. G. (2004). Evolution of seasonal ecological niches in the *Passerina* buntings (Aves: Cardinalidae). *Proceedings of the Royal Society of London B*, **271**, 1151–1157.

Monserud, R. A. and Leemans, R. (1992). Comparing global vegetation maps with kappa statistics. *Ecological Modelling*, **62**, 275–293.

New, M., Hulme, M. and Jones, P. D. (1999). Representing twentieth century space-time climate variability. Part 1: development of a 1961–90 mean monthly terrestrial climatology. *Journal of Climate*, **12**, 829–856.

New, M., Hulme, M. and Jones, P. D. (2000). Representing twentieth century space-time climate variability. Part 2: development of 1901–96 monthly grids of terrestrial surface climate. *Journal of Climate*, **13**, 2217–2238.

Nix, H. (1986). A biogeographic analysis of Australian elapid snakes. In *Atlas of Elapid Snakes of Australia*, ed. R. Longmore. Canberra: Bureau of Flora and Fauna, pp. 4–15.

Pearce, J. and Ferrier, S. (2000). An evaluation of alternative algorithms for fitting species' distribution models using logistic regression. *Ecological Modelling,* **128**, 128–147.

Pearman, P. B., Guisan, A., Broennimann, O. and Randin, C. F. (2007). Niche dynamics in space and time. *Trends in Ecology and Evolution,* **23**, 149–158.

Pearman, P. B., Randin, C. F., Broennimann, O. et al. (2008). Prediction of plant species' distributions across six millennia. *Ecology Letters,* **11**, 357–369.

Pearson, R.G. (2007). *Species' Distribution Modeling for Conservation Educators and Practitioners: Synthesis.* http://ncep.amnh.org. New York, NY: American Museum of Natural History.

Pearson, R. G., Raxworthy, C. J., Nakamura, M. and Peterson, A. T. (2007). Predicting species' distributions from small numbers of occurrence records: a test case using cryptic geckos in Madagascar. *Journal of Biogeography,* **34**, 102–117.

Peterson, A. T. (2003). Predicting the geography of species' invasions via ecological niche modeling. *The Quarterly Review of Biology,* **78**, 419–433.

Peterson, A. T. and Vargas, N. (1993). Ecological diversity in scrub jays, *Aphelocoma coerulescens.* In *The Biological Diversity of Mexico: Origins and Distribution,* ed. T. P. Ramamoorthy, R. Bye and J. Fa. New York, NY: Oxford University Press, pp. 309–317.

Peterson, A. T. and Vieglais, D. A. (2001). Predicting species' invasions using ecological niche modeling: new approaches from bioinformatics attack a pressing problem. *BioScience,* **51**, 363–371.

Peterson, A. T., Soberón, J. and Sánchez-Cordero, V. (1999). Conservation of ecological niches in evolutionary time. *Science,* **285**, 1265–1267.

Phillips, S. J., Dudík, M. and Shapire, R. E. (2004). A maximum entropy approach to species' distribution modeling. In *Proceedings of the 21st International Conference on Machine Learning.* New York, NY: ACM Press, pp. 655–662.

Phillips, S. J., Anderson, R. P. and Schapire, R. E. (2006). Maximum entropy modeling of species' geographic distributions. *Ecological Modeling,* **190**, 231–259.

Puliam, H. R. (2000). On the relationship between niche and distribution. *Ecology Letters,* **3**, 349–361.

Raxworthy, C. J., Ingram, C. M., Rabibisoa, N. and Pearson, R. G. (2007). Applications of ecological niche modeling for species' delimitation: a review and empirical evaluation using day geckos (*Phelsuma*) from Madagascar. *Systematic Biology,* **56**, 907–923.

Rödder, D. (2009). 'Sleepless in Hawaii': does anthropogenic climate change enhance ecological and socioeconomic impacts of the alien invasive *Eleutherodactylus coqui* Thomas, 1966 (Anura: Eleutherodactylidae)? *North-Western Journal of Zoology,* **5**, 16–25.

Rödder, D. and Böhme, W. (2009). Who is who? Comparison of the advertisement calls of two East African sister species of *Hyperolius* (Anura: Hyperoliidae). *Salamandra,* **45**, 180–185.

Rödder, D. and Dambach, J. (2010). Review: modelling future trends of relict species. In *Relict Species: Phylogeography and Conservation Biology,* ed. J. C. Habel and T. Assmann, New York, NY: Springer, pp. 373–382.

Rödder, D. and Lötters, S. (2009). Niche shift or niche conservatism? Climatic properties of the native and invasive range of the Mediterranean Housegecko *Hemidactylus turcicus*. *Global Ecology and Biogeography,* **18**, 674–687.

Rödder, D. and Weinsheimer, F. (2009). Will future anthropogenic climate change increase the potential distribution of the alien invasive Cuban treefrog (Anura: Hylidae)? *Journal of Natural History,* **43**, 1207–1217.

Rödder, D., Solé, M. and Böhme, W. (2008). Predicting the potential distribution of two alien invasive housegeckos (Gekkonidae: *Hemidactylus frenatus, Hemidactylus mabouia*). *North-Western Journal of Zoology,* **4**, 236–246.

Rödder, D., Schmidtlein, S., Veith, M. and Lötters, S. (2009). Alien invasive Slider turtle in unpredicted habitat: a matter of niche shift or predictors studied? *PLoS ONE,* **4**, e7843.

Rödder, D., Schlüter, A. and Lötters, S. (2010). Is the 'Lost World' lost? High endemism of amphibians and reptiles on South American tepuis in a changing climate. In *Relict Species: Phylogeography and Conservation Biology,* ed. J. C. Habel and T. Assmann. New York, NY: Springer, pp. 401–416.

Sax, D. F., Stachowicz, J. J., Brown, J. H. et al. (2008). Ecological and evolutionary insights from species invasions. *Trends in Ecology and Evolution,* **22**, 465–471.

Schiøtz, A. (1999). *Treefrogs of Africa*. Frankfurt: Chimaira.

Soberón, J. (2007). Grinnellian and Eltonian niches and geographic distributions of species. *Ecology Letters,* **10**, 1115–1123.

Soberón, J. and Peterson, A. T. (2004). Biodiversity informatics: managing and applying primary biodiversity data. *Philosophical Transactions of the Royal Society of London B,* **359**, 689–698.

Soberón, J. and Peterson, A. T. (2005). Interpretation of models of fundamental ecological niches and species' distributional areas. *Biodiversity Informatics,* **2**, 1–10.

Stockwell, D. B. R. and Noble, I. R. (1992). Introduction of sets of rules from animal distribution data: a robust and informative method of data analysis. *Math and Computers in Simulation,* **33**, 385–390.

Stockwell, D. and Peters, D. (1999). The GARP modeling system problems and solutions to automated spatial prediction. *International Journal of Geographical Information Science,* **13**, 143–158.

Swets, K. (1988). Measuring the accuracy of diagnostic systems. *Science,* **240**, 1285–1293.

Thuiller, W. (2003). BIOMOD: optimising predictions of species distributions and projecting potential future shifts under global change. *Global Change Biology,* **9**, 1352–1362.

Thuiller, W., Albert, C., Araújo, M. B. et al. (2008). Predicting global change impacts on plant species' distributions: future challenges. *Perspectives in Plant Ecology, Evolution and Systematics,* **9**, 137–152.

Tilman, D., May, R. M., Lehman, C. L. and Nowak, M. A. (1994). Habitat destruction and the extinction debt. *Nature,* **371**, 65–66.

Veith, M., Kosuch, J., Rödel, M. O. et al. (2009). Multiple evolution of sexual dichromatism in Afrotropical reed frogs. *Molecular Phylogenetics and Evolution,* **51**, 388–393.

Waltari, E. and Guralnick, R. P. (2009). Ecological niche modelling of montane mammals in the Great Basin, North America: examining past and present connectivity of species across basins and ranges. *Journal of Biography,* **36**, 148–161.

Waltari, E., Hijmans, R. J., Peterson, A. T. et al. (2007). Locating Pleistocene refugia: comparing phylogeographic and ecological niche model predictions. *PLoS ONE,* **7**, 1–11.

Warren, D. L., Glor, R. E. and Turelli, M. (2008). Environmental niche equivalency versus conservatism: quantitative approaches to niche evolution. *Evolution,* **62**, 2868–2883.

Wiens, J. J. and Graham, C. H. (2005). Niche conservatism: integrating evolution, ecology, and conservation biology. *Annual Review of Ecology and Systematics,* **36**, 519–539.

Wisz, M. S., Hijmans, R. J., Peterson, A. T. et al. (2008). Effects of sample size on the performance of species' distribution models. *Diversity and Distributions,* **14**, 763–773.

12

Biogeography of *Cyclamen*: an application of phyloclimatic modelling

C. Yesson

Institute of Zoology, Zoological Society of London, UK

A. Culham

School of Biological Sciences and *The Walker Institute for Climate Change, University of Reading, UK*

Abstract

Cyclamen is a genus of popular garden plant, protected by Convention on International Trade in Endangered Species (CITES) legislation. Many of its species are morphologically and phenologically adapted to the seasonal climate of the Mediterranean region. Most species occur in geographic isolation and will readily hybridise with their sister species when brought together. We investigate the biogeography of *Cyclamen* and assess the impact of palaeogeography and palaeoclimate change on the distribution of the genus. We use techniques of phyloclimatic modelling (combining ecological niche modelling and phylogenetic character optimisation) to investigate the heritability of climatic preference and

Climate Change, Ecology and Systematics, ed. Trevor R. Hodkinson, Michael B. Jones, Stephen Waldren and John A. N. Parnell. Published by Cambridge University Press.

to reconstruct ancestral niches. Conventional and phyloclimatic approaches to biogeography are compared to provide an insight into the historic distribution of *Cyclamen* species and the potential impact of climate change on their future distribution. The predicted climate changes over the next century could see a northward shift of many species' climatic niches to places outside their current ranges. However, such distribution changes are unlikely to occur through natural ant-based dispersal, so conservation measures are likely to be required.

12.1 Introduction

12.1.1 *Cyclamen*: present-day status and distribution

Cyclamen L. is a genus of *c.* 20 species in the family Myrsinaceae. Its species are perennial herbs, having distinctive flowers with reflexed petals, that are often scented, and winter blooming. These characteristics make *Cyclamen* a popular garden plant. Its popularity has prompted many studies on the group, including cytology (Bennett and Grimshaw, 1991; Anderberg, 1994), hybridisation (Gielly et al., 2001; Grey-Wilson, 2003) and phenology (Debussche et al., 2004). There are several phylogenetic studies based on morphological and molecular data (Anderberg et al., 2000; Clennett, 2002; Compton et al., 2004; Yesson et al., 2009). Although these studies present similar findings for many sister species pairings, they do not agree on the complete phylogenetic topology. Some uncertainty remains regarding subgeneric relationships and species delimitation.

Cyclamen is a phenologically interesting genus; in any month of the year, at least one species can be found flowering somewhere (Grey-Wilson, 2003), which is an unusual trait for such a small group, limited to boreal, seasonal climates around the Mediterranean basin. Its highest diversity is found in Turkey (11 species) and Greece (at least five species – Culham et al., 2009) (Fig 12.1). The seasonal winter-wet, summer-dry climate of this region is thought to be an important factor in the speciation of *Cyclamen*. Many species are adapted to a Mediterranean-type climate and die back to an underground organ during the dry summer months (Debussche et al., 2004).

As with most predominantly European species, *Cyclamen* species are under pressure from habitat reduction. All species of *Cyclamen* are listed in Appendix II of the CITES of Wild Fauna and Flora (www.cites.org), and thus receive some level of protection from wild collection. If protection of the species is to be effective, it is important to know how rapidly changing climate might affect *Cyclamen* in the wild.

12.1.2 Ecological niche models

It has long been accepted that climate creates boundaries to species distribution (Ricklefs and Latham, 1992; Inouye, 2000; Martínez-Meyer et al., 2004a; Peterson

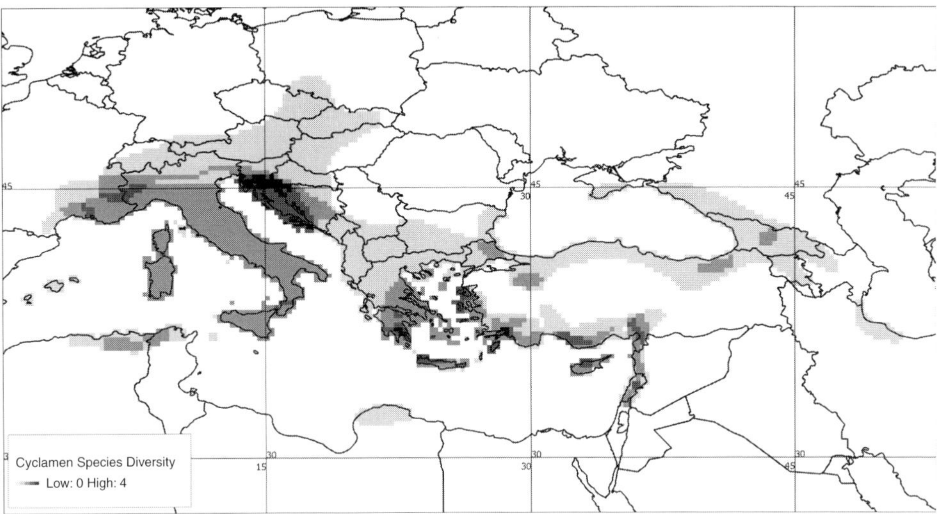

Figure 12.1 *Cyclamen* species diversity. Grid shows decimal degrees. Distribution data from Grey-Wilson (2003).

et al., 2005). There are many methodological applications that attempt to model the environmental preferences of species and use these to establish the environmental limits of species distribution (Nix, 1986; Guisan and Zimmermann, 2000; Guisan and Thuiller, 2005; Elith et al., 2006; Phillips et al., 2006). While individual techniques have been subject to criticism, the general technique of environmental niche modelling has been used widely for predicting species distributions (such as for invasive species - Peterson, 2003), predicting ancestral areas of extant species (Graham et al., 2004), and predicting species most likely to be threatened by climate change (Culham and Yesson, Chapter 10).

12.1.3 An evolutionary perspective

There have been successful efforts to extend the scope of these models to estimate historical geographic distributions by examining the models for extant species in relation to palaeoclimatic data from the Pleistocene (Hugall et al., 2002; Bonaccorso et al., 2006). This technique of historical area prediction has been tested with reference to the fossil record, and shown to predict fossil distributions successfully (Martínez-Meyer et al., 2004a). These studies demonstrate the long-term stability of species' climate preference. A logical next step is to look at longer-term stability of climate preferences on evolutionary/geological timescales. There is some evidence supporting phylogenetic niche conservatism and ecological niche heritability (Ackerly, 2003; Wiens and Donoghue, 2004; Hoffmann, 2005). Peterson et al. (1999) suggest that bioclimatic envelopes are statistically more similar among sister species in a range of animal taxa and that they are conserved across evolutionary time. This contention was supported by Martínez-Meyer et al. (2004b), who used species-level bioclimatic models to predict effectively the distribution of sister species of birds.

12.1.4 Ancestral areas

Ancestral niches have been modelled successfully by combining climatic preference data with phylogenetic trees using techniques of ancestral state reconstruction (Yesson and Culham, 2006a). The technique combining phylogenetic reconstruction and niche modelling is termed 'phyloclimatic modelling'. Yesson and Culham (2006a) recreated ancestral niches for lineages of *Drosera* (Droseraceae) and used a dated molecular phylogeny to select appropriate time-frames to examine these models within palaeoclimate reconstructions. They then estimated ancestral areas for *Drosera* lineages within the late Miocene of Australia and New Zealand.

Historically there has been a lot of interest in identifying ancestral areas and areas of prehistoric species diversity (Page, 1988; Morrone and Crisci, 1995). Perhaps the most popular technique for ancestral area reconstruction has been dispersal–vicariance analysis (DIVA – Ronquist, 1997). For DIVA, 'speciation is assumed to subdivide the ranges of widespread species into vicariant components; the optimal ancestral distributions are those that minimise the number of implied dispersal and extinction events' (Ronquist, 1997). DIVA and phyloclimatic modelling present alternative techniques for ancestral area reconstruction, but as yet there has been no comparison of these techniques.

Here we present an example of phyloclimatic modelling on *Cyclamen*, to gain a better understanding of ecological and evolutionary patterns for the past and future. The ancestral areas estimated by phyloclimatic modelling and DIVA are presented for comparison.

12.2 Future distribution of *Cyclamen*

Given the climatic specialisation of *Cyclamen* species, it is important to know how predicted climate change will affect these plants. Yesson and Culham (2006b) examined the climatic preferences of 21 *Cyclamen* species based on present-day distribution data, using annual and seasonal variations of climatic variables of temperature and precipitation. They found, using a randomisation test on the quantitative convergence index, that 8 out of 14 climatic variables displayed significant phylogenetic conservancy, but that within the genus climate specialisation ran from widely tolerant generalists (*Cyclamen coum* Mill. and *C. hederifolium* Alt.) through to Mediterranean specialists (*C. creticum* (Dörfl.) Hildebr. and *C. cyprium* Kotschv). These data were used to develop ecological niche models for each species using BIOCLIM (Busby, 1991) and Maxent (Phillips et al., 2006) algorithms. The majority of niches were isolated from each other, suggesting that species are climatically isolated, with the exception of the climatically tolerant species, whose niches were found to wholly encompass those of the more specialist species.

These niches were projected into a mid-severity future climate scenario for 2050 (scenario A2c), to examine whether areas presently occupied by *Cyclamen* species will be climatically suitable in the future. Except for the few species showing wide climatic tolerance, Yesson and Culham (2006b) found that for the BIOCLIM models the majority of species would be under severe threat from climate change, with 11 species predicted to have no climatically suitable area within their present range in 2050 (Fig 12.2). They noted a positive correlation of present-day range with proportion of area lost. The Maxent models (Fig 12.2) were less dramatic, but still predicted significant area loss for many species and indicated no relationship between present-day range and proportion of area lost. However, the models both indicate that areas outside the native ranges of many species are, or will become, climatically suitable.

There is some empirical evidence to support changes in geographic distribution of *Cyclamen* resulting from recent climate change (e.g. Fig 12.3). The climatically

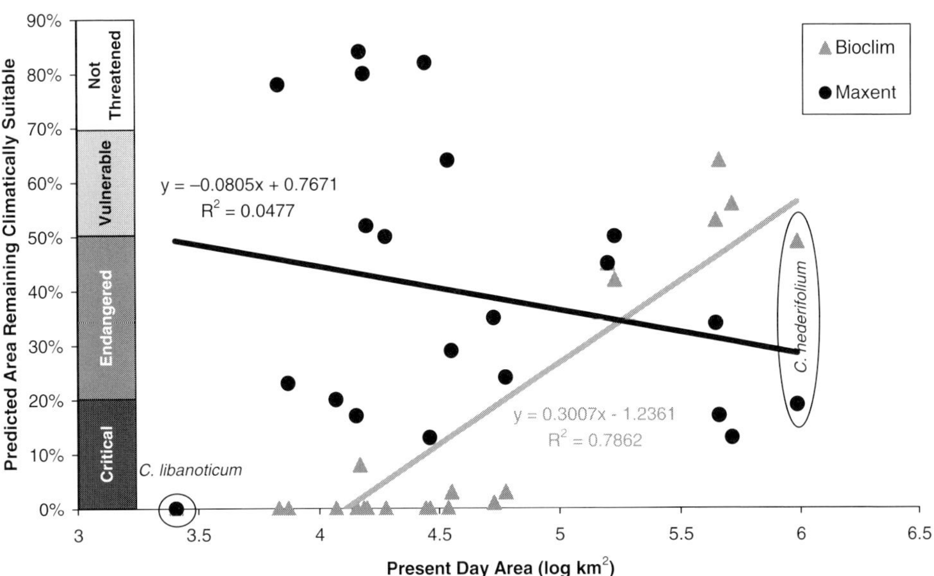

Figure 12.2 Projected area loss for *Cyclamen* species in 2050 using Maxent and BIOCLIM niche models. Lines show linear regression, with line equations and r^2 values adjacent. Bars to the right of the y-axis signify risk status based on International Union for Conservation of Nature (IUCN) classifications. Reproduced with permission from Yesson and Culham (2006a).

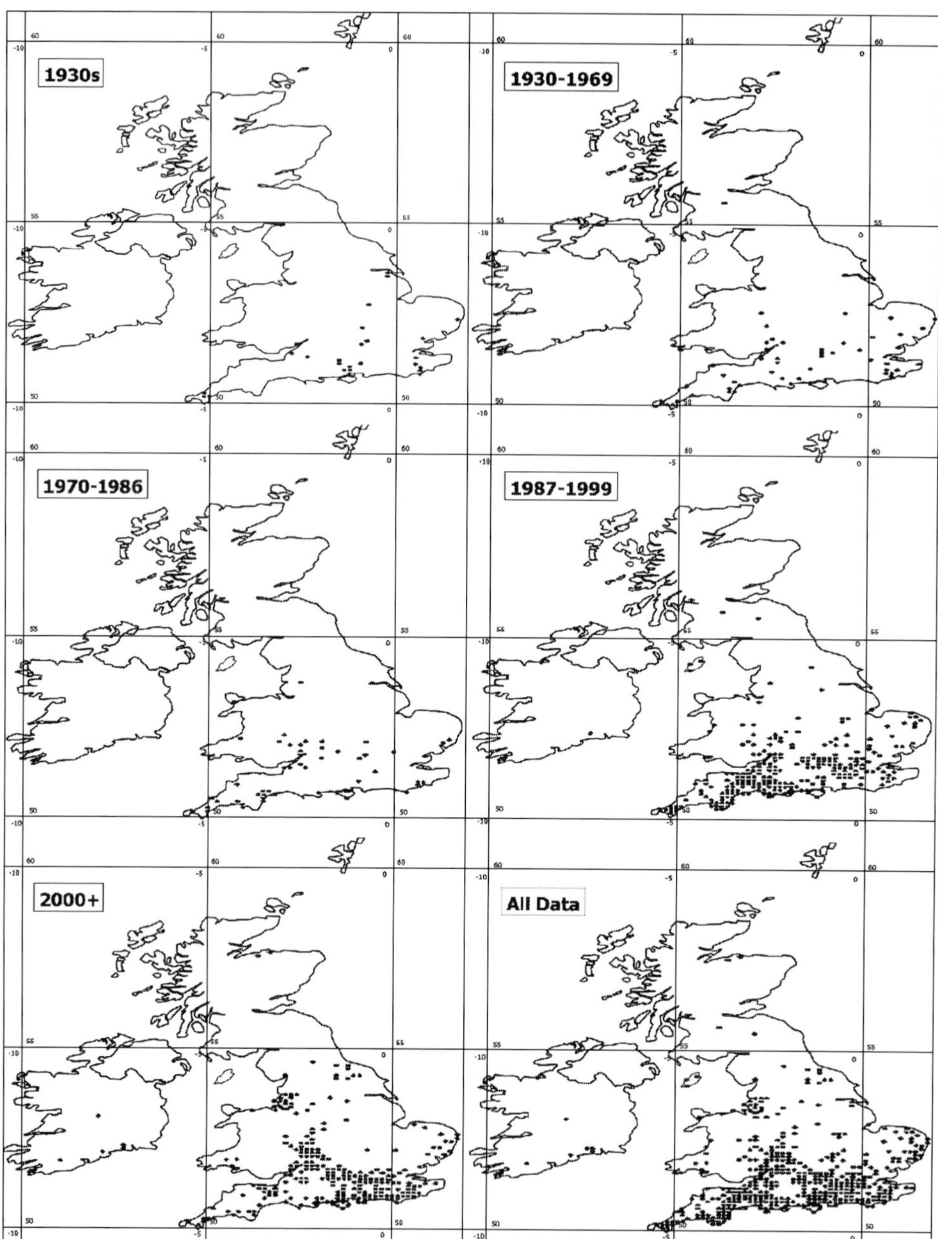

Figure 12.3 Distribution data recorded for *C. hederifolium* in the British Isles. New records over the past century (data from www.bsbimaps.org.uk). Grid shows decimal degrees.

tolerant species *C. hederifolium*, although native in central and southern Europe only, has spread northwards over the past few decades of warming (Stace, 1997). The niche models, based on native distribution patterns, suggest that northern Europe should be climatically suitable for this species. This is confirmed by Botanical Society of the British Isles (BSBI) distribution data, which indicate that *C. hederifolium* has become naturalised and has spread its range northwards in the UK over the past century (Fig 12.3). This northward migration shows distributional changes of the scale of hundreds of kilometres within several decades and is unlikely to be due to natural dispersal, as ant-based seed dispersal events are typically 0–10 m per year (Ness et al., 2004). Plant migration rates of 10–40 km per 100 years have been proposed for some tree species (Davis and Shaw, 2001; McLachlan et al., 2005). It is therefore considered highly unlikely that the ant-dispersed seeds of *Cyclamen* would exceed this migration rate. However, *C. hederifolium* and *C. coum* are highly popular garden plants, and gardeners may have unwittingly spread these plants throughout habitats that could become important areas for conservation as native areas become climatically unsuitable.

Yesson and Culham (2006b) tested their categorisations of extinction risk from a phylogenetic perspective, but reported no significant phylogenetic pattern of phylogenetic conservancy (Fig 12.4A). However, this result is somewhat dependent upon the choice of phylogeny. Figure 12.4B shows a revised topology (Yesson et al., 2009), based on increased sampling for both subspecific taxa and molecular character data. Superficially, there appears to be a pattern of conservancy of risk, but this is marginally short of being statistically significant ($0.05 < p < 0.1$) using a randomisation test of phylogenetic conservancy (Yesson and Culham, 2006b). The expectation that extinction risk based on climatic niches should show phylogenetic conservatism follows directly from the findings of phylogenetic conservatism of climatic characteristics, but for *Cyclamen* this is not the case.

12.3 Past distribution and biogeography of *Cyclamen*

The phylogenetic heritability of climatic preference and extinction risk for *Cyclamen* suggests that evolutionary history is an important factor in understanding the impact of climate change. Climatic conditions are inextricably linked with location, so understanding biogeography is an important step in this process.

Cyclamen species are currently restricted in their distribution. Their combined ranges cover approximately 2.25 million km^2, which is about the area of Western Australia (Fig 12.1), and many species have overlapping ranges; 38% of the quarter-degree squares contain more than one species. However, the majority of range overlap is accounted for by the few wide-ranging species (*C. coum*, *C. hederifolium*,

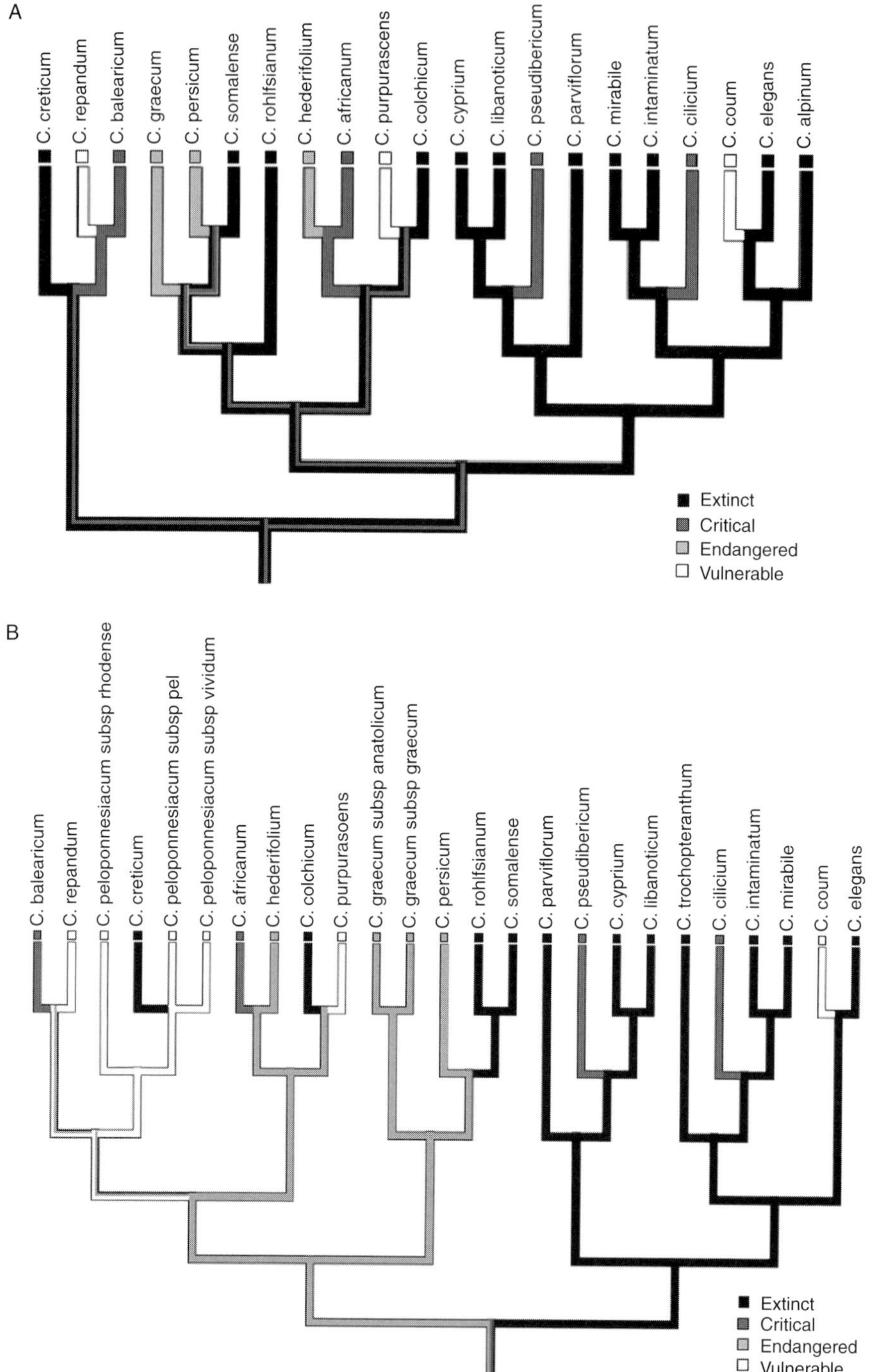

Figure 12.4 Parsimony optimisation of extinction risk for *Cyclamen*. Parsimony optimisation of extinction risk based on examination of models within 2050 scenario for BIOCLIM niche models (Yesson and Culham, 2006a). (A) Species-level phylogeny from Yesson and Culham (2006a). (B) Alternative topology, including subspecific sampling (Yesson et al., 2009). Risk categories are based on the proportion of area lost using the IUCN Red List categories. Note that risk value classification for species and subspecies is the same.

Table 12.1 Vicariance events proposed for *Cyclamen* that have been identified with reported geological events. From Yesson et al. (2009).

Epoch	Clade area 1	Clade area 2	Geological event
Mid–Late Miocene	*C. mirabile*–*C. parviflorum* Eastern Europe and Asia	*C. creticum*–*C.graecum* Western Europe	East/West European Divergence (Oberprieler, 2005)
Early–Mid Pliocene	*C. purpurascens* + *C. colchicum* Eastern Europe	subgen. *Psilanthum* Western Europe	East/West European Divergence (Oberprieler, 2005)
Late Miocene	*C. somalense* Somalia	*C. rohlfsianum* + *C. persicum* North Africa	Formation of Sahara (Douady et al., 2003)
Mid Pliocene	*C. hederifolium* Europe	*C. africanum* Africa	Loss of Tyrrhenian land bridge (Estabrook, 2001)

C. purpurascens Mill. and *C. repandum* Sm.), and no more than four species can be found in any one quarter-degree square.

Yesson et al. (2009) examined *Cyclamen* distribution patterns from a phylogenetic perspective and found that no pair of sister species overlapped in range, with the exception of *C. balearicum* Wilk. and *C. repandum* in a small area of southern France. Such allopatry, coupled with the ease of hybridisation between closely related species, implies a pattern of allopatric speciation in the evolutionary development of *Cyclamen*.

Assuming that allopatric speciation enables identification of patterns of vicariance and dispersal that best fit the observed distributions and reconstructed phylogeny, Yesson et al. (2009) used DIVA to identify 19 dispersal and 11 vicariance events in the evolutionary development of *Cyclamen*. Notably, there were four vicariance events coinciding with reported geological patterns elsewhere in the literature (Table 12.1). However, the majority of proposed vicariance events were not coupled with recognised geological patterns. One explanation for this might be climatic differentiation, which could help to explain the distinct climatic niches of extant species. The DIVA reconstruction was unable to discriminate between any of the areas of the present distribution at the root of the phylogenetic tree, and therefore the ancestral area for *Cyclamen* was estimated as the full extent of all extant *Cyclamen* species.

However, the niche model for the ancestral *Cyclamen* developed by Yesson and Culham (2006a), together with a phyloclimatic modelling approach, can be used to provide an alternative hypothesis of ancestral area (see Culham and Yesson, Chapter 10). This model was projected into a palaeoclimate reconstruction of the

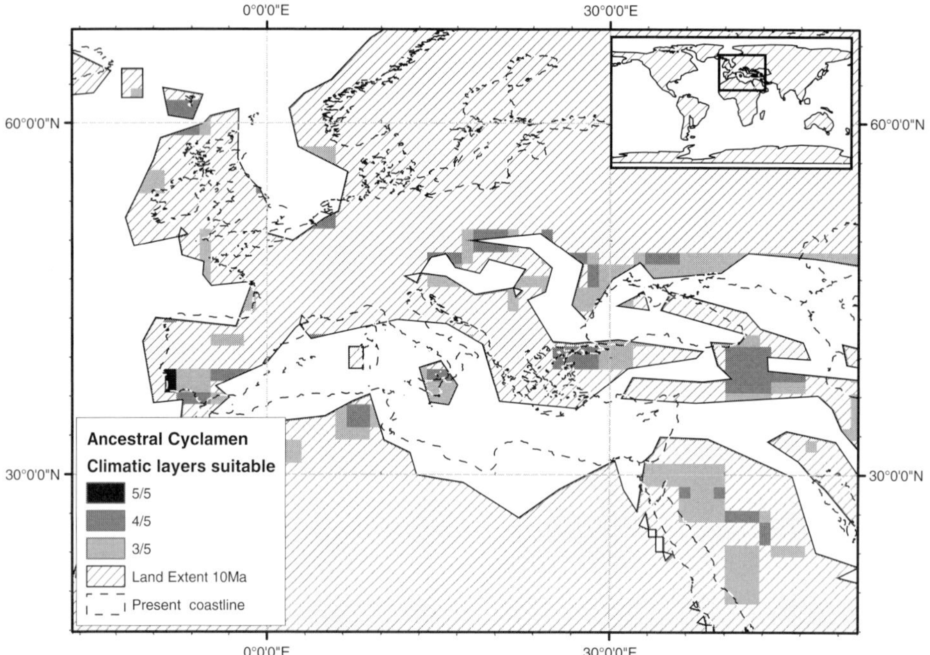

Figure 12.5 Area prediction for the ancestral *Cyclamen* in the late Miocene around the Mediterranean basin. Niche model developed by Yesson and Culham (2006a), projected using the BIOCLIM 'OR' methodology (Piñeiro et al., 2007), implemented in openModeller as the 'Envelope Score' algorithm. Reproduced with permission from Yesson (2008).

mid–late Miocene, which is the age of the ancestral lineage determined by the molecular dating of Yesson et al. (2009). Figure 12.5 demonstrates the climatically suitable areas for the ancestral *Cyclamen* for the mid-Miocene, and shows that a very small proportion of the extant range for all *Cyclamen* species would have been climatically suitable; these lie in what is now northern Turkey, North Africa and southern Iberia. In this case, the areas selected are a subset of the areas selected by DIVA. However, the geography of the Mediterranean at this time was very different from the present (Krijgsman, 2002). For example, the Italian peninsula was not formed at this time. To understand why the different methods vary we need to discuss their properties.

12.4 Potential of ancestral area reconstruction based on the reconstruction of ancestral niche

DIVA is a popular method for ancestral area reconstruction (Ronquist, 1997; Sanmartin, 2003; Oberprieler, 2005). The first step is to partition observed distributions into areas of endemism (Ronquist, 1997). These areas are chosen as being limited by geographical boundaries that may have acted as barriers to dispersal (Oberprieler, 2005). This partitioning is a somewhat arbitrary process, relying on the subjective decision of the individual researcher, and may present problems of circular logic if assumptions of dispersal ability are made prior to an analysis of dispersal. DIVA uses a cost-based model of dispersal and vicariance to reconstruct present-day observed areas onto internal nodes of a phylogeny (Ronquist, 1997). Clearly, if the true ancestral area was outside the present range, then DIVA cannot select the true area. Neither can such an approach discriminate between geographic regions that did not exist at the time of ancestral evolution. Furthermore, such analysis takes no account of the environment of the area at the time, and though the area selected may be climatically suitable for some species at present, it may not have been suitable during relevant periods of the past.

Ancestral area selection based on the reconstruction of the ancestral niche is a viable alternative to DIVA. It can construct independent hypotheses of potential distribution. As such it can be used as a complementary technique to refine area selection. It is a data-driven approach that does not rely on the subjective preselection of areas. Nor is it restricted analytically in the number of areas that it can consider. The underlying assumption is that of niche persistence and heritability.

Current research in ancestral area selection is focused on integrating probabilities of dispersal, but these methods still require initial partitioning of areas. For example, the program Lagrange uses a likelihood model to reconstruct ancestral areas and integrates dispersal probabilities by pre-assigning these probabilities based on hypotheses of the presence or absence of geographic barriers at any given time (Ree et al., 2005).

12.5 Conclusions

Phyloclimatic modelling does not integrate models of dispersal. This is an avenue for further development. One approach might be to track continuously, or at least at frequent intervals, the niches over the evolutionary timescales, which could theoretically show the potential for gradual migration through climatically suitable areas or reject such a theory by showing the need for long-distance dispersal. Unfortunately, we do not yet have continuous palaeoclimate reconstructions of appropriate resolution spanning the millions of years required for such analysis. This means that we are always likely to see predictions for different timeframes that do not spatially overlap, which presents difficulties in deciding between models of gradual migration or long-distance dispersal. Furthermore, the appearance of intermediate areas with suitable environments does not discount the hypothesis that a long-distance dispersal event could have bypassed this area.

Such a hypothesis could only be rejected by the discovery of fossil evidence in the intermediate areas, but the paucity of the fossil record often makes such testing impossible.

Climate change has already affected the distribution of *Cyclamen* species, and this seems likely to continue if predicted rates of warming prove accurate. Although niche modelling has met some criticism (see Rödder et al., Chapter 11), it is still a useful tool in the understanding, the prediction and ultimately the amelioration of the negative impacts of climate change. We have seen that phylo-climatic modelling approaches that take into account evolutionary perspectives can provide a deeper understanding of environmental niches, and how these can or cannot change over time. They can also be used to assess extinction risk. When we have insufficient data for traditional methods of extinction-risk estimation, phylogenetic relatedness to species of known risk might be employed as a useful proxy.

In the case of *Cyclamen* there is no significant pattern of phylogenetic relatedness, but if a new species was discovered in section *Gyropheobe*, then it would seem appropriate to regard it as being at high risk, as most closely related species are at high risk. This may provide a useful interim rapid assessment until a more rigorous process can be accomplished.

Acknowledgements

We wish to thank the Biotechnology and Biological Sciences Research Council for funding the BioDiversity World project (45/BEP17792), the University of Reading for funding the first author's PhD, and numerous colleagues for feedback and discussion of our ideas over the past five years.

References

Ackerly, D. D. (2003). Community assembly, niche conservatism, and adaptive evolution in changing environments. *International Journal of Plant Sciences*, **164**, S165–S184.

Anderberg, A. A. (1994). Phylogeny and subgeneric classification of *Cyclamen* L. (Primulaceae). *Kew Bulletin*, **49**, 455–467.

Anderberg, A. A., Trift, I. and Källersjö, M. (2000). Phylogeny of *Cyclamen* L. (Primulaceae): evidence from morphology and sequence data from the internal transcribed spacers of nuclear ribosomal DNA. *Plant Systematics and Evolution*, **220**, 147–160.

Bennett, S. T. and Grimshaw, J. M. (1991). Cytological studies in *Cyclamen* subg. *Cyclamen* (Primulaceae). *Plant Systematics and Evolution*, **176**, 135–143.

Bonaccorso, E., Koch, I. and Peterson, A. T. (2006). Pleistocene fragmentation of

Amazon species' ranges. *Diversity and Distributions,* **12**, 157–164.

Busby, J. R. (1991). BIOCLIM: a bioclimatic analysis and prediction system. In *Nature Conservation: Cost Effective Biological Surveys and Data Analysis*, ed. C. R. Margules and M. P. Austin. Melbourne: CSIRO, pp. 64–68.

Clennett, J. C. B. (2002). An analysis and revision of *Cyclamen* L. with emphasis on subgenus *Gyrophoebe* O. Schwarz. *Botanical Journal of the Linnean Society*, **138**, 473–481.

Compton, J. A., Clennett, J. C. B. and Culham, A. (2004). Nomenclature in the dock. Overclassification leads to instability: a case study in the horticulturally important genus *Cyclamen* (Myrsinaceae). *Botanical Journal of the Linnean Society*, **146**, 339–349.

Culham, A., Denney, M., Jope, M. and Moore, P. (2009). A new species of *Cyclamen* from Crete. *Cyclamen*, **33**, 12–15.

Davis, M. B. and Shaw, R. G. (2001). Range shifts and adaptive responses to Quaternary climate change. *Science*, **292**, 673–679.

Debussche, M., Garnier, E. and Thompson, J. D. (2004). Exploring thc causes of variation in phenology and morphology in Mediterranean geophytes: a genus-wide study of *Cyclamen*. *Botanical Journal of the Linnean Society*, **145**, 469–484.

Douady, C. J., Catzeflis, F., Raman, J., Springer, M. S. and Stanhope, M. J. (2003). The Sahara as a vicariant agent, and the role of the Miocene climatic events, in the diversification of the mammalian order Macroscelidae (elephant shrews). *Proceedings of the National Academy of Sciences of the USA*, **100**, 8325–8330.

Elith, J., Graham, C. H., Anderson, R. P. et al. (2006). Novel methods improve prediction of species' distributions from occurrence data. *Ecography*, **29**, 129–151.

Estabrook, G. F. (2001). Vicariance or dispersal: the use of natural historical data to test competing hypotheses of disjunction on the Tyrrhenian coast. *Journal of Biography*, **28**, 95–103.

Gielly, L., Debussche, M. and Thompson, J. D. (2001). Geographic isolation and evolution of Mediterranean endemic *Cyclamen*: insights from chloroplast trnL (UAA) intron sequence variation. *Plant Systematics and Evolution*, **230**, 75–88.

Graham, C. H., Ron, S. R., Santos, J. C., Schneider, C. J. and Moritz, C. (2004). Integrating phylogenetics and environmental niche models to explore speciation mechanisms in dendrobatid frogs. *Evolution*, **58**, 1781–1793.

Grey-Wilson, C. (2003). *Cyclamen: a Guide for Gardeners, Horticulturalists and Botanists*. London: Batsford.

Guisan, A. and Thuiller, W. (2005). Predicting species distribution: offering more than simple habitat models. *Ecology Letters*, **8**, 993–1009.

Guisan, A. and Zimmermann, N. E. (2000). Predictive habitat distribution models in ecology. *Ecological Modelling*, **135**, 147–186.

Hoffmann, M. H. (2005). Evolution of the realized climatic niche in the genus *Arabidopsis* (Brassicaceae). *Evolution*, **59**, 1425–1436.

Hugall, A., Moritz, C., Moussalli, A. and Stanisic, J. (2002). Reconciling paleodistribution models and comparative phylogeography in the wet tropics rainforest land snail *Gnarosophia bellendenkerensis*

(Brazier 1875). *Proceedings of the National Academy of Sciences of the USA*, **99**, 6112–6117.

Inouye, D. W. (2000). The ecological and evolutionary significance of frost in the context of climate change. *Ecology Letters*, **3**, 457–463.

Krijgsman, W. (2002). The Mediterranean: mare nostrum of earth sciences. *Earth and Planetary Science Letters*, **205**, 1–12.

Martínez-Meyer, E., Peterson, A. T. and Hargrove, W. W. (2004a). Ecological niches as stable distributional constraints on mammal species, with implications for Pleistocene extinctions and climate change projections for biodiversity. *Global Ecology and Biogeography*, **13**, 305–314.

Martínez-Meyer, E., Peterson, A. T. and Navarro-Siguenza, A. G. (2004b). Evolution of seasonal ecological niches in the passerina buntings (Aves: Cardinalidae). *Proceedings of the Royal Society of London B*, **271**, 1151–1157.

McLachlan, J. S., Clark, J. S. and Manos, P. S. (2005). Molecular indicators of tree migration capacity under rapid climate change. *Ecology*, **86**, 2088–2098.

Morrone, J. J. and Crisci, J. V. (1995). Historical biogeography: introduction to methods. *Annual Review of Ecology and Systematics*, **26**, 373–401.

Ness, J. H., Bronstein, J. L., Andersen, A. N. and Holland, J. N. (2004). Ant body size predicts dispersal distance of ant-adapted seeds: implications of small-ant invasions. *Ecology*, **85**, 1244–1250.

Nix, H. A. (1986). A biogeographic analysis of Australian elapid snakes. In *Atlas of Elapid Snakes of Australia*, ed. R. Longmore. Canberra: Australian Government Publishing Service, pp. 4–15.

Oberprieler, C. (2005). Temporal and spatial diversification of circum-Mediterranean Compositae-Anthemideae. *Taxon*, **54**, 951–966.

Page, R. D. M. (1988). Quantitative cladistic biogeography constructing and comparing area cladograms. *Systematic Zoology*, **37**, 254–270.

Peterson, A. T. (2003). Predicting the geography of species' invasions via ecological niche modeling. *Quarterly Review of Biology*, **78**, 419–433.

Peterson, A. T., Soberón, J. and Sánchez-Cordero, V. (1999). Conservation of ecological niches in evolutionary time. *Science*, **285**, 1265–1267.

Peterson, A. T., Tian, H., Martínez-Meyer, E. et al. (2005). Modeling distributional shifts of individual species and biomes. In *Climate Change and Biodiversity*, ed. T. E. Lovejoy and L. J. Hannah. New Haven, CT: Yale University Press, pp. 211–229.

Phillips, S. J., Anderson, R. P. and Schapire, R. E. (2006). Maximum entropy modeling of species geographic distributions. *Ecological Modelling*, **190**, 231–259.

Piñeiro, R., Aguilar, J. F., Munt, D. D. and Feliner, G. N. (2007). Ecology matters: Atlantic–Mediterranean disjunction in the sand-dune shrub *Armeria pungens* (Plumbaginaceae). *Molecular Ecology*, **16**, 2155–2171.

Ree, R. H., Moore, B. R., Webb, C. O. and Donoghue, M. J. (2005). A likelihood framework for inferring the evolution of geographic range on phylogenetic trees. *Evolution*, **59**, 2299–2311.

Ricklefs, R. E. and Latham, R. E. (1992). Intercontinental correlation of geographical ranges suggests stasis in ecological traits of relict genera of temperate perennial herbs. *American Naturalist*, **139**, 1305–1321.

Ronquist, F. (1997). Dispersal–vicariance analysis: a new approach to the quantification of historical biogeography. *Systematic Biology*, **46**, 195–203.

Sanmartin, I. (2003). Dispersal vs. vicariance in the Mediterranean: historical biogeography of the Palearctic Pachydeminae (Coleoptera, Scarabaeoidea). *Journal of Biogeography*, **30**, 1883–1897.

Stace, C. (1997). *New Flora of the British Isles*, 2nd edn. Cambridge: Cambridge University Press.

Wiens, J. J. and Donoghue, M. J. (2004). Historical biogeography, ecology and species richness. *Trends in Ecology and Evolution*, **19**, 639–644.

Yesson, C. (2008). Investigating plant diversity in Mediterranean climates. Unpublished PhD thesis, University of Reading.

Yesson, C. and Culham, A. (2006a). Phyloclimatic modelling: combining phylogenetics and bioclimatic modelling. *Systematic Biology*, **55**, 785–802.

Yesson, C. and Culham, A. (2006b). A phyloclimatic study of *Cyclamen*. *BMC Evolutionary Biology*, **6**, 72.

Yesson, C., Toomey, N. H. and Culham, A. (2009). *Cyclamen*: time, sea and speciation biogeography using a temporally calibrated phylogeny. *Journal of Biogeography*, **36**, 1234–1252.

13

Cenozoic climate changes and the demise of Tethyan laurel forests: lessons for the future from an integrative reconstruction of the past

F. RODRÍGUEZ-SÁNCHEZ AND J. ARROYO

Departamento de Biología Vegetal y Ecología, Universidad de Sevilla, Spain

Abstract

Climate on earth has always been changing. Despite decades of investigation, our limited knowledge of the ecological and evolutionary effects of climate changes often translates into uncertain predictions about the impact of future climates on biodiversity. Integrative biogeographical approaches using palaeobotanical, phylogenetic and niche-based species distribution models, when permitted by data availability, may provide valuable insights to address these key questions. Here we combine palaeobotanical and phylogeographical information with hindcast modelling of species distribution changes to reconstruct past range dynamics and differentiation in the bay laurel (*Laurus* spp., Lauraceae), an emblematic relict tree from the subtropical laurel forests that

Climate Change, Ecology and Systematics, ed. Trevor R. Hodkinson, Michael B. Jones, Stephen Waldren and John A. N. Parnell. Published by Cambridge University Press.

thrived in Tethyan realms during most of the Tertiary period. We provide plausible examples of climate-driven migration, extinction and persistence of populations and taxa, and discuss the factors that influence niche conservatism or adaptation to changing environments. Finally, we discuss the likely impacts of the predicted climate change on laurophyllous taxa in the Mediterranean and Macaronesia.

13.1 Introduction

The reconstruction of the evolutionary history and distribution of plants has been based primarily on the information supplied by two relatively independent research fields: palaeobotany and phylogenetics. More recently, it has also relied on statistical modelling approaches to hindcast species distributions on geological timescales. Palaeobotany has long relied on the description of fossils, the resolution power of microscopes and fossil sampling in cores. Isotope dating, and other approaches for absolute timing of events, then led to a methodological revolution in this field (Stewart and Rothwell, 1993). Concurrently, the field of phylogenetics has developed quickly, particularly in the past two decades, because of its strong conceptual framework, the generation of huge amounts of new data and new methods to analyse these data (Felsenstein, 2004). Despite both fields sharing a number of aims, there has been little contact between them, and their research agendas have remained mostly separated. There is now recognition that a better understanding of the history of living organisms and their ecological and evolutionary processes can be accomplished through the integration of independent evidence. This has fuelled a number of studies combining both palaeobotanical and molecular data (e.g. Petit et al., 2002; Magri et al., 2006; López de Heredia et al., 2007 – see also Posadas et al., 2006, for a review of past and current approaches in historical biogeography). Furthermore, the use of habitat suitability or species distribution models (Guisan and Zimmermann, 2000; Guisan and Thuiller, 2005) to hindcast the palaeodistributions of species has become a new source of biogeographical inference. This new tool is especially powerful when combined with fossil data (Martínez-Meyer and Peterson, 2006; Pearman et al., 2008a), phylogenetic information (Hugall et al., 2002) or both (Cheddadi et al., 2006); the integrative approach appears particularly useful for solving long-standing questions for biogeographers (Donoghue and Moore, 2003; Posadas et al., 2006). In general, the reliability of biogeographical reconstructions increases when supported by different types of data (Cleland, 2001); disagreement between data types may uncover neglected processes or sources of error, and may also highlight new conceptual advances and research questions (Givnish and Renner, 2004; Pulquerio and Nichols, 2007).

Currently, one of the most active research topics in biogeography deals with the ecological and evolutionary effects of climate change. A range of studies has examined the role of climate change in shaping the evolutionary differentiation, extinction and distribution of taxa (see Jansson and Dynesius, 2002, and Parmesan, 2006, for reviews). Despite the existence of some clear-cut patterns (e.g. Svenning, 2003), there are still wide gaps in our knowledge of the actual processes by which climate change affects the ecology and evolution of species, and how different species and/or populations may react to them. This limited understanding translates into considerable uncertainty when trying to predict species' persistence, migration and/or extinction under future climate change (Thuiller, 2004; Araújo et al., 2005). The analysis of past events can certainly provide solid evidence for or against hypotheses on the response of taxa to future scenarios of climate change. However, the enormous potential of historical approaches for assessing such hypotheses has not yet been achieved, perhaps because of the scarcity of systematic data for most taxa and regions, but also because of the traditional bias towards experimental hypothesis testing (Cleland, 2001).

The origin and diversification of Mediterranean flora has long interested botanists and biogeographers (Quézel, 1985; Thompson, 2005). Despite the diversity of methodological approaches (analyses of floristic diversity, palaeobotanical, phylogenetic and/or phylogeographical reconstructions) and the range of studied taxa, our knowledge of Mediterranean biogeography is biased towards recently evolved taxa and recent processes, while the history of ancient taxa and the processes that moulded such patterns remain largely unknown (Petit et al., 2005). Among these ancient plants, laurophyllous taxa (e.g. *Laurus nobilis* L., *Prunus lusitanica* L., *P. laurocerasus* L. and *Rhododendron ponticum* L.) are particularly interesting, as they represent members of one of the ancestral vegetation types prior to the establishment of the current Mediterranean vegetation, now dominated by sclerophyllous trees and shrubs, malacophyllous shrubs and annual plants (Thompson, 2005). They are the few survivors of the once extensive laurel forests that covered southern Eurasia and North Africa from the early Cenozoic until the Miocene (Mai, 1989). Profound long-term climate change (Fig 13.1) provoked the gradual southward retreat of these laurel forests, and finally the extinction of most of their constituent species (Kovar-Eder et al., 2006; Utescher et al., 2007). The few extant laurophyllous taxa in the Mediterranean basin are currently restricted to particular regions of mild climate, while some species become more frequent in the adjacent Macaronesian region (Azores, Canary Islands, Cape Verde Islands and Madeira). Only one species, the bay laurel (*Laurus nobilis*, Lauraceae), has an extensive range throughout the Mediterranean, whereas the congeneric and closely related *L. azorica* (Seub.) Franco is present mostly in Macaronesia. Furthermore, *Laurus* appears relatively frequently in the fossil record since the late Oligocene to early Miocene. Therefore, *Laurus* represents an unparalleled case study to explore the

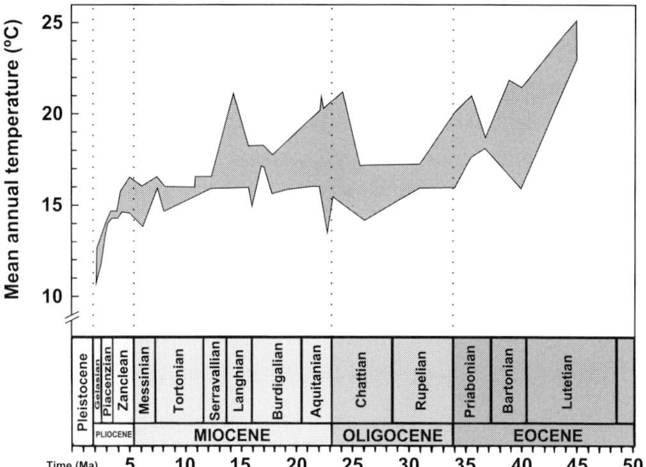

Figure 13.1 Variation of mean annual temperatures in central Europe during the last 45 million years, as inferred from the composition of palaeofloras. The marked trend towards cooler conditions was mainly driven by lower winter temperatures, while summer temperatures remained relatively constant (i.e. increased seasonality). Modified from Mosbrugger et al. (2005).

effects of past climate change and other palaeogeographical events on the range dynamics of plant species and their diversification throughout the Mediterranean over an extended (geological) timescale.

Here we integrate the analysis of the fossil record with palaeodistribution models and phylogeographical analyses of *Laurus* to reconstruct the climate-driven demise of Tethyan laurel forests, and of *Laurus* in particular. The specific aims of this integrative approach are: (1) to infer the processes of climate-driven migration and extinction of Tethyan laurel species; (2) to understand how some species managed to persist and even thrive in largely unsuitable environments; and (3) to investigate the ecological and genetic differentiation patterns within *Laurus* populations and species in the context of Mediterranean–Macaronesian biogeography. Finally, we apply the knowledge gained from these analyses of past events to predict the likely impacts of climate change on the fate of these Mediterranean relict taxa.

13.2 Palaeobotanical evidence of the rise and demise of the Eurasian–Tethyan laurel forests

The widespread discoveries of plant fossil material whose morphological features closely resemble extant tropical–subtropical species impelled early biogeographers to postulate the presence of extensive evergreen laurophyllous forests in Eurasia during most of the Palaeogene and early Neogene (Mai, 1989, 1995). These forests were dominated by taxa of tropical affinities, such as Lauraceae (e.g. *Lindera, Litsea, Ocotea*), evergreen Fagaceae (e.g. *Trigonobanalopsis*), Myrtaceae (e.g. *Rhodomyrtophyllum*), Vitaceae and Arecaceae, all requiring warm and humid climates (Mai, 1989, 1995; Collinson and Hooker, 2003). Although the earliest fossils of these subtropical or 'paratropical' taxa go back to the Cretaceous (Coiffard et al., 2007), Eurasian laurel forests apparently reached their maximum diversity and distribution during the early Eocene, a period of warm and humid global climate (Fig 13.1 - Zachos et al., 2001; Mosbrugger et al., 2005). At that time, laurel forests occupied a wide latitudinal belt from North Africa to northern Europe, although the richest assemblages occurred around the Tethys Sea (Utescher and Mosbrugger, 2007).

From the late Eocene, laurel forests experienced a gradual retreat from northern latitudes and a southward migration that is linked to a long-term global cooling (Fig 13.1). Concomitantly, deciduous Arcto-Tertiary vegetation expanded its range southward in northern Europe, while sclerophyllous taxa became locally dominant at more arid sites in southwestern Europe and the easternmost Mediterranean (Axelrod, 1975; Collinson and Hooker, 2003). Nonetheless, laurel forests were still relatively widespread in central Europe and the Mediterranean basin until the middle Miocene (Fig 13.2 - Mai, 1989; Erdei et al., 2007; Utescher et al., 2007). Northernmost satellite populations of laurophyllous taxa such as *Rhododendron, Laurophyllum* and *Sassafras* have even been recovered recently from Miocene deposits in Iceland (Denk et al., 2005).

The harsh climatic cooling (Fig 13.1) that occurred since the late Miocene, simultaneously with regional increases in aridity, triggered the southward retreat and the eventual extinction of most laurophyllous taxa. Several Lauraceae were still present in scattered, relict populations at the Pannonian (Erdei et al., 2007),

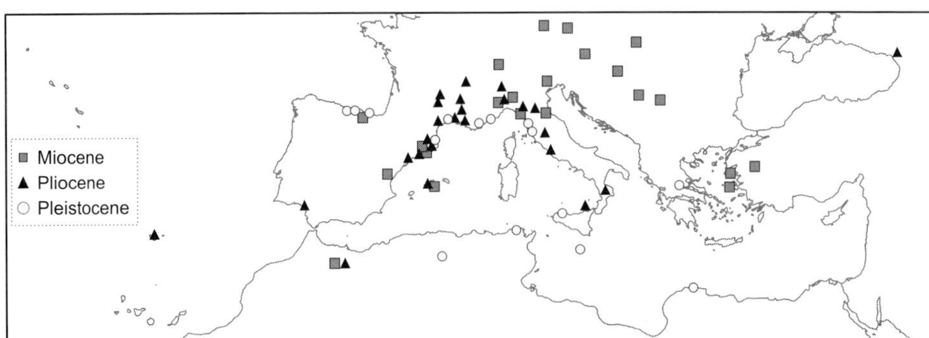

Figure 13.2 Location of Neogene fossil records attributed to *Laurus*. The figure depicts the southward range retreat experienced by Eurasian laurel forests since the mid-Miocene. Redrawn from Rodríguez-Sánchez et al. (2009).

Mediterranean and Black Sea basins (Kovar-Eder et al., 2006) during the lower Pliocene, but only *Laurus* was able to overcome this climatic deterioration and the subsequent Pleistocene glaciations, and thereby persist to the present in relatively moist locations of the Mediterranean and Black Sea basins. *Laurus* and some other Lauraceae species were able to survive in Macaronesian archipelagos (Santos, 1990), where they arrived probably between the Miocene and Pliocene (Axelrod, 1975). The strong regional extinction of laurophyllous taxa that occurred in continental Europe did not occur in eastern Asia or in North America; these regions represented suitable long-term refugia for a significant number of species (Latham and Ricklefs, 1993). Today, excluding a few temperate species, all extant Lauraceae have a tropical–subtropical distribution (Rohwer et al., 1993), and *Laurus* is the only one present in western Eurasia and northern Africa.

Although the fossil record enables a rough reconstruction of the Neogene demise of Tethyan laurel forests, the spatiotemporal patterns of range retreat and extinction are still largely unknown for most taxa. Uncertainty stems mainly from the scarcity of palaeobotanical data for most areas (particularly Macaronesia and the southern Mediterranean) and time periods (e.g. Quaternary). Given the poor pollen preservation of the Lauraceae (Ferguson, 1974), palaeobotanical data are mostly restricted to macrofossils. Furthermore, the limited sampling effort, and particularly its spatial heterogeneity, prevent robust inferences on colonisation dates, local extinctions and other aspects of the range dynamics of species. Other sources of uncertainty are inherent to palaeobotanical data, such as the likely misidentification of poorly conserved fossils, their limited usefulness to infer the distribution and abundance of low-density species, or the difficulty of detecting local extinctions of taxa followed by recolonisations from fragmentary records. Despite these limitations, palaeobotanical data are a fundamental source of information for reconstructing the past. Moreover, inferences provided by palaeobotanical data are likely to broaden when integrated with other tools for biogeographical reconstructions, namely molecular phylogeographics and species distribution models (Cheddadi et al., 2006; Magri et al., 2006). More thorough sampling should therefore be undertaken.

13.3 Reconstructing Plio–Pleistocene *Laurus* distribution shifts through niche-based species distribution models

Species distribution models comprise a diverse group of statistical techniques that relate the geographical distribution of organisms to particular features of the environment, such as climate and soil type, usually with the aim of predicting species distributions at different places or time stages (Guisan and Zimmermann,

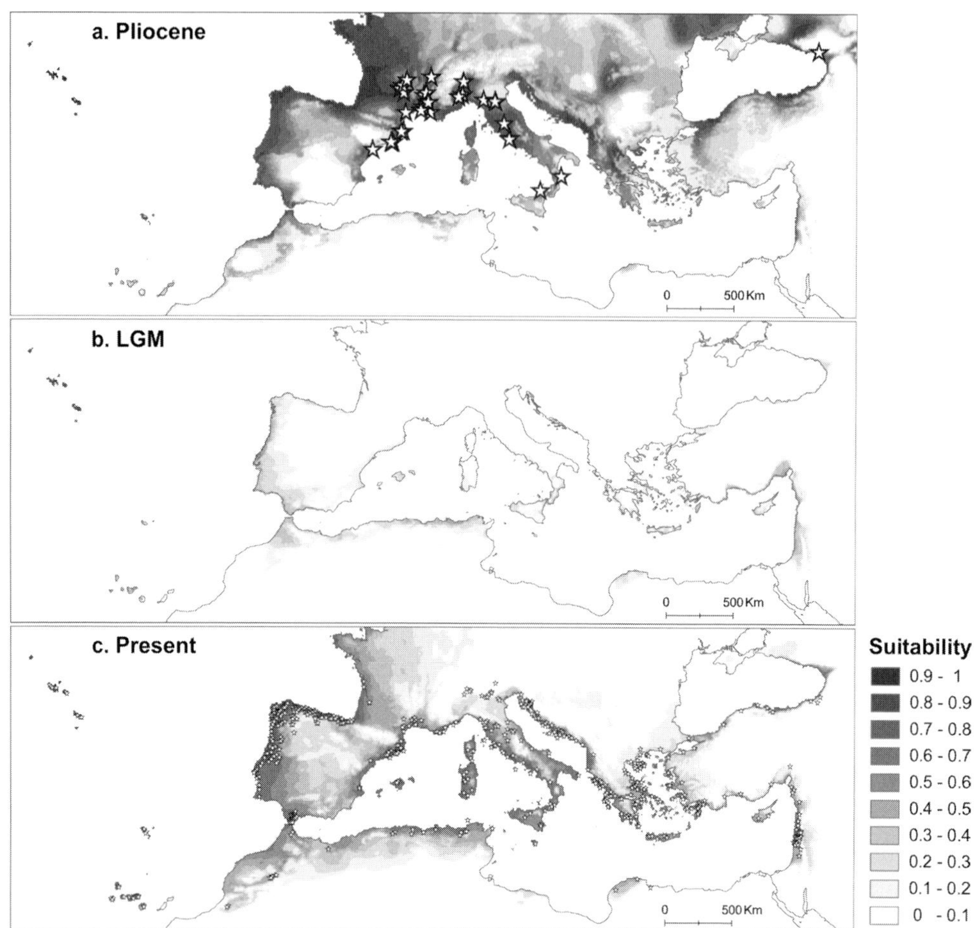

Figure 13.3 Climatically suitable areas for *Laurus* at the mid-Pliocene (3 million years ago), Last Glacial Maximum (LGM; *c.* 21 000 years ago) and present. Habitat suitability inferred from models of *Laurus*' climatic niche projected onto palaeoclimatic data (Rodríguez-Sánchez and Arroyo, 2008). Actual occurrences of *Laurus* for the Pliocene and present are shown by white stars.

2000; Pearson and Dawson, 2003). When successfully fitted to data for a particular time period, these models can be used to predict future (forecasting) or past (hind-casting) species distributions as a function of changing climates or environments (e.g. Pearman et al., 2008a). Rodríguez-Sánchez and Arroyo (2008) employed such an approach to reconstruct the range dynamics of *Laurus* for the last three million years. The scarcity of palaeobotanical data for this period limits the assessment of hypotheses about distribution shifts or location of glacial refugia. Their fossil-validated model predictions (Fig 13.3) documented a process of severe *Laurus* range retreat driven by the overall cooling and, secondarily, by the increase of thermal and rainfall (drought) seasonality since the end of the Tertiary. At the mid-Pliocene climatic optimum, the potential distribution of *Laurus* was much wider than in later periods, favoured by a warm and relatively humid climate (Haywood et al., 2000). The onset of the Pleistocene glaciations drastically reduced the areas suitable for the persistence of *Laurus*, particularly at northern latitudes (central and southeastern Europe), causing the extinction of temperate relict populations. During the largely unsuitable glacial periods, *Laurus* may have persisted at multiple but small and scattered refugia distributed across the Macaronesian archipelagos, Transcaucasia and the Mediterranean basin (Fig 13.3b), with the Iberian peninsula, North Africa, southern Italy, the Balkan–Aegean region and the Near East as the main glacial refugia. At present (interglacial conditions), *Laurus* inhabits coastal and relatively moist areas in the Mediterranean and Macaronesia. On a coarse geographical scale, the current geographical range of *Laurus* seems to be constrained mostly by climate: cold temperatures impede range expansion towards northern latitudes and inland ranges, while drought stress sets its southern range limits (Rodríguez-Sánchez and Arroyo, 2008). Therefore, climate change seems to have exerted a strong influence on the range dynamics of *Laurus* over geological time.

As an alternative approach to these geographical projections, more mechanistic insights into the fate of *Laurus* species confronted with climate change can be obtained by mapping the climatic requirements of *Laurus*, as estimated by species niche models, onto an environmental space chart, together with the 'realised climates' or range of climatic conditions available at each time stage (Fig 13.4). For a species to persist, there must exist at least some overlap between the climatic requirements of the species and the range of available environmental conditions over time (Jackson and Overpeck, 2000). When there is no overlap, the species must either adapt to the new environment or else migrate to areas conserving suitable climates in order to avoid extinction. In this framework, species niche modelling analyses using two relevant climatic variables and data on the current distribution of *Laurus* showed that its climatic requirements are skewed towards warmer and moister conditions in relation to the available environment (Fig 13.4). Independent experimental evidence confirms the sensitivity of *Laurus* to cold temperatures (Larcher, 2000; González-Rodríguez et al., 2005), which can damage photosynthetic organs and kill seeds or even adult individuals (Giacobbe, 1939; Takos, 2001). Water availability throughout the year also plays a relevant role in limiting distribution, excluding *Laurus* plants from dry sites (those with actual to potential evapotranspiration ratios below 0.5 – Fig 13.5). These climatic requirements are consistent with the subtropical affiliation of *Laurus* (Larcher, 2005). Although here we can only depict part of its climatic niche (that constrained within the range of present climates), both palaeobotanical and experimental data confirm that *Laurus* can inhabit areas with climates warmer than present, as long

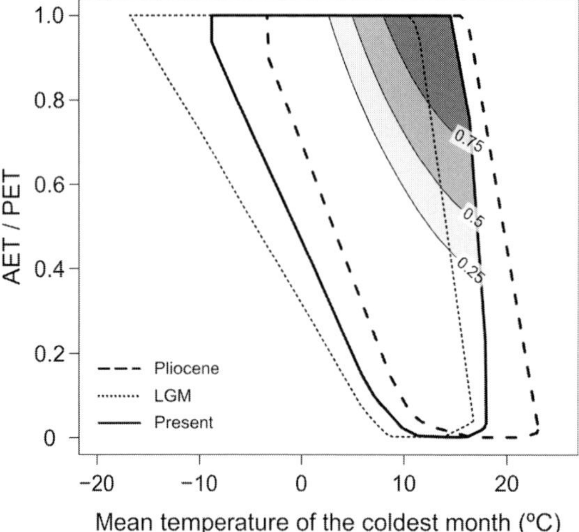

Figure 13.4 Convex hulls enclosing the range of climatic conditions available at the mid-Pliocene, LGM and present for the two climatic variables mostly determining the distribution of *Laurus*. Variables are mean temperature of the coldest month and the actual to potential evapotranspiration ratio (AET/PET). Shaded areas represent relative suitability of climatic conditions for *Laurus* (on a 0–1 scale). The latter was estimated using maximum entropy niche modelling (Phillips et al., 2006) and data on present *Laurus* occurrences and climate (see Rodríguez-Sánchez and Arroyo, 2008). Averaged predictions after 100 random model replicates are shown. Note that *Laurus* requires warm and moist climates (even more than that currently available), and that glacial conditions are largely unsuitable for the species, as shown by the small overlap between the available climatic conditions at the LGM and relative suitability for *Laurus*.

as they also offer high water availability (Konis, 1949; Utescher et al., 2007). Thus, the warm and moist mid-Pliocene climate appeared more suitable for *Laurus* than any other period in the last three million years, as shown by the higher overlap between *Laurus* requirements and the available climate. By contrast, this overlap became minimal during maximum glacial conditions (Fig 13.4). The cold and relatively dry climates of Pleistocene glaciations provoked regional extinction of populations (mainly at northern latitudes) but not the (complete) extinction of the species, as there was still some overlap between the requirements of *Laurus* and the available climatic conditions at some geographical areas (those depicted as glacial refugia in Fig 13.3b).

It is worth noting the close match between the climatic response curves of *Laurus* at the mid-Pliocene and at present (Fig 13.5), which suggests that the climatic requirements of *Laurus* have remained virtually unchanged over millions of years. This marked niche conservatism (evolutionary maintenance of ancestral

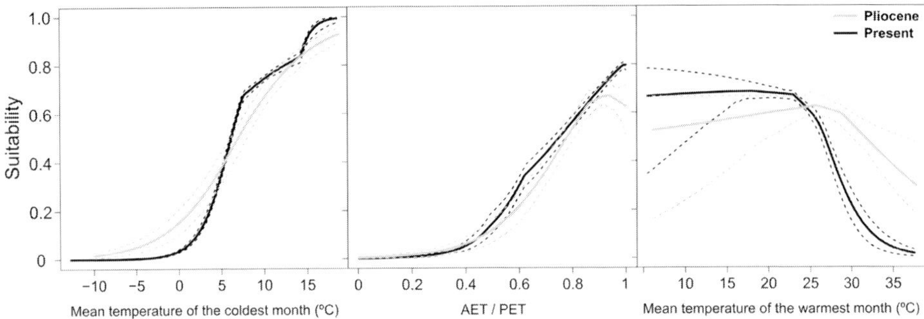

Figure 13.5 Fitted relationships between climatic variables and suitability for *Laurus*, as estimated for the mid-Pliocene and the present. Variables are mean temperature of the coldest month, actual to potential evapotranspiration ratio, and mean temperatures of the warmest month, in decreasing order of importance. The large overlap between the response curves for the two periods points to a high conservatism of *Laurus*' climatic requirements for several million years. Dotted lines enclose the 95% confidence intervals obtained through replicating model fitting 100 times after randomly partitioning the occurrences data set into training and test locations (Rodríguez-Sánchez and Arroyo, 2008). Note that other sources of uncertainty, such as likely biases in palaeoclimatic data, are not accounted for.

ecological characteristics – Wiens and Graham, 2005) has forced largely deterministic (hence predictable) range dynamics of *Laurus* over time in relation to changing climate (see also Svenning, 2003). In fact, the current distribution of *Laurus* could be successfully predicted from a rough estimate of the environments occupied by the species in the mid-Pliocene (Rodríguez-Sánchez and Arroyo, 2008). Although rarely assessed in this kind of study, confirmation of niche stasis over time provides support to palaeodistributions reconstructed through niche-based models (Pearman et al., 2008b).

13.4 Molecular footprints of past range dynamics in *Laurus*

One of the most frequently used approaches for reconstructing the biogeographical history of lineages is molecular phylogeography (Avise, 2000). There has been a proliferation of phylogeographical studies in both animal and plant groups, often aimed at locating refugia and reconstructing species distribution shifts in response to Pleistocene glaciations (Hewitt, 2004). In angiosperms, phylogeographical studies are often based on the analysis of chloroplast DNA (cpDNA) polymorphisms, as they offer advantages compared with those from nuclear DNA markers, such as moderate variation, non-recombination and smaller effective population size. They also generally show maternal inheritance, which means that patterns of cpDNA variation are not directly influenced by pollen flow among populations (Comes and Kadereit, 1998). Thus, patterns of genetic relatedness of populations based on cpDNA markers should arise exclusively from successful seed dispersal events and hence reflect the history of colonisations and range retreats experienced by the species (Avise, 2000). Alternatively, a number of studies have used nuclear DNA markers, usually to compensate for the low variability of cpDNA, or have worked at the level of whole genomes using markers such as those generated by amplified fragment length polymorphism (AFLP – Vos et al., 1995; Mueller and Wolfenbarger, 1999). The existence of contrasting phylogeographical patterns obtained by both plastid and nuclear markers has sometimes revealed interesting evolutionary events and has thus contributed to a broader picture of the history of taxa (e.g. Arnold et al., 1991).

To date, two studies have attempted a phylogeographical analysis of *Laurus* over most of its distribution range. Firstly, Arroyo-García et al. (2001) used AFLP analyses to assess the genetic relatedness between populations from the western and central Mediterranean (*L. nobilis*) and part of Macaronesia (*L. azorica*). Although their analysis did not consider phylogenetic relationships of the species and populations, they found a clear geographical structure of genetic similarity. Somewhat surprisingly, they found a close similarity between western and southern Iberian populations and those in Macaronesia (even though they are geographically separated and even considered as different species), whereas there was a strong divergence between this western (Iberian and Macaronesian) group and the central Mediterranean (French and Italian) *L. nobilis* populations. Therefore, the Pyrenees seem to have acted as a significant (bio)geographical barrier for *Laurus* over recent historical periods, while some genetic connection, through either pollen or seed, must have occurred among distant western Mediterranean populations. Nevertheless, sampling was limited to the western half of the *Laurus* range and excluded some areas such as northern Africa, which are crucial in order to explain the colonisation of Macaronesia by *Laurus* and, more generally, the historical range dynamics of *Laurus* throughout the western Mediterranean.

A second study (Rodríguez-Sánchez et al., 2009) attempted a historical reconstruction of the past range dynamics of *Laurus* over the whole Mediterranean basin and Macaronesia. This time, *Laurus* populations were sampled throughout the complete geographical range of the genus (from the Azores to the Caucasus, and from the northern Mediterranean basin to the Moroccan Anti-Atlas and the Canary Islands). Overall cpDNA variation was low, but two cpDNA regions (*trnK-matK*, *trnT-trnD*) were sufficiently informative and unambiguous. The use of both phylogeographical and phylogenetic analyses (maximum parsimony

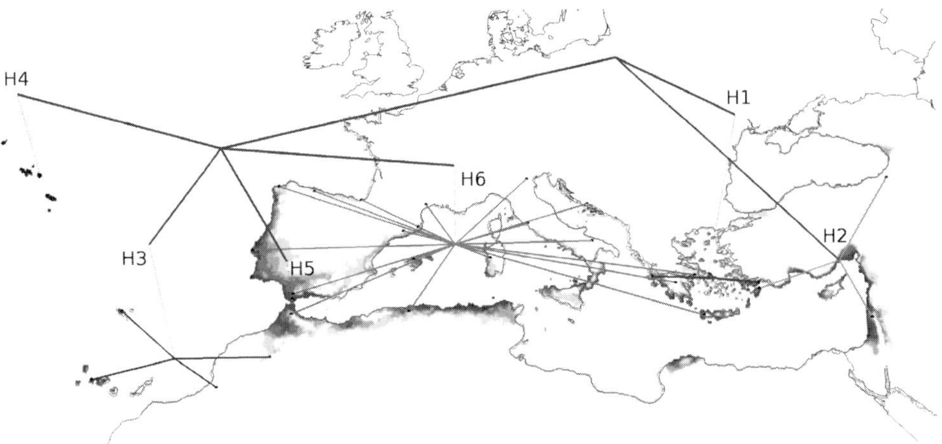

Figure 13.6 Geophylogeny showing the phylogenetic tree (red) and the geographical distribution of the six distinct *Laurus* haplotypes (H1 to H6). After Rodríguez-Sánchez et al. (2009). Shaded areas represent likely glacial refugia for *Laurus*, as predicted by niche-based distribution models (Rodríguez-Sánchez and Arroyo, 2008). See Kidd and Liu (2008) for further details on geophylogenies. *See colour plate section.*

and Bayesian inference) allowed the assessment of relationships among populations, which gave several insights into the differentiation and colonisation patterns of the species. Firstly, *Laurus* was found to be a monophyletic group, an important prerequisite for subsequent analyses of its past migration history. Secondly, six distinct haplotypes were found across its geographical range (Fig 13.6), but, in general, variation was low. This low genetic variability may stem from a number of different causes, but the most likely explanations are the long lifespan of *Laurus* and the subsequent slow generation turnover of its populations, and a historically low mutation rate in the whole lineage of Lauraceae (Chanderbali et al., 2001; Willis and Niklas, 2004; Petit and Hampe, 2006; Smith and Donoghue, 2008). On the other hand, the dioecious breeding system of the species might have promoted haplotype sorting, as separate sexes diminish the effective population size of cpDNA (Cruzan and Templeton, 2000). Thirdly, the six haplotypes were grouped into three lineages, two of which (consisting of one haplotype each) were distributed along the eastern Mediterranean (Fig 13.6). The third clade contained four haplotypes, three of them being exclusive of the westernmost part of the range (Macaronesia, Morocco and southwestern Iberia). Therefore, *Laurus* follows an east-to-west diversification pattern that seems to be common in other ancient Mediterranean lineages, and is consistent with Tethyan–Mediterranean palaeogeographical dynamics (Oosterbroek and Arntzen, 1992; Petit et al., 2005).

13.5 Integrating the fossil record with palaeodistribution models and molecular phylogeography to reconstruct the Neogene history of *Laurus*

Although an absolute time framework could not be provided, Rodríguez-Sánchez et al. (2009) exploited the relatively abundant fossil record of *Laurus* (Fig 13.2) and a previous reconstruction of *Laurus* palaeodistributions based on bioclimatic models (Rodríguez-Sánchez and Arroyo, 2008) in order to provide an integrative interpretation of its past range dynamics. An ancient lineage splitting in the eastern Mediterranean was proposed, followed by a westward migration and a secondary diversification in the western Mediterranean and Macaronesia (four out of six haplotypes). This westerly diversification is probably ancient too, as supported by the relatively abundant Miocene and Pliocene fossils of *Laurus* in Iberia, France and Italy (Fig 13.2), suggesting an early arrival in these areas. Despite the range restrictions imposed by Pleistocene glaciations (Fig 13.3), most populations were able to persist at scattered locations along the Mediterranean and in Macaronesia, as shown by the presence of singular haplotypes at several putative refugia (Fig 13.6). The hypothesis of multiple glacial refugia receives further support from the geographical genetic structure determined by Arroyo-García et al. (2001). However, we cannot exclude the extinction of additional lineages during the Pleistocene. The expansion of one of the derived haplotypes (H6) throughout most of the Mediterranean might be a consequence of relatively recent dispersal processes, probably linked to interglacial periods and even to human-assisted migrations.

Unfortunately, the scarcity of fossils in North Africa and Macaronesia and the low overall genetic variability of the populations preclude further inferences on the dates and migration routes towards and within Macaronesian archipelagos. Nonetheless, the recent finding of a further new haplotype in the Atlas region of North Africa (Rodríguez-Sánchez, unpublished data) supports the hypothesis that this area was an important long-term refugium for *Laurus*. This region has also probably acted as a stepping stone in the migration of *Laurus* to Macaronesia, as has been shown for many other taxa (Vargas, 2007).

13.6 Taxonomic differentiation in *Laurus* and other pre-Mediterranean lineages

The low rate of molecular differentiation and high morphological stasis shown by *Laurus* over geological time are also features in other Tertiary relict taxa from the Mediterranean and elsewhere (Milne and Abbott, 2002). Two species have been considered traditionally within the genus *Laurus* (*L. nobilis* in the Black Sea and Mediterranean basins and *L. azorica* in Macaronesia), but it may be better to treat them as a single monophyletic group in light of previous morphological reassessments (Ferguson, 1974; Marques and Sales, 1999) and recent molecular studies (Arroyo-García et al., 2001; Rodríguez-Sánchez et al., 2009). Thus, our former sketch of *Laurus'* historical biogeography, as a monophyletic lineage, is valid for the whole genus. The formal taxonomic description of both species is based mostly on continuous, overlapping traits, such as leaf shape and tomentum, and many specimens collected within the alleged disjunct range of the species cannot be unequivocally ascribed to one of the two species (Arroyo-García et al., 2001). Macaronesian *Laurus* was originally described in the mid-nineteenth century as two separate species: *Persea azorica* Seub. from the Azores and *Laurus canariensis* Webb and Berthelot from Madeira and the Canary Islands (see Franco, 1960, who combined both taxa into a single species, *L. azorica*). Later, some *L. azorica* populations were reported from the Middle Atlas and the Anti-Atlas on the African continent (Barbero et al., 1981). *Laurus nobilis* is found hundreds of kilometres from here, and spreads across the rest of the Mediterranean and Black Sea basins. *Laurus nobilis* and *L. azorica* present higher morphological variability within their geographical ranges than amongst them, as shown by a morphometric study based on herbarium specimens in which intrapopulation variability was not considered (Marques and Sales, 1999; see also Giacomini and Zaniboni, 1946). A study across the full range of *Laurus*, incorporating morphological variability within and amongst populations, is thus desirable to ascertain the precise differences, if any, amongst populations and species.

Several hypotheses can be posed regarding this enormous morphological variability. Firstly, the isolated location of most populations, certainly more pronounced in the Macaronesian archipelagos, may have favoured some stochastic morphological differentiation among populations. In particular, founder events may have been more important than genetic drift, given the apparent long generation times of *Laurus*. Hence, a wide range of ancestral phenotypic variation may have existed among populations throughout the Mediterranean and these may have undergone subsequent random local extinction and colonisation events to generate a pattern similar to what is currently observed. Nevertheless, we cannot discard the role of phenotypic plasticity or the possible adaptive significance of population modal values. These should be specifically tested, at least through classical reciprocal transplant experiments. Additional insight may come from ecological niche models. Although there is significant niche overlap (Warren et al., 2008) between both species (actual overlap of 0.75, on a 0–1 scale, while values of niche overlap according to a proper null model span from 0.68 to 0.70), some ecological segregation between the two species also appears to exist (Fig 13.7). In particular, *L. nobilis* is found in colder (i.e. more continental) habitats than *L. azorica*,

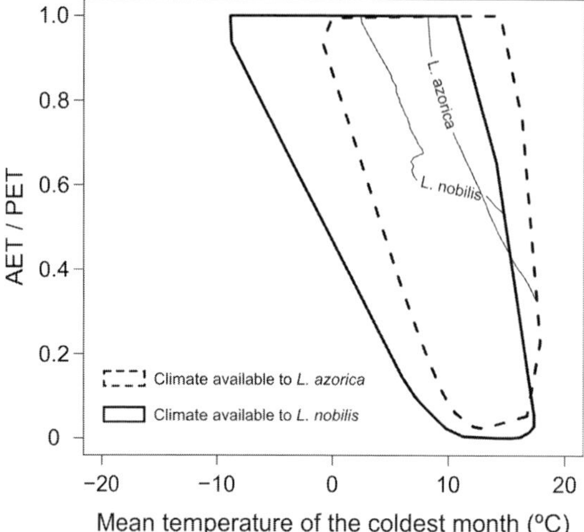

Figure 13.7 Range of current climatic conditions available to *L. nobilis* and *L. azorica*, and the portion of their climatic space occupied by each species. *L. nobilis* is found in colder habitats, but these are also much more common within its distribution area. Warm and moist habitats are optimal for both species. Climatic niches were estimated by maximum entropy niche modelling (see Fig 13.4). For better visibility, environmental suitability values have been thresholded for each species using the maximisation of sensitivity plus specificity criteria (Liu et al., 2005).

which may reflect a better adaptive performance of the former to Pleistocene glaciations that were more intense in continental ranges. In fact, ecophysiological studies have shown higher tolerance to cold stress in continental (*L. nobilis*) populations (Larcher, 2000) than in Macaronesian (*L. azorica*) ones (González-Rodríguez et al., 2005). Nevertheless, this apparent habitat segregation might also result from processes of competitive exclusion in cold habitats or from intense historical land-use change in Macaronesian islands (Parsons, 1981). Thus, further work is needed to explore these hypotheses and ultimately understand the ecological attributes of the two *Laurus* 'species'.

Caryological studies (Ehrendorfer et al., 1968) have found variable ploidy levels within *Laurus* that may help explain its infrageneric morphological differentiation. However, so far the weight of available evidence suggests that *Laurus* should be considered a monospecific genus as proposed by Marques and Sales (1999), and any attempt to split it into new species, such as the recent proposal of *L. novocanariensis* (Rivas-Martínez et al., 2002), should be based on rigorous and comprehensive research.

Most Mediterranean woody taxa (e.g. families Cistaceae, Fabaceae and Lamiaceae) show high phylogenetic diversification and endemism levels as a consequence of adaptive radiation, hybridisation and/or polyploidy processes. Recent work on rockroses (*Cistus*) has shown strong molecular and taxonomic divergence across the Mediterranean, which probably occurred during the Plio–Pleistocene (Guzmán and Vargas, 2005). An even stronger diversification has been found in Mediterranean *Ulex* spp. (De Castro et al., 2002; Pardo et al., 2004; Cubas et al., 2005). In contrast, there is low taxonomic diversification in *Laurus* and other pre-Mediterranean lineages of subtropical origin, irrespective of their current status as relicts (e.g. *L. nobilis*, *Prunus lusitanica*, *Rhododendron ponticum*) or as widespread species in typical Mediterranean forests and shrublands (e.g. *Myrtus*, *Olea*, *Phillyrea*, *Pistacia* – Herrera, 1992; Milne and Abbott, 2002). The wide phylogenetic span of this pre-Mediterranean guild, with several families represented, does not support phylogenetic relatedness as a possible explanation for their shared low diversification. Instead, their longevities and strong survival capacity against severe disturbance and their subsequent slow generation turnover, together with their frequent vegetative reproduction, emerge as more likely factors (Willis and Niklas, 2004; Smith and Donoghue, 2008). Additionally, genetic constraints, long-term habitat stability and/or historical gene flow among populations may also have precluded differentiation or even adaptation to changing environments (Milne and Abbott, 2002; Ackerly, 2003). This low differentiation and diversification capacity could partly explain the relatively high extinction rates of subtropical taxa at the Plio–Pleistocene boundary (Svenning, 2003). Clearly, more combined molecular, palaeobotanical and ecological evidence is needed to ascertain the causes of the morphological stasis and low speciation rates of Tertiary lineages.

13.7 A cautionary tale for the future of laurel forests in the Mediterranean and Macaronesia

The insights obtained from our integrative reconstruction of the past range dynamics of *Laurus* in relation to historical climate change can tentatively be used to predict the likely effects of future climate changes on *Laurus* and other laurophyllous taxa with similar climatic requirements. All current climatic models coincide in predicting warmer temperatures and an overall reduction in rainfall (i.e. higher drought stress) in the Mediterranean basin (Christensen et al., 2007), with similar predictions for the Canary Islands (Sperling et al., 2004). Although laurophyllous taxa could initially benefit from a warmer climate, a marked increase in water deficit would be detrimental (Rodríguez-Sánchez and Arroyo, 2008). This impact would be stronger in southern Mediterranean *Laurus* populations, which currently harbour most of the genetic diversity of the species. In contrast, the milder climate predicted for western and central Europe would favour the spread of *Laurus* to those regions because predicted warmer minimum temperatures would no longer be limiting. Field observations indicate an ongoing northward range expansion of thermophilous taxa to this area (Walther, 2003), consistent with previous predictions (Parmesan, 2006). In the particular case of *Laurus*, this northward expansion might be hastened by humans (van der Veken et al., 2008) because *Laurus* is widely used as a cultivated and ornamental plant. Similarly, the authors have detected reduced growth and have recorded reproductive failures caused by dry years in Mediterranean populations from southern Iberia and North Africa (unpublished data). Although long lifespan and resprouting ability confer strong demographic resilience, a sustained trend towards more arid climates would certainly increase the risk of local extinction.

13.8 Conclusions: adaptation, niche conservatism and the evolutionary effects of climate changes

There has been an increase in theoretical and empirical studies trying to foresee the ecological and evolutionary consequences of future climate change, and historical approaches are central to these efforts. They allow researchers to assess competing hypotheses on the effects of varying past climate regimes with actual data obtained from the past. Within this framework, the climate-driven apogee and demise of Tethyan laurel forests during the Cenozoic represents an invaluable case study to explore timely topics of paramount biogeographical relevance. Firstly, it is important to know what ecological factors and/or intrinsic species features determine different resiliences to extinction in different taxa confronted with the same process of climate change. In this regard, it is particularly important to know why *Laurus* was able to endure the Plio–Pleistocene climate changes while all other European Lauraceae went extinct. In a landmark paper, Svenning (2003) showed that the final fate of European tree species (extinction, relictual persistence or wide distribution) after the Plio–Pleistocene transition was mostly determined by their previous climatic requirements. Comparative data from extant Lauraceae in the Canary Islands indeed suggest that *Laurus* is more freeze-tolerant and less drought-sensitive than confamilial taxa (Gandullo et al., 1992; González-Rodríguez et al., 2005), which could help explain its successful persistence in the Mediterranean basin. Secondly, the history of *Laurus* provides an illustrative case of how some plant species manage to persist despite largely unsuitable and changing climates. Although climatic cooling since the late Tertiary provoked the extinction of its once thriving northernmost populations, the existence of several climatically suitable areas through the Quaternary, albeit small and isolated, enabled the long-term persistence of the species.

A large number of studies have shown that the maintenance of ancestral ecological features of taxa may play a determinant role in the responses of species to climate change. If niche conservatism occurs, rather than adaptation to the new environment, then species must track suitable climates by migration or go extinct (Huntley and Webb, 1989; Jackson and Overpeck, 2000). Several causes have been proposed to explain the lack of adaptation in natural ecological scenarios, including: (1) lack of sufficient genetic variation; (2) genetic constraints such as pleiotropy; (3) presence of gene flow, which hinders possible local adaptive processes; and (4) the existence of stabilising selection on the climatic niches of species (Wiens and Graham, 2005). The last of these could be promoted by community assembly processes or the ability of species to track suitable habitats (Ackerly, 2003; de Mazancourt et al., 2008). Furthermore, while rapid evolution or adaptation in short-lived plant populations seems likely (Willis and Niklas, 2004; Franks et al., 2007), such processes might be more limited in tree populations, due to the longevity of most tree species and their long generation cycles (Petit and Hampe, 2006). This would reduce their opportunities to adapt to changing climates at a suitable pace (Jump and Peñuelas, 2005). On the other hand, the typically high survival rates of adult trees and their long lifespan (of up to several thousand years) makes them, in principle, more resilient to extinction in transient periods of unsuitable climate (Brubaker, 1986; Bond and Midgley, 2001).

Future studies should focus on the role of the many factors and processes that determine either niche conservatism or adaptive responses of species, as these will ultimately decide the fate of taxa under changing climates. In this sense, historical approaches that integrate information from phylogenetic and phylogeographical analyses, detailed studies of the fossil record, and palaeodistribution statistical models will provide fruitful insights. Furthermore, when combined with experimental, theoretical and contemporary ecological approaches, they will certainly improve our knowledge of the ecological and evolutionary effects of climate change.

Acknowledgements

We thank D. Lunt, A. Haywood, P. J. Valdes, and the developers of WorldClim and the Palaeoclimate Modelling Intercomparison Project (PMIP) for providing climate data, and S. Phillips and D. Warren for kind assistance in terms of software. We are also grateful to the many people who helped to locate and sample *Laurus* populations, and to P. Vargas, B. Guzmán and A. Valido for collaborating on the *Laurus* phylogeography work. F. Ojeda, T. R. Hodkinson, J. S. Carrión and an anonymous reviewer revised and improved previous versions of the manuscript. Financial support was provided through research contracts with GIASA, and grants from PAIDI (P07-RNM-02869 and P09-RNM-5280) and MEC (BOS2003–07924-CO2–01, CGL2006–13847-CO2–01). The first author was supported by a predoctoral FPU studentship from MEC (AP2002-3730).

References

Ackerly, D. D. (2003). Community assembly, niche conservatism, and adaptive evolution in changing environments. *International Journal of Plant Sciences*, **164**, S165-S184.

Araújo, M. B., Whittaker, R. J., Ladle, R. J. and Erhard, M. (2005). Reducing uncertainty in projections of extinction risk from climate change. *Global Ecology and Biogeography*, **14**, 529-538.

Arnold, M. L., Buckner, C. M. and Robinson, J. J. (1991). Pollen-mediated introgression and hybrid speciation in Louisiana irises. *Proceedings of the National Academy of Sciences of the USA*, **88**, 1398-1402.

Arroyo-García, R., Martínez-Zapater, J. M., Fernández Prieto, J. A. and Álvarez-Arbesu, R. (2001). AFLP evaluation of genetic similarity among laurel populations (*Laurus* L.). *Euphytica*, **122**, 155-164.

Avise, J. (2000). *Phylogeography: the History and Formation of Species*. Cambridge, MA: Harvard University Press.

Axelrod, D. I. (1975). Evolution and biogeography of Madrean-Tethyan sclerophyll vegetation. *Annals of the Missouri Botanical Garden*, **62**, 280-334.

Barbero, M., Benabid, A., Peyre, C. and Quézel, P. (1981). Sur la presence au Maroc de *Laurus azorica* (Seub.) Franco. *Anales del Jardín Botánico de Madrid*, **37**, 467-472.

Bond, W. J. and Midgley, J. J. (2001). Ecology of sprouting in woody plants: the persistence niche. *Trends in Ecology and Evolution*, **16**, 45-51.

Brubaker, L. B. (1986). Responses of tree populations to climatic change. *Vegetatio*, **67**, 119-130.

Chanderbali, A. S., van der Werff, H. and Renner, S. S. (2001). Phylogeny and historical biogeography of Lauraceae: evidence from the chloroplast and nuclear genomes. *Annals of the Missouri Botanical Garden*, **88**, 104-134.

Cheddadi, R., Vendramin, G. G., Litt, T. et al. (2006). Imprints of glacial refugia in the modern genetic diversity of *Pinus sylvestris*. *Global Ecology and Biogeography*, **15**, 271-282.

Christensen, J. H., Hewitson, B., Busuioc, A. et al. (2007). Regional climate projections. In *Climate Change 2007: the Physical Science Basis. Contribution of Working Group I to the Fourth Assessment Report of the Intergovernmental Panel on Climate Change*, ed. S. Solomon, D. Qin, M. Manning et al. Cambridge: Cambridge University Press, pp. 847-940.

Cleland, C. (2001). Historical science, experimental science, and the scientific method. *Geology*, **29**, 987-990.

Coiffard, C., Gómez, B. and Thevenard, F. (2007). Early Cretaceous angiosperm invasion of western Europe and major environmental changes. *Annals of Botany*, **100**, 545-553.

Collinson, M. and Hooker, J. (2003). Paleogene vegetation of Eurasia: framework for mammalian faunas. *Deinsea*, **10**, 41-83.

Comes, H. P. and Kadereit, J. W. (1998). The effect of Quaternary climatic changes on plant distribution and evolution. *Trends in Plant Science*, **3**, 432-438.

Cruzan, M. B. and Templeton, A. R. (2000). Paleoecology and coalescence: phylogeographic analysis of hypotheses from the fossil record. *Trends in Ecology and Evolution*, **15**, 491–496.

Cubas, P., Pardo, C. and Tahiri, H. (2005). Genetic variation and relationships among *Ulex* (Fabaceae) species in southern Spain and northern Morocco assessed by chloroplast microsatellite (cpSSR) markers. *American Journal of Botany*, **92**, 2031–2043.

De Castro, O., Cozzolino, S., Jury, S. L. and Caputo, P. (2002). Molecular relationships in *Genista* L. sect. *Spartocarpus* Spach (Fabaceae). *Plant Systematics and Evolution*, **231**, 91–108.

de Mazancourt, C., Johnson, E. and Barraclough, T. G. (2008). Biodiversity inhibits species' evolutionary responses to changing environments. *Ecology Letters*, **11**, 380–388.

Denk, T., Grimsson, F. and Kvaček, Z. (2005). The Miocene floras of Iceland and their significance for late Cainozoic North Atlantic biogeography. *Botanical Journal of the Linnean Society*, **149**, 369–417.

Donoghue, M. J. and Moore, B. R. (2003). Toward an integrative historical biogeography. *Integrative and Comparative Biology*, **43**, 261–270.

Ehrendorfer, F., Krendl, F., Habeler, E. and Sauer, W. (1968). Chromosome numbers and evolution in primitive angiosperms. *Taxon*, **17**, 337–353.

Erdei, B., Hably, L., Kazmer, M., Utescher, T. and Bruch, A. A. (2007). Neogene flora and vegetation development of the Pannonian domain in relation to palaeoclimate and palaeogeography. *Palaeogeography, Palaeoclimatology, Palaeoecology*, **253**, 115–140.

Felsenstein, J. (2004). *Inferring Phylogenies*. Sunderland, MA: Sinauer Associates.

Ferguson, D. K. (1974). On the taxonomy of recent and fossil species of *Laurus* (Lauraceae). *Botanical Journal of the Linnean Society*, **68**, 51–72.

Franco, J. A. (1960). Lauráceas macaronésicas. *Anais do Instituto Superior de Agronomia de Lisboa*, **23**, 89–104.

Franks, S. J., Sim, S. and Weis, A. E. (2007). Rapid evolution of flowering time by an annual plant in response to a climate fluctuation. *Proceedings of the National Academy of Sciences of the USA*, **104**, 1278–1282.

Gandullo, J. M., Bañares, A., Blanco, A. et al. (1992). *Estudio Ecológico de la Laurisilva Canaria*. Madrid: ICONA.

Giacobbe, A. (1939). Richerche geografiche et ecologiche sul *Laurus nobilis* L. *Archivio Botanico*, **15**, 33–82.

Giacomini, V. and Zaniboni, A. (1946). Osservazioni sulla variabilitá del '*Laurus nobilis* L.' nel bacino del Lago di Garda. *Archivio Botanico*, **22**, 1–16.

Givnish, T. J. and Renner, S. S. (2004). Tropical intercontinental disjunctions: Gondwana breakup, immigration from the boreotropics, and transoceanic dispersal. *International Journal of Plant Sciences*, **165**, S1-S6.

González-Rodríguez, A. M., Jiménez, M. S. and Morales, D. (2005). Seasonal and intraspecific variation of frost tolerance in leaves of three Canarian laurel forest tree species. *Annals of Forest Science*, **62**, 423–428.

Guisan, A. and Thuiller, W. (2005). Predicting species distribution: offering more than simple habitat models. *Ecology Letters*, **8**, 993–1009.

Guisan, A. and Zimmermann, N. E. (2000). Predictive habitat distribution models in ecology. *Ecological Modelling*, **135**, 147–186.

Guzmán, B. and Vargas, P. (2005). Systematics, character evolution, and biogeography of *Cistus* L. (Cistaceae) based on ITS, *trnL-trnF*, and *matK* sequences. *Molecular Phylogenetics and Evolution*, **37**, 644–660.

Haywood, A. M., Sellwood, B. W. and Valdes, P. J. (2000). Regional warming: Pliocene (3 Ma) paleoclimate of Europe and the Mediterranean. *Geology*, **28**, 1063–1066.

Herrera, C. M. (1992). Historical effects and sorting processes as explanations for contemporary ecological patterns: character syndromes in Mediterranean woody plants. *American Naturalist*, **140**, 421–446.

Hewitt, G. M. (2004). Genetic consequences of climatic oscillations in the Quaternary. *Philosophical Transactions of the Royal Society of London B*, **359**, 183–195.

Hugall, A., Moritz, C., Moussalli, A. and Stanisic, J. (2002). Reconciling paleodistribution models and comparative phylogeography in the Wet Tropics rainforest land snail *Gnarosophia bellendenkerensis* (Brazier 1875). *Proceedings of the National Academy of Sciences of the USA*, **99**, 6112–6117.

Huntley, B. and Webb, T. (1989). Migration: species response to climatic variations caused by changes in the earth's orbit. *Journal of Biogeography*, **16**, 5–19.

Jackson, S. T. and Overpeck, J. T. (2000). Responses of plant populations and communities to environmental changes of the late Quaternary. *Paleobiology*, **26**, 194–220.

Jansson, R. and Dynesius, M. (2002). The fate of clades in a world of recurrent climatic change: Milankovitch oscillations and evolution. *Annual Review of Ecology and Systematics*, **33**, 741–777.

Jump, A. S. and Peñuelas, J. (2005). Running to stand still: adaptation and the response of plants to rapid climate change. *Ecology Letters*, **8**, 1010–1020.

Kidd, D. M. and Liu, X. (2008). GEOPHYLOBUILDER 1.0: an ArcGis extension for creating 'geophylogenies'. *Molecular Ecology Resources*, **8**, 88–91.

Konis, E. (1949). The resistance of maquis plants to supramaximal temperatures. *Ecology*, **30**, 425–429.

Kovar-Eder, J., Kvacek, Z., Martinetto, E. and Roiron, P. (2006). Late Miocene to early Pliocene vegetation of southern Europe (7–4 Ma) as reflected in the megafossil plant record. *Palaeogeography, Palaeoclimatology, Palaeoecology*, **238**, 321–339.

Larcher, W. (2000). Temperature stress and survival ability of Mediterranean sclerophyllous plants. *Plant Biosystems*, **134**, 279–295.

Larcher, W. (2005). Climatic constraints drive the evolution of low temperature resistance in woody plants. *Journal of Agricultural Meteorology*, **61**, 189–202.

Latham, R. and Ricklefs, R. (1993). Continental comparisons of temperate-zone tree species diversity. In *Species Diversity in Ecological Communities: Historical and Geographical Perspectives*, ed. R. E. Ricklefs and D. Schluter, Chicago, IL: University Press, pp. 294–314.

Liu, C., Berry, P. M., Dawson, T. P. and Pearson, R. G. (2005). Selecting thresholds of occurrence in the prediction of species distributions. *Ecography*, **28**, 385–393.

López de Heredia, U., Carrión, J. S., Jiménez, P., Collada, C. and Gil, L. (2007). Molecular and palaeoecological evidence for multiple glacial refugia for evergreen oaks on the Iberian Peninsula. *Journal of Biogeography*, **34**, 1505–1517.

Magri, D., Vendramin, G. G., Comps, B. et al. (2006). A new scenario for the Quaternary history of European beech populations: palaeobotanical evidence and genetic consequences. *New Phytologist*, **171**, 199–221.

Mai, D. H. (1989). Development and regional differentiation of the European vegetation during the Tertiary. *Plant Systematics and Evolution*, **162**, 79–91.

Mai, D. H. (1995). *Tertiäre Vegetationsgeschichte Europas: Methoden und Ergebnisse.* Jena: Fischer.

Marques, A. R. and Sales, F. (1999). *Laurus* L., um elemento arcaico na flora da Macaronésia. *V Jornadas de Taxonomia Botânica*, Lisboa.

Martínez-Meyer, E. and Peterson, A. T. (2006). Conservatism of ecological niche characteristics in North American plant species over the Pleistocene-to-recent transition. *Journal of Biogeography*, **33**, 1779–1789.

Milne, R. I. and Abbott, R. J. (2002). The origin and evolution of Tertiary relict floras. *Advances in Botanical Research*, **38**, 281–314.

Mosbrugger, V., Utescher, T. and Dilcher, D. L. (2005). Cenozoic continental climatic evolution of Central Europe. *Proceedings of the National Academy of Sciences of the USA*, **102**, 14964–14969.

Mueller, U. G. and Wolfenbarger, L. L. (1999). AFLP genotyping and fingerprinting. *Trends in Ecology and Evolution*, **14**, 389–394.

Oosterbroek, P. and Arntzen, J. W. (1992). Area-cladograms of circum-Mediterranean taxa in relation to Mediterranean palaeogeography. *Journal of Biogeography*, **19**, 3–20.

Pardo, C., Cubas, P. and Tahiri, H. (2004). Molecular phylogeny and systematics of *Genista* (Leguminosae) and related genera based on nucleotide sequences of nrDNA (ITS region) and cpDNA (*trnL-trnF* intergenic spacer). *Plant Systematics and Evolution*, **244**, 93–119.

Parmesan, C. (2006). Ecological and evolutionary responses to recent climate change. *Annual Review of Ecology, Evolution, and Systematics*, **37**, 637–669.

Parsons, J. J. (1981). Human influences on the pine and laurel forests of the Canary Islands. *Geographical Review*, **71**, 253–271.

Pearman, P. B., Randin, C. F., Broennimann, O. et al. (2008a). Prediction of plant species distributions across six millennia. *Ecology Letters*, **11**, 357–369.

Pearman, P. B., Guisan, A., Broennimann, O. and Randin, C. F. (2008b). Niche dynamics in space and time. *Trends in Ecology and Evolution*, **23**, 149–158.

Pearson, R. G. and Dawson, T. P. (2003). Predicting the impacts of climate change on the distribution of species: are bioclimate envelope models useful? *Global Ecology and Biogeography*, **12**, 361–371.

Petit, R. J. and Hampe, A. (2006). Some evolutionary consequences of being a tree. *Annual Review of Ecology, Evolution, and Systematics*, **37**, 187–214.

Petit, R. J., Brewer, S., Bordacs, S. et al. (2002). Identification of refugia and post-glacial colonisation routes of European white oaks based on chloroplast DNA and fossil pollen evidence. *Forest Ecology and Management*, **156**, 49–74.

Petit, R. J., Hampe, A. and Cheddadi, R. (2005). Climate changes and tree phylogeography in the Mediterranean. *Taxon*, **54**, 877–885.

Phillips, S. J., Anderson, R. P. and Schapire, R. E. (2006). Maximum entropy modeling of species geographic distributions. *Ecological Modelling*, **190**, 231–259.

Posadas, P., Crisci, J. V. and Katinas, L. (2006). Historical biogeography: a review of its basic concepts and critical issues. *Journal of Arid Environments*, **66**, 389–403.

Pulquerio, M. J. F. and Nichols, R. A. (2007). Dates from the molecular clock: how wrong can we be? *Trends in Ecology and Evolution*, **22**, 180–184.

Quézel, P. (1985). Definition of the Mediterranean region and the origin of its flora. In *Plant Conservation in the Mediterranean Area*, ed. C. Gómez-Campo. Dordrecht: Dr. W. Junk Publishers, pp. 9–24.

Rivas-Martínez, S., Díaz, T. E., Fernández-González, F. et al. (2002). Vascular plant communities of Spain and Portugal: addenda to the syntaxonomical checklist of 2001. *Itinera Geobotanica*, **15**, 5–922.

Rodríguez-Sánchez, F. and Arroyo, J. (2008). Reconstructing the demise of Tethyan plants: climate-driven range dynamics of *Laurus* since the Pliocene. *Global Ecology and Biogeography*, **17**, 685–695.

Rodríguez-Sánchez, F., Guzmán, B., Valido, A., Vargas, P. and Arroyo, J. (2009). Late Neogene history of the laurel tree (*Laurus* L., Lauraceae) based on phylogeographical analyses of Mediterranean and Macaronesian populations. *Journal of Biogeography*, **36**, 1270–1281.

Rohwer, J. G., Kubitzki, K. and Bittrich, V. (1993). Lauraceae. In *The Families and Genera of Vascular Plants*, ed. K. Kubitzki. Berlin: Springer, pp. 366–390.

Santos, A. (1990). *Evergreen Forests in the Macaronesian Region*. Strasbourg: Council of Europe.

Smith, S. A. and Donoghue, M. J. (2008). Rates of molecular evolution are linked to life history in flowering plants. *Science*, **322**, 86–89.

Sperling, F. N., Washington, R. and Whittaker, R. J. (2004). Future climate change of the subtropical North Atlantic: implications for the cloud forests of Tenerife. *Climatic Change*, **65**, 103–123.

Stewart, W. and Rothwell, G. (1993). *Paleobotany and the Evolution of Plants*. New York, NY: Cambridge University Press.

Svenning, J. C. (2003). Deterministic Plio-Pleistocene extinctions in the European cool-temperate tree flora. *Ecology Letters*, **6**, 646–653.

Takos, I. A. (2001). Seed dormancy in bay laurel (*Laurus nobilis* L.). *New Forests*, **21**, 105–114.

Thompson, J. D. (2005). *Plant Evolution in the Mediterranean*. Oxford: Oxford University Press.

Thuiller, W. (2004). Patterns and uncertainties of species' range shifts under climate change. *Global Change Biology*, **10**, 2020–2027.

Utescher, T. and Mosbrugger, V. (2007). Eocene vegetation patterns reconstructed from plant diversity: a global perspective. *Palaeogeography, Palaeoclimatology, Palaeoecology*, **247**, 243–271.

Utescher, T., Erdei, B., Francois, L. and Mosbrugger, V. (2007). Tree diversity in the Miocene forests of western Eurasia. *Palaeogeography, Palaeoclimatology, Palaeoecology*, **253**, 226–250.

van der Veken, S., Hermy, M., Vellend, M., Knapen, A. and Verheyen, K. (2008). Garden plants get a head start on climate change. *Frontiers in Ecology and the Environment*, **6**, 212–216.

Vargas, P. (2007). Are Macaronesian islands refugia of relict plant lineages? A molecular survey. In *Phylogeography in Southern European Refugia: Evolutionary Perspectives on the Origins and Conservation of European Biodiversity*, ed. S. J. Weiss and N. Ferrand. Berlin: Springer, pp. 297–314.

Vos, P., Hogers, R., Bleeker, M. et al. (1995). AFLP: a new technique for DNA fingerprinting. *Nucleic Acids Research*, **23**, 4407–4414.

Walther, G. R. (2003). Plants in a warmer world. *Perspectives in Plant Ecology, Evolution and Systematics*, **6**, 169–185.

Warren, D. L., Glor, R. E. and Turelli, M. (2008). Environmental niche equivalency versus conservatism: quantitative approaches to niche evolution. *Evolution*, **62**, 2868–2883.

Wiens, J. J. and Graham, C. H. (2005). Niche conservatism: integrating evolution, ecology, and conservation biology. *Annual Review of Ecology, Evolution, and Systematics*, **36**, 519–539.

Willis, K. J. and Niklas, K. J. (2004). The role of Quaternary environmental change in plant macroevolution: the exception or the rule? *Philosophical Transactions of the Royal Society of London B*, **359**, 159–172.

Zachos, J., Pagani, M., Sloan, L., Thomas, E. and Billups, K. (2001). Trends, rhythms, and aberrations in global climate 65 Ma to present. *Science*, **292**, 686–693.

14

The impact of climate change on the origin and future of East African rainforest trees

L. W. Chatrou

Nationaal Herbarium Nederland and *Biosystematics Group, Wageningen University, the Netherlands*

J. J. Wieringa

Nationaal Herbarium Nederland and *Biosystematics Group, Wageningen University, the Netherlands*

T. L. P. Couvreur

Institut de Recherche pour le Développement, Montpellier, France

Abstract

East African rainforests are characterised by a high percentage of endemic species. The occurrence of Annonaceae in the area conforms to this pattern. We review the historical biogeography of species of this family endemic to East Africa,

Climate Change, Ecology and Systematics, ed. Trevor R. Hodkinson, Michael B. Jones, Stephen Waldren and John A. N. Parnell. Published by Cambridge University Press.

in the light of episodes of climate change during the Tertiary. Based on herbarium specimen data, and using a phyloclimatic modelling approach, we identify the environmental variables that are associated with the origin of East African endemics of the genus *Monodora*. We discuss the possible responses of *Monodora* to future trends, based on inferences from past evolutionary changes linked to climatic transitions.

14.1 Introduction

Ecological changes due to the process of global warming and climate change are increasingly documented (Hannah et al., 2005; Lovett et al., 2005a; Lewis, 2006). Species distributions have shifted, and changes in ethology and phenology have caused the disruption of synchrony in plant–insect and predator–prey interactions (Parmesan, 2006). One of the tools applied to study the effect of climate change on organisms is species distribution modelling (Heikkinen et al., 2006; Beaumont et al., 2007). These models, also named bioclimatic envelopes or bioclimatic niches (see Kearney, 2006, for a discussion of species distribution modelling and terms involved), reflect the potential distribution of a species that is predicted on the basis of the relationships between species absence/presence, or presence-only, data, and environmental parameters of areas in which these species occur.

A species distribution model therefore is an extrapolation that disregards historic constraints that have prevented a species from fully exploring its potential distribution. For example, Fig 14.1 shows the predicted area of occurrence of *Monodora tenuifolia* Benth., produced by the BIOCLIM approach (Busby, 1991), as implemented in DIVA-GIS 5.2 (Hijmans et al., 2005). *Monodora tenuifolia* is distributed in West and Central Africa, from Guinea in the west to the eastern part of the Democratic Republic of Congo (Couvreur, 2009). As can be seen, the results of the modelling procedure indicate that some parts of the current distribution area in Ivory Coast, Ghana, Togo, Nigeria and Cameroon are designated as having excellent suitability for this species. Yet some parts of East Africa that are outside the current distribution area are also found to have suitability for the occurrence of *M. tenuifolia*. On a larger spatial scale, one can even identify areas in the Neotropics and the Malay archipelago with a climate that would support this species reasonably well, although this is not shown here.

Why *Monodora* and its allies never ended up in the Neotropics and Asia, despite a suitable bioclimatic niche, is revealed when the biogeography of the family is taken into account (Richardson et al., 2004). They simply were not present at the source areas at the times when cross-continental dispersal was taking place (Wiens and Donoghue, 2004). Discussion of large-scale biogeography, however, is outside the scope of this chapter. Rather, we focus on the evolutionary history of

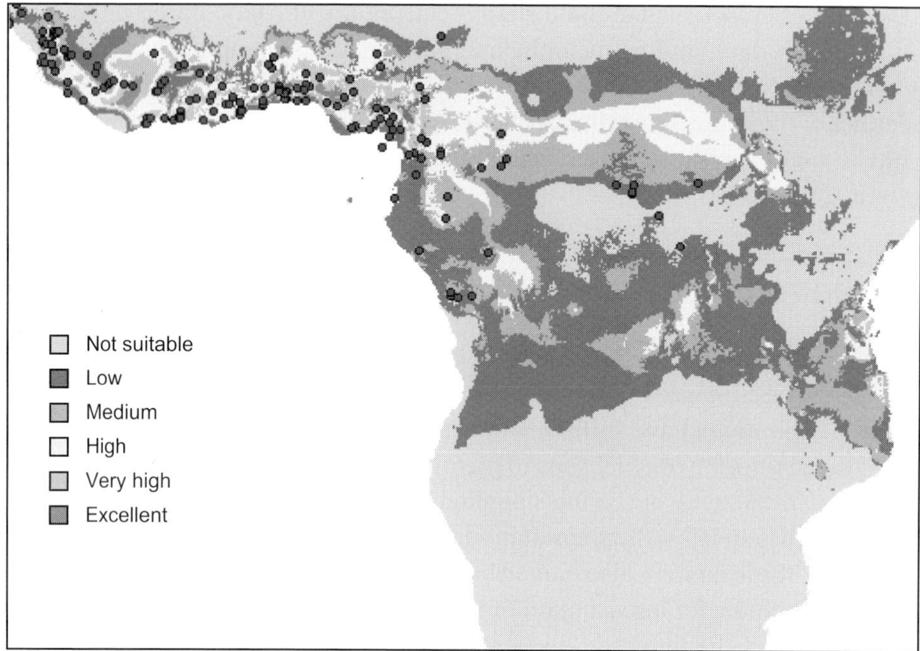

Figure 14.1 Species distribution model for *Monodora tenuifolia* Benth. Based on 106 species occurrence localities (represented by blue circles) and BIOCLIM extrapolation of 19 climate variables as indicated in Table 14.2. *See colour plate section.*

Monodora, a genus comprising species restricted to African rainforests, and on the relationship between its evolutionary history and climate change. We have a good understanding of the phylogenetic and biogeographic patterns underlying the present-day distribution of *Monodora* in Africa (Couvreur et al., 2008; Couvreur, 2009). Species of *Monodora* are endemic to the forests either in West and Central Africa, or in East Africa, with no overlap. This distribution pattern is typical for African Annonaceae in general, as only a handful of species in this family have a distribution encompassing the two major African rainforest blocks (Couvreur et al., 2006), which have been separated since the formation of the African Plateau during the Miocene. Dispersal opportunities for tropical elements between these blocks have been limited ever since. The separation of forest blocks during the Miocene was the last one of a series of geologically or climatically induced vicariance events. These large-scale events involved the fragmentation followed by the re-expansion of rainforest areas. These expansions and contractions have acted as a species pump that may in part explain the high species diversity of hotspots such as the Eastern Arc Mountains and Coastal Forests of Tanzania and Kenya (Couvreur et al., 2008; Chatrou et al., 2009). Such a scenario of repeated contraction and expansion of ranges, and subsequent allopatric speciation, is well known, but in connection with much more recent timescales, such as the Pleistocene (e.g. Bowie et al., 2006).

Couvreur et al. (2008), however, using dated molecular phylogenetic trees of African Annonaceae, demonstrated that the origin of species endemic to the Eastern Arc Mountains and Coastal Forests of Tanzania and Kenya is traceable to three periods extending into the Oligocene. Each of these periods coincides with documented geological or climatological events that likely caused aridification of the African continent, and hence the contraction of rainforest ranges.

Evidence of ancient diversifications is also noticeable in other extant floras and faunas. During the mid-Miocene, vicariant speciation occurred in many unrelated plant lineages in southern Australia (Crisp and Cook, 2007). The age of some species of Neotropical seasonally dry forests similarly was dated to mid-Miocene times (Pennington et al., 2004). As a last example, in the Mediterranean, the majority of all lineages in *Cyclamen* accumulated during the Miocene–Pliocene (Yesson et al., 2009; Yesson and Culham, Chapter 12).

These papers exemplify the fact that we may only have started to understand how climate change has been driving speciation on longer timescales than, for example, the relatively short Pleistocene period. Inferred phylogenies published in these papers doubtlessly bear the signature of past climatic transitions and/or geological events. Given the timescales involved, however, it is hard to arrive at exact correlations between climate change and speciation events, let alone mechanistic links between them. The estimation of branch lengths ('substitutional noise' – Sanderson and Doyle, 2001), the quality of the fossil calibrations (e.g. Near and Sanderson, 2004), the quality of taxon sampling (Pirie et al., 2005) and the methodology used to accommodate rate variation (Linder et al., 2005) all are possible sources of error. The magnitude of the confidence intervals of molecular dating studies generally exceeds the time periods during which species will be faced with changing environmental conditions that are likely to modify the bioclimatic envelopes of species (Tilman and Lehman, 2001).

Despite this disparity in timescales and rate of change between past (evolutionary) and present-day (ecological) climatic changes, it is worthwhile exploring the evolutionary history of climatic preferences. Climate parameters can be optimised onto a phylogenetic tree, and given a well-supported and well-resolved phylogenetic tree this will indicate the evolutionary stability or lability of climatic preferences of species (Nogués-Bravo, 2009). This integration of distributional data, climate data, phylogenetics and molecular dating has been termed 'phyloclimatic modelling' (Yesson and Culham, 2006; Culham and Yesson, Chapter 10). In this chapter we will: (1) identify the main climatic variables that correlate with the distribution of *Monodora* species; (2) trace the evolution of these variables from the crown node of *Monodora* up to the terminal species; and (3) discuss the possible effects of climate change in the light of the biogeographic and phyloclimatic history of *Monodora*.

14.2 Materials and methods

A taxonomic revision of *Monodora* has been published recently (Couvreur, 2009). Data on species' occurrences were extracted from the herbarium specimen database (BRAHMS – Filer, 2008) of the National Herbarium of the Netherlands. A total of 1167 specimens of *Monodora* are available that are both identified down to the species level by the monographer and are georeferenced. Nine specimens of *M. zenkeri* Engl. and Diels were not included as this species is not represented in the species-level phylogeny published by Couvreur et al. (2008). The *Monodora* specimens include duplicates, i.e. multiple specimens of the same species collected at the same locality. These duplicates were removed, resulting in a total of 498 unique occurrence data points for 13 species. The number of unique occurrence localities ranges from two (*M. hastipetala* Couvreur) to 106 (*M. tenuifolia*) (Table 14.1).

Current environmental variables were downloaded from www.worldclim.org at 2.5 minute resolution. These 19 variables are listed in Table 14.2. We assessed all species distribution models using Maxent version 3.3.1 (Phillips et al., 2006). Maxent has been demonstrated to perform well when two conditions apply that are both relevant to the present study, namely: (1) the availability of solely presence-only occurrence data (Elith et al., 2006), and (2) the presence of a small number of point localities (Wisz et al., 2008). Given a set of point localities and environmental variables over geographical space, Maxent estimates the predicted distribution for each species. It does so by finding the distribution of maximum entropy (the distribution closest to uniform) under the constraint that the expected value (expectation) of each feature (derived from the environmental variables) for the estimated distribution matches its empirical average over a sample of locations (Phillips et al., 2006). The resulting distribution model is a relative probability distribution over all grid cells in the geographical area of interest. It expresses the relative probability of the occurrence of a species in a grid cell as a function of the values of the environmental variables in that grid cell. Maxent runs were performed for each of the 13 species with the following options: auto features, random test percentage = 25, maximum iterations = 500. A jackknife test was performed to measure the importance of the environmental variables. During this test three different models are created. The first type of model excludes each variable in turn and creates a species distribution model using the remaining variables. Next, models are created using each variable on its own. Finally, a model is created using all variables, as before.

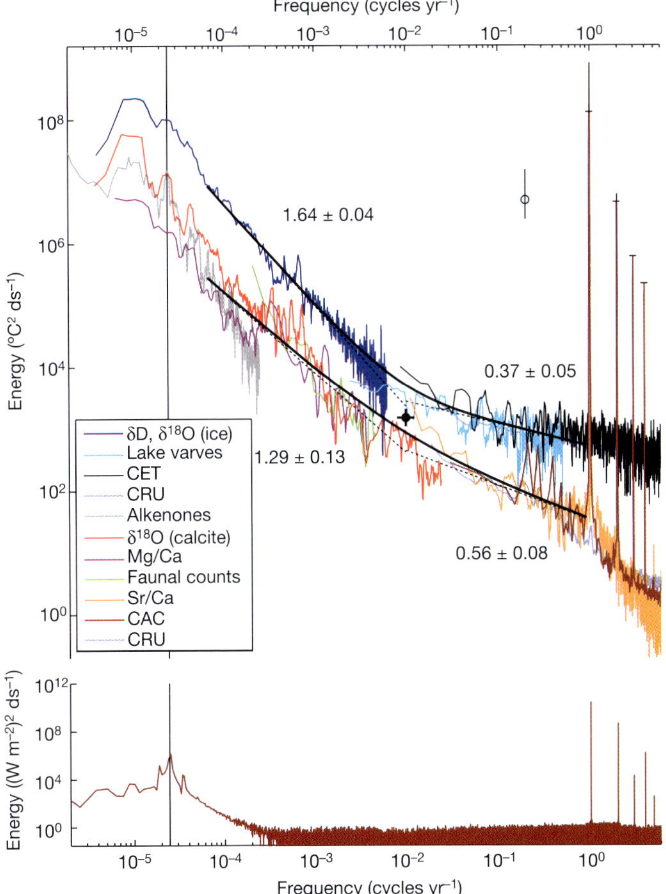

Figure 2.1 Patchwork spectral estimate using instrumental and proxy records of surface temperature variability. The more energetic spectral estimate is from high-latitude continental records and the less energetic estimate from tropical sea-surface temperatures. Power-law estimates for 1.1–100-year and 100–15 000-year periods are listed along with standard errors, and are indicated by the dashed lines. The sum of the power laws fitted to the long and short period continuum is indicated by the black curve. The vertical line segment indicates the approximate 95% confidence interval, where the circle indicates the background level. The mark at 1/100 yr indicates the region midway between the annual and Milankovitch periods. At the bottom is the spectrum of insolation at 65° N sampled monthly over the past million years plus a small amount of white noise. The vertical black line indicates the 41-kyr obliquity period. Reproduced with permission from Huybers and Curry (2006).

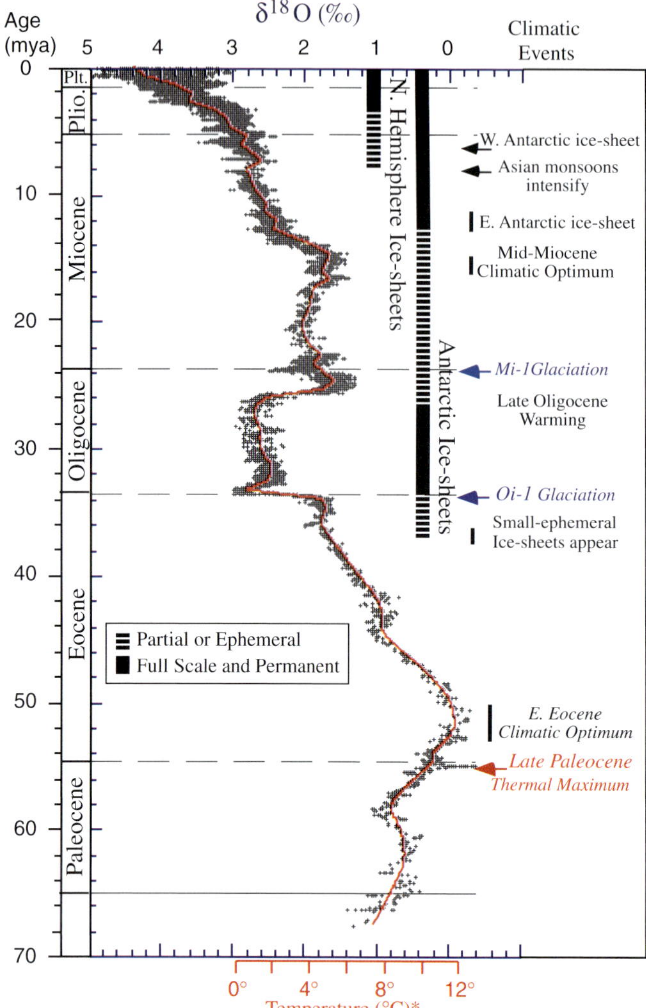

Figure 2.2 Global deep-sea oxygen isotope records for the Cenozoic era, based on data compiled from more than 40 ocean drilling sites. The raw data were smoothed using a five-point running mean and curve fitted with a locally weighted mean. The oxygen isotope temperature scale was computed for an ice-free ocean, and thus only applies to the time preceding the onset of large-scale glaciation on Antarctica (35 mya). From the early Oligocene to the present, much of the variability in the oxygen isotope record reflects changes in Antarctica and northern hemisphere ice volume. The vertical bars provide a rough qualitative representation of ice volume in each hemisphere relative to the Last Glacial Maximum (21 000 years ago), with the dashed bar representing periods of minimal ice coverage (less than 50%), and the full bar representing close to maximum ice coverage (more than 50% of present). Some key tectonic and biotic events are listed as well. Adapted from Zachos et al. (2001).

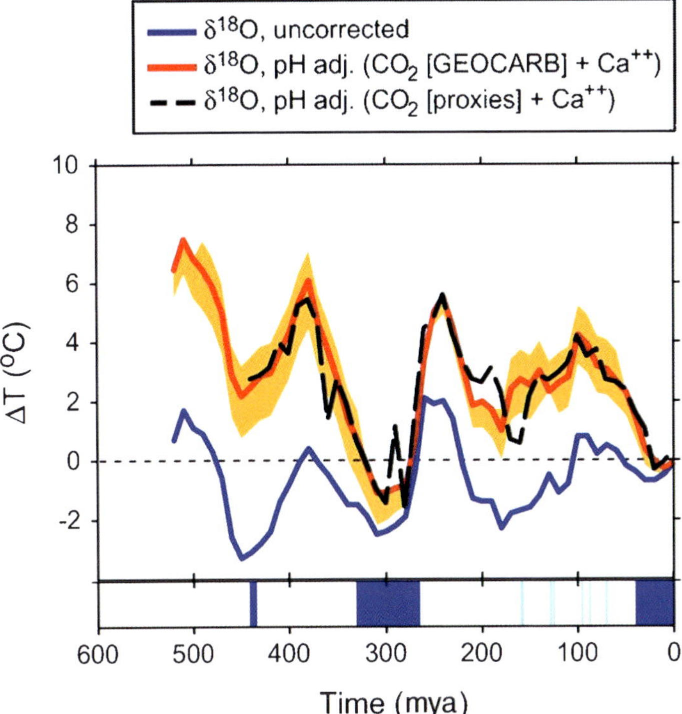

Figure 2.3 Shallow marine carbonate oxygen isotope record over the Phanaerozoic aeon. The blue curve corresponds to temperature deviations relative to today calculated by Shaviv and Veizer (2003). In the two remaining curves, data from the blue curve have been adjusted for pH effects due to changes in seawater calcium ion concentration and CO_2 based either on model (GEOCARB) or proxy reconstructions. Blue bands in the strip along the bottom indicate icehouse intervals, with extensive, permanent continental ice sheets (dark blue) or cool climates, with modest, ephemeral ice sheets (light blue). Adapted from Royer et al. (2004).

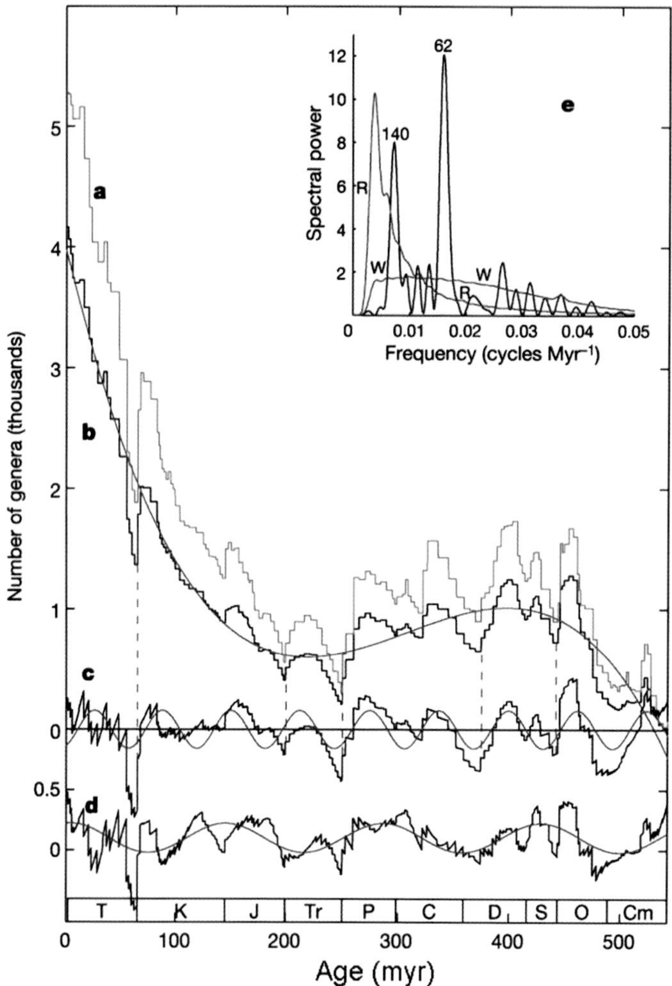

Figure 4.2 Number of marine animal genera against age. (a) All data; (b) singletons and poorly dated genera removed; (c) detrended to show the 62-million-year (myr) cycle; (d) detrended to show the 140 myr cycle; and (e) Fourier spectrum of (c), showing the 62 myr and 140 myr cycles rising beyond the spectral background (W and R). Reproduced from Rohde and Muller (2005) with permission of Macmillan Publishers Ltd.

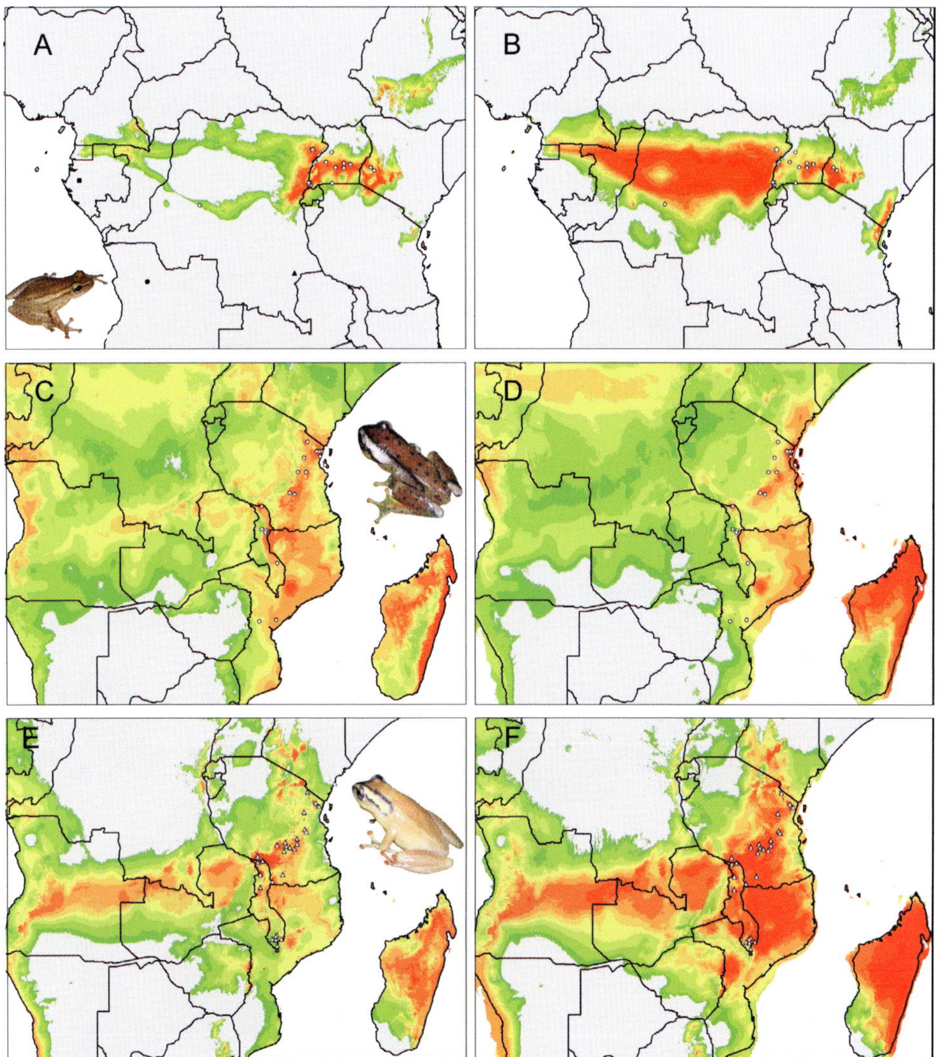

Figure 11.6 CEMs for three African reed frog species (*Hyperolius*). See Fig 11.5 for names and known distributions. CEMs computed with Maxent for different African reed frog species under current climate conditions (A, C, E) and palaeoclimate conditions suggested for the Last Glacial Maximum, *c.* 21 000 years bp (B, D, F, considering past sea-level fluctuation). White circles indicate presence data for species used for modelling; other symbols used in (A) represent type localities of nominal species currently treated as junior synonyms of *H. cinnamomeoventris* (filled square, *H. fimbriolatus* and *H. olivaceus*; filled circle, *H. cinnamomeoventris* sensu stricto and *H. tristis*; filled triangle, *H. wittei*; grey circle, *H. ituriensis*).

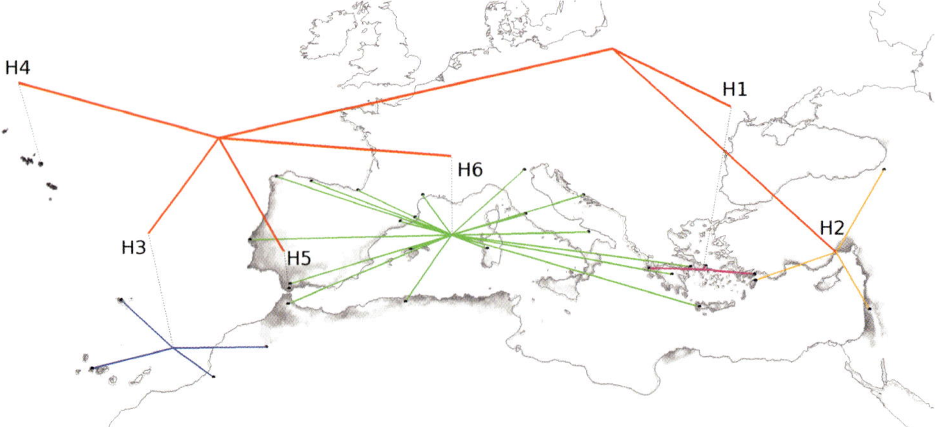

Figure 13.6 Geophylogeny showing the phylogenetic tree (red) and the geographical distribution of the six distinct *Laurus* haplotypes (H1 to H6). After Rodríguez-Sánchez et al. (2009). Shaded areas represent likely glacial refugia for *Laurus*, as predicted by niche-based distribution models (Rodríguez-Sánchez and Arroyo, 2008). See Kidd and Liu (2008) for further details on geophylogenies.

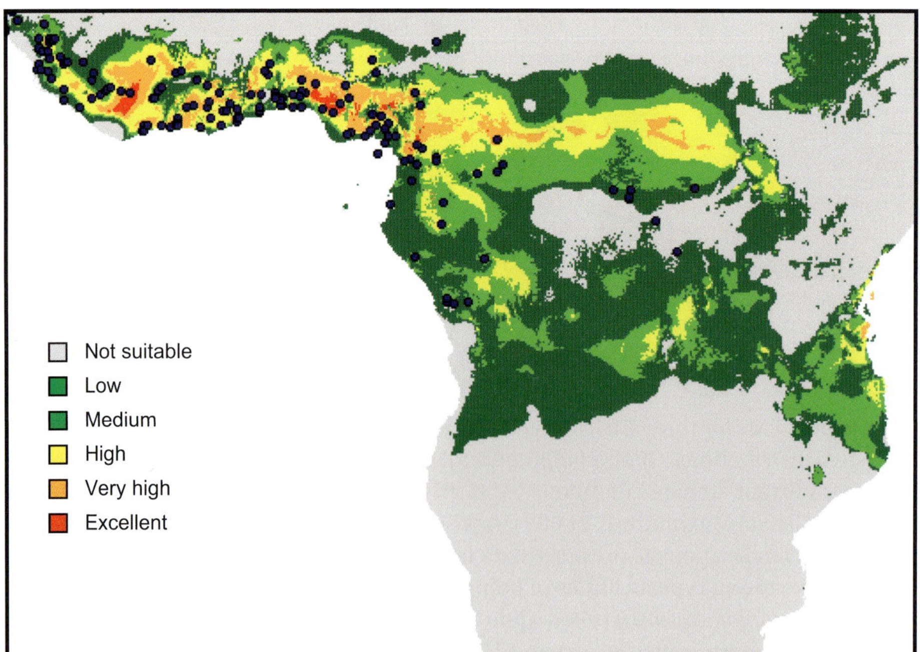

Figure 14.1 Species distribution model for *Monodora tenuifolia* Benth. Based on 106 species occurrence localities (represented by blue circles) and BIOCLIM extrapolation of 19 climate variables as indicated in Table 14.2.

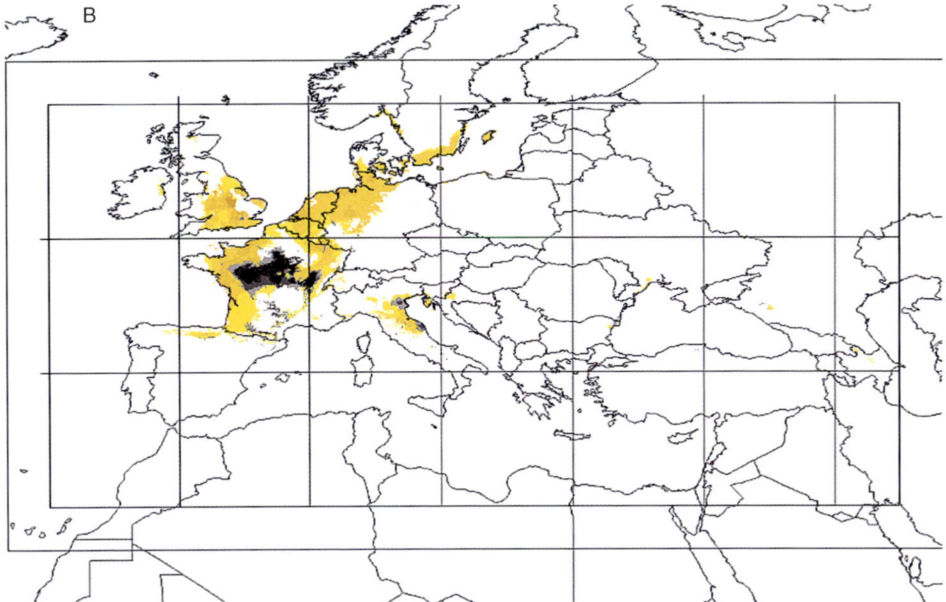

Figure 15.2 (B) Maxent distribution of the potential niche of *F. angustifolia* hybrids under a global warming scenario of doubled CO_2.

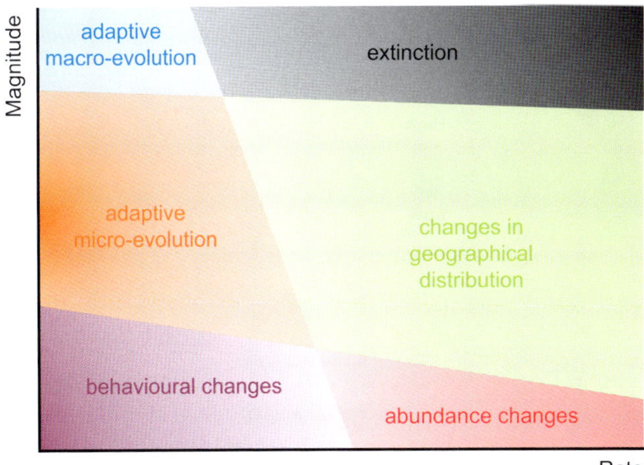

Figure 16.1 Species' responses to climatic change. Species exhibit six general types of response, each of which predominates under particular combinations of rate and magnitude of change, although the boundaries between their areas of predominance are 'fuzzy' rather than abrupt. Changes in geographical distribution are species' principal response to relatively rapid climatic changes of a wide range of magnitudes.

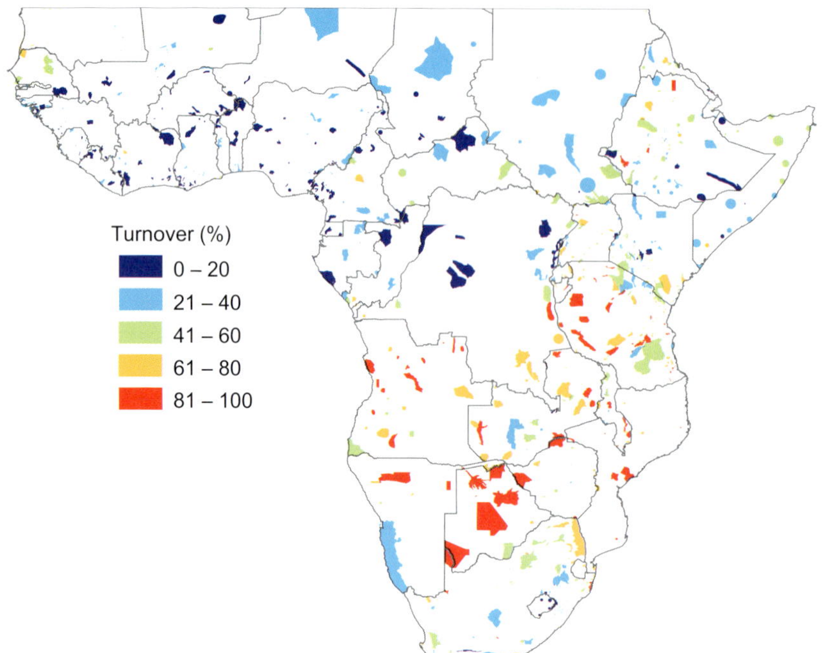

Figure 16.3 Mean potential turnover of priority species by 2085. Mean of potential turnover (%) of priority species in each IBA simulated for the climatic conditions projected by the three GCMs.

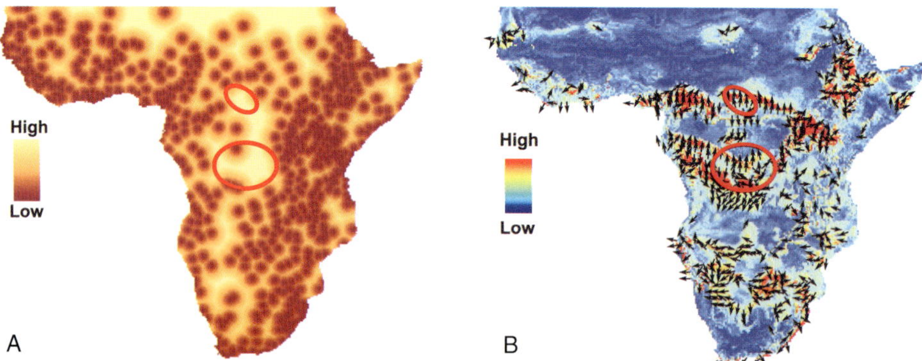

Figure 16.4 Identifying priority areas for network enhancement. (A) Shading indicates the distance between the centre of each cell of a 0.25° longitude × latitude grid and the centre of its nearest neighbour in the IBA network (maximum 669 km, minimum < 1 km). (B) Shading indicates the absolute number of priority species for which each cell of a 0.25° longitude × latitude grid is potentially newly climatically suitable in 2085, having been simulated as unsuitable in 2055. The numbers are ensemble means for three GCM projections (maximum 100, minimum 0). Arrows indicate the mean direction of movement required for species to colonise 1° grid cells that are simulated as potentially newly climatically suitable from the nearest previously suitable grid cell (arrows are plotted only for grid cells in the upper quartile with respect to the number of potential colonising species). The two red ellipses on each panel highlight areas where the present network of IBAs has particularly low connectivity but that potentially will be newly suitable in 2085 for a large number of species that will thus be attempting to shift their ranges across the area.

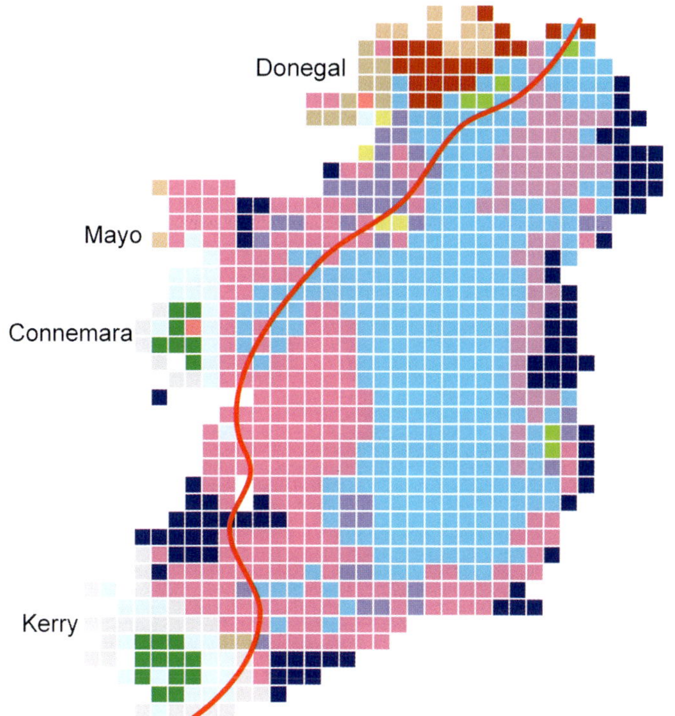

Figure 21.1 Map of Irish bioclimatic regions, with regions defined by the MONARCH project (Berry et al., 2005). Each group of coloured squares represents a different bioclimatic region. The mountains of Kerry and Connemara are within a distinct bioclimatic region, different from that of the mountains of Donegal. The bold line represents a value of 20 on the index of climatic oceanicity (after Averis et al., 2004). This is calculated as the mean annual number of wet days (> 1 mm of rain) divided by the range of monthly mean temperatures in °C.

Figure 21.2 Northern hepatic mat vegetation. Relatively ungrazed *Calluna vulgaris* heath, on Errigal, Co. Donegal, Ireland, supporting northern hepatic mat vegetation. Photo: R. Hodd.

Figure 21.3 *Salix herbacea*, a widespread arctic-montane species, on Dooish, Co. Donegal, Ireland. Photo: R. Hodd.

Figure 21.4 Montane heath, dominated by the moss *Racomitrium lanuginosum*, on the Slieve Mish mountains, Co. Kerry, Ireland. Photo: R. Hodd.

Figure 21.5 Northern hepatic mat vegetation on Errigal, Co. Donegal, with *Bazzania pearsonii* ssp. *hutchinsiae* and *Scapania ornithopodioides* prominent. Photo: R. Hodd.

Table 14.1 Species used in this study, classified by geographical area. The number of data points (species' occurrence localities) is shown in the third column. The fourth column gives the value for the area under the curve (AUC) of the receiver operating characteristic (ROC) plot, produced by the respective Maxent run for each species. The fifth column gives the four climatic variables (as in Table 14.2) that contribute most to the Maxent species distribution model. The variable shown in bold is the single most important predictor of the species distribution model, based on the jackknife procedure implemented in Maxent.

Species	Distribution	No. of data points	AUC ± standard deviation	Variable
M. angolensis Welw.	West–Central Africa	81	0.985 ± 0.003	BIO4 (24.5%) BIO12 (15.9%) **BIO7** (14.4%) BIO1 (11.9%)
M. crispata Engl.	West–Central Africa	24	0.999 ± 0.001	**BIO4** (27.2%) BIO19 (24.0%) BIO13 (17.4%) BIO14 (7.5%)
M. laurentii De Wild.	West–Central Africa	13	0.990 ± 0.001	BIO3 (48.9%) **BIO7** (11.9%) BIO4 (9.6%) BIO16 (8.5%)
M. myristica (Gaertn.) Dunal	West–Central Africa	101	0.990 ± 0.002	BIO4 (33.2%) **BIO7** (21.5%) BIO12 (16.7%) BIO 14 (6.9%)
M. tenuifolia Benth.	West–Central Africa	106	0.992 ± 0.001	**BIO4** (55.2%) BIO16 (10.2%) BIO11 (6.8%) BIO19 (6.7%)
M. undulata (P. Beauv.) Couvreur	West-Central Africa	28	0.993 ± 0.006	BIO19 (37.3%) **BIO4** (18.5%) BIO15 (10.4%) BIO13 (7.9%)
M. carolinae Couvreur	East Africa	4	0.999 ± < 0.0001	BIO4 (38.2%) BIO19 (26.1%) **BIO7** (23.7%) BIO 2 (10.8%)

Table 14.1 (*cont.*)

Species	Distribution	No. of data points	AUC ± standard deviation	Variable
M. globiflora Couvreur	East Africa	6	0.997 ± 0.001	**BIO4** (45.7%) BIO19 (18.9%) BIO2 (8.1%) BIO10 (7.3%)
M. grandidieri Baill.	East Africa	52	0.995 ± 0.002	BIO3 (29.7%) **BIO4** (23.7%) BIO7 (11.6%) BIO12 (8.1%)
M. hastipetala Couvreur	East Africa	2	1.000 ± 0.001	**BIO7 (51.6%)** BIO19 (24.3%) BIO2 (13.5%) BIO3 (5.9%)
M. junodii Engl. and Diels	East Africa	55	0.990 ± 0.002	BIO3 (43.7%) **BIO4** (13.5%) BIO16 (8.4%) BIO14 (7.1%)
M. minor Engl. and Diels	East Africa	16	0.998 ± 0.001	**BIO4** (27.9%) BIO3 (20.1%) BIO7 (18.1%) BIO19 (11.6%)
M. stenopetala Oliv.	East Africa	4	0.929 ± 0.003	BIO11 (70.4%) BIO19 (14.7%) BIO8 (3.5%) **BIO1** (0.5%)

Based on the jackknife test and the analysis of variable contribution, which form part of the standard output from Maxent, the most significant environmental variables were identified. Values for all environmental variables, and for all species occurrence points, were exported using DIVA-GIS 5.2 (Hijmans et al., 2005). Mean values of these variables for each of the species were optimised as a continuous character using square change parsimony optimisation, as implemented in Mesquite (Maddison and Maddison, 2006). The topology of the molecular phylogenetic tree was taken from Couvreur (2009) and includes all species of *Monodora* except *M. zenkeri*. Couvreur et al. (2008) dated the crown node of *Monodora* to the Miocene, with an age of 8.4 million years, with a 95% highest posterior density interval of 4.7–12.2 million years.

Table 14.2 Environmental variables used in this study. The variables represent annual trends (e.g. mean annual temperature, annual precipitation), seasonality (e.g. annual range in temperature and precipitation) and extreme or limiting environmental factors (e.g. temperature of the coldest and warmest month, and precipitation of the wet and dry quarters). From www.worldclim.org/bioclim.

Abbreviation	Description
BIO1	Annual mean temperature
BIO2	Mean diurnal range (mean of monthly (max temp – min temp))
BIO3	Isothermality (BIO2/BIO7) (× 100)
BIO4	Temperature seasonality (standard deviation × 100)
BIO5	Max temperature of warmest month
BIO6	Min temperature of coldest month
BIO7	Temperature annual range (BIO5 – BIO6)
BIO8	Mean temperature of wettest quarter
BIO9	Mean temperature of driest quarter
BIO10	Mean temperature of warmest quarter
BIO11	Mean temperature of coldest quarter
BIO12	Annual precipitation
BIO13	Precipitation of wettest month
BIO14	Precipitation of driest month
BIO15	Precipitation seasonality (coefficient of variation)
BIO16	Precipitation of wettest quarter
BIO17	Precipitation of driest quarter
BIO18	Precipitation of warmest quarter
BIO19	Precipitation of coldest quarter

14.3 Results and discussion

With the exception of *M. hastipetala* Couvreur, the statistics produced by Maxent indicate that all species were represented by sufficient data points (i.e. presence localities) to adequately assess the species distribution models. The area under the curve (AUC) values are at, or close to, the maximum value of 1.0 (indicating perfect model fit), with small standard deviations (Table 14.1). The AUC value of

0.929 ± 0.003 returned for *M. stenopetala* Oliv. indicates that the model has performed accurately even with only four presence localities. It is unclear exactly how many data points are needed for meaningful modelling, and for assessing how much regularisation is needed to avoid overfitting in cases where the number of data points is small (Phillips et al., 2006). Our results are certainly encouraging for the implementation of species distribution modelling in the case of tropical plant groups generally characterised by species with few recorded localities, in turn correlated with a limited representation in herbaria.

We identified the main contributing environmental variables using two indicators. The first was the percentage that each of the 19 variables has contributed to the respective species distribution models. We arbitrarily chose to list the four variables contributing most to the Maxent species distribution models, as for all species these percentages added up to roughly three-quarters of the total (Table 14.1). The second was the results of the Maxent jackknife procedure, which identifies the single most important environmental variable that contributes to the species distribution model (Table 14.1). The performance of individual environmental variables cannot be deduced from the first measure (percentage contribution) as these percentages indicate performance of variables in concert with the remaining 18 variables. Only during the jackknife procedure are the individual performances teased out. Thus, the best single predictor of the species distribution model is not always equal to the variable with the highest percentage contribution (e.g. *M. angolensis* Welw. in Table 14.1). It should be noted that for none of the species distribution models did the best single predicting variable vastly outperform the other variables. In other words, differences among each of the 19 environmental variables, in their ability to generalise the species distribution model on their own, were not large in any of the 13 species.

A consistent suite of variables was identified containing important defining factors for the distribution of *Monodora* species: temperature seasonality (BIO4), temperature annual range (BIO7), precipitation of driest month (BIO17) and precipitation of coldest quarter (BIO19). The obvious question of whether these variables demonstrate a phylogenetic pattern within the genus then arises. In the *Monodora* molecular phylogeny reconstruction, the West–Central African species and the East African species make up two reciprocally monophyletic groups. The mean values of the four variables given above were plotted by square change parsimony optimisation (Fig 14.2). This clearly indicates that the East African species tolerate higher temperature seasonality and larger annual range of temperature, and receive less precipitation during both the driest month and the coldest quarter (Fig 14.2). This pattern is especially obvious with regard to the latter two variables (Fig 14.2C,D). Moreover, all four variables have evolved to values characteristic for the East African clade at deeper nodes. In other words, the preference for smaller amounts of rainfall during the driest month and the coldest quarter are stable features from an evolutionary point of view. Couvreur et al. (2008) demonstrated that the timing of the first diversification (the crown node) of *Monodora* corresponds to the formation of the African Plateau during the Miocene, an era in which rainforests in East Africa became greatly reduced and were replaced by open woodland and grassland (Morley, 2000).

From the phylogenetic trees shown in Fig 14.2 we cannot deduce whether East African species adapted to drier conditions, or rather whether West–Central species adapted to more humid climates. Character optimisation will remain ambiguous because of the nature of the early split in the *Monodora* phylogeny (reciprocal monophyly). One way to resolve this ambiguity would be to expand taxon sampling to more inclusive clades and to incorporate East African endemics that diverged during episodes of climatic changes before the late-Miocene, which saw the origin of *Monodora*. In the early Oligocene and the early Miocene, Africa witnessed two more periods of aridification, with concomitant contraction of rainforest areas. Couvreur et al. (2008) demonstrated temporal congruence between these two periods of aridification and the diversification of lineages of Annonaceae endemic to East Africa.

There is evidence to support the hypothesis that the East African species shifted to a derived suite of climatic preferences, and that the West–Central African species, in contrast, did not. Firstly, Annonaceae in essence comprise species that are adapted to rainforests. Strong correlations between occurrence and temperature and precipitation have been shown for Annonaceae in a Neotropics-wide survey (Punyasena et al., 2008). As species from the Neotropics are present throughout the family phylogeny (Richardson et al., 2004; Chatrou et al., unpublished), it is likely that these correlations that have been demonstrated for the Neotropics extend to the Palaeotropical Annonaceae as well. Secondly, East Africa used to have a more humid climate, like that of West Africa, with aridification slowly but progressively intensifying since the mid-Tertiary (Bobe, 2006). Thus, past conditions would have favoured species adapted to high humidity, such as *Monodora*, across Africa. After the separation of the two rainforest blocks, conditions in East Africa provided an opportunity to adapt to a drier climate, resulting in speciation. This window of opportunity has not been exploited by all genera, however. Some widespread West–Central African genera are absent from East Africa (e.g. *Piptostigma*). This might have been the result of a failure to adapt to these drier climates, or, as stated above, it might simply underline the fact that the geographical range of ancestral species of *Piptostigma* was not affected by vicariance events that did affect other ancestral species (Wiens and Donoghue, 2004).

The recurring diversification of East African endemics during the Tertiary, the slow rate of net speciation rate compared to other areas in Africa (South African Cape, Cameroon to Gabon), and the evolutionary stability of key environmental

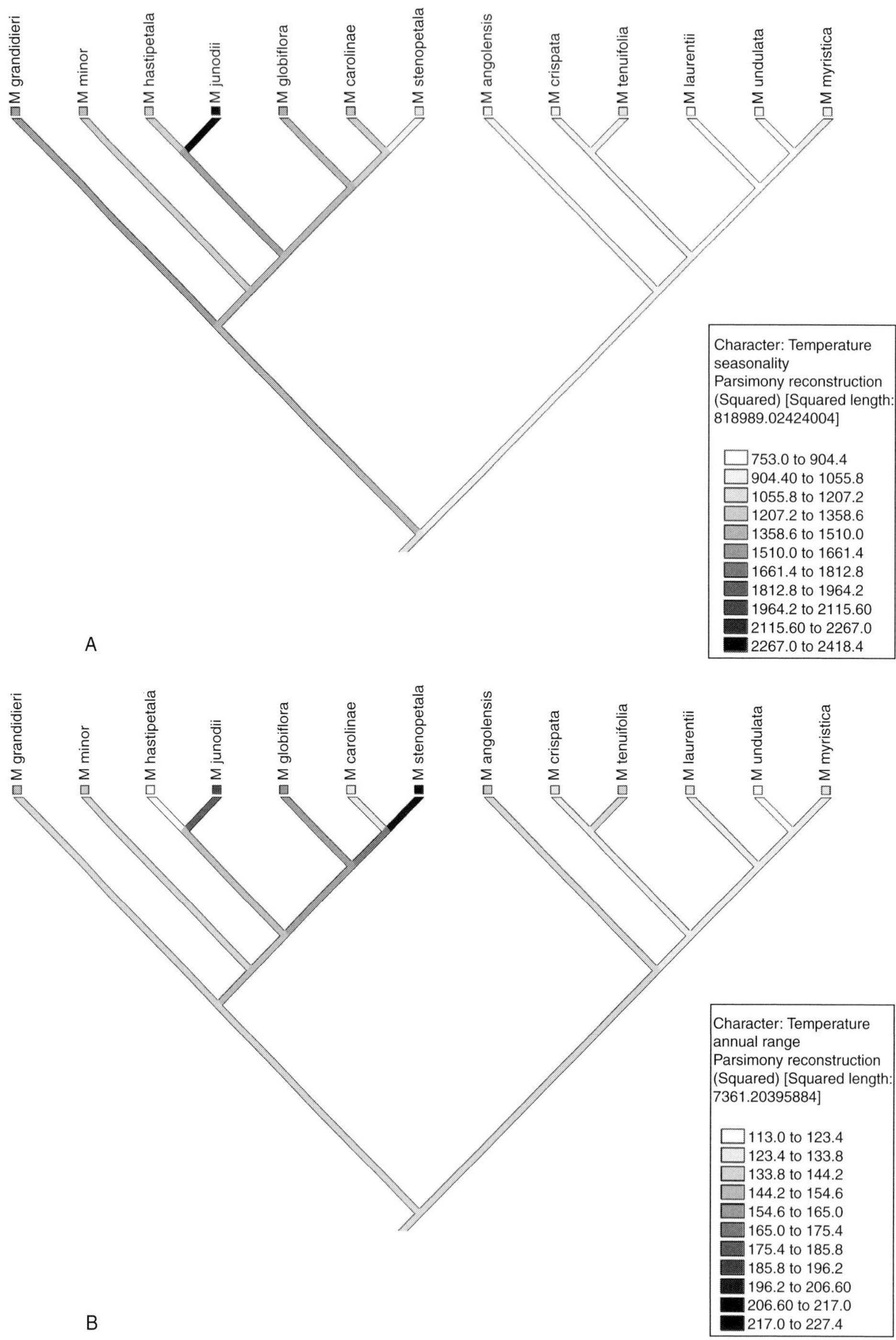

Figure 14.2 Square change parsimony optimisation of the mean of four climatic variables: (A) temperature seasonality; (B) temperature annual range; (C) precipitation of coldest quarter; (D) precipitation of driest month.

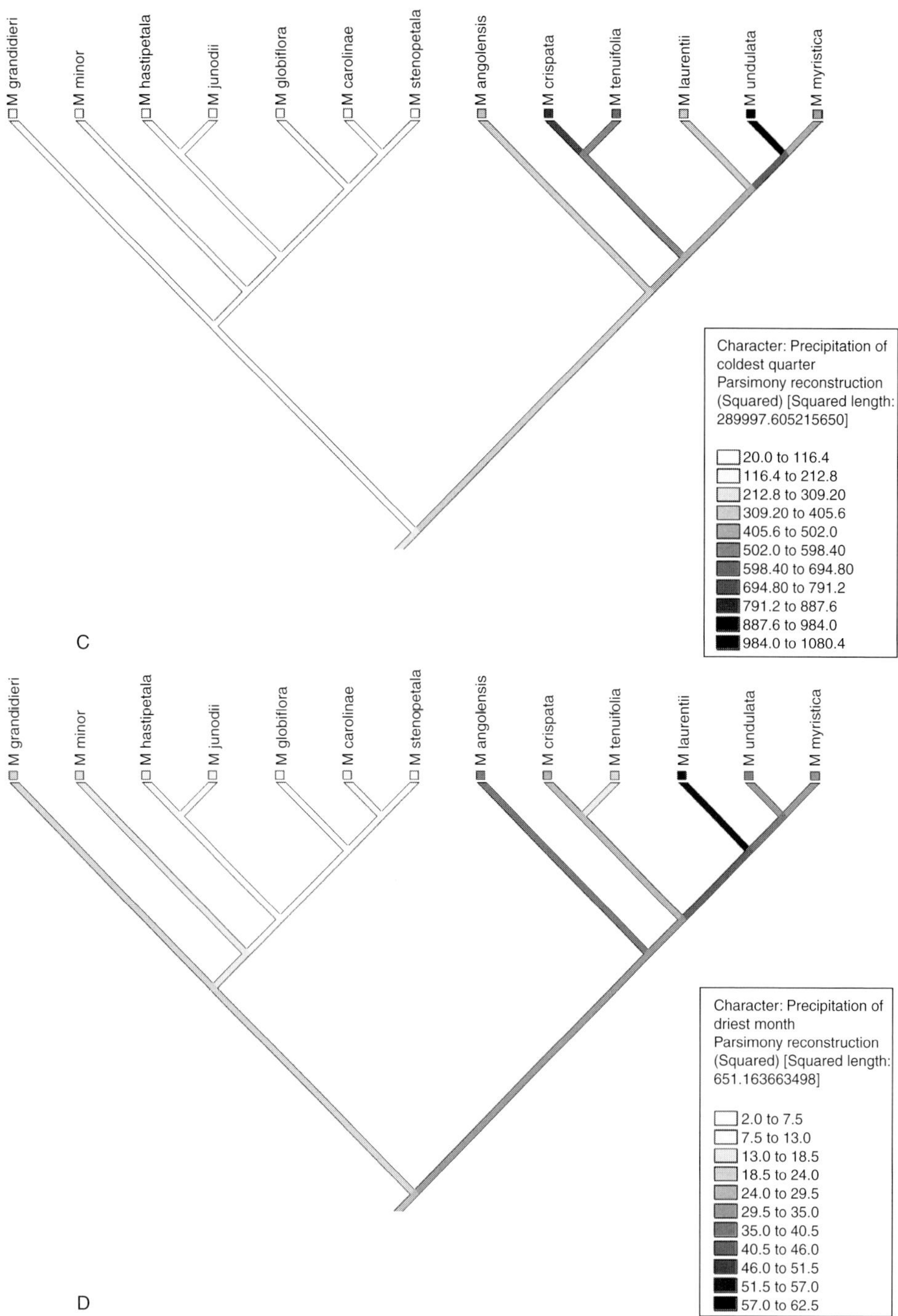

Figure 14.2 (*cont.*)

variables, all point towards the long-term ecological stability of East Africa (Lovett et al., 2005b; Mumbi et al., 2008). This pattern ties in well with the importance of phylogenetic niche conservatism for allopatric speciation (Wiens, 2004; Wiens and Donoghue, 2004; Graham et al., 2006). The retention of ancestral traits, and limited adaptive capacity to occupy habitats different from rainforests, would explain why Tertiary vicariance events produced the clear East–West disjunct distribution patterns in African Annonaceae, and in African plants in general.

But are we not oversimplifying matters by extending the explanation of niche conservatism, plausibly relevant for earlier splits, to more recent timescales and adopting it to explain speciation patterns within *Monodora*? Our analyses of climatic variables and the optimisation of the most significant of them seem to support the importance of niche conservatism. This would be corroborated by the environmental stability of East African rainforests over long timescales (Lovett et al., 2005b; Mumbi et al., 2008). Yet climate is known to be highly variable, even on small spatial scales. It remains to be determined whether the niche conservatism would be maintained on a finer scale than that applied here. The evidence for ecological constancy seems contradicted by information on the ecology of East African species already available from herbarium specimens. There is considerable variation in altitudinal distribution among the species, with *M. globiflora* Couvreur, for example, being confined to altitudes of 1700–2000 m, whereas most species have a wider altitudinal range at middle elevations, up to 900 m. A similar picture appears in relation to the vegetation type from which species have been collected. *Monodora minor* Engl. and Diels has been collected in a variety of habitats, from coastal rainforests to thickets. *Monodora stenopetala* has been found only in thickets and woodlands, challenging the generalisation that Annonaceae occurrence co-varies with temperature and precipitation. One might argue that these differences have not been supported by our analysis. Notwithstanding, answering questions on the apparent contrast between conservatism and adaptation seems to be difficult using rather coarse geographic information systems (GIS)-based techniques, and would require the analysis of ecophysiological traits of the species.

The long-term environmental stability of the East African rainforests could imply a relatively mild impact of future climatic changes on East African plants (Lovett et al., 2005b). The evolutionary stability of key environmental variables in the distribution models, when supported after more rigorous analyses than the ones we present here, would add to that encouraging prospect. Still, the quantity of future climatic changes and their impact are to a large extent undetermined. Climate change will have a clear impact on distribution of species through changes in amounts and seasonality of precipitation, temperature and atmospheric chemistry. Apart from studying the effect of changes in individual parameters on the distribution of species, the interactions between these factors in particular pose key issues for analysis and policy making. Even in the case of precipitation only, there are many uncertainties in our knowledge of African climates, with consequent limitations on the establishment of future scenarios (Hulme et al., 2005). In light of the multidimensionality of species distribution models, it is conceivable that small changes in multiple relevant environmental variables would have as much impact as a single major change (Tilman and Lehman, 2001).

Acknowledgements

The authors are grateful to the editors for the opportunity to contribute to this volume, and to the reviewers for comments. Wageningen University is acknowledged for financial support to the first author to attend the Systematics Association conference on Climate Change and Systematics.

References

Beaumont, L. J., Pitman, A. J., Poulsen, M. and Hughes, L. (2007). Where will species go? Incorporating new advances in climate modelling into projections of species' distributions. *Global Change Biology*, **13**, 1368–1385.

Bobe, R. (2006). The evolution of arid ecosystems in eastern Africa. *Journal of Arid Environments*, **66**, 564–584.

Bowie, R. C. K., Fjeldså, J ., Hackett, S. J., Bates, J. M. and Crowe, T. M. (2006). Coalescent models reveal the relative roles of ancestral polymorphism, vicariance, and dispersal in shaping phylogeographical structure of an African montane forest robin. *Molecular Phylogenetics and Evolution*, **38**, 171–188.

Busby, J. R. (1991). BIOCLIM: a bioclimatic analysis and prediction system. In *Nature Conservation: Cost Effective Biological Surveys and Data Analysis*, ed. C. R. Margules and M. P. Austin. Melbourne: CSIRO, pp. 64–68.

Chatrou, L. W., Couvreur, T. L. P. and Richardson, J. E. (2009). Spatio-temporal dynamism of hotspots enhances plant diversity. *Journal of Biogeography*, **36**, 1628–1629.

Couvreur, T. L. P. (2009). Monograph of the syncarpous African genera *Isolona* and *Monodora* (Annonaceae). *Systematic Botany Monographs*, **87**, 1–150.

Couvreur, T. L. P., Gereau, R. J., Wieringa, J. J. and Richardson, J. E. (2006). Description of four new species of *Monodora* and *Isolona* (Annonaceae) from Tanzania and an overview of Tanzanian Annonaceae diversity. *Adansonia (Paris)*, **28**, 243–266.

Couvreur, T. L. P., Chatrou, L. W., Sosef, M. S. M. and Richardson, J. E. (2008). Molecular phylogenetics reveal multiple Tertiary vicariance origins of the African rain forest trees. *BMC Biology*, **6**, 54.

Crisp, M. D. and Cook, L. G. (2007). A congruent molecular signature of vicariance across multiple plant lineages. *Molecular Phylogenetics and Evolution*, **43**, 1106–1117.

Elith, J., Graham, C. H., Anderson, R. P. et al. (2006). Novel methods improve prediction of species' distributions from occurrence data. *Ecography,* **29**, 129–151.

Filer, D. L. (2008). BRAHMS Version 6. Oxford: Department of Plant Sciences, University of Oxford.

Graham, C. H., Moritz, C. and Williams, S. E. (2006). Habitat history improves prediction of biodiversity in rainforest fauna. *Proceedings of the National Academy of Sciences of the USA,* **103**, 632–636.

Hannah, L., Midgley, G., Hughes, G. and Bomhard, B. (2005). The view from the Cape: extinction risk, protected areas, and climate change. *BioScience,* **55**, 231–242.

Heikkinen, R. K., Luoto, M., Araújo, M. B. et al. (2006). Methods and uncertainties in bioclimatic envelope modelling under climate change. *Progress in Physical Geography,* **30**, 751–777.

Hijmans, R. J., Guarino, L., Jarvis, A. et al. (2005). DIVA-GIS, version 5. www.diva-gis.org.

Hulme, M., Doherty, R., Ngara, T. and New, M. (2005). Global warming and African climate change: a reassessment. In *Climate Change and Africa*, ed. P. S. Low. Cambridge: Cambridge University Press, pp. 29–40.

Kearney, M. (2006). Habitat, environment and niche: what are we modelling? *Oikos,* **115**, 186–191.

Lewis, S. L. (2006). Tropical forests and the changing earth system. *Philosophical Transactions of the Royal Society of London B,* **361**, 439–450.

Linder, H. P., Hardy, C. R. and Rutschmann, F. (2005). Taxon sampling effects in molecular clock dating: an example from the African Restionaceae. *Molecular Phylogenetics and Evolution,* **35**, 569–582.

Lovett, J. C., Midgley, G. F. and Barnard, P. (2005a). Climate change and ecology in Africa. *African Journal of Ecology,* **43**, 167–169.

Lovett, J. C., Marchant, R., Taplin, J. and Küper, W. (2005b). The oldest rainforests in Africa: stability or resilience for survival and diversity? In *Phylogeny and Conservation*, ed. A. Purvis, J. L. Gittleman and T. Brooks. Cambridge, UK: Cambridge University Press, pp. 198–229.

Maddison, W. P. and Maddison, D. R. (2006). Mesquite: a modular system for evolutionary analysis, version 1.11. http://mesquiteproject.org.

Morley, R. J. (2000). *Origin and Evolution of Tropical Rain Forests.* Chichester: Wiley.

Mumbi, C. T., Marchant, R., Hooghiemstra, H. and Wooller, M. J. (2008). Late Quaternary vegetation reconstruction from the Eastern Arc Mountains, Tanzania. *Quaternary Research,* **69**, 326–341.

Near, T. J. and Sanderson, M. J. (2004). Assessing the quality of molecular divergence time estimates by fossil calibrations and fossil-based model selection. *Philosophical Transactions of the Royal Society of London B,* **359**, 1477–1483.

Nogués-Bravo, D. (2009). Predicting the past distribution of species' climatic niches. *Global Ecology and Biogeography,* **18**, 521–531.

Parmesan, C. (2006). Ecological and evolutionary responses to recent climate change. *Annual Review of Ecology, Evolution, and Systematics,* **37**, 637–669.

Pennington, R. T., Lavin, M., Prado, D. E. et al. (2004). Historical climate change and speciation: neotropical seasonally dry forest plants show patterns of both Tertiary and Quaternary diversification. *Philosophical Transactions of the Royal Society of London B,* **359**, 515–537.

Phillips, S. J., Anderson, R. P. and Schapire, R. E. (2006). Maximum entropy modeling of species' geographic distributions. *Ecological Modelling,* **190**, 231–259.

Pirie, M. D., Chatrou, L. W., Erkens, R. H. J. et al. (2005). Phylogeny reconstruction and molecular dating in four Neotropical genera of Annonaceae: the effect of taxon sampling in age estimation. In *Plant Species-Level Systematics: New Perspectives on Pattern and Process*, ed. F. T. Bakker, L. W. Chatrou, B. Gravendeel and P. B. Pelser. Ruggell, Liechenstein: A. R. G. Gantner Verlag, pp. 149–174.

Punyasena, S. W., Eshel, G. and McElwain, J. C. (2008). The influence of climate on the spatial patterning of Neotropical plant families. *Journal of Biogeography,* **35**, 117–130.

Richardson, J. E., Chatrou, L. W., Mols, J. B. et al. (2004). Historical biogeography of two cosmopolitan families of flowering plants: Annonaceae and Rhamnaceae. *Philosophical Transactions of the Royal Society of London B,* **359**, 1495–1508.

Sanderson, M. J. and Doyle, J. A. (2001). Sources of error and confidence intervals in estimating the age of angiosperms from *rbcL* and 18S rDNA data. *American Journal of Botany,* **88**, 1499–1516.

Tilman, D. and Lehman, C. (2001). Human-caused environmental change: impacts on plant diversity and evolution. *Proceedings of the National Academy of Sciences of the USA,* **98**, 5433–5440.

Wiens, J. J. (2004). Speciation and ecology revisited: phylogenetic niche conservatism and the origin of species. *Evolution,* **58**, 193–197.

Wiens, J. J. and Donoghue, M. J. (2004). Historical biogeography, ecology and species richness. *Trends in Ecology and Evolution,* **19**, 639–644.

Wisz, M. S., Hijmans, R. J., Li, J. et al. (2008). Effects of sample size on the performance of species distribution models. *Diversity and Distributions,* **14**, 763–773.

Yesson, C. and Culham, A. (2006). Phyloclimatic modeling: combining phylogenetics and bioclimatic modeling. *Systematic Biology,* **55**, 785–802.

Yesson, C., Toomey, N. H. and Culham, A. (2009). *Cyclamen*: time, sea and speciation biogeography using a temporally calibrated phylogeny. *Journal of Biogeography,* **36**, 1234–1252.

15

Hybridisation, introgression and climate change: a case study of the tree genus *Fraxinus* (Oleaceae)

M. Thomasset

School of Natural Sciences, Trinity College Dublin and *Kinsealy Research Centre, Teagasc, Dublin, Ireland*

J. F. Fernández-Manjarrés

CNRS and *Université Paris-Sud XI, Orsay,* and *AgroParisTech, Paris, France*

G. C. Douglas

Kinsealy Research Centre, Teagasc, Dublin, Ireland

N. Frascaria-Lacoste

CNRS and *Université Paris-Sud XI, Orsay,* and *AgroParisTech, Paris, France*

T. R. Hodkinson

School of Natural Sciences, Trinity College Dublin, Ireland

Climate Change, Ecology and Systematics, ed. Trevor R. Hodkinson, Michael B. Jones, Stephen Waldren and John A. N. Parnell. Published by Cambridge University Press.

Abstract

The distribution of potential hybrid zones depends largely on climate, habitat quality and historical biogeographic factors including dispersal and local extinctions. Global climate change can produce more favourable conditions for certain species to survive in areas that were previously unsuitable for their growth and/or their reproduction, and it may therefore change the potential for their hybridisation with closely related taxa. This chapter discusses general issues of plant hybridisation and invasiveness in the context of global climate change and presents a case study of hybrid ash trees (*Fraxinus excelsior* × *F. angustifolia*) that are mostly geographically separated in their natural range by climate but can have large hybrid zones. In general, both species are temporally separated by flowering times, which occur in early winter for *F. angustifolia* and in early spring for *F. excelsior.* In Ireland, introduced alien ash (*F. angustifolia, F. excelsior* × *F. angustifolia* hybrids, and non-native *F. excelsior*) can be found growing in sympatry with native *F. excelsior* populations. It is not known whether alien ash will hybridise with native populations or how climate change, principally in temperature and precipitation, will influence their hybridisation and invasiveness potential. We firstly examine the climate presently associated with known hybrid zones for ash in continental Europe and in Ireland. We then evaluate if a double CO_2 global warming scenario (2 × CO_2, CCM3 model) would provide improved climatic conditions for hybrids in Ireland and elsewhere. We also present results of phenological observations, which show that flowering periods of alien trees in two Irish plantations overlap with adjacent native trees. Nevertheless, peak pollen release dates for the alien and native populations at these two plantation sites remain separated by at least six weeks.

15.1 Introduction: climate change and hybridisation

Climate and biodiversity have interacted closely since the first appearances of life on earth, *c.* 3.8 billion years ago (Lenton, 2004; IPCC, 2007). For example, changes in temperature, precipitation patterns or prevalence of extreme events can have an impact on species in many ways (Root et al., 2003). They may influence their distribution by altering the suitability of an area for growth and survival. They may create favourable conditions for alien species to survive in areas that were previously unsuitable for their growth and/or reproduction. Conversely, climate change might render an area unsuitable for the survival of a species. Therefore, it can influence the chance that individuals will hybridise, the chance that their progeny will survive and the chance of introgression of genes among populations.

There are numerous examples of how the geographical range of a species has moved because of climate change, either over geographical timescales or over the most recent human-induced warming period (Tsukada, 1982; Mahoney, 2004; Kuchta and Tan, 2005; Schönswetter et al., 2005). For instance, fossil, palaeoecological and DNA evidence has shown how distributions of species have changed during the most recent glacial episodes (Hewitt, 2000; Kullman, 2008; Liepelt et al., 2009). Furthermore, there is evidence that species are already shifting their ranges in response to current climate change (Parmesan et al., 2005; Root et al., 2005; Lavergne et al., 2006), and that some species are consequently facing extinction (Parmesan, 2006; Pauli et al., 2006; Foden et al., 2007). There is also evidence that hybrid zones are shifting. For example, Buggs (2007) referenced examples in animals (ticks, salamanders, crickets and lizards) where climate fluctuations might have produced hybrid-zone movement. Climate change may also alter the timing of developmental events (phenology) or species' behaviour (Menzel and Fabian, 1999; Visser and Holleman, 2001; Fitter and Fitter, 2002; White et al., 2003). Such changes can have a profound effect on the reproductive biology of species and their evolution.

Climate change may also influence hybridisation among closely related species or populations within species. Hybridisation refers to the mating between individuals of two populations (be they determined as separate species or not) distinguishable by one or more heritable characters (Arnold, 1997). If hybrid progenies backcross with one or more of their parent taxa, the phenomenon is termed introgression. Introgression was first defined as 'the infiltration of germplasm from one species into another, through repeated backcrossing of hybrids with parental species' (Anderson and Hubricht, 1938).

There has been much discussion in the scientific literature about the potential role of hybridisation in evolution. Some biologists believe that hybridisation has no major long-term effect on the evolution of species (Wagner, 1970). However, most biologists now recognise hybridisation as an important evolutionary phenomenon; it is particularly well documented in plants (Stace, 1975; Ellstrand et al., 1999; Ellstrand and Schierenbeck, 2000; Rieseberg et al., 2003; Arnold, 2004; Rieseberg and Willis, 2007). Hybridisation may lead to speciation, increased variability (in hybrid swarm situations), reinforcement of reproductive isolation, or species convergence (Harrison, 1993). Hybridisation can also provide a source of evolutionary novelty (Anderson and Stebbins, 1954). Experimental work on hybridisation in natural populations by Anderson and Hubricht (1938) and Anderson and Stebbins (1954) showed that hybridisation and introgression could be favoured by selection, and both contribute to the development of adaptation of populations to the environment.

Several reviews exist on the topic of hybridisation (Arnold, 1997; Rieseberg, 1997; Rieseberg and Carney, 1998) but few studies have attempted to discuss the influence of climate change on this phenomenon. This chapter aims to highlight the influence of climate change (mainly temperature and precipitation) on hybridisation and introgression and provides a case study demonstrating the influence of climate change on the distribution and phenology of two closely related species within the genus *Fraxinus* (Oleaceae) in western Europe.

15.2 Consequences of climate change on hybridisation

15.2.1 Hybrid zones and climate change

Hybridisation can result in a diverse range of evolutionary outcomes. Interbreeding between two different taxa can lead to hybrid-zone formation, introgression, hybrid speciation or species' isolation and divergence. Hybrid zones can occur when previously allopatric taxa or populations come into contact (Lowe et al., 2004). Therefore, climate change can bring previously isolated populations together in geographical space. Hybrid zones can be short-lived (ephemeral) when the hybrids, and genes transferred via them, are removed by selection and have no lasting effect on the parental populations via introgression. However, hybrid zones can also be long-lasting and of more consequence to the evolution of the plant populations because there is permanent introgression of genes among species (Anderson and Hubricht, 1938).

A consequence of hybridisation may be the disappearance of the hybrid zone by fusion of different species into one, a process also called 'speciation in reverse' (Taylor et al., 2006). Hybridisation may also lead to the formation of new species, and this process has been documented for the origin of many species, especially in plants (Rieseberg, 1997; Otto and Whitton, 2000; Ramsey and Schemske, 2002). Hybrids that are isolated by reproductive barriers can become fertile new species that are reproductively isolated from their parental species (Avise, 2004). There are two main mechanisms of such hybrid speciation. The first one that can be detected easily is allopolyploidy (Levin, 2002). Winge (1917) was the first to report that hybridisation, accompanied by duplication of chromosome sets, can create a new stable species in plants. The hybrid population could then become isolated from the parental taxa. In the second mechanism, hybridisation can create new species without a change in ploidy level, and this is known as homoploid hybrid speciation. Some annual species of the genus *Helianthus* provide well-studied examples of this mode of speciation (Rieseberg, 2006), as do species of *Tragopogon* (Soltis and Soltis, 2009).

Hybridisation may lead to the extinction of one of the parental species, particularly in the case of rare species with small population size (Levin et al., 1996).

It may also be a mechanism of facilitating the invasion of introduced species. Pollen-mediated gene flow from exotic plant species can result in the introgression of exotic genes into native gene pools and lead to the dilution or contamination of 'pure' native species (Raybould and Gray, 1994; Rhymer and Simberloff, 1996; Schierenbeck et al., 2005; Chapman and Burke, 2006; Currat et al., 2008). Such gene flow might be a real threat for many native plant species around the world and could be exacerbated by climate change, especially if hybrids are more competitive and reproductively successful (Vila et al., 2000).

However, interbreeding between distinct species or populations can result in enhanced genetic diversity (Rieseberg et al., 1999) and novel genetic combinations (Anderson, 1954; Arnold, 2004; Seehausen, 2004). Hybrids can have phenotypes that are extreme variants to those of either parental line, a phenomenon called transgressive segregation (Rieseberg et al., 1999). By generating individuals that exceed parental phenotypic values, transgressive segregation could give hybrid individuals the opportunity to access new ecological niches such as those that may be created through changing climatic conditions. Furthermore, when hybrids are fertile, repeated backcrossing can result in introgression and the exchange of genetic material from one species to another (Anderson and Stebbins, 1954). This may increase genetic diversity, providing raw material for novel adaptations required for new environments. Favourable alleles may spread efficiently from one species or population to another through introgression (Rieseberg and Burke, 2001). Therefore small populations, which often show a low genetic diversity due to the effects of genetic drift and inbreeding (Young et al., 1996), may well benefit from influxes of new genetic material through hybridisation and introgression. In the context of global warming, high genetic diversity may help species to respond to climatic variation and adapt to the new conditions. For example, Savolainen et al. (2007) showed that the potential for tree adaptation to climate change depends on genetic variation, dispersal and establishment rate.

15.2.2 Influence of climate change on barriers to hybridisation

The degree of gene flow between two populations, or the probability of hybrid formation between two taxa, is influenced by reproductive isolation mechanisms (Levin, 1978). These can be divided into two types, based on whether they act before (prezygotic) or after (postzygotic) fertilisation (Rieseberg and Carney, 1998).

Postzygotic barriers act after zygote formation by decreasing viability or fertility of the hybrids. They can occur via several mechanisms, including genomic incompatibility, hybrid inviability or sterility. Hybrid inviability also includes situations where hybrids are viable but ill-adapted for survival. Postzygotic mechanisms can be either intrinsic or extrinsic (Rieseberg and Carney, 1998). Intrinsic postzygotic mechanisms are the consequence of genetic incompatibility or negative epistasis between alleles from different species (Dobzhansky, 1937). Intrinsic postzygotic isolation can occur at different levels: non-viability of the first-generation hybrids (F1), sterility of F1 hybrids and hybrid depression that can affect viability and fertility beyond the first hybrid generation (Coyne and Orr, 2004). Such factors are not likely to be greatly influenced by climate change. However, extrinsic postzygotic isolation, or exogenous selection, could be influenced significantly by climate. In such cases, hybrids can show phenotypes which are ill-adapted to the parental environment (Coyne and Orr, 2004; Lowe et al., 2004). Under future climatic scenarios, the ecological tolerance of a species might change. For example, Ross and Harrison (2006) showed that egg viability and survival over different winter climate regimes plays an important role in determining the distribution of two cricket species (*Gryllus firmus* Scudder S. H., 1902, and *G. pennsylvanicus* Burmeister, 1838) and the position of their hybrid zone.

Premating or prezygotic isolation mechanisms limit hybrid formation before fertilisation and can, again, be due to intrinsic and extrinsic factors. An incompatibility of reproductive organs (mechanical isolation) can act as a barrier. For example, within many insects, the male and female copulatory organs of closely related species do not fit together, preventing sperm transfer (Sota and Volger, 2001). Intrinsic isolation may also be due to gametic isolation, either by gametic incompatibility (e.g. intrinsic defects causing physiological or biochemical recognition of gametes during fertilisation - Lewis and Crowe, 1958), or by gametic competition, also, in plants, called pollen competition (see Howard, 1999).

Prezygotic barriers to hybridisation are arguably more likely to be influenced by climate change. They can be temporal (including phenological) or ecological in origin. Temporal isolation is related to a time delay of reproductive events. It can refer to reproductive seasons as well as the timing of the release of gametes. In plant species, this mechanism involves different flowering periods (Broeck et al., 2003; Gerard et al., 2006a, 2006b; Pascarella, 2007; Donnelly et al., Chapter 8) or a delay in the different steps of flowering: flowering peak or period of pollination (Schluter, 2001; Weis and Kossler, 2004). Species with overlapping flowering time have a greater opportunity to interact. The timing of reproductive events is thus an important mechanism for maintaining species coexistence in diverse plant communities (Rathcke and Lacey, 1985). Recent studies have shown a change of flowering time (mainly earlier spring phenology) in several angiosperms, which is significantly correlated with global warming (Fitter and Fitter, 2002; Menzel et al., 2006).

15.3 Detection of hybridisation

While some studies have used only molecular markers to identify and characterise hybridisation (Khasa et al., 2005; Valbuena-Carabanã et al., 2005; Coyer et al.,

2007; Mebert, 2008; Wachowiak and Prus-Glowacki, 2008), it has often been necessary to combine chromosome, molecular DNA and morphological markers for that purpose (Perron and Bousquet, 1997; Hodkinson et al., 2002; Kelleher et al., 2005; Ebina and Otho, 2006; Curtu et al., 2007; Conceição et al., 2008). Hybrids with a level of ploidy different from their parents can be detected readily by examination of their chromosome number (Carroll and Borrill, 1965). Molecular cytogenetic approaches such as fluorescent in-situ hybridisation (FISH) and genome in-situ hybridisation (GISH) have also proven useful in this respect (Leitch and Bennett, 1997; Hodkinson et al., 2002).

When chromosome studies are not suitable or convenient, hybrids have generally been identified based on morphological or ecological attributes (Stace, 1975). However, Allendorf et al. (2001) argued that detection of hybrids based on morphological traits can be difficult. Hybrids do not necessarily present intermediate morphology, because they often express a mosaic of parental phenotypes. Moreover, in the presence of a population of hybrids with varying numbers of generations of backcrossing with parental species, or among hybrids (hybrid swarms), distinguishing between hybrids and parental taxa becomes a challenging task.

Molecular DNA markers can provide fast and efficient tools to study hybridisation and are often the method of choice for hybridisation investigations. A variety of methods including allozymes, restriction fragment length polymorphism (RFLP), microsatellites (SSRs) or DNA sequence analysis are now available to study species' boundaries and hybrids (Lowe et al., 2004). While uniparentally inherited markers (i.e. chloroplast DNA or mitochondrial DNA) can be used to study differential parental contribution, biparental inherited markers can be used to observe genetic differentiation between species or to examine the genetic profile of hybrids which combine the specific markers of their putative parents.

15.4 Natural hybridisation in the genus *Fraxinus*: a case study

15.4.1 Hybridisation in *Fraxinus*

The genus *Fraxinus* L. is one of 24 genera in the olive family, Oleaceae (Wallander and Albert, 2000). *Fraxinus excelsior* L. (common ash) and *F. angustifolia* Vahl. (narrow-leaved ash) are native tree species of western Europe, with important economic value for the production of timber (Horgan et al., 2004). Despite the fact that they are generally geographically isolated because of different climate preferences, they are also known to form viable hybrids (Gerard et al., 2006c) and their distributions overlap in several areas, potentially producing different hybrid zones (Fig 15.1). As climate is predicted to change towards warmer conditions, the most southern *Fraxinus* species, *F. angustifolia*, could, for example, have the potential

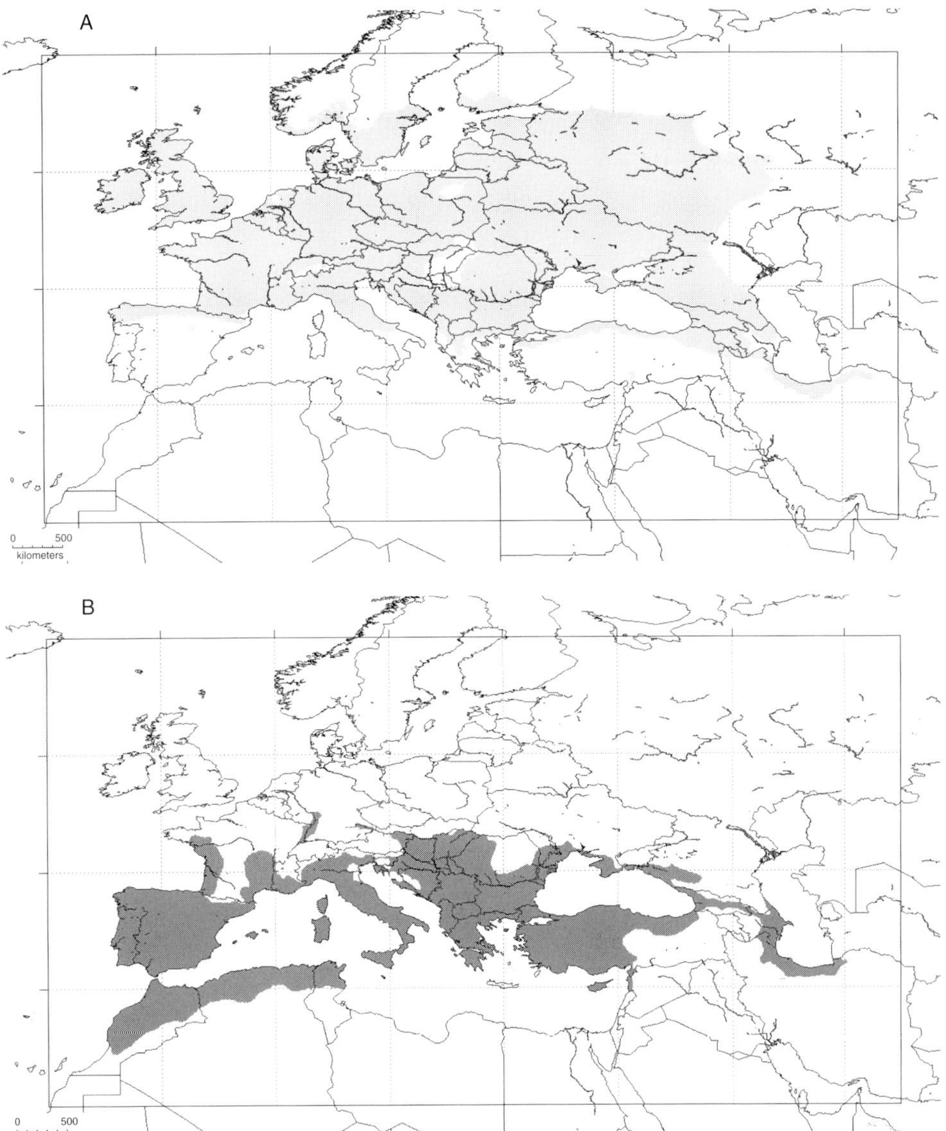

Figure 15.1 General distribution of *F. excelsior* and *F. angustifolia*. (A) Distribution map of *F. excelsior*, plotted with the shape file provided with permission by EUFORGEN. (B) Approximate distribution of *F. angustifolia*, based on records of our own research, literature and public databases (www.gbif.org).

to shift its range northwards and increase the potential hybridisation zone with *F. excelsior*.

Here we review our current knowledge about hybridisation in ash and analyse present and future climate data at the hybrid zones of *F. angustifolia* and

F. excelsior to predict the potential impact of climate warming on these species. We focus on the potential future climate of Ireland under a global warming scenario (2 × CO_2 climatic conditions, CCM3 model) to evaluate whether accidentally introduced hybrids could find a climate resembling that of their geographical origin. Finally, we present data on the phenology of the two species and their hybrids to assess how much their flowering periods currently overlap in Ireland when they are grown together artificially in plantations.

15.4.2 Extent of hybridisation and introgression in *Fraxinus*

Fraxinus excelsior is mainly present in central and northern Europe and is gradually replaced towards the Mediterranean basin by *F. angustifolia* (Fig 15.1). The present-day distribution of these two species indicates a contact zone where hybridisation has been suspected for many years (Picard, 1983). In France, the extent of hybridisation between the two species has been analysed, and evidence for hybridisation between them obtained from: (1) plants showing intermediate morphological characteristics (Rameau et al., 1989); (2) experimental crosses which show that the two parental species are compatible (Morand-Prieur et al., 2002; Raquin et al., 2002); (3) molecular markers (Jeandroz et al., 1996; Heuertz et al., 2001, 2004); and (4) a combination of molecular and physiological (based on seed dormancy) markers (Gerard et al., 2006c).

Molecular analyses show that *Fraxinus* chloroplast DNA haplotypes largely overlap in western Europe, confirming the occurrence of historical interspecies gene flow (Heuertz et al., 2006). The lack of a clear separation between plastid haplotypes according to species suggests that hybridisation and introgression have been widespread. Analyses with nuclear microsatellites and multivariate morphological characters suggested differential patterns of hybridisation in the Loire and the Saône valleys in France (Fernández-Manjarrés et al., 2006). The mild climatic conditions in the Loire appear to promote introgression of morphological characters and molecular markers of *F. angustifolia* into *F. excelsior*, while the most continental areas of the Saône appear to allow mostly molecular introgression between species.

15.4.3 Climate and hybridisation *in Fraxinus*

The distribution of the two species and their hybrids is linked to ecological variables; climate is a particularly important factor in determining hybridisation potential (Fernández-Manjarrés et al., 2006). Within a sympatric population in the Loire valley, Gerard et al. (2006a) assessed the role of floral phenology in restricting gene flow at a local scale. Reproductive events mainly occurred between co-flowering trees, and pollen flow appeared to be asymmetric. For early-flowering hybrids, male and female reproductive success was high, and they were producing more flowers and fruits than later-flowering hybrids. Moreover, the authors found relatively high selfing rates, indicating that hybrids may have reproductive assurance resulting in higher fitness in this type of intermediate ecotone.

Physiological studies also show that the two species have different responses and tolerances to summer temperature stresses, as *F. excelsior* relies on malate (Patonnier et al., 1999; Marigo et al., 2000) and *F. angustifolia* on mannitol accumulation (Oddo et al., 2002) to cope with water deficiencies. However, the growth of *F. angustifolia* depends greatly on water availability, a feature not seen as markedly in *F. excelsior* (Marigo et al., 2000). In winter, frost damages the flowers of *Fraxinus* and limits its most northerly distribution. However, *F. angustifolia* can survive as an adult tree in areas colder than its native range, as it is often seen in parks throughout Europe, although it seldom sets and disperses seeds successfully. Indeed, the capability of *F. angustifolia* to produce hybrids with *F. excelsior* in hybrid zones is variable across years, as frost patterns can induce a late second flowering period of *F. angustifolia*, allowing the pollen of this species to potentially sire seeds of *F. excelsior* (personal observation).

15.4.4 Alien ash in northwest Europe

Ireland is a good place to study the potential expansion (invasion) of *F. angustifolia* to more northerly and western regions of Europe. *Fraxinus angustifolia* is not native to Ireland but has been planted either deliberately or accidentally, principally by foresters and horticulturalists. There are therefore several study sites with a mix of ash species that can be used to study the potential impact that invading *F. angustifolia*, or hybrid populations, will have on Irish *F. excelsior* as a result of range shifts arising from global warming. These sites came into existence by accident, because the Irish government introduced substantial planting grants in 1992 for farmers to plant hardwood trees. Today, there are at least 100 afforested sites in which suspected alien ash trees were planted between 1992 and 2000.

Given the extent of introduced plants in Ireland, one central question regarding the potential invasiveness of hybrid material is whether future Irish climate will change to resemble the climate of continental *F. angustifolia*. If that is the case, hybrid individuals and any introduced *F. angustifolia* will have increased chances of establishing, hybridising and introgressing with native Irish ash. To investigate this, we simulated the present and potential niche distribution of *F. angustifolia* from the hybrid zones of the Loire and Saône valleys using maximum entropy models (Maxent – Phillips et al., 2006) of species distributions. A large-scale analysis of *F. angustifolia* and *F. excelsior* (Fernández-Manjarrés et al., submitted) indicates that six variables are highly explanatory of their distribution: (1) temperature of the coldest month; (2) temperature of the warmest month; (3) annual temperature variance; (4) precipitation of the warmest quarter; (5) precipitation of the coldest quarter; and (6) precipitation variability.

We used 706 geolocalised localities obtained from our own research and the public database GBIF (www.gibf.org) for which the six climate variables data listed above were extracted. The climate database (Hijmans et al., 2005) has a spatial resolution of 2.5 minutes and contains climate data obtained from several world meteorological stations covering the period 1950–2000. Adequacy of the distribution model produced by Maxent was verified using the area under the curve (AUC) of a receiver operating characteristic (ROC) plot (Fielding and Bell, 1997). Firstly, the Maxent model was run with default settings (50% of the data used at random to train the model, convergence threshold 10–5, maximum iterations 500 and automatic regularisation value) to simulate the distribution of hybrids for the present climate. Next, this present-day distribution was projected by Maxent onto a global warming scenario. The global warming model used was a high-resolution mid to strong case scenario of doubled CO_2 (2 × CO_2, CCM3 model) and other greenhouse gases from pre-industrial times to the year 2100 (e.g. CO_2 concentration of 710 ppmv and CH_4 concentration of 2538 ppbv) (Duffy et al., 2003) downloaded from www.diva-gis.org/climate.htm. This represents a 2–3 °C increment in global mean temperatures.

Simulation results for current climate for the two hybrid zones (Loire and Saône) indicated that the potential habitat for hybrids was relatively restricted in these areas (Fig 15.2A). However, projection of the hybrid habitat under the future global warming scenario suggested that weather variables suitable for the development of *F. angustifolia* and hybrids would be widespread in northwestern coastal Europe, including the southeast of Britain. Nevertheless, a potential suitable habitat for *F. angustifolia* or hybrids would not be significantly present in Ireland (Fig 15.2B); but note the small area of suitability around Dublin, possibly due to local warming effects.

Hence, neither the Loire-type climate nor the Mediterranean-type climate seems to be a likely scenario for Ireland, a projection validated with observations and several other models (Giorgi et al., 2004; Räisänen et al., 2004; Alcamo et al., 2007). For example, temperatures are increasing more in winter than in summer and mean winter precipitation is increasing in most of the Atlantic and northern Europe (Jones and Moberg, 2003). Hence, the likelihood of an overall favourable climate for *F. angustifolia* and hybrids in Ireland remains low, at least under the current state of knowledge.

The present analysis, however, does not take into account how *F. excelsior* will respond to climate change, or if phenology of the two species would be modified if winters become milder in northwestern Europe. If *F. excelsior* flowers earlier and *F. angustifolia* flowers later in the winter season, the risk of introgression into Irish plantations would be increased. Most field observations show that during mild winters, *F. angustifolia* completes its flowering in late December/early January, long before the spring flowering of *F.* excelsior (Jato et al., 2004). However, this

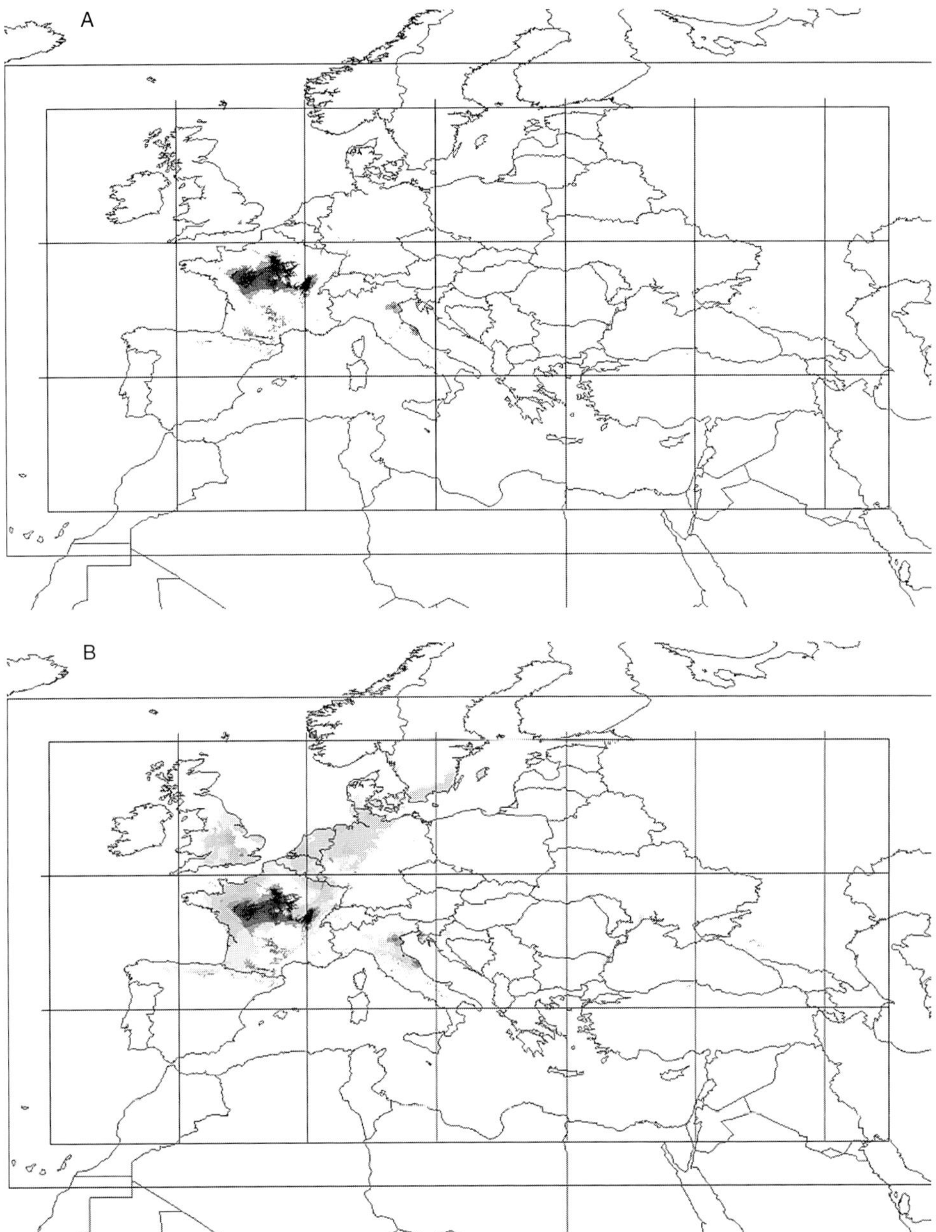

Figure 15.2 Potential niche distributions of *Fraxinus* hybrid zone under current conditions and a future climate scenario. (A) Maxent simulation of the potential niche distribution of *F. angustifolia* hybrid populations under current climate conditions (average 1950–2000, see text for details); AUC of the simulation = 0.997. (B) Maxent distribution of the potential niche of *F. angustifolia* hybrids under a global warming scenario of doubled CO_2. *See colour plate section.*

needs to be monitored in Ireland for a number of years before definite conclusions can be drawn. Similarly, our analysis would be insufficient to summarise adequately the other types of hybrids that are likely to exist across the large distribution range; but to the best of our knowledge, the provenances introduced to Ireland are likely to be from the Saône and Alsace regions.

15.4.5 Phenology and climate change in *Fraxinus*

We also wanted to evaluate the risk of hybridisation and introgression of non-native *Fraxinus* with native ash populations under current conditions. We wanted to know if prezygotic barriers to hybridisation existed between *F. excelsior* and non-*excelsior* types. Highly appropriate study sites for such research are available in Ireland because plantations containing non-native *F. angustifolia* and their hybrids exist that are adjacent to native *F. excelsior* woodlands. These trees are now mature enough to flower.

We recorded the phenology of trees during spring 2008 at two alien ash sites, Greenan in County Wicklow (8 ha planted in 1996) and Kildalkey in County Meath (20 ha planted in 1999). Thirty-two trees (Greenan) and 24 trees (Kildalkey) that had previously been selected on the basis of their morphological characteristics as '*excelsior* type', '*angustifolia* type' or 'hybrid type' were visited to record flowering phenology. Recordings were also made on 30 native *F. excelsior* trees growing around each plantation. Observations of flowering phenology were recorded from January to May 2008, every 8–10 days. Flower phenology was scored using a scale of 0 to 5 (0 for the dormant flower bud stage to 5 for the formation of seed stage – Fig 15.3).

At the Greenan site, 93% of native trees showed full flowering, whereas 62% of the alien trees flowered. Our observations showed that the native trees flowered later than the alien ash. The beginning of the pollen release was earlier for the selected hybrid type and *angustifolia* type trees inside the plantation compared to the native trees (Fig 15.4A). However, the trees that were selected as *excelsior* type showed a pattern similar to the native trees. The peak time of flowering for the selected trees of hybrid type and *angustifolia* type was around 10 March, whereas for native *F. excelsior* the peak was at the end of March (Fig 15.4A). A considerable overlap was found to exist in the flowering times of the native trees and the imported trees.

At the Kildalkey site, the proportion of flowering for the native trees was 97%, compared to 63% for the imported trees. As at Greenan, we observed that the beginning of the pollen release and the moment where the female became receptive was earlier in the selected hybrid type trees than in the native trees (Fig 15.4B). However, in one out of 30 native *F. excelsior* trees, the flower buds started to flush at the end of February, producing a fully exposed inflorescence (stage 2.5 – Fig 15.3).

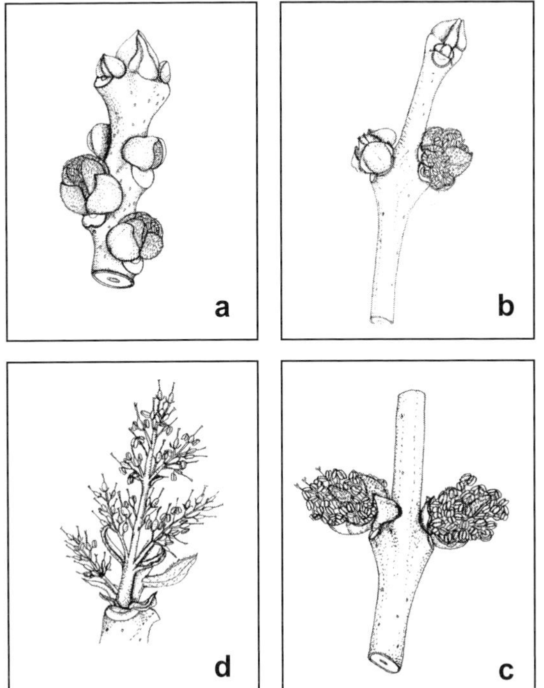

Figure 15.3 Stages and score notation of *Fraxinus* inflorescences and flowers: (a) stage 1 and 2 – flushing starts and first inflorescences are visible but are mainly covered by the bracts; (b) stage 2.5 – bracts fallen, completely visible inflorescences; (c) stage 3 – first pollen release and stigmas are receptive; (d) stage 4 – end of pollen release.

The timing of peak flowering for the alien trees and the native trees of *F. excelsior* was approximately the same (i.e. from the end of March to the middle of April).

Therefore the flowering periods (when pollen is released and stigmas are receptive) of the alien *angustifolia* type trees and native trees were found to overlap for approximately six weeks. This indicated the potential for pollen transfer and interbreeding between the native trees and the aliens. Our results do not project what will happen under future climate scenarios but do demonstrate that synchronous flowering can occur with a high proportion of the material under current conditions.

Fitter and Fitter (2002) showed that of six pairs of species, all of which form natural hybrids, four showed more synchronous flowering under present-day conditions than in the past, promoting the probability of hybridisation. Change in spring phenology in response to climate warming across the northern hemisphere, including Ireland, is especially well documented (Parmesan and Yohe, 2003; Root et al., 2003; Donnelly et al., 2006). In temperate climates, temperature is one of the

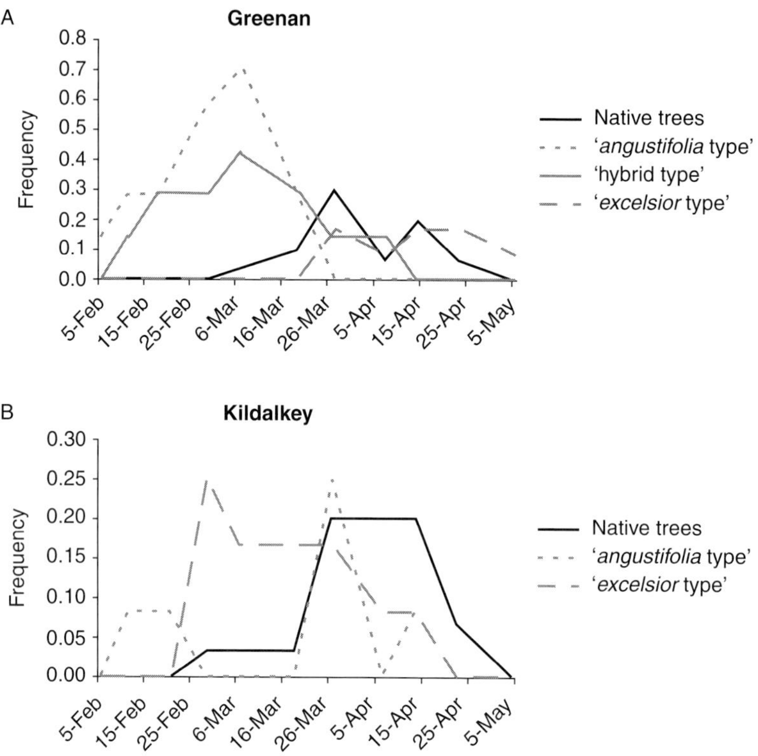

Figure 15.4 Frequency of trees reaching flowering stage 2.5 at (A) the Greenan site and (B) the Kildalkey site. Stage 2.5 represents the stage just before pollen release and receptivity of the stigmas.

most important factors influencing the timing of bud dormancy release and tree flushing (Jato et al., 2004; Vitasse et al., 2009). For *F. angustifolia*, Jato et al. (2004) have shown that temperature during the months prior to flowering determines the date of the pollen season. Maximum temperatures in December are critical to accumulate the required amount of heat to break dormancy, and temperatures below 0 °C in December will lead to a delay in flowering time. In *F. excelsior* the timing of spring leaf flush is also influenced by winter chilling and warm temperature (Cottignies, 1986; Jouve et al., 2007). Given predicted increases in winter temperature under global warming, *F. excelsior* could have an earlier start to its flowering. In Ireland, under future climatic predictions, both species may start flowering earlier, but the period of overlap found between the alien trees and the native trees may still be maintained. However, *F. excelsior* was not found to flower significantly earlier during the last 30 years in a study by van Vliet et al. (2002). Moreover, *F. angustifolia* is highly sensitive to frost (Jato et al., 2004), and an early start will expose it to increased risk of frost damage. In our case, frost damage was observed on some early-flowering alien trees, and the buds were dying before pollen was released (data not shown). If winter frosts become more prevalent, hybridisation between native trees and alien trees may become less likely.

15.5 Conclusions

Few studies have examined the impact of climate change on hybridisation and introgression. We have reviewed several ways in which climate can influence hybridisation in plants. The most significant impacts are likely to involve range shifts of parental species/populations and changes in phenology. We have presented a case study of the tree genus *Fraxinus* showing that the future climate in Ireland is not likely to be more favourable than the present to *F. angustifolia* and its hybrids with *F. excelsior*. However, the evolution of the phenology of *F. excelsior* and the alien trees is unknown under future climatic conditions in Ireland. Overlap in flowering time was found between alien and native trees, leading to the opportunity for interbreeding. More studies on gene flow using molecular markers and seed germination capacity are in progress.

Acknowledgements

Thanks to the Teagasc Walsh Fellowship Scheme for funding the studentship of M. Thomasset and COFORD for funding the research. Special thanks to John McNamara, Kinsealy Research Centre, Teagasc, for his technical support, and to Atchara Teerawatananon for drawings of the flowering states in *Fraxinus*.

References

Alcamo, J., Floerke, M. and Maerker, M. (2007). Future long-term changes in global water resources driven by socio-economic and climatic changes. *Hydrological Sciences*, **52**, 247–275.

Allendorf, F. W., Leary, R. F., Spruell, P. and Wenburg, J. K. (2001). The problems with hybrids: setting conservation guidelines. *Trends in Ecology and Evolution*, **16**, 613–622.

Anderson, E. and Hubricht, L. (1938). Hybridization in *Tradescantia*. III. The evidence for introgressive hybridization. *American Journal of Botany*, **25**, 396–402.

Anderson, E. and Stebbins, J. (1954). Hybridization as an evolutionary stimulus. *Evolution*, **8**, 378–388.

Arnold, M. L. (1997). *Natural Hybridization and Evolution*. Oxford: Oxford University Press.

Arnold, M. L. (2004). Natural hybridization and the evolution of domesticated, pest and disease organisms. *Molecular Ecology*, **13**, 997–1007.

Avise, J. C. (2004). *Molecular Markers, Natural History, and Evolution*, 2nd edn. Sunderland, MA: Sinauer Associates.

Broeck, A. V., Cox, K., Quataert, P., van Bockstaele, E. and van Slycken, J. (2003). Flowering phenology of *Populus nigra* L., *P. nigra* cv. *italica* and *P.* × *canadensis* Moench. and the potential for natural hybridisation in Belgium. *Silvae Genetica*, **52**, 280–283.

Buggs, R. J. A. (2007). Empirical study of hybrid zone movement. *Heredity*, **99**, 301–312.

Carroll, C. P. and Borrill, M. (1965). Tetraploid hybrids from crosses between diploid and tetraploid *Dactylis* and their significance. *Genetica*, **36**, 65–82.

Chapman, M. A. and Burke, J. M. (2006). Letting the gene out of the bottle: the population genetics of genetically modified crops. *New Phytologist*, **170**, 429–443.

Conceição, A. de S., de Queiroz, L. and Borba, E. (2008). Natural hybrids in *Chamaecrista* sect. *Absus* subsect. *Baseophyllum* (Leguminosae–Caesalpinioideae): genetic and morphological evidence. *Plant Systematics and Evolution*, **271**, 19–27.

Cottignies, A. (1986). The hydrolysis of starch as related to the interruption of dormancy in the ash bud. *Journal of Plant Physiology*, **123**, 373–380.

Coyer, J. A., Hoarau, G., Stam, W. T. and Olsen, J. L. (2007). Hybridization and introgression in a mixed population of the intertidal seaweeds *Fucus evanescens* and *F. serratus*. *Journal of Evolutionary Biology*, **20**, 2322–2333.

Coyne, J. A. and Orr, H. A. (2004). *Speciation*. Sunderland, MA: Sinauer Associates.

Currat, M., Ruedi, M., Petit, R. J. and Excoffier, L. (2008). The hidden side of invasions: massive introgression by local genes. *Evolution*, **62**, 1908–1920.

Curtu, A. L., Gailing, O. and Finkeldey, R. (2007). Evidence for hybridization and introgression within a species-rich oak (*Quercus* spp.) community. *BioMed Central Evolutionary Biology*, **7**, 218–233.

Dobzhansky, T. (1937). *Genetics and the Origin of Species*. New York, NY: Cambridge University Press.

Donnelly, A, Salamin, N. and Jones, M. B. (2006). Changes in tree phenology: an indicator of spring warming in Ireland? *Biology and Environment, Proceedings of the Royal Irish Academy*, **106B**, 49–56.

Duffy, P. B., Govindasamy, B., Lorio, J. P. et al. (2003). High-resolution simulations of global climate, part 1: present climate. *Climate Dynamics*, **21**, 371–390.

Ebina, T. and Otho, K. (2006). Morphological characters and PCR-RFLP markers in the interspecific hybrids between *Bactrocera carambolae* and *B. papayae* of the *B. dorsalis* species complex (Diptera: Tephritidae). *Research Bulletin of the Plant Protection Service, Japan*, **42**, 23–34.

Ellstrand, N. C. and Schierenbeck, K. A. (2000). Hybridization as a stimulus for the evolution of invasiveness in plants? *Euphytica*, **148**, 35–46.

Ellstrand, N. C., Prentice, H. C. and Hancock, J. F. (1999). Gene flow and introgression from domesticated plants into their wild relatives. *Annual Review of Ecology and Systematics*, **30**, 539–563.

Fernández-Manjarrés, J. F., Gerard, P. R., Dufour, J., Raquin, C. and Frascaria-

Lacoste, N. (2006). Differential patterns of morphological and molecular hybridization between *Fraxinus excelsior* L. and *Fraxinus angustifolia* Vahl (Oleaceae) in eastern and western France. *Molecular Ecology,* **15**, 3245–3257.

Fielding, A. H. and Bell, J. F. (1997). A review of methods for the assessment of prediction errors in conservation presence/absence models. *Environmental Conservation,* **24**, 38–49.

Fitter, A. H. and Fitter, R. S. R. (2002). Rapid changes in flowering time in British plants. *Science,* **296**, 1689–1691.

Foden, W., Midgley, G. F., Hughes, G. O. et al. (2007). A changing climate is eroding the geographical range of the Namib Desert tree Aloe through population declines and dispersal lags. *Diversity Distribution,* **13**, 645–653.

Gerard, P. R., Fernández-Manjarrés , J. F. and Frascaria-Lacoste, N. (2006a). Temporal cline in a hybrid zone population between *Fraxinus excelsior* L. and *Fraxinus angustifolia* Vahl. *Molecular Ecology,* **15**, 3655–3667.

Gerard, P. R., Klein, E. K., Austerlitz, F., Fernández-Manjarrés, J. F. and Frascaria-Lacoste, N. (2006b). Assortative mating and differential male mating success in an ash hybrid zone population. *BMC Evolutionary Biology,* **6**, 96.

Gerard, P. R., Fernández-Manjarrés, J. F., Bertolino, P. et al . (2006c). New insights in the recognition of the European ash species *Fraxinus excelsior* L. and *Fraxinus angustifolia* Vahl as useful tools for forest management. *Annals of Forest Science,* **63**, 733–738.

Giorgi, F., Bi, X. and Pal, J. (2004). Mean interannual and trends in a regional climate change experiment over Europe. II: climate change scenarios (2071–2100). *Climate Dynamics,* **23**, 839–858.

Harrison, R. G. (1993). *Hybrid Zones and the Evolutionary Process.* New York, NY: Oxford University Press.

Heuertz, M., Hausman, J. F., Tsvetkov, I., Frascaria-Lacoste, N. and Vekemans, X. (2001). Assessment of genetic structure within and among Bulgarian populations of the common ash (*Fraxinus excelsior* L.). *Molecular Ecology,* **10**, 1615–1623.

Heuertz, M., Hausman, J. F., Hardy, O. J. et al. (2004). Nuclear microsatellites reveal contrasting patterns of genetic structure between western and southeastern European populations of the common ash (*Fraxinus excelsior* L.). *Evolution,* **58**, 976–988.

Heuertz, M., Carnevale, S., Fineschi, S. et al. (2006). Chloroplast DNA phylogeography of European ashes, *Fraxinus* sp. (Oleaceae): roles of hybridization and life history traits. *Molecular Ecology,* **15**, 2131–2140.

Hewitt, G. (2000). The genetic legacy of the Quaternary ice ages. *Nature,* **405**, 907–913.

Hijmans, R. J., Cameron, S. E., Parra, J. L., Jones, P. G. and Jarvis, A. (2005). Very high resolution interpolated climate surfaces for global land areas. *International Journal of Climatology,* **25**, 1965–1978.

Hodkinson, T. R., Chase, M. W., Takahashi, C. et al. (2002). The use of DNA sequencing (ITS and trnL-F), AFLP, and fluorescent in situ hybridization to study allopolyploid *Miscanthus* (Poaceae). *American Journal of Botany,* **89**, 279–286.

Horgan, T., Keane, M., Mccarthy, R., Lally, M. and Thompson, D. (2004).

A Guide to Forest Tree Species Selection and Silviculture in Ireland. Dublin: COFORD.

Howard, D. J. (1999). Conspecific sperm and pollen precedence and speciation. *Annual Review of Ecology and Systematics*, **30**, 109–132.

Intergovernmental Panel on Climate Change (IPCC) (2007). *Climate Change 2007: Impacts, Adaptation and Vulnerability. Contribution of Working Group II to the Fourth Assessment Report of the Intergovernmental Panel on Climate Change*, ed. M. Parry, O. Canziani, J. Palutikof, P. van der Linden and C. Hanson. Cambridge: Cambridge University Press.

Jato, V., Rodriguez-Rajo, F. J., Dacosta, N. and Aira, M. J. (2004). Heat and chill requirements of *Fraxinus* flowering in Galicia (NW Spain). *Grana*, **43**, 217–223.

Jeandroz, S., Frascaria-Lacoste, N. and Bousquet, J. (1996). Molecular recognition of the closely related *Fraxinus excelsior* and *F. oxyphylla* (Oleaceae) by RAPD markers. *Forest Genetics*, **3**, 237–242.

Jones, P. D. and Moberg, A. (2003). Hemispheric and large scale surface air temperature variations: an extensive revision and an update to 2001. *Journal of Climate*, **16**, 206–223.

Jouve, L., Jacques, D., Douglas, G. C., Hoffmann, L. and Hausman, J. F. (2007). Biochemical characterization of early and late bud flushing in common ash (*Fraxinus excelsior* L.). *Plant Science*, **172**, 962–969.

Kelleher, C. T., Hodkinson, T. R., Douglas, G. C. and Kelly, D. L. (2005). Species distinction in Irish populations of *Quercus petraea* and *Q. robur*: morphological versus molecular analyses. *Annals of Botany*, **96**, 1237–1246.

Khasa, D., Pollefeys, P., Navarro-Quezada, A., Perinet, P. and Bousquet, J. (2005). Species-specific microsatellite markers to monitor gene flow between exotic poplars and their natural relatives in eastern North America. *Molecular Ecology Notes*, **5**, 920–923.

Kuchta, S. R. and Tan, A. M. (2005). Isolation by distance and post-glacial range expansion in the rough-skinned newt, *Taricha granulosa*. *Molecular Ecology*, **14**, 225–244.

Kullman, L. (2008). Early postglacial appearance of tree species in northern Scandinavia: review and perspective. *Quaternary Science Reviews*, **27**, 2467–2472.

Lavergne, S., Molina, J. and Debussche, M. (2006). Fingerprints of environmental change on the rare Mediterranean flora: a 115-year study. *Global Change Biology*, **12**, 1466–1478.

Leitch, I. J. and Bennett, M. D. (1997). Polyploidy in angiosperms. *Trends in Plant Science*, **2**, 470–476.

Lenton, T. M. (2004). The coupled evolution of life and atmospheric oxygen. In *Evolution on Planet Earth*, ed. L. J. Rothschild and A. Lister. London: Academic Press, pp. 35–53.

Levin, D. A. (1978). The origin of isolating mechanisms in flowering plants. *Evolutionary Biology*, **11**, 185–317.

Levin, D. A. (2002). *The Role of Chromosomal Change in Plant Evolution*. Oxford: Oxford University Press.

Levin, D. A., Francisco-Ortega, J. and Jansen, R. K. (1996). Hybridization and the extinction of rare plant species. *Conservation Biology*, **10**, 10–16.

Lewis, D. and Crowe, L. K. (1958). Unilateral interspecific incompatibility in flowering plants. *Heredity*, **12**, 233–256.

Liepelt, S., Cheddadi, R., de Beaulieu, J. L. et al. (2009). Postglacial range expansion and its genetic imprints in *Abies alba* Mill.: a synthesis from palaeobotanic and genetic data. *Review of Palaeobotany and Palynology,* **153**, 139–149.

Lowe, A., Harris, S. and Ashton, P. (2004). *Ecological Genetics: Design, Analysis, and Application.* Oxford: Blackwell.

Mahoney, M. J. (2004). Molecular systematics and phylogeography of the *Plethodon elongatus* species group: combining phylogenetic and population genetic methods to investigate species history. *Molecular Ecology,* **13**, 149–166.

Marigo, G., Peltier, J. P., Girel, J. and Pautou, G. (2000). Success in the demographic expansion of *Fraxinus excelsior* L.. *Trees,* **15**, 1–13.

Mebert, K. (2008). Good species despite massive hybridization: genetic research on the contact zone between the watersnakes *Nerodia sipedon* and *N. fasciata* in the Carolinas, USA. *Molecular Ecology,* **17**, 1918–1929.

Menzel, A. and Fabian, P. (1999). Growing season extended in Europe. *Nature,* **397**, 659.

Menzel, A., Sparks, T. H., Estrella, N. et al. (2006). European phenological response to climate change matches the warming pattern. *Global Change Biology,* **12**, 1969–1976.

Morand-Prieur, M. E., Brachet, S., Rossignol, P., Dufour, J. and Frascaria-Lacoste, N. (2002). A generalized heterozygote deficiency assessed with microsatellites in French common ash populations. *Molecular Ecology,* **11**, 377–385.

Oddo, E., Saiano, F., Alonzo, G. and Bellini, E. (2002). An investigation of the seasonal pattern of mannitol content in deciduous and evergreen species of the *Oleaceae* growing in northern Sicily. *Annals of Botany,* **90**, 239–243.

Otto, S. P. and Whitton, J. (2000). Polyploid incidence and evolution. *Annual Review of Genetics,* **34**, 401–437.

Parmesan, C. (2006). Ecological and evolutionary responses to recent climate change. *Annual Review of Ecology, Evolution, and Systematics,* **37**, 637–669.

Parmesan, C. and Yohe, G. (2003). A globally coherent fingerprint of climate change impacts across natural systems. *Nature,* **421**, 37–42.

Parmesan, C., Gaines, S., Gonzalez, L. et al. (2005). Empirical perspectives on species borders: from traditional biogeography to global change. *Oikos,* **108**, 58–75.

Pascarella, J. B. (2007). Mechanisms of prezygotic reproductive isolation between two sympatric species, *Gelsemium rankinii* and *G. sempervirens* (Gelsemiaceae), in the southeastern United States. *American Journal of Botany,* **94**, 468–476.

Patonnier, M. P., Peltier, J. P. and Marigo, G. (1999). Drought-induced increase in xylem malate and mannitol concentrations and closure of *Fraxinus excelsior* L. stomata. *Journal of Experimental Botany,* **50**, 1223–1229.

Pauli, H., Gottfried, M., Reiter, K., Klettner, C. and Grabherr, G. (2006). Signals of range expansions and contractions of vascular plants in the high Alps: observations (1994–2004) at the GLORIA master site Schrankogel, Tyrol, Austria. *Global Change Biology,* **13**, 147–156.

Perron, M. and Bousquet, J. (1997). Natural hybridization between black spruce and red spruce. *Molecular Ecology,* **6**, 725–734.

Phillips, S. J., Anderson, R. P. and Schapire, R. E. (2006). Maximum entropy modeling of species geographic distributions. *Ecological Modelling,* **190**, 231–259.

Picard, J. F. (1983). A propos du frêne oxyphylle, *Fraxinus angustifolia* Vahl. *Forêt-Entreprise,* **83**, 2–4.

Räisänen, J., Hansson, U., Ullerstig, A. et al. (2004). European climate in the late 21st century: regional simulations with two driving global models and two forcing scenarios. *Climate Dynamics,* **22**, 13–31.

Rameau, J. C., Mansion, D. and Dumé, G. (1989). *Flore forestière Française, guide écologique illustré.* Paris: Institut pour le Développement Forestier.

Ramsey, J. and Schemske, D. W. (2002). Neopolyploidy in flowering plants. *Annual Review of Ecology and Systematics,* **33**, 589–639.

Raquin, C., Jung-Muller, B., Dufour, J. and Frascaria-Lacoste, N. (2002). Rapid seedling obtaining from European ash species *Fraxinus excelsior* L. and *Fraxinus angustifolia* Vahl. *Annals of Forest Sciences,* **59**, 219–224.

Rathcke, B. and Lacey, E. P. (1985). Phenological patterns of terrestrial plants. *Annual Review of Ecology and Systematics,* **16**, 179–214.

Raybould, A. F. and Gray, A. J. (1994). Will hybrids of genetically modified crops invade natural communities? *Trends in Ecology and Evolution,* **9**, 85–89.

Rhymer, J. M. and Simberloff, D. (1996). Extinction by hybridization and introgression. *Annual Review of Ecology and Systematics,* **27**, 83–109.

Rieseberg, L. H. (1997). Hybrid origins of plant species. *Annual Review of Ecology and Systematics,* **28**, 359–389.

Rieseberg, L. H. (2006). Hybrid speciation in wild sunflowers. *Annals of Missouri Botanical Garden,* **93**, 34–48.

Rieseberg, L. H. and Burke, J. M. (2001). The biological reality of species: gene flow, selection, and collective evolution. *Taxon,* **50**, 47–67.

Rieseberg, L. H. and Carney, S. E. (1998). Plant hybridization. *New Phytologist,* **140**, 599–624.

Rieseberg, L. H. and Willis, J. H. (2007). Plant speciation. *Science,* **317**, 910–914.

Rieseberg, L. H., Archer, M. A. and Wayne, R. K. (1999). Transgressive segregation, adaptation and speciation. *Heredity,* **83**, 363–372.

Rieseberg, L. H., Raymond, O., Rosenthal, D. M. et al. (2003). Major ecological transitions in wild sunflowers facilitated by hybridization. *Science,* **301**, 1211–1216.

Root, T. L., Price, J. T., Hall, K. R., Schneider, S. H. and Rosenzweig, C. (2003). Fingerprints of global warming on wild animals and plants. *Nature,* **421**, 57–60.

Root, T. L., Macmynowski, D. P., Mastrandrea, M. D. and Schneider, S. H. (2005). Human-modified temperatures induce species changes: joint attribution. *Proceedings of the National Academy of Sciences of the USA,* **102**, 7465–7469.

Ross, C. L. and Harrison, R. G. (2006). Viability selection on overwintering eggs in a field cricket mosaic hybrid zone. *Oikos,* **115**, 53–68.

Savolainen, O., Pyhajarvi, T. and Knurr, T. (2007). Gene flow and local adaptation in trees. *Annual Review of Ecology, Evolution, and Systematics,* **38**, 595–619.

Schierenbeck, K. A., Symonds, V. V., Gallagher, K. G. and Bell, J. (2005). Genetic variation and phylogeographic analyses of two species of *Carpobrotus* and their hybrids in California. *Molecular Ecology,* **14**, 539–547.

Schluter, D. (2001). Ecology and the origin of species. *Trends in Ecology and Evolution,* **16**, 372–379.

Schönswetter, P., Stehlik, I., Holderegger, R. and Tribsch, A. (2005). Molecular evidence for glacial refugia of mountain plants in the European Alps. *Molecular Ecology,* **14**, 3547–3555.

Seehausen, O. (2004). Hybridization and adaptive radiation. *Trends in Ecology and Evolution*, **19**, 198–207.

Soltis, P. S. and Soltis, D. E. (2009). The role of hybridization in plant speciation. *Annual Review of Plant Biology,* **60**, 561–588.

Sota, T. and Volger, A. P. (2001). Incongruence of mitochondrial and nuclear gene trees in the Carabid beetles *Ohomopterus*. *Systematic Biology,* **50**, 39–59.

Stace, C. A. (1975). *Hybridization and the Flora of the British Isles*. London: Academic Press.

Taylor, E. B., Boughman, J. W., Groenenboom, M. et al. (2006). Speciation in reverse: morphological and genetic evidence of the collapse of a three-spined stickleback (*Gasterosteus aculeatus*) species pair. *Molecular Ecology,* **15**, 343–355.

Tsukada, M. (1982). Late-Quaternary shift of fagus distribution. *Journal of Plant Research,* **95**, 203–217.

Valbuena-Carabanã, M., Gonzàlez-Martinez, S. C., Sork, V. L. et al. (2005). Gene flow and hybridisation in a mixed oak forest (*Quercus pyrenaica* Willd. and *Quercus petraea* (Matts.) Liebl.) in central Spain. *Heredity,* **95**, 457–465.

van Vliet, A. J. H., de Groot, R. S., Overeem, A., Jacobs, A. F. G. and Spieksma, F. T. M. (2002). The influence of temperature and climate change on the timing of pollen release in the Netherlands. *International Journal of Climatology,* **22**, 1757–1767.

Vila, M., Weber, E. and Antonio, C. M. D. (2000). Conservation implications of invasion by plant hybridization. *Biological Invasions,* **2**, 207–217.

Visser, M. E. and Holleman, L. J. M. (2001). Warmer springs disrupt the synchrony of oak and winter moth phenology. *Proceedings of the Royal Society of London B,* **268**, 289–294.

Vitasse, Y., Delzon, S., Dufrêne, E. et al. (2009). Leaf phenology sensitivity to temperature in European trees: do within-species populations exhibit similar responses? *Agricultural and Forest Meteorology,* **149**, 735–744.

Wachowiak, W. and Prus-Glowacki, W. (2008). Hybridisation processes in sympatric populations of pines *Pinus sylvestris* L., *P. mugo* Turra and *P. uliginosa* Neumann. *Plant Systematics and Evolution,* **271**, 29–40.

Wagner, W. H. J. (1970). Biosystematics and evolutionary noise. *Taxon,* **19**, 146–151.

Wallander, E. and Albert, V. A. (2000). Phylogeny and classification of Oleaceae based on *rps16* and *trnL-F* sequence data1. *American Journal of Botany,* **87**, 1827–1841.

Weis, A. E. and Kossler, T. M. (2004). Genetic variation in flowering time induces phenological assortative mating: quantitative genetic methods applied to *Brassica rapa*. *American Journal of Botany,* **91**, 825–836.

White, M. A., Brunsell, N. and Schwartz, M. D. (2003). Vegetation phenology in global change studies. In *Phenology: an Integrative Environmental Science*, ed. M. D.

Schwartz. Dordrecht: Kluwer, pp. 453–466.

Winge, Ö. (1917). The chromosomes: their number and general importance. *Comptes Rendues des Travaux du Laboratoire Carlesberg,* **13**, 131–275.

Young, A., Boyle, T. and Brown, T. (1996). The population genetic consequences of habitat fragmentation for plants. *Trends in Ecology and Evolution,* **11**, 413–418.

Section 4

Conservation

16

Assessing the effectiveness of a protected area network in the face of climatic change

B. HUNTLEY

School of Biological and Biomedical Sciences, Durham University, UK

D. G. HOLE

School of Biological and Biomedical Sciences, Durham University, UK and *Science and Knowledge Division, Conservation International, Arlington, VA, USA*

S. G. WILLIS

School of Biological and Biomedical Sciences, Durham University, UK

Abstract

Climatic change is expected to result in changes in species' distributions. However, current networks of protected areas, designed to conserve biodiversity, have been designated and designed on the basis of a paradigm of long-term stability of

Climate Change, Ecology and Systematics, ed. Trevor R. Hodkinson, Michael B. Jones, Stephen Waldren and John A. N. Parnell. Published by Cambridge University Press.

species' geographical distributions. As a result, these networks may not be effective in conserving biodiversity in a world with rapidly changing climatic conditions. We investigate this using as a model system the 1679 bird species breeding in sub-Saharan Africa and the network of 803 Important Bird Areas (IBAs) designated in the region by BirdLife International. Using climatic envelope models fitted to species' present distributions and the current climate, species' present and potential future occurrences in IBAs were simulated. The results show that the current network has the potential to maintain most species throughout the present century. However, they also indicate that this outcome depends upon substantial potential species turnover in many IBAs. This is only likely if the connectivity of the current network is enhanced substantially in key areas, and will also depend upon sympathetic management of the wider landscape, so as to enhance its permeability, and appropriate management of individual sites, taking into account their role in the overall network.

16.1 Introduction

It is now generally accepted that anthropogenic activities have resulted in global climatic changes over the past century (Trenberth et al., 2007); they may even have done so over several millennia (Ruddiman, 2003). It is also now generally accepted that the recent rapid increase in global mean temperature will continue throughout the present century and perhaps even beyond, although the magnitude and rate of warming beyond *c.* 2050 will depend upon the amounts of greenhouse gases emitted over the coming decades (Meehl et al., 2007). Although these recent and forecast rapid climatic changes have focused attention upon how species and ecosystems respond to changes in climate, it is important to recognise that climatic change is not a new phenomenon, but has always been a feature of the earth system (Jansen et al., 2007). The past million or so years of the Quaternary period, in particular, have been characterised by global climatic fluctuations at timescales ranging from centennial to multi-millennial (GRIP Members, 1993; McDermott et al., 2001; EPICA Community Members, 2004; Allen et al., 2007). Although species exhibit a range of responses to climatic changes, depending upon their rate and magnitude (Fig 16.1), the species forming present ecosystems have survived the rapid large-magnitude climatic changes of the Quaternary principally by shifting their distributions in response to the changes in climate (Huntley and Webb, 1989).

Climatic conditions during the period of recorded human history, however, and even throughout the time since the early Holocene when agriculture was first developed, have been relatively stable when compared to the last interglacial (Fronval and Jansen, 1997) or to the Quaternary as a whole. Climate has of course

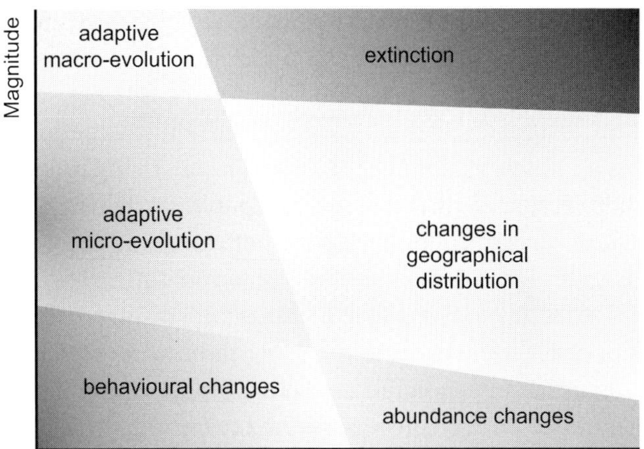

Figure 16.1 Species' responses to climatic change. Species exhibit six general types of response, each of which predominates under particular combinations of rate and magnitude of change, although the boundaries between their areas of predominance are 'fuzzy' rather than abrupt. Changes in geographical distribution are species' principal response to relatively rapid climatic changes of a wide range of magnitudes. *See colour plate section.*

varied, even over the past millennium (Juckes et al., 2007), but the magnitude and rate of these variations generally have been small and slow compared either to pre-Holocene variations or to recent and forecast future changes resulting from anthropogenic greenhouse gas emissions. This relative stability has conditioned the expectations of human society with respect to the magnitude and rate of climatic changes and, in particular, has resulted in an approach to biodiversity conservation that is rooted in an assumption that the global environment is essentially static, and that, as a result, so too are the geographical distributions of species and of the ecosystems of which they are components.

This static view is reflected by the fundamental dependence of global biodiversity conservation strategies upon the establishment of networks of protected areas and other designated sites that, by their very nature, occupy particular fixed geographical locations. Given the recent and expected future rapid, large-magnitude climatic changes, however, it is now apparent that many such sites will in future no longer provide suitable conditions for the species and/or ecosystems for whose protection they were established or designated (Araújo et al., 2004; Hannah et al., 2007). As a result, both governmental and non-governmental bodies responsible for biodiversity conservation have recognised the urgent need to assess how effective such site networks may be in conserving biodiversity in a world with rapidly changing climatic conditions and to develop appropriate biodiversity conservation strategies that will facilitate ecological adaptation to these changing conditions (Hopkins et al., 2007; Huntley, 2007; Hannah, 2008).

We used the birds breeding in sub-Saharan Africa, and the network of Important Bird Areas (IBAs) identified and designated in that region by BirdLife International (Fishpool and Evans, 2001), as a model system to explore this issue. Using models describing the relationship between each bird species' geographical distribution and current climate, along with scenarios of projected climate for three intervals during the present century, we simulated the current and potential future pattern of occurrence of each species in the IBAs. We then synthesised the results of these simulations in various ways in order to investigate the potential future effectiveness of this continent-wide network of designated areas in supporting the elements of biodiversity it was designed to protect, given projected twenty-first-century climatic changes. We also explored ways to identify potential gaps in the network and areas that potentially will increase in value in the future but that do not form part of the current network.

16.2 Methods

16.2.1 Species distribution modelling

In order to simulate the potential future distributions of birds in sub-Saharan Africa, including their potential occurrences in IBAs, we first fitted climatic envelope models relating each species' present distribution to the present climate. Although such models, and especially their underlying assumptions, have been debated (Gaston, 2003; Pearson and Dawson, 2003), and their reliability as a basis for predicting species' potential responses to climatic change has been questioned by some authors (Davis et al., 1998; Beale et al., 2008), several studies have demonstrated their robustness. They have been shown, for example, successfully to predict species distributions in regions (Beerling et al., 1995) and for times (Hijmans and Graham, 2006) independent of those for which data were used to fit the models. They also have been shown successfully to hindcast changes in species abundance over recent decades, both for rare species near their range margins (Green et al., 2008) and more extensively for a range of common species (Gregory et al., 2009). Furthermore, the method that we apply, of fitting climatic response surfaces (CRSs) (Huntley et al., 1989, 1995, 2007): (1) makes no assumptions about the general form of the relationship between a species' probability of occurrence and the climatic variables; (2) uses a locally weighted regression fitting approach (Cleveland and Devlin, 1988) that allows for differences in the relative importance of variables as determinants of the species' range limits in different parts of the climatic space; (3) takes into account the role of interactions between the climatic variables in determining the species' range limits, and also allows for the varying nature and importance of these interactions in different parts of the climatic space; (4) requires that an a priori selection of a limited set of climatic variables is made, precluding the 'shotgun' approach often adopted in fitting climatic envelopes using other methods, in which numerous variables are included, many of which are highly correlated and few of which have plausible or distinctive mechanistic roles in limiting a species' distribution; and (5) has been shown to give models with a comparable goodness of fit to those obtained using methods advocated by various authors as being amongst the best, for example generalised additive models (Doswald et al., 2009; Hole et al., 2009). Additionally, the behaviour of the CRS models when extrapolated into regions of climatic space outside the domain to which they are fitted is conservative and well understood, in comparison with the variable and often unpredictable behaviour of many widely applied techniques (Thuiller, 2004; Pearson et al., 2006).

Goodness of fit of the CRS models was assessed using the area under the curve (AUC) of a receiver operating characteristic (ROC) plot (Fielding and Bell, 1997), models being validated using k-fold partitioning (k = 100 models for each species, each being fitted to a 70% random sample of observations and its goodness of fit assessed using the remaining 30% of observations). Having identified the four-variable model giving the highest mean k-fold AUC for each species, the full model for those variables (i.e. that fitted to all the observations) was used to simulate both that species' potential future distribution (Huntley et al., 2006) and its potential present and future occurrence in individual IBAs. The threshold probability of occurrence used to discriminate simulated presences and absences was determined for each species' model as the minimum probability of occurrence that gave the maximum value of Cohen's kappa (Cohen, 1960), kappa being evaluated for probability of occurrence increments of 0.001.

Potential impacts upon species distributions were assessed using three measures that compare different properties of the simulated potential future and present distributions (Huntley et al., 2007, 2008):

(1) Relative range extent (R) was calculated by expressing the number of potentially suitable grid cells in the future as a proportion of the number of grid cells simulated as suitable under present climatic conditions.
(2) Potential range overlap (O) was calculated by expressing the number of grid cells simulated as suitable under both present and future climatic conditions as a proportion of the number of grid cells simulated as suitable under present climatic conditions.
(3) Range displacement (D) was calculated as the geodesic distance between the centroid of the potential future distribution and the centroid of the distribution simulated for present climatic conditions.

16.2.2 Data and variables

The bird species distribution data used recorded the presence or absence of the 1679 species breeding in the region for the 1963 cells of a 1° longitude × latitude grid (i.e. *c.* 110 × 110 km in regions near the equator) (Burgess et al., 1998; Brooks et al., 2001). Avian taxonomy followed the BirdLife International checklist (www.birdlife.org/datazone/species/taxonomy.html). Modelling was impractical for those 71 species with fewer than five recorded presences. The 1608 species for which models were fitted included 815 species, referred to hereafter as 'priority species', falling into one of those categories whose presence triggered the designation of a site as an IBA. These categories are: (1) globally threatened species, i.e. those assigned to one of the classes Critically Endangered, Endangered or Vulnerable, as used by the International Union for the Conservation of Nature (IUCN) to compile its Red Lists (www.iucnredlist.org) (BirdLife International, 2000); (2) range-restricted species; and (3) biome-restricted species (Fishpool and Evans, 2001).

Mean monthly temperature and precipitation data for the present climate, as well as mean elevation, were obtained for a 2.5′ longitude × latitude (i.e. *c.* 5 km) grid from the WorldClim database (Hijmans et al., 2005). Mean monthly cloudiness data were interpolated to the same 2.5′ grid using thin plate spline surfaces (Hutchinson, 1989) fitted to a compilation of mean monthly cloudiness data from meteorological stations (Leemans and Cramer, 1991) and relating mean monthly cloudiness to longitude, latitude and elevation. A series of bioclimatic variables was calculated from these climatic data for use in fitting CRS models relating species distributions to the present climate. Bioclimatic values for the cells of the 1° grid were estimated as the means of the values for the second and third quartiles of the relevant 2.5′ grid cells ranked by elevation; omitting 2.5′ grid cells in the first and fourth quartiles avoided distortion of the 1° grid cell values by inclusion in calculating the mean of values from limited areas of extreme high or low elevation atypical of the grid cell as a whole.

The bioclimatic variables used were selected on the basis that plausible mechanistic roles could be identified whereby they could act, whether directly or indirectly, to limit species distributions. The seven variables used were as follows:

(1) coldest month mean temperature, a proxy for the lower temperature threshold tolerated by the species, whether acting directly through the species' own physiology or indirectly by determining the availability of, for example, food or habitat;

(2) warmest month mean temperature, a proxy for the species' upper temperature threshold;

(3) an estimate of the annual ratio of actual to potential evapotranspiration;

(4) wet season duration;

(5) dry season duration;

(6) wet season intensity;

(7) dry season intensity.

Whereas the third variable provides an overall measure of moisture availability integrated throughout the year, the last four variables reflect different aspects of the seasonality of rainfall, and hence moisture availability, that is a predominant feature especially of the tropical seasonal climates prevailing throughout large parts of the region and that determines the annual cycle of activity of many bird species breeding in these regions. Definitions of these last four variables, and details of how they were calculated, are given by Huntley et al. (2006), while Huntley et al. (1995) describe the calculation of the other three variables (see also Prentice et al., 1992). Four models were fitted for each species using the first three variables plus each in turn of the last four. We could identify no biologically plausible basis for fitting models using five or more variables, including two or more of the last four variables, and in any case the additional explanatory power even of the fourth variable was often small.

16.2.3 IBAs

In order to use the CRS models to simulate species' potential occurrences in IBAs, an estimate of the present climate of each IBA was required. Digitised and georeferenced outlines of 803 IBAs in sub-Saharan Africa, excluding those on surrounding islands, were provided by BirdLife International. Using the ArcInfo GIS (ESRI, 1998), these outlines were overlaid onto the 2.5′ grid in order to identify those cells of this grid partially or completely within each IBA. Given the very wide range of areas of individual IBAs, and the considerable range of elevation spanned by some, the 2.5′ grid cells relating to each IBA were stratified if the IBA extended over > 1° of longitude or latitude, or if the difference in mean elevation between the lowest and highest 2.5′ grid cells exceeded 600 m. For IBAs extending over > 1° of longitude or latitude the 2.5′ grid cells were first stratified using the cells of the 1° grid. For IBAs, or for sections of larger IBAs stratified using the 1° grid, within which the range of mean elevations of the 2.5′ grid cells exceeded 600 m, the 2.5′ grid cells were stratified into elevation bands, the number of these bands being one plus the integer portion of the result of dividing the range of mean elevations in metres by 600. Finally, bioclimatic values were estimated for each stratum, or for the IBA as a whole where no stratification had been required, using the same procedure as for the 1° grid cells (i.e. as the mean of the values for the second and third quartiles of the relevant 2.5′ grid cells ranked by elevation). The 600 m threshold for elevation range was determined on the basis of the mean temperature lapse rate (*c.* 6 °C km^{-1}) and the magnitude of the uncertainty typical of interpolated mean monthly temperatures (*c.* ± 1 °C). Using this threshold resulted in elevation bands spanning 300–600 m, with mean temperature of adjacent bands typically differing by an amount comparable to or greater than the inherent uncertainty in the interpolated values (1.8–3.6 °C).

Applying the CRS models, and threshold probability of occurrence values, obtained using 1° data to simulate species' potential occurrence in IBAs, many of which are much smaller in extent than the cells of the 1° grid, represents a substantial downscaling. In order to assess the reliability of the downscaled simulations, we obtained from BirdLife International species inventories from 64 of the IBAs for which the level of monitoring was considered sufficient that inventories based upon observations were likely to be largely, albeit never entirely, complete. Simulated species inventories were obtained by projecting each species' presence or absence for the IBA, or for its component strata, using the CRS model; where an IBA had been stratified, simulated presence of a species in any one stratum was taken as simulated occurrence in the IBA. Various measures, including the Hansen-Kuiper discriminant (or true skill statistic (TSS) – Allouche et al., 2006), were used to assess how well our simulated species inventories for these IBAs matched those recorded.

In order to identify gaps in the current IBA network, we calculated geodesic distances between the centre of each cell of a 0.25° grid extending across the study area and the centroid of each IBA. We then identified the nearest neighbour IBA for each grid cell and recorded its geodesic distance from the grid cell. Areas where these distances were relatively large would then represent gaps in the network. Areas of potential future value not included in the present network were identified by simulating and comparing the present and potential future suitability of each 0.25° grid cell for priority species, calculating for each cell the number of priority species for which it was simulated to become newly suitable. Bioclimatic values for the 0.25° cells were estimated following the same procedure as for the 1° cells. Using the potential future and simulated present complements of priority species for the cells of the 1° grid, we also evaluated the predominant direction of movement required for species to achieve the range shifts required in order to occupy newly suitable grid cells. To do this we calculated for each priority species for which a grid cell was newly suitable the azimuth of the geodesic path from the nearest grid cell simulated to be suitable under the previously prevailing conditions. We then calculated for each grid cell the mean azimuth for those priority species for which it was newly suitable (Hole et al., in press).

16.2.4 Future climates

Projected future climatic conditions were estimated for three 30-year periods during the present century centred on 2025, 2055 and 2085. Three projections were made for each period using the results from transient simulations made by three general circulation models (GCMs) for an intermediate scenario of future greenhouse gas emissions (SRES B2a – see Nakicenovic and Swart, 2000). The three GCMs selected were: HadCM3 (Gordon et al., 2000); ECHAM4 (Roeckner et al., 1996); and GFDL-R30 (Knutson et al., 1999). These three GCMs were chosen firstly because they each have an equilibrium sensitivity for global mean temperature close to the mean for the nine models, including these three, included in the Intergovernmental Panel on Climate Change (IPCC) *Third Assessment Report*, and secondly because each represents one of the three groups into which these nine GCMs fall with respect to projected increase in global mean precipitation (Cubasch et al., 2001). Of the three, GFDL-R30 is in the 'wet' group, HadCM3 in the 'intermediate' group and ECHAM4 in the 'dry' group. Given the importance of moisture availability in determining major ecological patterns and species distributions in sub-Saharan Africa, and the greater uncertainty surrounding GCM projections of precipitation than those of temperature, it was important that the selection of GCMs used enabled us to accommodate the uncertainty in precipitation changes within our ensemble projections of potential climatic change impacts. For each GCM and time period, anomalies were calculated, relative to the mean values simulated by the GCM for 1961–90, for the 30-year means of monthly mean temperature and precipitation for each GCM grid cell. Temperature anomalies were additive, whereas those for precipitation were multiplicative. Spline surfaces were fitted to these anomalies and used to interpolate their values to the cells of the 2.5′ grid; the interpolated anomalies were then applied to the values for the present climate. Finally, projected future values for the bioclimatic variables were calculated for the cells of the 2.5′ grid and used to estimate projected future values for these variables for the cells of the 1° grid and for the IBAs, using the same procedures as were used to estimate their present values.

16.3 Results

16.3.1 Species distribution models

Models were fitted for the 1608 species with five or more recorded presences for the 1° grid. Goodness of fit analyses of the CRS models revealed that the overall median *k*-fold partitioned mean AUC for the 1608 species was 0.950, with the majority of species modelled (86%) having *k*-fold partitioned mean AUC values > 0.9 ('high' performance – Swets, 1988) and only four species with *k*-fold partitioned mean AUC < 0.7 ('low' performance). Figure 16.2 illustrates the performance of models for two exemplar species, one widespread (*Merops hirundineus* A. A. H. Lichtenstein, 1793 – swallow-tailed bee-eater) and the other a biome-restricted

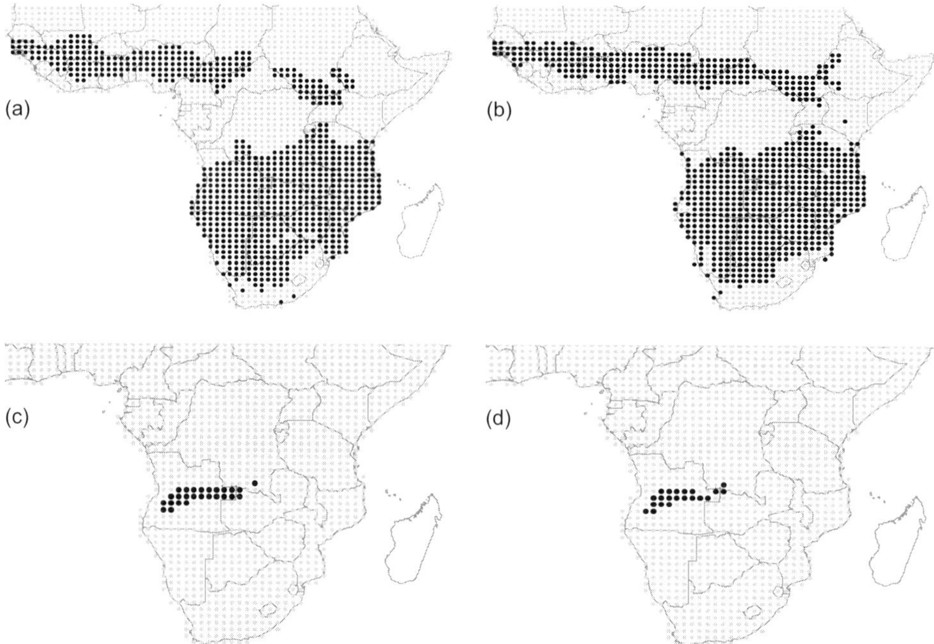

Figure 16.2 Climatic response surface (CRS) models for two exemplar species: (a, c) recorded distributions, and (b, d) distributions simulated by CRS models, illustrating model performance. (a) and (b) show the widespread *Merops hirundineus* (swallow-tailed bee-eater) (mean *k*-fold AUC 0.955); (c) and (d) show the biome-restricted *Ploceus temporalis* (Bocage's weaver) (mean *k*-fold AUC 0.998).

species (*Ploceus temporalis* (Bocage, 1880) – Bocage's weaver) found only in the Zambesian miombo woodland biome, principally in Angola.

16.3.2 IBA species inventories

Species inventories that were likely to be close to being complete were obtained for 64 IBAs representing five countries. Comparison of these with the inventories of species simulated by the CRS models potentially to occupy those IBAs on the basis of their climatic characteristics alone revealed a generally good match (Table 16.1), although the simulated inventories included larger numbers of species than had been recorded. However, the median sensitivity of the simulations, a measure of the success of the models in simulating the presence of species recorded as present, was 0.882, indicating that the majority of model errors were errors of commission (i.e. simulated presence of species not recorded as present). Furthermore, the median value of 0.721 for the TSS, a measure that takes into account both errors of commission and errors of omission, indicates a generally very good overall match between the simulated and recorded inventories.

Table 16.1 Comparison of observed and simulated IBA species inventories.

Country	IBAs with 'complete' recorded inventories	Median sensitivity	Median TSS
Gabon	7	0·777	0·705
Ghana	7	0·803	0·700
Malawi	9	0·849	0·688
Uganda	10	0·887	0·551
Zambia	31	0·920	0·786
Overall	64	0·882	0·718

The simulated presence of species not recorded from the IBAs is not unexpected, given the relatively small size of most IBAs relative to the 1° grid cells for which data were used to fit the CRS models. Species are increasingly likely to be absent from smaller areas as a result of lack of suitable habitat, as has been shown by modelling species distributions at various spatial grains using habitat variables as well as bioclimatic variables in the models (Luoto et al., 2007). Given, in addition, that many IBAs comprise only one or a small number of habitat types that often are relatively restricted in occurrence in the wider landscape, it is to be expected that species inventories simulated on the basis only of the suitability of the climate will include species for which the IBA does not provide suitable habitat.

16.3.3 Potential impacts of projected future climatic changes

At the level of individual species, projected future climatic changes are expected to result in changes in distribution, many species' range boundaries potentially shifting by hundreds of kilometres by the end of the present century (Huntley et al., 2006). Across all 1608 modelled species and all three GCMs the ensemble mean value of D by the year 2085 is 592 km, although extreme values are in excess of 4000 km (HadCM3: min = 9.2 km, mean = 629.7 km, max = 4678.5 km; GFDL-R30: min = 5.0 km, mean = 489.8 km, max = 3796.9 km; ECHAM4: min = 6.7 km, mean = 656.9 km, max = 4806.3 km). These very large shifts in potential range location are accompanied by an overall reduction in the relative extent of species' ranges. Median R for 2085 is 0.902 for the 793 species not in priority categories and only 0.738 for the 815 priority species, indicating a significantly greater potential impact on priority species (Mann-Whitney U: $z = -4$, d.f. = 1524, $p < 0.001$). Furthermore, median O for 2085 is 0.563 for the species not in priority categories and only 0.315 for the priority species, again indicating a significantly greater potential impact on priority species ($t = 12.6$, d.f. = 1524, $p < 0.001$). These potential shifts and changes in extent of species distributions are reflected by substantial potential changes in the species inventories simulated for IBAs. Across all the IBAs, the median turnover for all species is potentially 10–13% by 2025, increasing to 20–26% by 2085, the ranges reflecting the results obtained using the future climatic conditions projected by the three GCMs. If only priority species are considered, median turnover values are much higher: 18–21% by 2025 and 35–45% by 2085.

Examination of the turnover values for individual IBAs reveals considerable variation, while mapping the ensemble mean values across the three GCMs (Fig 16.3) reveals a degree of systematic geographic patterning of this variation. High turnover, for example, is more prevalent amongst the IBAs of the southern tropical zone, with low turnover, in contrast, being characteristic of many IBAs in the Guinea–Congo forest region. The achievement of any turnover, but especially of the higher rates of turnover, will require many species to shift their distributions, often over long distances, across the wide landscape between IBAs. If these movements are achieved, however, then the network has the potential to support in future the majority of the bird species of sub-Saharan Africa. Depending upon which GCM's projection of future climate is used, for 88–92% of priority species at least one of the

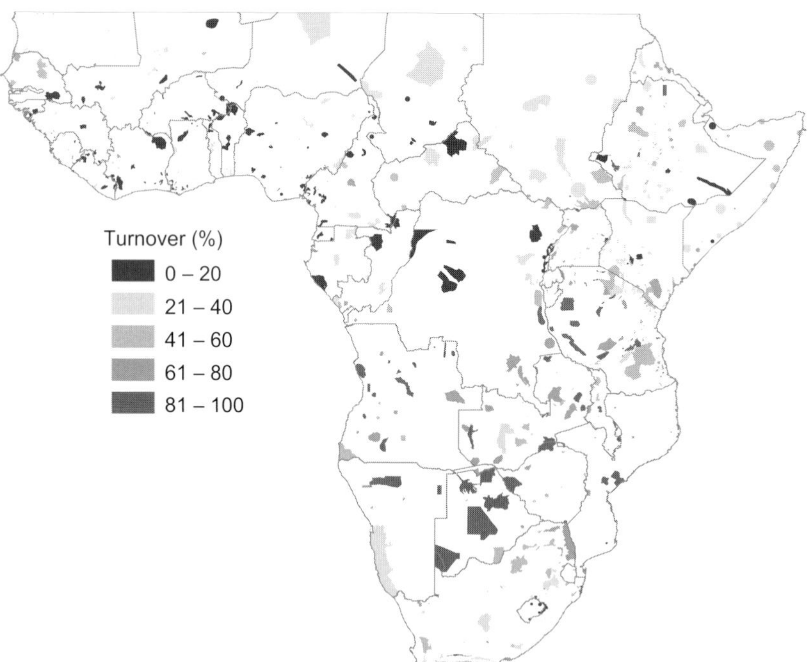

Figure 16.3 Mean potential turnover of priority species by 2085. Mean of potential turnover (%) of priority species in each IBA simulated for the climatic conditions projected by the three GCMs. *See colour plate section.*

IBAs simulated to offer suitable climatic conditions in 2085 is amongst those for which the species contributes to the basis for the site's current designation. At least one IBA in the network is simulated to offer suitable conditions in 2085 for a further 8–11% of priority species, although none of the IBAs for which these species contribute to the current basis for designation continues to offer suitable conditions. Only for *c.* 1% of species are none of the IBAs simulated to offer suitable climatic conditions in 2085. In other words, even if species are unable to achieve the range shifts required to adapt to climatic change, the network potentially may still support most of the sub-Saharan avifauna at the end of the present century.

These encouraging figures, however, must be balanced against the simulation of 51–55% of IBAs in which priority species contribute to the current basis for designation no longer providing suitable climatic conditions for those species by 2085, as well as the potential marked reduction in range extent for priority species, to a median of only 74% of present range extent. Thus, if species were unable to achieve the range shifts required for adaptation, the number of IBAs in which each priority species is present would decrease by about half. Furthermore, even if species did fully achieve their potential range shifts, there would still be an overall decrease in species richness of individual IBAs throughout the network as a result of a reduction in the number of IBAs in which each species is simulated potentially to occur. Expressing the number of IBAs simulated as suitable in 2085 as a proportion of those simulated as suitable at present gives median values, across the three GCM scenarios, of a 15% reduction in the number of suitable IBAs for priority species and a 10% reduction for species not in priority classes.

Figure 16.4A illustrates the geodesic distance between each cell of a 0.25° grid and its nearest neighbour IBA in the current network. While IBAs are relatively dense in some parts of the study area, the map highlights some areas where the current network is extremely sparse, with nearest neighbouring IBAs > 500 km distant, and thus distances of > 1000 km between IBAs. The potential future value of areas is shown in Fig 16.4B, which illustrates, as an ensemble mean across the three GCMs, the number of priority species for which each 0.25° grid cell is newly suitable in 2085. Also shown are arrows indicating, for those 1° grid cells in the upper quartile with respect the number of priority species for which they are newly suitable, the mean direction of movement required for priority species to colonise newly suitable grid cells from the nearest previously suitable grid cell.

16.4 Discussion

Although previous authors have highlighted the potential for climatic change to result in species moving out of protected areas in which they occur at present (Araújo et al., 2004; Hannah et al., 2007), our simulations of priority species'

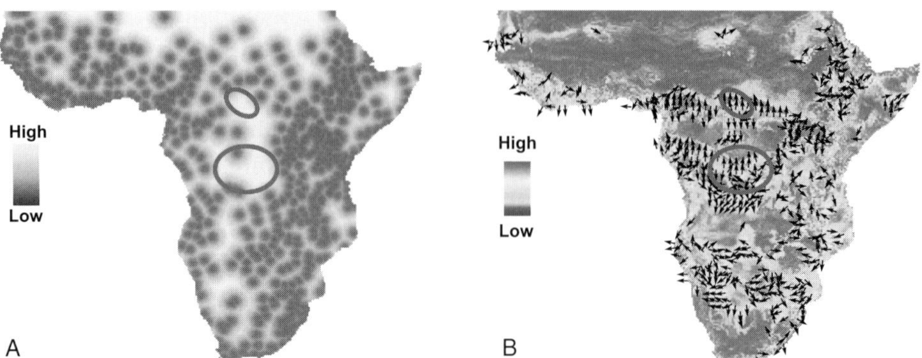

Figure 16.4 Identifying priority areas for network enhancement. (A) Shading indicates the distance between the centre of each cell of a 0.25° longitude × latitude grid and the centre of its nearest neighbour in the IBA network (maximum 669 km, minimum < 1 km). (B) Shading indicates the absolute number of priority species for which each cell of a 0.25° longitude × latitude grid is potentially newly climatically suitable in 2085, having been simulated as unsuitable in 2055. The numbers are ensemble means for three GCM projections (maximum 100, minimum 0). Arrows indicate the mean direction of movement required for species to colonise 1° grid cells that are simulated as potentially newly climatically suitable from the nearest previously suitable grid cell (arrows are plotted only for grid cells in the upper quartile with respect to the number of potential colonising species). The two red ellipses on each panel highlight areas where the present network of IBAs has particularly low connectivity but that potentially will be newly suitable in 2085 for a large number of species that will thus be attempting to shift their ranges across the area. *See colour plate section.*

potential future patterns of occurrence in the sites of a continent-wide network of designated sites, for a range of projections of climatic changes resulting from a moderate greenhouse gas emissions scenario, indicate that this network has the potential to sustain almost all of the species it was designed to protect, at least until the late twenty-first century. However, this positive result must be tempered by the marked reduction in the number of sites in the network simulated to be climatically suitable for each individual species, and especially for individual priority species. The simulated potential overall reduction in the number of IBAs suitable for individual priority species, and especially the simulated overall reduction in their range extents, indicate an increased risk of their eventual extinction. A reduction in range extent almost inevitably will result in a reduction in population, increasing the species' vulnerability to extinction (Thomas et al., 2004); a reduction in the number of suitable IBAs will similarly increase the species' vulnerability.

Of even greater importance, however, is the dependence of this positive outcome upon the ability of species to shift their ranges in response to climatic change, and hence to occupy IBAs that become newly climatically suitable as a result of climatic change. Across all 1608 species modelled, the ensemble mean simulated shift in potential range centroid by 2085 was 592 km, with maximum shifts approaching or exceeding 4000 km depending upon the GCM considered. Such large potential range shifts result in mean overlaps of species' potential future ranges with their present ranges of around half for non-priority species and of less than a third for priority species. They also result in a substantial potential turnover of species within IBAs by 2085, especially of priority species. The achievement of such turnover, and of the underlying potential range shifts, will depend upon the connectivity of sites in the network and the extent to which species are able to cross, and/or make use of habitat patches in, the intervening wider landscape.

Our assessment of the connectivity of the present network (Fig 16.4A) reveals that for substantial parts of the continent the nearest IBA is > 500 km distant, a distance substantially greater than the dispersal distances many species, even many bird species, are capable of attaining, especially if the intervening terrain is inhospitable. While this in itself need not necessarily represent a major problem if, for example, few species potentially will be expanding their ranges into these areas or the areas between IBAs have little human land use and thus present no particular difficulties for species moving through them, this is unlikely to be the case for all of the areas where the current network has low connectivity. Figure 16.4B shows the number of species for which each cell of a 0.25° grid potentially will become newly suitable by 2085, and also illustrates, for those 1° grid cells in the upper quartile with respect to the number of potential colonists, the mean direction of movement of the ranges of those potential colonists. The two red ellipses on the panels of Fig. 16.4 highlight two examples of areas where the present IBA network has low connectivity but that are in the upper quartile with respect to potential colonising species by 2085. These areas, into and across which large numbers of species potentially will be shifting their ranges this century, are examples of areas where augmentation of the current IBA network is likely to be necessary if these potential range changes are to be achieved. Without such augmentation of the network it is likely that realised rates of species turnover within IBAs will be less than the simulated potential rates and, as a result, that the future effectiveness of the network will be compromised.

The approach that we have developed and applied has general utility. It provides a basis for similar assessments of other networks and of their effectiveness in conserving the biodiversity of other taxonomic groups. In order to apply this approach, however, it is necessary to have some understanding of the bioclimatic variables that are likely to determine the distributions of members of the taxonomic group to be examined, as well as a sound taxonomy for members of the group and reasonably reliable information about each species' geographical distribution. Although such data are becoming available for vertebrates and higher plants in most areas of the globe, it is likely to be some time before sufficient data are available to permit such an approach to be applied to most invertebrate groups, bryophytes or lichens in any other than the most intensively studied geographical regions.

Although much can thus be learned from the judicious application of current models, further work is needed to go beyond the current generation of climatic envelope models. In particular, the development of models that incorporate the dynamics of the population and dispersal processes upon which species depend to realise potential range changes will allow exploration of the spatial and temporal dynamics of these potential range changes (Barnard and Thuiller, 2008; Huntley et al., 2010). Nonetheless, the development and implementation of biodiversity conservation strategies that will be effective in a world with rapidly changing climatic conditions is urgent and cannot await the 'last word' in model development. These strategies, however, must be adaptive (Sutherland, 2006), being subject to review and modification in the light of evidence as to their effectiveness, as well as in response to new evidence of the rates and magnitudes of climatic changes and new insights into the potential nature and rates of species' responses to these changes. Our results provide an illustration of how current models can be applied to assess the likely future effectiveness of a network of protected areas, and also to highlight areas where the network is in most urgent need of augmentation if its future effectiveness is to be maximised. Our results show that the model network we examined, that of the IBAs in sub-Saharan Africa, potentially will be effective in sustaining avian biodiversity throughout the present century despite the impacts of anthropogenic climatic change. Many of the sites in this network, however, do not at present have any formal protection; the effectiveness of the network is likely to be seriously compromised if such protection is not implemented as a matter of urgency throughout the network. Our results also highlight that this effectiveness is likely to be compromised if urgent steps are not taken to augment the network in some areas where it currently has very low connectivity, and draw attention to the need to manage the wider landscape in ways that are more sympathetic to wildlife and that will enhance its permeability (Huntley, 2007). Finally, as we discuss elsewhere (Hole et al., in press), they provide the basis for developing management strategies for individual sites in a network that take account of each site's potential future role in the overall network rather than being developed in isolation for each individual site, as is often the case at present.

Acknowledgements

The research reported was supported by a contract from the Royal Society for the Protection of Birds. We are grateful to Carsten Rahbek for making available the one-degree gridded species distribution data for sub-Saharan Africa. Literally hundreds of ornithologists, birdwatchers, conservationists and others interested in birds and the wider environment have contributed information upon which the African Important Bird Areas network is based, and continue to do so. We also thank the BirdLife Africa Partnership, UNEP-World Conservation Monitoring Centre, Paul Britten, Graeme Buchanan, Richard Dean, Tim Dodman, Kenna Kelly, Pete Leonard and Ian Tarplee for assistance in assembling IBA GIS polygons. Individual IBA species data were kindly provided by Derek Pomeroy, Herbert Tushabe and Françoise Dowsett-Lemaire.

References

Allen, J. R. M., Long, A. J., Ottley, C. J., Pearson, D. G. and Huntley, B. (2007). Holocene climate variability in northernmost Europe. *Quaternary Science Reviews,* **26**, 1432–1453.

Allouche, O., Tsoar, A. and Kadmon, R. (2006). Assessing the accuracy of species distribution models: prevalence, kappa and the true skill statistic (TSS). *Journal of Applied Ecology,* **43**, 1223–1232.

Araújo, M. B., Cabeza, M., Thuiller, W., Hannah, L. and Williams, P. H. (2004). Would climate change drive species out of reserves? An assessment of existing reserve-selection methods. *Global Change Biology,* **10**, 1618–1626.

Barnard, P. and Thuiller, W. (2008). Global change and biodiversity: future challenges. *Biology Letters,* **4**, 553–555.

Beale, C. M., Lennon, J. J. and Gimona, A. (2008). Opening the climate envelope reveals no macroscale associations with climate in European birds. *Proceedings of the National Academy of Sciences of the USA,* **105**, 14908–14912.

Beerling, D. J., Huntley, B. and Bailey, J. P. (1995). Climate and the distribution of *Fallopia japonica*: use of an introduced species to test the predictive capacity of response surfaces. *Journal of Vegetation Science,* **6**, 269–282.

BirdLife International (2000). *Threatened Birds of the World*. Barcelona and Cambridge: Lynx Edicions and BirdLife International.

Brooks, T., Balmford, A., Burgess, N. et al. (2001). Toward a blueprint for conservation in Africa. *BioScience,* **51**, 613–624.

Burgess, N. D., Fjeldså, J. and Rahbek, C. (1998). Mapping the distributions of Afrotropical vertebrate groups. *Species,* **30**, 16–17.

Cleveland, W. S. and Devlin, S. J. (1988). Locally weighted regression: an approach to regression analysis by local fitting. *Journal of the American Statistical Association,* **83**, 596–610.

Cohen, J. (1960). A coefficient of agreement for nominal scales. *Educational and Psychological Measurements,* **20**, 37–46.

Cubasch, U., Meehl, G. A., Boer, G. J. et al. (2001). Projections of future climate change. In *Climate Change 2001: the Scientific Basis,* ed. J. T. Houghton, Y. Ding , D. J Griggs et al. Cambridge: Cambridge University Press, pp. 525–582.

Davis, A. J., Jenkinson, L. S., Lawton, J. H., Shorrocks, B. and Wood, S. (1998). Making mistakes when predicting shifts in species range in response to global warming. *Nature,* **391**, 783–786.

Doswald, N., Willis, S. G., Collingham, Y. C. et al. (2009). Potential impacts of climatic change on the breeding and non-breeding ranges and migration distance of European *Sylvia* warblers. *Journal of Biogeography*, **36**, 1194–1208.

Environmental Systems Research Institute (ESRI) (1998). ArcInfo, version 7.2.1. Redlands, CA: ESRI.

European Project for Ice Coring in Antarctica (EPICA) Community Members (2004). Eight glacial cycles from an Antarctic ice core. *Nature*, **429**, 623–628.

Fielding, A. H. and Bell, J. F. (1997). A review of methods for the assessment of prediction errors in conservation presence/absence models. *Environmental Conservation*, **24**, 38–49.

Fishpool, L. D. C. and Evans, M. I. (2001). *Important Bird Areas in Africa and Associated Islands: Priority Sites for Conservation*. Barcelona: BirdLife International and Lynx Edicions.

Fronval, T. and Jansen, E. (1997). Eemian and early Weichselian (140–60 ka) paleoceanography and paleoclimate in the Nordic seas with comparisons to Holocene conditions. *Paleoceanography*, **12**, 443–462.

Gaston, K. J. (2003). *The Structure and Dynamics of Geographic Ranges*. Oxford: Oxford University Press.

Gordon, C., Cooper, C., Senior, C. A. et al. (2000). The simulation of SST, sea ice extents and ocean heat transports in a version of the Hadley Centre coupled model without flux adjustments. *Climate Dynamics*, **16**, 147–168.

Green, R. E., Collingham, Y. C., Willis, S. G. et al. (2008). Performance of climate envelope models in retrodicting recent changes in bird population size from observed climatic change. *Biology Letters*, **4**, 599–602.

Greenland Ice-core Project (GRIP) Members (1993). Climate instability during the last interglacial period recorded in the GRIP ice core. *Nature*, **364**, 203–207.

Gregory, R. D., Willis, S. G., Jiguet, F. et al. (2009). An indicator of the impact of climatic change on European bird populations. *PLoS ONE*, **4**, e4678.

Hannah, L. (2008). Protected areas and climate change. In *Year in Ecology and Conservation Biology 2008*, ed. R. S. Ostfeld and W. H. Schlesinger. Oxford: Blackwell, pp. 201–212.

Hannah, L., Midgley, G., Andelman, S. et al. (2007). Protected area needs in a changing climate. *Frontiers in Ecology and the Environment*, **5**, 131–138.

Hijmans, R. J. and Graham, C. H. (2006). The ability of climate envelope models to predict the effect of climate change on species distributions. *Global Change Biology*, **12**, 2272–2281.

Hijmans, R. J., Cameron, S. E., Parra, J. L., Jones, P. G. and Jarvis, A. (2005). Very high resolution interpolated climate surfaces for global land areas. *International Journal of Climatology*, **25**, 1965–1978.

Hole, D. G., Willis, S. G., Pain, D. J. et al. (2009). Projected impacts of climate change on a continent-wide protected area network. *Ecology Letters*, **12**, 420–431.

Hole, D. G., Huntley, B., Arinaitwe, J. et al. (in press). Towards a management framework for key biodiversity networks in the face of climatic change. *Conservation Biology*.

Hopkins, J. J., Allison, H. M., Walmsley, C. A., Gaywood, M. and Thurgate, G. (2007). *Conserving Biodiversity in a Changing Climate: Guidance on Building Capacity to Adapt.* London: Department for Environment, Food and Rural Affairs.

Huntley, B. (2007). *Climatic Change and the Conservation of European Biodiversity: Towards the Development of Adaptation Strategies.* Strasbourg: Council of Europe, Convention of the Conservation of European Wildlife and Natural Habitats.

Huntley, B. and Webb, T. (1989). Migration: species' response to climatic variations caused by changes in the earth's orbit. *Journal of Biogeography,* **16**, 5–19.

Huntley, B., Bartlein, P. J. and Prentice, I. C. (1989). Climatic control of the distribution and abundance of beech (*Fagus* L.) in Europe and North America. *Journal of Biogeography,* **16**, 551–560.

Huntley, B., Berry, P. M., Cramer, W. P. and McDonald, A. P. (1995). Modelling present and potential future ranges of some European higher plants using climate response surfaces. *Journal of Biogeography,* **22**, 967–1001.

Huntley, B., Collingham, Y. C., Green, R. E. et al. (2006). Potential impacts of climatic change upon geographical distributions of birds. *Ibis,* **148**, 8–28.

Huntley, B., Green, R. E., Collingham, Y. C. and Willis, S. G. (2007). *A Climatic Atlas of European Breeding Birds.* Barcelona: Lynx Edicions.

Huntley, B., Collingham, Y. C., Willis, S. G. and Green, R. E. (2008). Potential impacts of climatic change on European breeding birds. *PLoS ONE,* **3**, e1439.

Huntley, B., Barnard, P., Altwegg, R. et al. (2010). Beyond bioclimatic envelopes: dynamic species' range and abundance modelling in the context of climatic change. *Ecography,* **33**, 621–626.

Hutchinson, M. F. (1989). *A New Objective Method for Spatial Interpolation of Meteorological Variables from Irregular Networks Applied to the Estimation of Monthly Mean Solar Radiation, Temperature, Precipitation and Windrun.* Technical Memo 89/5. Canberra: CSIRO Division of Water Resources.

Jansen, E., Overpeck, J., Keith, R. B. et al. (2007). Paleoclimate. In *Climate Change 2007: the Physical Science Basis. Contribution of Working Group I to the Fourth Assessment Report of the Intergovernmental Panel on Climate Change,* ed. S. Solomon, D. Qin, M. Manning et al. Cambridge: Cambridge University Press, pp. 433–497.

Juckes, M. N., Allen, M. R., Briffa, K. R. et al. (2007). Millennial temperature reconstruction intercomparison and evaluation. *Climate of the Past,* **3**, 591–609.

Knutson, T. R., Delworth, T. L., Dixon, K. W. and Stouffer, R. J. (1999). Model assessment of regional surface temperature trends (1949–1997). *Journal of Geophysical Research-Atmospheres,* **104**, 30981–30996.

Leemans, R. and Cramer, W. (1991). *The IIASA Database for Mean Monthly Values of Temperature, Precipitation and Cloudiness of a Global Terrestrial Grid.* Laxenburg: International Institute for Applied Systems Analysis (IIASA).

Luoto, M., Virkkala, R. and Heikkinen, R. K. (2007). The role of land cover in bioclimatic models depends on spatial resolution. *Global Ecology and Biogeography,* **16**, 34–42.

McDermott, F., Mattey, D. P. and Hawkesworth, C. (2001). Centennial-scale Holocene climate variability revealed by a high-resolution speleothem $\delta^{18}O$ record from SW Ireland. *Science,* **294**, 1328–1331.

Meehl, G. A., Stocker, T. F., Collins, W. D. et al. (2007). Global climate projections. In *Climate Change 2007: the Physical Science Basis. Contribution of Working Group I to the Fourth Assessment Report of the Intergovernmental Panel on Climate Change,* ed. S. Solomon, D. Qin, M. Manning et al. Cambridge: Cambridge University Press, pp. 747–845.

Nakicenovic, N. and Swart, R. (2000). *Emissions Scenarios. Special Report of the Intergovernmental Panel on Climate Change.* Cambridge: Cambridge University Press.

Pearson, R. G. and Dawson, T. P. (2003). Predicting the impacts of climate change on the distribution of species: are bioclimate envelope models useful? *Global Ecology and Biogeography,* **12**, 361–371.

Pearson, R. G., Thuiller, W., Araújo, M. B. et al. (2006). Model-based uncertainty in species range prediction. *Journal of Biogeography,* **33**, 1704–1711.

Prentice, I. C., Cramer, W., Harrison, S. P. et al. (1992). A global biome model based on plant physiology and dominance, soil properties and climate. *Journal of Biogeography,* **19**, 117–134.

Roeckner, E., Oberhuber, J. M., Bacher, A., Christoph, M. and Kirchner, I. (1996). ENSO variability and atmospheric response in a global coupled atmosphere-ocean GCM. *Climate Dynamics,* **12**, 737–754.

Ruddiman, W. F. (2003). The anthropogenic greenhouse era began thousands of years ago. *Climatic Change,* **61**, 261–293.

Sutherland, W. J. (2006). Predicting the ecological consequences of environmental change: a review of the methods. *Journal of Applied Ecology,* **43**, 599–616.

Swets, J. A. (1988). Measuring the accuracy of diagnostic systems. *Science,* **240**, 1285–1293.

Thomas, C. D., Cameron, A., Green, R. E. et al. (2004). Extinction risk from climate change. *Nature,* **427**, 145–148.

Thuiller, W. (2004). Patterns and uncertainties of species' range shifts under climate change. *Global Change Biology,* **10**, 2020–2027.

Trenberth, K. E., Jones, P. D., Ambenje, P. et al. (2007). Observations: surface and atmospheric climate change. In *Climate Change 2007: the Physical Science Basis. Contribution of Working Group I to the Fourth Assessment Report of the Intergovernmental Panel on Climate Change,* ed. S. Solomon, D. Qin, M. Manning et al. Cambridge: Cambridge University Press, pp. 235–336.

17

Documenting plant species in a changing climate: a case study from Arabia

M. Hall and A. G. Miller

Centre for Middle Eastern Plants, Royal Botanic Garden Edinburgh, UK

Abstract

Plant taxonomy must re-evaluate its outputs in order to be part of an effective response to climate change. Traditional taxonomic works, such as floras and monographs, are not appropriate tools for plant conservation and monitoring programmes. Such outputs need to be more widely supplemented with practical, field-based publications (field guides), which are more suited to providing rapid species identifications in the field. This chapter argues that to be as effective and as inclusive as possible, plant field guides need to be based on images rather than text. Using recent case studies from the Arabian Peninsula, we present a series of practical methods for documenting plant species using digital photography and assess the advantages and disadvantages of digital image-based identification.

Climate Change, Ecology and Systematics, ed. Trevor R. Hodkinson, Michael B. Jones, Stephen Waldren and John A. N. Parnell. Published by Cambridge University Press.

17.1 Introduction: current Arabian climate

The latest climate projections for the Arabian region predict significant change by the end of the twenty-first century. According to Dawson (2007), under a low-emissions scenario (B2a), across much of the region, the mean winter temperature is predicted to have increased by 3 °C and the mean summer temperature by up to 4 °C in 2070–99. In the same period, under a high-emissions scenario (A1f), predictions suggest that the mean winter temperature will have increased by 5 °C across much of the region. The mean summer temperature is likely to increase by up to 6 °C in the south and 7 °C in the north of Arabia. Under both high- and low-emissions scenarios, the projected rainfall levels between 2070 and 2099 are predicted to be in the category range of 'no change or a reduction in rainfall by up to 0.5 mm per day'. Maximally, this direct reduction in precipitation could equal as much as 180 mm per year.

With significantly elevated temperatures and decreased precipitation predicted by the end of the century, there is a strong possibility that the available moisture across much of Arabia will be radically reduced. In the already water-stressed environments of the Arabian region, any such reduction in available water is likely to further restrict plant growth. However, despite these recent regional predictions of climate change, there are still a number of uncertainties with modelling future Arabian climates. Much of Arabia's topography is mountainous, and the most diverse plant habitats are found in mountainous regions. To date, the climate of these regions is still not adequately represented in large-scale climatic models. As the most recent Intergovernmental Panel on Climate Change (IPCC) report acknowledges, the spatial resolution of global climatic models is not able to accurately reproduce the complex terrain and land cover of mountainous regions, even though these are critical factors in generating the mountain climate (IPCC, 2007). In addition, any projected changes in precipitation in mountainous topography may well be unreliable because the links between topography and precipitation are not adequately represented in most global climatic models.

Not only is the complexity of modelling mountain systems a problem, but the lack of reliable, spatially appropriate data for rainfall, temperature and other important precipitation sources such as mists is also a significant limiting factor in modelling future Arabian climates on scales appropriate for assessing impacts on plant life. As well as these deficiencies in climatic data, there is a lack of detailed distribution data for many plant species. These data are required to formulate accurate predictions of the impacts of climate change on local biodiversity. The Arabian Peninsula is clearly vulnerable to climate change, but climate/ecological projections are limited because of problems in modelling techniques and an absence of adequate modelling data. However, in an environment where the availability of sufficient water is the principal limiting factor for plant growth, a change in mean precipitation is likely to have significant effects.

One particular study highlights the potential vulnerability of plant species to changes in received precipitation. In the southwest of Saudi Arabia, the Asir Mountains are home to 7600 km^2 of *Juniperus procera* Hochst. ex Endl. (juniper) woodland (NCWCD and JICA, 2007). Over the last 15 years, there has been a noticeable and marked process of 'die-back' across a number of discrete *Juniperus* populations. This process is markedly different from the normal ageing and decay of *Juniperus* individuals and is characterised by the synchronised decay and death of the majority of the individuals in a population (Fig 17.1). From repeated fieldwork observations over a number of years (A. G. Miller, unpublished field data), it is clear that the areas of *Juniperus* woodland which suffer die-back are also the areas that receive reduced mists during the winter period. This contrasts sharply with the occurrence of healthy *Juniperus* populations in the regions of the Asir Mountains that still receive regular mists; it appears that the change in the local climate of the Asir Mountains (whether anthropogenic or not) results in die-back. Currently, a research programme using bioindicators is being established to examine the causal factors.

Figure 17.1 Large-scale die-back of *Juniperus procera* woodlands in the Asir Mountains, Saudi Arabia, is associated with climate change.

The decline in *J. procera* populations in Saudi Arabia is an example of the potential for climate change to increase the extinction risk for a species at both a national and a regional level. Globally, recent modelling studies have predicted that between 18% and 35% of all species will become extinct by 2050 as a result of anthropogenic climatic change (Thomas et al., 2004). Predictions of global extinction risk for plants estimate similar levels of loss. The Global Strategy for Plant Conservation estimates that up to 30% of plant species are at risk of extinction (SCBD, 2002). Regionally, within the Arabian Peninsula, over 10% of the assessed endemic species for the International Union for the Conservation of Nature Red List is currently classified as Endangered or Critically Endangered. Any reduction in future water availability to plants is highly likely to increase the extinction pressures on the species listed in these categories.

Climate change is not only a threat at the species level. In Arabia there are a number of particularly vulnerable habitats that are most likely to suffer from reductions in available soil moisture. One of the most striking examples of a vulnerable habitat is the valley forest found in the western escarpment mountains of Yemen and Saudi Arabia (Hall et al., 2008). Patches of valley forest are located in isolated valleys between 500 and 1000 m on Jabal Raymah, Jabal Melhan and Jabal Bura, in Wadi Liyah in Khawlan Ash Sham, in the Haraz Mountains and on Jabal Fayfa (Wood, 1997; Al-Turki, 2004). These valleys are characterised by a west or southwest aspect and receive locally high levels of orographic rainfall in the spring and late summer. While the high local rainfall contributes to the existence of vegetation along the wadi channel (riparian forest), in a similar fashion to the *Juniperus* woodland, the persistence of vegetation occurring on the lower catchment slopes is likely to be aided by the occurrence of winter mists.

The largest area of valley forest in Yemen is found on Jabal Bura. This area has recently received legislative protection in order to conserve the closed forest vegetation. While there are no species that are entirely restricted to the valley forest habitat, it is also an important site for a number of plant and animal species that are endemic to Arabia. Notable plant endemics are the legumes (Fabaceae), *Acacia johnwoodii* Boulos, *Abrus bottae* Deflers and *Ormocarpum yemenense* J. B. Gillett. Important faunal endemics include the Yemen monitor lizard *Varanus yemenensis* Böhme, Joger and Schätti, 1989 and the bright blue agamid lizard *Acanthocercus adramitanus* (Anderson, 1986). Another very important feature of the valley forest is the occurrence of several plant species that are almost entirely restricted to this habitat in Arabia such as *Stereospermum kunthianum* Cham., *Mimusops laurifolia* (Forssk.) Friis and *Piliostigma thonningii* (Schumach.) Milne-Redh.

With any reduction in local precipitation and/or the occurrence of winter mists, this rare closed forest habitat, the plant and animal endemics, and the regionally rare species will all be exposed to heightened extinction pressures. To ensure the survival of this vegetation and these species in Arabia, such change would necessitate the implementation of local in-situ conservation management and monitoring programmes, or if local conditions deteriorated greatly, the implementation of ex-situ conservation initiatives. The necessity of responding to climate change with conservation planning (however successful it may be) provides an important basis for asking a number of relevant questions in a discussion of climate change, systematics and ecology. In a situation where Arabian species and habitats are threatened with extinction, how should plant taxonomy respond? How can botanists specialising in the Arabian region contribute to the conservation of habitats, species and populations that are especially vulnerable to a drying climate?

17.2 Taxonomy contributing to conservation

Mace (2004) has elucidated the link between systematics and the conservation of species:

> Taxonomy and conservation go hand in hand. We cannot necessarily expect to conserve organisms that we cannot identify, and our attempts to understand the consequences of environmental change and degradation are compromised fatally if we cannot recognize and describe the interacting components of natural ecosystems.

To maintain this clear link between conservation and identification, taxonomy must provide ways of identifying species. The correct, efficient identification of a species is the simple, yet fundamental, factor in the initial field observations (of populations, subspecies and species) which indicate that conservation action is required (Mace, 2004). The accurate identification of plant and animal species is necessary for the cycle of conservation activities to commence. In the UK this fact has recently been acknowledged by a House of Lords report into the 'taxonomic impediment' (House of Lords, 2008 – see also Wheeler et al., 2004). For effective conservation to continue, the report recommends a renewed commitment to producing field guides – from users, funders, host institutions and publishers. The reason for this is simple: it is difficult to know which species are in decline if you do not know to which species the individuals in front of you belong.

Quick, reliable identification is not only fundamental to conservation management, it is also crucial for conducting monitoring programmes that aim to assess the occurrence of any directional change in the status of a population or species (Hill et al., 2005). Misidentification is a significant source of bias in sampling, and can lead to monitoring programmes yielding skewed results (Hill et al., 2005).

Despite this importance (both in the UK and in Arabia, and in more biodiversity-rich regions of the world), the provision of accessible identification tools has been given a low priority. This is despite warnings that the limited availability of usable field guides, field keys and other identification tools greatly hinders the assessment of biodiversity (Heywood, 1995).

One of taxonomy's most important contributions to the global climate change crisis must be the provision and widespread dissemination of accessible tools that allow the correct identification of species. Reliable, efficient and accurate species identifications are essential for any programme aimed at conserving species vulnerable to extinction through climate change (and other, synergistic, factors such as habitat loss). They are particularly important for indicator taxa, which may be used to select important areas for biodiversity conservation (Larsen et al., 2009). In addition, correct species identifications are also crucial for compiling the comprehensive plant distribution data that are necessary for accurate modelling of species' responses to climate change. Sufficient data must underpin the models that predict the impacts of climate change on biodiversity (Guisan et al., 2007). Without such data, the accuracy of predicted climate change impacts is significantly reduced.

From both a philosophical and a practical perspective it is no longer acceptable to rely solely on a handful of taxonomic specialists to undertake species identification and distribution mapping (Hill et al., 2005). In Arabia, there are literally only a handful of individuals capable of recording the presence and absence of all the species in a particular habitat. In order to gather accurate and comprehensive species distributions, monitor the changes in plant distributions and population health, and enact management plans for species and habitats under threat, teams of people, rather than individuals, are required. These teams need to be composed of people who are capable of accurately identifying a large number of plant species in the field.

This need is particularly acute during the conservation assessment and monitoring phases of protected area programmes. In the establishment of a protected area network for Saudi Arabia, the administering agency, the Saudi Wildlife Commission (SWC), currently relies heavily on a handful of specialists to collect distribution data, to assess the status of plant species and populations, and to informally monitor changes in plant habitats. Yet because of the vast area of reserves such as Jabal Aja' (2203 km^2), it is necessary to systematically collect adequate data on the occurrence and status of hundreds of plant and animal species, and this requires that the teams of staff assigned to conserve these areas are able to identify and collect data on these species (Llewellyn, 2008). Without such coordinated efforts the collection of adequate distribution data will be practically impossible.

To date, 15 areas, covering almost 4% of the country (an area twice the size of Switzerland), are listed by the SWC as protected areas. A further 27 are proposed, which would bring the total under protection up to 10% of the area of the country. Startlingly, out of this total of 42 protected areas, only 19 are wholly or partially inventoried. Currently, all the georeferenced herbarium records for Saudi Arabia do not provide sufficient distribution data for climate modelling purposes. Much more detailed fieldwork is necessary in order to provide both basic inventories and the species distribution data which are necessary to predict the impact of climate change on these vast protected areas (Guisan et al., 2007).

Having identified this fundamental need for field identifications, we must seriously question whether the current products of plant taxonomy are fit for purpose. Floras, monographs and phylogenetic analyses are the major outputs of taxonomic institutions. Although phylogenetic studies have begun to eclipse the task of flora writing, major flora projects such as the Flora of Arabia, Flora Malesiana and Flora Zambesiaca are still in progress. As Frodin (2001) notes, the main purpose of these floras is to exist as tools for communication, providing basic information and identification. Yet, although it is still fundamental to taxonomic research in the twenty-first century, the flora in its traditional form is not suited to providing quick, reliable identifications in the field for non-specialists. This staple of taxonomic endeavour is not suited to empowering non-taxonomists with the taxonomic knowledge required to implement and conduct species surveys, monitoring programmes or habitat conservation initiatives.

17.3 Flora problems, flora solutions

The initial problem for a non-specialist reader is that floras are exclusive. They are predominantly written in technical taxonomic language, which the average non-taxonomist, including ecologists and park rangers, finds frustratingly time-consuming and difficult to comprehend. To take an Arabian example, a description of *Juniperus phoenicea* L. leaves in the *Flora of the Arabian Peninsula* (Miller and Cope, 1996) reads: 'Mature leaves rhomboid ovate, *c.* 1 mm. Long, closely imbricate with an oblong depressed gland on the dorsal surface; margins denticulate, scarious.' Although plant taxonomists may view this as a concise description, the use of such taxonomic language is undoubtedly a barrier to communication with non-taxonomists. If potential readers persevere and manage to bridge this initial communication barrier, they may soon discover other problems with using floras. They may become aware that the account of a species contains information that is unnecessary for the process of identification, such as nomenclature and specimen data. Many descriptive terms are either ambiguous or poorly defined; as is the case with many descriptions of colour and texture. Although sympathetic readers may well be aware that this is simply a function of having to reduce visual morphological data into written language, this does not make their task any easier. Neither does the fact that the synoptic keys provided lack many vegetative characters, because the account is actually a description of dried specimens and not living plants in the field.

In summary, a fundamental problem with floras is their orientation. While floras may be designed for identification, they are not designed for non-specialists in biodiverse countries who wish to identify living plants in their natural habitats. In their current incarnations, floras act more as working tools for taxonomists and as repositories of taxonomic information. Their lack of design for fieldwork in biodiverse regions extends into the fact that they are too large to use in the field, they are produced very slowly, they often lack the language of the target country, they are too expensive for the majority of interested parties and therefore they have limited circulation.

As Wheeler (2004) notes, taxonomy should make good use of digital technologies, such as online databases, hand-held identification tools and digital photography, in order to respond to the demands of the twenty-first century. In response to the imminent threat of climate change and the taxonomic impediment, we advocate flora projects employing digital technologies to produce photographic field guides aimed at providing swift, effective, accurate plant identification for non-specialists. Moving towards image-based identification tools, which can be easily translated to the internet, immediately opens up access to botanical identification. This should be a fundamental part of the systematic community's contribution to mitigating the effects of global climate change.

In a global context, there are an increasing number of photographic guides aimed at providing accurate field identifications (Gardner et al., 2000; Hawthorne and Gyakari, 2006; Krishen, 2006). In an Arabian context, there are few photographic field guides to the plants of the region. Existing guides suffer from a number of problems similar to the problems of the flora, problems that stem from the fact that they are not designed for systematically identifying plants in the field. Guides such as *The Comprehensive Guide to the Wildflowers of the UAE* (Jongbloed, 2003) and *Field Guide to the Wild Plants of Oman* (Pickering and Patzelt, 2008) are in effect very concise floras supplemented with pictures. In many instances these pictures often serve aesthetic rather than diagnostic purposes and as such do not provide a reliable identification. Although there is a move towards incorporating digital images in botanical guides, this use of non-diagnostic images is common. Existing guides still rely on textual description rather than digital images to provide identification. Another noticeable limitation is the absence of simple keys, or the use of keys which rely entirely on flower colour, a rarely observed character (Pickering and Patzelt, 2008). A lack of clear circumscription of the plants treated by these guides is also a problem for generating confidence in identifications.

17.4 Arabian-region field guide project

Aware of the regional need for photographic field guides, the Centre for Middle Eastern Plants at the Royal Botanic Gardens Edinburgh has been conducting preliminary research into a series of photographic guides in conjunction with partner institutions in the Arabian region (e.g. SWC). The basic aim of this research project is to use digital photography to produce a series of field guides that use images rather than text for identification. This shift from using an exclusive, language-based approach allows non-specialists to identify plants in the field. A parallel aim is to produce guides that transfer basic taxonomic knowledge to countries where such knowledge needs strengthening. This will not only increase the capacity to generate new biological data, it will also empower countries in the region to address their national biodiversity commitments under the Convention on Biological Diversity and to implement monitoring and management strategies in the face of a changing environment (Samper, 2004).

The most prominent feature of the planned guides will be maximum use of diagnostic digital photographs. This will serve two functions: to immediately render the guides accessible to non-specialists and non-English speakers, and to remove much of the ambiguity and subjectivity of text-based accounts. High-quality photographs of living plants retain the maximum morphological information content, thereby increasing the accuracy of identification. The images in the prototype versions of this field guide focus on all diagnostic characters. Importantly for field identification, these include images of diagnostic vegetative characters and of the plant's ecology, including any variations in these characters (Fig 17.2). Diagnostic support will be provided by an internet database of photographs alongside more detailed descriptions and keys. Unlike existing photo guides, images will provide the primary means of identification. Text will be kept to an absolute minimum, which removes problems of translation in an Arabic-speaking region.

The 'way into' the photographic depictions of each species will again differ from the standard taxonomic use of dichotomous keys. Key structure will differ depending on the group of plants being treated, but all the keys in the Arabian field guides will be visual, using diagnostic images from photographs taken in the field. Rather than dichotomising, the keys will display a collection of diagnostic characters, with all species in a particular group grouped on the same page for comparison. While this may not be possible for many regions of the world, in an Arabian context such visual keys provide a quick and efficient way of identification (Miller and Morris, 2004). Another prominent feature of such keys is that they can be easily used in conjunction with any automated identification technology based upon DNA barcoding (Chase et al., 2007).

Figure 17.2 Variation in vegetative characters can be easily demonstrated using digital images.

Originally the proposed guides will focus on clearly circumscribed groups (e.g. trees of Saudi Arabia) and focused areas (e.g. plants of Jabal Qaraqir). This narrow focus minimises the guide size, enabling it to be small enough for field research, and ensures that the information is tailored to specific user needs. Initially the guides will be paper-based; however, with the rapid development of hand-held mobile phone technology and mobile photograph-viewing platforms, future field guides will become increasingly electronic. Electronic guidebooks for mobile phones will be founded upon a web-accessible image database. Similar in nature to projects such as the Virtual Field Herbarium (http://herbaria.plants.ox.ac.uk), the Arabian-region field database will differ by featuring interactive keys and regional mapping software with geo-tagged images. With mobile phones, cameras and GPS systems becoming increasingly sophisticated, it is likely that mobile guidebook users will soon also be able to submit field photographs to the database and geo-tag images of their own. As in South America, where a similar approach to field guides is being developed (T. Pennington, personal communication, 2009), the proliferation of mobile-phone ownership will make these electronic photo guides highly accessible.

Although there are many advantages to using images as the basis of field identification guides, preliminary work in Yemen and Saudi Arabia has revealed a number of disadvantages with this approach. A fundamental problem is related to the number of work hours. As with any taxonomic endeavour, accumulating sufficient high-quality digital images requires significant time and effort. While recognising the preference of users for photographic guides, Lawrence and Hawthorne (2006) highlight the increased publication costs of printing full-colour field guides. For flora projects that are already under-resourced, the commitment to producing digital identification guides is a substantial investment that requires additional funding for the extra staff, fieldwork and equipment involved. Unless this funding is forthcoming, flora projects will struggle to produce these crucial resources.

More technical difficulties occur with an increased reliance on images for taxonomic information. Foremost, in paper-based publication at least, is the problem of representing a range of plant forms, such as a variety of leaf shapes, within the restricted space of the page. Additionally, there is difficulty in accurately identifying cryptic taxa and species associations using digital images. For example, genera in the Compositae (such as *Senecio*) are notoriously difficult to identify because the characters used for identification are hard to see in the field (see Bernado, Chapter 18). Another significant disadvantage is that focusing on compiling a digital image database will be to the detriment of collecting herbarium specimens, and ultimately to the detriment of taxonomic studies in the Arabian region.

In an effort to overcome some of these technical drawbacks, preliminary work in Arabia has focused on developing appropriate fieldwork methods for compiling digital image databases. By conducting rapid digital profiling of all available characters on an individual plant, we have in effect begun to create digital plant specimens of living plants, which are numbered in the same sequential system as other records. The creation of digital specimens during fieldwork periods makes the best use of fieldwork time, because unlike digitised herbarium specimens these field specimens can be used for both photographic guides and taxonomic work. Many of these digital specimens require dissections to be performed in the field in the same way as they would be done in the herbarium, particularly for difficult characters and cryptic species (Fig 17.3). While these take time, the advantage of field dissections coupled with digital photography is that they enable characters to be retained (and easily accessed) that would otherwise be lost in taking traditional herbarium material. In addition, as Knees et al. (2007) have demonstrated, photographs of these dissections are able to provide excellent colour illustrations of plant species, at a fraction of the time and cost of traditional illustrations (see also N. Simpson's digital botanical illustrations at http://nikisimpson.co.uk).

The problem of heavy investment in fieldwork time (e.g. to photograph species that rarely flower) can also be overcome by conducting photographic training

Figure 17.3 Critical characters can be displayed on digital plant specimens by performing dissections in the field.

courses for local researchers to target key taxa. By creating partnerships with local institutions and interested groups, not only does the field guide become a collaborative project, but also the digital image database is filled at a much faster rate and at a much lower cost. Even in the early stages of its compilation, the digital specimen database will be made available online for the Arabian region, and will prove to be a significant resource for botanical studies. Once active, this digital identification resource will also be readily available on any hand-held electronic device with internet access, making interactive digital field guides a possibility even with current technology. Along with a series of interactive keys and descriptions, the digital database will be of great use for local taxonomic studies. Perhaps more importantly, it will provide an accessible tool for making accurate identifications of plants in the field. This basic skill will be critical for assessing, monitoring and modelling the impact of climatic change on the Arabian flora.

17.5 Conclusions

Research into plant field guides is of vital importance for a systematic understanding of, and response to, climatic change. The capacity to identify plants is fundamental to recording their distribution. These distribution data, along with the climatic, habitat and edaphic data, are crucial for modelling the future impacts of climate change on biodiversity, and ultimately on all aspects of human society (e.g. climate envelope modelling – see Rödder et al., Chapter 11). While the taxonomic community is charged with providing tools for plant identification, the major outputs (floras) are not designed to provide field identifications.

The major advantage of producing both print and electronic photographic guides to the plants of the Arabian region is their potential to provide rapid, accurate identifications in the field. Transmitting the capacity to identify plant species and record the status of plant populations is critical for the longevity of Arabian conservation programmes aimed at protecting plant species and habitats. Without training large numbers of people to undertake reliable identification of plant species, the task of monitoring species and habitat responses to climatic change will be extremely difficult. Passing on identification skills to large numbers of park wardens, parataxonomists, conservationists and resource managers is also vital to any serious effort to accumulate sufficient, accurate plant distribution data.

With this in mind, another major advantage of photographic field guides is that they can be used easily as the basis of training courses in botanical research and plant identification. Across the region, the preliminary field-guide research is providing a major resource for a series of training programmes (including training for Nature Iraq's Key Biodiversity Areas Programme) aimed at building capacity in plant taxonomy, for taxonomy students, ecologists, environmental impact assessors and conservationists. The use of photographic guides for training and capacity building points us towards the real importance of research into well-designed field guides (Samper, 2004).

The need for biological data in modelling and in conservation monitoring programmes clearly links the production of accessible field guides and efforts to predict, and mitigate, the impacts of climate change. Yet, while it is relatively simple for a taxonomic institution to highlight these theoretical links, it is currently far harder to manifest such links in the stream of research funding and research prioritisation. Even though policy makers and funding agents are paying increasing attention to the increasingly fashionable topic of 'climate change impact', in Arabia at least there is little awareness of the need to collect large volumes of adequate data on which rational predictions can be made. In an Arabian context, funding for the collection of climatic, topographic and biological data needs sufficient increases to provide credible data for conservation, monitoring and modelling programmes. Increases in funding are particularly needed for training large numbers of people in biological species identification. Yet unfortunately the current situation is unlikely to change until the theoretical, practical and financial links are made between the realities of field observation and data collection, and the overwhelming need to mitigate the impacts of climate change.

References

Al-Turki, R. A. (2004). A prelude to the study of the flora of Jabal Fayfa in Saudi Arabia. *Kuwait Journal of Science and Engineering*, **31**, 77–145.

Chase, M. W., Cowan, R. S., Hollingsworth, P. M. et al. (2007). A proposal for a standard protocol to barcode all land plants. *Taxon*, **56**, 295–299.

Dawson, T. P. (2007). Potential impacts of climate change in the Arabian Peninsula. In *Proceedings, International Conference on Desertification, 12–16 May 2007.* Kuwait: Kuwait Institute for Scientific Research (KISR).

Frodin, D. G. (2001). *Guide to the Standard Floras of the World*, 2nd edn. Cambridge: Cambridge University Press.

Gardner, S., Sidisunthorn, P. and Anusarnsunthorn, V. (2000). *A Field Guide to Forest Trees of Northern Thailand*. Bangkok: Kobfai Publication Project.

Guisan, A., Graham, C.H., Elith, J. and Huettmann, F. (2007). Sensitivity of predictive species distribution models to change in grain size. *Diversity and Distribution*, **13**, 332–340.

Hall, M., Al-Khulaidi, A. W., Miller, A. G., Scholte, P. and Al-Qadasi, A. H. (2008). Arabia's last forests under threat: plant biodiversity and conservation in the valley forest of Jabal Bura'a (Yemen). *Edinburgh Journal of Botany*, **65**, 113–135.

Hawthorne, W. and Gyakari, N. (2006). *Photoguide for the Forest Trees of Ghana: a Tree-Spotter's Field Guide for Identifying the Largest Trees.* Oxford: Oxford Forestry Institute.

Heywood, V. H., ed. (1995). *Global Biodiversity Assessment*. United Nations Environment Programme. Cambridge: Cambridge University Press.

Hill, D., Fasham, M., Tucker, G., Shewry, M. and Shaw, P., eds. (2005). *Handbook of Biodiversity Methods*. Cambridge, Cambridge University Press.

House of Lords (2008). *Systematics and Taxonomy: Follow-up* (S. Sutherland, chair). 5th report of Session 2007–08, House of Lords Science and Technology Committee, HL162. London: Stationery Office.

Intergovernmental Panel on Climate Change (IPCC) (2007). *Climate Change 2007: Synthesis Report. Contribution of Working Groups I, II and III to the Fourth Assessment Report of the Intergovernmental Panel on Climate Change*, ed. R. K. Pachauri and A. Reisinger. Geneva: IPCC.

Jongbloed, M. (2003). *The Comprehensive Guide to the Wildflowers of the UAE.* Abu Dhabi: Environmental Research and Wildlife Development Agency.

Knees, S. G., Laser, S., Miller, A. G. and Patzelt, A. (2007). A new species of *Barleria* (Acanthaceae) from Oman. *Edinburgh Journal of Botany*, **64**, 107–112.

Krishen, P. (2006). *Trees of Delhi: a Field Guide*. Delhi: Dorling Kindersley.

Larsen, F. W., Bladt, J. and Rahbek, C. (2009). Indicator taxa revisited: useful for conservation planning? *Diversity and Distributions,* **15**, 70–79.

Lawrence, A. and Hawthorne, W. (2006). *Plant Identification: Creating User-Friendly Field Guides for Biodiversity Management*. London: Earthscan.

Llewellyn, O. (2008). *Protected Sites of NW Saudi Arabia*. NCWCD Report.

Mace, G. M. (2004). The role of taxonomy in species conservation. *Philosophical*

Transactions of the Royal Society of London B, **359**, 711–719.

Miller, A. G. and Cope, T. A. (1996). *Flora of the Arabian Peninsula and Socotra.* Edinburgh: Edinburgh University Press.

Miller, A. G. and Morris, M. (2004). *Ethnoflora of Socotra.* Edinburgh: Royal Botanic Garden, Edinburgh.

National Commission for Wildlife Conservation and Development (NCWCD) and Japan International Cooperation Agency (JICA) (2007). *The Joint Study Project on the Conservation of Juniper Woodland in the Kingdom of Saudi Arabia. Final Report.* Saudi Arabia: NCWCD.

Pickering, H. and Patzelt, A. (2008). *Field Guide to the Wild Plants of Oman.* Kew: Royal Botanic Gardens.

Samper, C. (2004). Taxonomy and environmental policy. *Philosophical Transactions of the Royal Society of London B,* **359**, 721–728.

Secretariat of the Convention on Biodiversity (SCBD) (2002). *Global Strategy for Plant Conservation.* Montreal: SCBD.

Thomas, C. D., Cameron, A., Green, R. E. et al. (2004). Extinction risk from climate change. *Nature,* **427**, 145–148.

Wheeler, Q. D. (2004). Taxonomic triage and the poverty of phylogeny. *Philosophical Transactions of the Royal Society of London B,* **359**, 571–583.

Wheeler, Q. D, Rave, P. H. and Wilson, E. O. (2004). Taxonomy: impediment or expedient? *Nature,* **303**, 285.

Wood, J. R. I. (1997). *A Handbook of the Yemen Flora.* Kew: Royal Botanic Gardens.

18

A critical appraisal of the meaning and diagnosability of cryptic evolutionary diversity, and its implications for conservation in the face of climate change

J. Bernardo

Department of Natural Resources, Cornell University, NY and *Southern Appalachian Biodiversity Institute, Roan Mountain, TN, USA*

Abstract

Accurate species delimitation is a foundational assumption of biological research. It is especially relevant to conservation, because species names are the currency for conservation policy. Cryptic species are species that are deeply genetically divergent from other such lineages, but that have escaped detection and description because they lack obvious morphological discontinuities. They are not necessarily closely related. Genetic data have revealed surprising amounts of cryptic diversity, which has provoked numerous criticisms concerning their taxonomic recognition and relevance to conservation. I critically examine these and other concerns in the context of a hypothetico-deductive framework (HDF) for species

Climate Change, Ecology and Systematics, ed. Trevor R. Hodkinson, Michael B. Jones, Stephen Waldren and John A. N. Parnell. Published by Cambridge University Press.

delimitation and conclude that they are unfounded. I explore links between taxonomy and systematics with respect to cryptic species recognition, claims about the relative usefulness of morphological versus genetic data for species delimitation, and the kinds of inferential errors that attach to the process of inferring species boundaries. The balance of the chapter shows that the description of cryptic diversity is an important enterprise and considers its implications for conservation biology, especially in the context of global warming.

18.1 Introduction

Biodiversity conservation is a multidisciplinary enterprise that seeks to preserve species diversity in the form of ecologically and evolutionarily viable populations. Importantly, there are deeper dimensions to biodiversity conservation beyond sustaining populations, including preservation of population genetic (Thorpe et al., 1995; Avise, 1996; Crozier, 1997) and phylogenetic (Allen et al., 2009) diversity, maintenance of the ecological and evolutionary processes that regulate diversity (Crandall et al., 2000; Templeton et al., 2001; Moritz, 2002; Mace and Purvis, 2008), and preservation of intact ecosystems, which comprise the matrix for these processes.

A great deal has been written about the way in which our understanding of the amount and distribution of biodiversity influences conservation assessments and practical conservation management of organisms and regions. Amongst the issues being debated are:

(1) whether and how phylogenetic distinctiveness, named or not (such as cryptic species), should be considered in defining conservation priorities, including hotspots (Sechrest et al., 2002; Pérez-Losada and Crandall, 2003; Agapow et al., 2004; Crozier et al., 2005; Avise, 2008; Ceballos and Ehrlich, 2009);
(2) the unevenness of information content about evolutionary distinctiveness represented by a given taxon rank across biodiversity (de Queiroz and Gauthier, 1992; Avise and Johns, 1999; Avise and Mitchell, 2007);
(3) how to assess the relative vulnerabilities of species to extinction (Bernardo et al., 2007; Deutsch et al., 2008);
(4) how to allocate scarce conservation resources to a problem whose scale is well beyond humanity's economic capacity to solve exhaustively (Meiri and Mace, 2007).

One issue pervades this debate, and that is the idea of how we define species conceptually (Huxley, 1940; Mayr 1940, 1942; Simpson, 1943, 1945; Donoghue, 1985; de Queiroz and Donoghue, 1988, 1990; Baum and Shaw, 1995; de Queiroz, 1998,

1999, 2005a, 2005b, 2006, 2007a, 2007b), and how we operationally delimit their genealogical and geographical boundaries (Sites and Crandall, 1997; Wiens and Penkrot, 2002; Sites and Marshall, 2003, 2004). Ecologists, evolutionary biologists, geneticists, physiologists and conservationists have been struggling with this issue since before Darwin's (1859) *Origin*. This discussion is critical for conservation biology, because enumeration of named taxa is the fundamental currency for conservation assessment (Fig 18.1), planning, species and habitat management, legal protections and allocation of limited economic resources (Daugherty et al., 1990, 1994; Agapow et al., 2004; Agapow, 2005; Crozier et al., 2005).

An especially challenging dimension of the species delimitation problem is the diagnosability of cryptic species, highly evolutionarily divergent lineages that share a very similar morphology and which are therefore difficult to distinguish based upon morphological discontinuities. A recent review revealed surprising levels of cryptic diversity distributed uniformly across both phylogeny and geography (Pfenniger and Schwenk, 2007). The phylogenetic and geographical

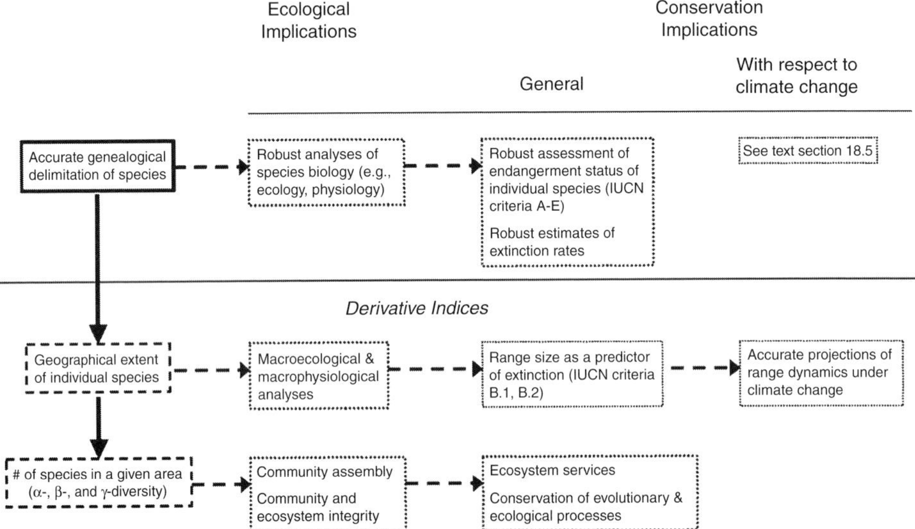

Figure 18.1 The fundamental place of accurate species delimitation in ecology and conservation biology. Accurate species delimitation is a premise of most other kinds of biology. At the level of individual species, species limits are relevant to studies of ecological, physiological and other species traits. Also, assessments of individual species' conservation status (e.g. IUCN, 2001) assume accurate species delimitation. Derivative indices based on counts and distributions of sets of species that form the basis of macroscale comparative ecology, community ecology and dynamics are also fundamentally dependent on accurate species delimitation. In conservation biology, species lists are an important assessment and management tool for identifying global biodiversity hotspots, guiding the design of reserves, and so on.

ubiquity of cryptic diversity reveals greater complexity of ecological and evolutionary processes in determining morphological evolution and generation of biodiversity than previously suspected. Moreover, it also indicates that local, regional and global distributions of biodiversity are grossly underestimated (e.g. Agapow et al., 2004) and biased towards morphologically distinct lineages, and that biodiversity hotspots and reserves are misaligned (Orme et al., 2005; Ceballos and Ehrlich, 2009). Nonetheless, there is a great deal of controversy emerging about the reality and meaning of cryptic diversity, and whether it should even be considered in conservation biology planning and assessments.

The problem of accurately characterising species' genealogical and geographical limits, and consequently patterns of species diversity, has grown in urgency in the face of climate change. While habitat loss has long been the key cause of global biodiversity loss, anthropogenic warming is now considered the dominant threat (Thomas et al., 2004). Long-standing conservation strategies, such as habitat protection, permit ecological recovery of species once protected, but the insidious threat from climate change is that it may impede or prohibit such ecological responses by species that are physiologically (Bernardo and Spotila, 2006) or evolutionarily (Hoffmann and Blows, 1993) constrained. Thus, most species, including many that are not presently of conservation concern will be affected in some way by climate change, including species located in pristine or protected habitats (Deutsch et al., 2008; Bernardo et al., unpublished data).

Organisms will respond in one of three ways to new or magnified stressors (e.g. thermal, salinity, oxygen, desiccation) caused by global change, or face extinction. Firstly, organisms can persist in altered native habitats via a plastic stress response, if the magnitude of the change is within their tolerance zone. Secondly, organisms can track suitable habitat, migrating from deteriorating sites that they can no longer tolerate into more suitable habitats. Thirdly, organisms may evolve their tolerances so as to be able to remain in their altered native habitats. Other evolutionary changes, such as enhanced dispersal behaviour that alters migratory potential (Dytham, 2009), are also possible, as are combinations of these responses. Hence the key to understanding the relative susceptibility of species to climate change is to understand their physiological tolerances, and migratory and evolutionary potential (Bernardo et al., 2007, unpublished data). In turn, the ability to assess these properties of species relies upon accurate species delimitation.

As an example, the emerging field of macrophysiology (Chown et al., 2004; Gaston et al., 2009) has elucidated mechanistic links between physiological tolerance and distribution at the local (Stillman, 2002, 2003, 2004; Bernardo and Spotila, 2006) and landscape levels (Chown and Gaston, 1999; Bernardo et al., 2007). Consequently, macrophysiology has inspired a novel, mechanistic approach to assessing the relative vulnerabilities of species to climate change (Stillman, 2003; Bernardo et al., 2007, unpublished data; Deutsch et al., 2008; Huey et al., 2009).

The potential of such mechanistic approaches to predict species' relative vulnerabilities to climate change rests on accurate characterisation of their genealogical and geographical boundaries.

This chapter is about the implications of cryptic biodiversity for conservation, particularly in the context of climate change. I first critically review the concept of cryptic species, including recent controversies, especially as they pertain to the challenge of conservation. I examine issues such as philosophies and operational criteria for species delimitation; inferential errors pertaining to species delimitation and their relationships to specific concerns that have been raised about cryptic species; and the emergent controversy about the relative phylogenetic informativeness of morphological versus molecular character data for inferring species boundaries. Using these insights, I examine unappreciated ways in which cryptic diversity affects conservation assessments and management, besides the issue of dynamic species lists that troubles many conservation planners.

18.2 Acknowledging, defining and delimiting cryptic species

A recent review (Pfenniger and Schwenk, 2007) found that cryptic species occur uniformly across phylogeny and geography and thus are not restricted to poorly studied regions or groups (see also Knowlton 1993, 2000; Bickford et al., 2006; Trontelj and Fiser, 2009). Therefore, the acknowledgement of the potential for their existence, as well as the detection and delimitation of cryptic species diversity, are not marginal issues for conservation biology, despite claims to the contrary (Chaitra et al., 2004; Isaac et al., 2004; Mace, 2004). Unfortunately, there has been little discussion of the implications of cryptic diversity for conservation beyond the issue of the stabilisation of species lists (Chaitra et al., 2004; Isaac et al., 2004). Other critical dimensions of cryptic species in conservation, such as assessment of individual species or species groups, prioritisation of species or areas for protection, and management strategies that necessarily cross geopolitical borders, have not been explored. The nexus of cryptic diversity and endangerment due to climate change has not yet been examined.

18.2.1 The foundational role of accurate species delimitation in biology, including conservation assessment and management

The issue of species uncertainty has profound conceptual and operational implications for all branches of biology. Some fields have confronted these issues while other fields have largely ignored them. Although on the surface many of these other implications do not appear to be directly relevant to conservation, many dependencies exist (Fig 18.1) and will be discussed throughout the chapter.

These include theoretical and empirical community ecology, which remain rooted in the principle of limiting similarity (MacArthur and Levins, 1967). There is no theory for how ecomorphologically similar, but evolutionarily distinct, species should be able to coexist locally, yet analyses of diverse organisms show this to be common (e.g. flatworms – Casu and Curini-Galletti, 2006; nematodes – Derycke et al., 2008; annelids – King et al., 2008; echinoderms – Boissin et al., 2008; Onychophora – Trewick, 1998; insects – Molbo et al., 2003; Hajibabaei et al., 2006; Roy et al., 2006; amphibians – Tilley et al., 1978; Tilley, 1981; Highton, 2000; Camp et al., 2000; Stuart et al., 2006; lizards – Toda et al., 2001; Gentile et al., 2009; mammals – Baker, 1984; Mayer and von Helversen, 2001; Yoder et al., 2002; LeDuc et al., 2008). Moreover, systematic underestimates of α-, β- and γ-diversity (Whittaker, 1972) introduce insidious error in comparative ecological analyses such as the number of interacting species, the sizes of local and regional species pools from which local assemblages are derived, and estimates of the level of niche diversification or niche saturation (Fig 18.1). Also, the degree to which parasites or phytophagous insects may be host generalists or specialists assumes accurate delimitation of both host and exploiter species, but recent studies of a range of such systems have found that generalists actually comprise assemblages of syntopic cryptic specialists (e.g. Molbo et al., 2003; Hebert et al., 2004; Kankare and Shaw, 2004; Smith et al., 2006; Burns et al., 2008; Smith et al., 2008).

Cryptic species have a wide range of implications for public health and medicine. These include accurate epidemiological analysis and management of pathogens, parasites and their vectors (Cruse et al., 2002; Pringle et al., 2005; Hemmerter et al., 2007), especially those infecting humans and economically important crop or livestock species (Kastin, 2006; Siddall et al., 2007).

Accurate species delimitation, including both genealogical and geographical limits, is a critical assumption of all aspects of comparative biology including macroecology (Brown, 1995; Gaston and Blackburn, 2000; Blackburn and Gaston, 2003), macrophysiology (Chown and Gaston, 1999, 2008; Chown et al., 2004; Gaston et al., 2009), and the comparative analysis of trait evolution. For instance, estimates of the physical distributions of species on the landscape such as range area and latitudinal limits are central to conservation biology, macroecology and macrophysiology (Fig 18.1), and cryptic diversity causes overestimates of these parameters (Riddle and Hafner, 1999). Patterns in biogeography are also fundamental to other areas of biology such as in the evolutionary analysis of range size.

Other implications of cryptic diversity relate to understanding evolutionary process and outcomes, such as quantitative analyses of speciation (Alizon et al., 2008) and extinction rates (Balmford et al., 2003), other analyses of the tempo and mode of evolutionary processes including the number of evolutionary innovations and reversals, the degree of linearity of character evolution (e.g. the overall correspondence between rates of morphological and genetic divergence – Omland, 1997), empirical estimates of lineage accumulation rates (McPeek and Brown, 2007; Adams et al., 2009) and conclusions of tests of niche filling models of diversification (Price, 1997; Harvey and Rambault, 2000; Freckleton et al., 2003). These areas are also pertinent to conservation in the face of climate change, because a key issue is whether species can evolve their physiological tolerances quickly enough to keep pace with environmental change (Hoffmann and Blows, 1993).

18.2.2 Emergent misconceptions and controversies about the importance of cryptic diversity in conservation biology

In the context of the pressing operational problem of assessing how global biodiversity will respond to climate change, the recognition of cryptic diversity has been an unexpected obstacle that has provoked a vigorous debate about what it is, why it matters and even whether it should be acknowledged in conservation assessments and management plans. Amongst the issues raised (Ferguson, 2002; Dunn, 2003; Lipscombe et al., 2003; Isaac et al., 2004; Mace, 2004; Wheeler, 2004, 2008; Meiri and Mace, 2007; Schlick-Steiner et al., 2007; Tattersall, 2007), and considered critically herein, are:

- that cryptic species are a recent phenomenon, arising as an artefact of new species concepts based on molecular phylogenetic analyses, and so are not relevant to conservation;
- that philosophical debates about species concepts, especially the phylogenetic species concept, are distorting traditional taxonomy and destabilising species lists used in conservation; and derivatively, that long-standing taxonomic arrangements should not be upset by phylogenetic analyses revealing cryptic lineage diversity;
- that genetic data are insufficient for delimiting evolutionarily distinctive units, although for some reason morphological features are deemed sufficient even in isolation from other types of data;
- that genetic data are causing taxonomic inflation, that is, unnecessary erection of species binomina for genetic groups that do not merit specific status, apart from their relevance to conservation;
- that recognition of cryptic diversity is dangerous because the increase in species numbers will diminish overall resources available for conservation; rather, that a polytypic species concept should be adopted;
- that only morphological data should be used to delimit species, otherwise it will be impossible to identify species in the field; and that even if genetic data do reveal highly divergent lineages, that one must nonetheless find corroborating

morphological data, even when the evolutionarily distinct groups have long escaped detection via examination of morphological features;

- that whatever the reason for recognising it, cryptic diversity is dangerous for conservation because it causes instability in species lists that are used for establishing patterns of endemism and hotspots for conservation effort.

All of these arguments pertain to the inferential process of delimiting species and supposed biases in this process. Thus, to evaluate them, a thorough and critical examination of the meaning and diagnosability of cryptic species is required.

18.3 Appraisal of cryptic diversity using objective inference and a hypothetico-deductive framework (HDF)

18.3.1 Cryptic species are not a new phenomenon

The accelerating pace of species discovery over the last three decades has been spurred by: rapidly evolving technologies for generating molecular genetic data; a renaissance of phylogenetic thinking coupled with expanding theory and methods for phylogenetic and population genetic inference (Felsenstein, 2004) and increasing availability of computational tools (Felsenstein, 2009); expanding and pluralistic views of species concepts; and heightened awareness of the biodiversity crisis caused by climate change, which has increased field effort aimed at species discovery.

Cryptic species are often mischaracterised as a recent phenomenon arising from the application of molecular genetic analyses and evolutionary or phylogenetic species concepts (e.g. Chaitra et al., 2004; Isaac et al., 2004). While it is true that the pace of discovery of cryptic diversity appears to be increasing as a result of applying molecular tools (Bickford et al., 2006), the recognition of cryptic species was clearly articulated by Darwin (1859):

> Those forms which possess in some considerable degree the character of species, but which are so closely similar to some other forms, or are so closely linked to them by intermediate gradations, that naturalists do not like to rank them as distinct species.

Darlington (1940), however, appears to have coined the term 'cryptic' species. It is also noteworthy that critical discussions about objective criteria for discovering cryptic species based on subtle morphological discontinuities, or discontinuities within other features such as genitalic morphology, life-history features or cytological characters long predate the use of molecular genetic tools for this purpose (e.g. Dobzhansky, 1937a; Lubischew, 1962; Kim et al., 1966).

18.3.2 Species as hypotheses in an HDF

Often lost in debates about species definitions and concepts, and more specifically in critiques of the application of new species concepts, is that species are hypotheses that are subject to falsification (see Bateman, Chapter 3). These hypotheses include phylogenetically explicit definitions of species boundaries, and of hierarchical, nested species groups such as sibling species and species complexes (Table 18.1). This approach is not universally applied, however, and explains some of the reluctance to admit molecular genetic data to the enterprise of species delimitation. Simpson (1943) discussed the failure to recognise species as hypotheses, challenging the prevailing view that morphologically defined taxonomic species are definitive. Rather, he described them as subjective inferences of an objective group, concluding:

> What is really done in classifying organisms is to base, on a series of specimens, a taxonomic species which is a subjective estimate of a morphological species, which in turn is a group of organisms so defined as to approximate a genetic species.

The explicit consideration of species as hypotheses has been advocated before (e.g. Sites and Crandall, 1997; Grube and Kroken, 2000; Highton, 2000; Sites and Marshall, 2003, 2004; Agnarsson and Kuntner, 2007), but the operational advantages of this approach in the context of an HDF have not been explored in detail. These advantages are particularly valuable in the enterprise of diagnosing cryptic diversity, and essential to understanding the criticisms about its recognition.

The treatment of species and species complexes as hypotheses within an HDF makes the process of species delimitation one of strong inference (Platt, 1964). This process, like any application of an HDF, is rendered objective by the establishment of two or more, a priori, alternative hypotheses (e.g. these populations do comprise a distinct species; these species do comprise a species complex). This clarifies the kinds of data required for discriminating between them, which are then used to designate a priori criteria in the form of probabilistic limits (Fig 18.2A). This construct admits multiple types of character data as separate criteria, or, more usefully, specifies a null model expectation for the degree of congruence between data partitions and thus an a priori critical value (e.g. between two different molecular markers, between a molecular marker and some morphological feature, or between morphology and behaviour).

These two attributes (competing a priori hypotheses and a priori criteria for hypothesis discrimination) confer objectivity upon the process of species delimitation. This has several advantages compared with other approaches. One is that species boundaries are subject to revision given new data (Fig 18.2A), a hallmark of the scientific method and strong inference (Platt, 1964). Hence many types of data can be brought to bear upon hypotheses of species boundaries, precluding

Table 18.1 Phylogenetically explicit definitions of species groupings.

Concept	Definition
Sibling species	Sister species (i.e. two descendent lineages that share a most recent common ancestor) that are also morphologically similar to such a degree that their distinct evolutionary trajectories cannot be readily recognised based upon quantitative analysis of morphological features alone, and whose morphological similarity is due to common descent (i.e. homology).
Cryptic species	Species that are morphologically similar to such a degree that their distinct evolutionary trajectories cannot be readily recognised based upon quantitative analysis of morphological features alone. The morphological similarity between or among species could be due to common descent (i.e. homologous character states), due to convergent evolution (i.e. homoplasious character states), or both.
Species complex	A monophyletic group containing more than two independently evolving lineages that are also morphologically similar to such a degree that their distinct evolutionary trajectories cannot be readily recognised based upon quantitative analysis of morphological features alone, and whose morphological similarity is due to common descent (i.e. homology).

data monotypy (the use of a single type of data to delimit species). Thus new data are welcomed under the HDF approach. Related to this point is that when multiple kinds of data are used as criteria, the levels of congruence or incongruence in establishing species boundaries can be objectively compared statistically. The second advantage is that, because this approach permits objective rejection of hypothesised species boundaries or species groups, it also precludes confirmation bias (Loehle, 1987). Thirdly, the objectivity inherent in a strong inference approach in turn contributes to uniformity and logical consistency across taxa, between workers studying the same taxa, and so on. The last advantage of treating species as hypotheses within an HDF is that it clarifies the distinct kinds of inferential errors, and their relationship to each other, which arise during hypothesis testing. In the context of species delimitation these errors are the probability of inferring that too many species exist, relative to how many actually exist (the 'true' but unknowable number), and the probability of failing to recognise species that actually do exist (Table 18.2). Avise (1989) gave examples of these errors, and Highton (2000) discussed them in general terms; I define them formally here in the context of inferential error in hypothesis testing of species

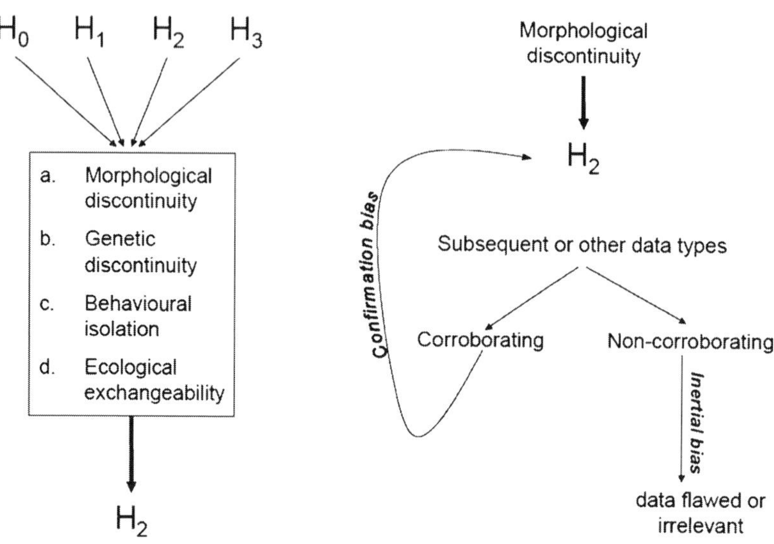

Figure 18.2 Schematic depiction of contrasting philosophical approaches to operationally delimiting species. (A) The HDF for inferring species limits. This approach admits both multiple alternative hypotheses (H_0–H_3) about species limits, which stem from species concepts, as well as multiple types of data, which are actually a priori criteria (criteria a–d in the box). This approach to species delimitation leads to strong inference by eliminating confirmation bias (via the articulation of competing hypotheses) and precludes data monotypy (by defining multiple criteria). (B) Inferential process under primacy or 'inertial' (Good, 1994) approaches for inferring species. Under this view species limits are set solely by historical precedence, typically based on the traditional use of morphological data. Subsequent or other data types are then compared post hoc. If these data are corroborative of defined species limits, they are accepted, leading to confirmation bias. If found to be non-corroborative, the data may in some circumstances be discredited or discarded as irrelevant, which is inertial bias. Although it is not illustrated, any data monotypy approach to inference, such as sole dependence upon DNA barcodes or other DNA-only approaches, follow a similar inferential path.

limits (Table 18.3). This clarification aids greatly in understanding some of the concerns that have been expressed about cryptic species, as well as in providing a logical basis for evaluating those concerns. Because nothing can be known with certainty in science, we rely on probabilistic statements, conditioned on statistical criteria. These criteria are constrained by the observer based upon the related, inherent errors that exist in statistical inference (Cohen, 1977; Toft and Shea, 1983).

Table 18.2 Decision table for hypothesis testing in studies of species delimitation, with associated probabilities. After Toft and Shea (1983).

	Decision after data analysis	
'Truth'	**Accept H_0**	**Reject H_0**
H_0 is true H_1 is false	**No error ($1 - \alpha$)**	**Type I error (α)** This is the probability of mistakenly rejecting a true H_0. In the context of species delimitation, it is the error of recognising more species than actually exist in nature.
H_0 is false H_1 is true	**Type II error (β)** This is the probability of mistakenly failing to reject a false H_0. In the context of species delimitation, it is the error of failing to detect or acknowledge species that really do exist.	**No error ($1 - \beta$)**

H_0 = evolutionarily non-independent lineages that have not, and will not, attain reproductive isolation (i.e. entities that are not distinct species).

H_1 = evolutionarily independent, monophyletic lineages progressing towards, or having attained, reproductive isolation (i.e. entities that are in fact distinct species).

As in other fields, specification of inferential error can guide study design for evaluating species limits in several ways. Firstly, the fact that these two types of inferential error are related to each other clarifies that they trade off (Fig 18.3), which at least can be acknowledged given a particular research design (Toft and Shea, 1983). Thus it is impossible to minimise simultaneously both kinds of error. As we will see below, these two types of error have been raised as specific concerns in the context of cryptic species recognition, although most critiques concern just one of them. Secondly, power analyses (Cohen, 1977; Toft and Shea, 1983) can be used to estimate sampling intensity required to reduce either type of error, although this approach has not yet been deliberately applied to the design of molecular genetic studies of species limits. Thirdly, understanding that these two types of inferential error exist, and are related to each other, permits critical evaluation of the extent of each kind of error by retrospective analyses of taxonomic trends in various groups. Finally, understanding that these two types of error are a property of any kind of logical inference clarifies that they are not a consequence of species concept, data type or other attributes of species delimitation per se, as has been repeatedly claimed.

Table 18.3 Distinct hypothesised errors in species delimitation and their relationships to specific types of inferential error.

Hypothesised bias	Definition and notes	References invoking	References refuting	Inferential error
Taxonomic inflation (TI) (i.e. taxonomic oversplitting)	A hypothesised error in species delimitation in which previously identified species are reclassified (under new species concepts or using molecular genetic data) into smaller and more numerous new taxa. Often an argument cited as evidence of TI is that such taxa are invalid because they are not recognisable based upon morphological discontinuities.	Avise 1989; Chaitra et al., 2004; Isaac et al., 2004; Tattersall, 2007	Agapow and Sluys, 2005; Padial and de la Riva, 2006; this chapter	Type I
Taxonomy–phylogeny gap (TPG) (i.e. taxonomic underdescription)	A hypothesised error in species delimitation in which cryptic species are revealed by systematic analyses but not formally named.	Avise, 1989; Highton, 1990, 2000; Dubois, 2003; Franz, 2005; Perez and Minton, 2008		Type II
Confirmation bias (CB)	A hypothesised error in species delimitation arising from the tendency, outside of an HDF, to accept only corroborative evidence of a historical species concept. CB can result from sampling designs that are constrained by existing taxonomy (Funk and Omland, 2003), and from the way in which new data are applied to testing species boundaries (Fig 18.2B).	This chapter		Type II
Inertial bias (IB)	A hypothesised error in species delimitation arising from the tendency, outside of an HDF, to disregard non-corroborative evidence that falsifies a historical species concept. IB can result from sampling designs that are constrained by existing taxonomy (Funk and Omland, 2003) and from the way in which new data are applied to testing species boundaries (Fig 18.2B).	Good, 1994; this chapter		Type II

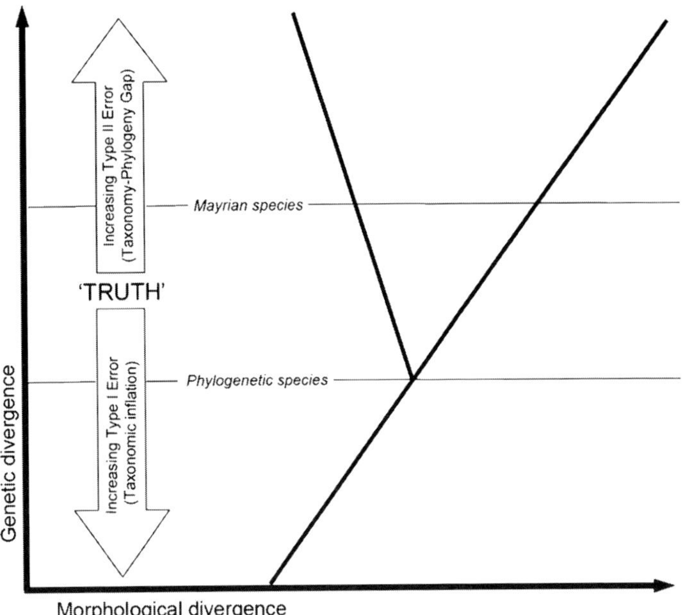

Figure 18.3 Underappreciated, interactive relationships between two types of inferential error and species concepts in species delimitation. If we consider the speciation process depicted as a branching event, the two types of inferential error are seen as increasing in magnitude in opposite directions away from the point at which the two lineages have attained 'true' species status. Thus the trade-off between the two kinds of inferential error is clear in that minimisation of one kind of error necessarily increases the other, except when both converge on the unknowable 'truth'. The two major species concepts recognise the bifurcation event differently. The phylogenetic species concept (PSC) regards the lineages at or shortly after the split as species (distinct evolutionary lineages), whereas the biological species concept (BSC) would not regard the two lineages as species until they have attained reproductive isolation, what I term Mayrian species. Both concepts have some uncertainty about them, but the two concepts can be seen as minimising alternative types of inferential error. The PSC minimises type II error (probability of failing to recognise distinct lineages) but in so doing has an increased type I error rate. Conversely, the BSC minimises type I error (probability of recognising too many species) but therefore accrues an increased type II error rate. Thus, the two species concepts constitute, in a practical sense, upper and lower confidence intervals about the inference of speciation.

In marked contrast to the HDF approach is a widespread but seldom acknowledged alternative approach described by Good (1994) as the 'inertial species' approach (Fig 18.2B). In this view, traditional taxonomic arrangements, typically based upon morphological characters, are considered robust simply because of their age (primacy). Indeed, many critics of molecular genetic data make this argument (Dunn, 2003; Bruce, 2005; Tattersall, 2007). These traditional arrangements thus are not viewed as hypotheses and most have never faced the scrutiny of rigorous tests with alternative data. Advocates of 'stable taxonomy', when confronted by new data, generally accept corroborative data (confirmation bias) but criticise or ignore non-corroborating data (inertial bias - Fig 18.2B, Table 18.3 - Ramey et al., 2006).

18.3.3 Relative informativeness of morphological versus molecular genetic data

Cryptic species are by definition not identifiable based upon morphology. It is not surprising, then, that the most vigorously disputed issue about the taxonomic recognition of cryptic species centres on the relative use of genetic and morphological data both for deciphering species boundaries and as criteria for formally naming them. This issue has many interrelated dimensions, including the relative informativeness of each kind of data, how each relates to various species concepts, and the practicality of the different types of data in species descriptions.

Supposed informative primacy of morphology

Critiques of cryptic species implicitly or explicitly contend that morphological characters are somehow a superior criterion for inferring species boundaries. In inferential terms, this argument is that morphological features are a more phylogenetically informative (less homoplasious) data partition than, say, molecular sequence data. I have always found this argument troubling. It is true that the morphospecies (or phenetic species) concept is as old as binomial nomenclature (Linnaeus, 1758), and that morphological discontinuities are easy for human perceptual systems to detect, which facilitates both field and museum identification. But neither priority nor intuitiveness renders morphology the sine qua non of species' evolutionary discontinuities, although many workers, including prominent conservation biologists, continue to stress this requirement (e.g. Bateman and DiMichele, 2003; Dunn, 2003; Lipscomb et al., 2003; Sites and Marshall, 2004; Wheeler, 2004, 2008; Will and Rubinoff, 2004; Meiri and Mace, 2007; Tattersall, 2007), even those who acknowledge that genetic data have their own resolving power (Beheregaray and Caccone, 2007; Schlick-Steiner et al., 2007).

Detailed conceptual arguments against the primacy of morphology were made by Simpson (1943, 1945), and the empirical evidence that morphology can mislead by both oversplitting and under-recognising diversity is beyond extensive. Morphology may not be more informative, or even as informative, as molecular genetic data about species limits, for two reasons. The first is that there is an intrinsic difference between morphological and molecular genetic characters. Compared with sequence data, which are genotypes, morphological features are phenotypes, and perforce contain an environmental contribution. The $Z = G + E + G \times E$ equation defines the simplest model of phenotypic variance illustrating the relationship of morphological and genetic data to variance components, in which each capital letter corresponds to a variance component (Z = total phenotypic variance; G = total genotypic variance; E = total environmental variance; and $G \times E$ = genotypic by environment interaction variance). Morphological characters, as phenotypes (Z), comprise both genotypic and environmental contributions, whereas molecular sequence data (G) reflect only genotypic variance. This difference has long been recognised in the context of phylogenetics and taxonomy (Simpson, 1945; Avise, 2000, 2007).

The extent to which environmental contributions to morphology influence supposedly diagnostic features seems to have been largely ignored by advocates of morphological primacy. This is surprising, because examples can be found across most metazoa, including plant growth forms (Clausen et al., 1940, 1948, 1958), coral shape and form (Miller and Benzie, 1997; Todd, 2008; Forsman et al., 2009), larval sea urchin size and shape (Strathmann et al., 1992), vertebrate body size and life histories (Bernardo, 1994; Bernardo and Reagan-Wallin, 2002; Bernardo and Agosta, 2003), molluscan shell architecture (Wullschleger and Jokela, 2002; Quesada et al., 2007; Perez and Minton, 2008), limb shape and size (Malhotra and Thorpe, 1997; Losos et al., 2001), and diverse aspects of skeletal and skull structure in fish and mammals (Dunham et al., 1979; Meyer, 1987; Bell and Foster, 1994; Myers et al., 1996). Thus, morphological data are often relatively noisier than molecular genetic markers. This does not mean that morphological features contain no evolutionary signal from the underlying genetic component of the phenotype (Linneaus got a lot right!), only that the signal-to-noise ratio of homologous (genotypic) signal to homoplasious (environmental and convergent) noise is a factor that must be evaluated empirically. This can only be done by comparison with other data.

A further reason why morphology may not be a more phylogenetically informative data partition than molecular genetic data (and may even be worse, that is, more homoplasious) is that it is not a neutral marker. Because morphology is the outward phenotype through which an organism interacts with its environment, it is subjected to the constant scrutiny of selection. Selection can act either to maintain an adaptive phenotype (stabilising selection leading to stasis – Estes and Arnold, 2007), or selection may disrupt morphology, leading to divergent forms capable of exploiting a wider range of resources (Schluter, 2000a, 2000b). Only if morphology were selectively neutral would it be expected to evolve in parallel to the rate of neutral genetic characters, but for many aspects of morphology this is unlikely. Thus, selection can drive morphology in different directions away from a close relationship with genetic divergence (J. Bernardo, unpublished data).

Utility of molecular genetic data in recognising species groups

Although it is not generally described in this way, Mayr's biological species concept (BSC) is in fact a genetic discontinuity criterion for species delimitation (Bradley and Baker, 2001; Baker and Bradley, 2006). Thus, genetic data yield unambiguous tests for reproductive isolation. Examples include fixed allelic differences between syntopic, cryptic salamander species (Tilley et al., 1978; Camp et al., 2000; Highton and Peabody, 2000) and syntopic lemurs (Yoder et al., 2002). Molecular genetic data have also resolved decades of century-old taxonomic flogging of groups that were found to be divergent using multiple other lines of evidence such as the tuatara (Daugherty et al., 1990) and well-documented morphological differences between forest- and savanna-dwelling African elephants recognised for more than a century (Roca et al., 2001).

Many recent critiques of genetic data in species delimitation imply that this criterion is an emergent property of modern genetic technology, but the idea is far older. Specifically, the epistemological distinction between genetic compared with morphological discontinuities as a source of evidence in species delimitation predates Mayr's BSC, and was explored in detail by Dobzhansky (1937a, 1937b, 1950) and his contemporaries (Mayr, 1940; Simpson, 1943, 1945). Simpson (1943) wrote: 'In modern taxonomy it is a basic concept that the species in nature is a genetic group.'

These earlier workers, in making the conceptual argument that deep genetic divergence perforce evidences a history of a lack of genetic exchange, did so without much empirical genetic data from natural populations. Nonetheless, this idea is now well substantiated by a range of empirical reviews (Avise and Aquadro, 1982; Thorpe, 1982; Johns and Avise, 1998; Avise and Walker, 1999; Highton, 2000; Bradley and Baker, 2001). Evidence of a historical lack of genetic exchange has come from divergence and reciprocal monophyly of mitochondrial DNA (mtDNA), deep divergence of nuclear sequences, fixed allelic differences in allozyme studies, extreme allele frequency differences at multiple allozyme loci and extreme departures from Hardy–Weinberg equilibrium. For instance, comparisons across major vertebrate classes of variability in mtDNA cytochrome-b divergence within species (Avise and Walker, 1999) and between species (Johns and Avise, 1998) find generally strong concordance of molecular partitions with previously inferred taxonomic boundaries in most groups, although there are strong differences in the level of divergence between species of birds compared to between species within other classes.

Finer-scale analyses of amphibians (Highton, 1990, 2000) and rodents and bats (Bradley and Baker, 2001) have also detected such discontinuities. In both cases, the authors have argued that such empirical approaches, calibrated by reference to the amount of genetic divergence exhibited by otherwise well-supported species, can provide an objective guide to recognising cryptic species, which are revealed as species having statistically significantly higher than expected 'intraspecific' genetic variation compared with the levels found across other species in the same genus (see also an analysis for fishes – McCune, 1997).

It is essential to note that the tool of molecular phylogenetic analyses in species delimitation cuts both ways. When implemented within an HDF, it can also powerfully reject hypothesised subspecies, species, species complexes and other increasingly nested hierarchical groups that were erected without strong statistical support from other data sets (Ramey et al., 2005, 2006). Some examples in which genetic data have reduced the number of evolutionarily significant units (ESUs), subspecies or supposedly endangered species include well-studied large eastern North American snakes (Burbrink et al., 2000; Burbrink, 2001, 2002) and lizards (Miles et al., 2002); charismatic megafauna including tigers (Cracraft et al., 1998) and some giraffe subspecies (Brown et al., 2007); some Madagascan primates (Groeneveld et al., 2009); the purportedly endangered and legally protected Preble's mouse (Ramey et al., 2005); two long known, common and ecologically divergent snipefishes (Robalo et al., 2009); and a supposedly endangered crayfish (Crandall et al., 2009). Lastly, recent studies employing molecular tools and historical museum samples have demonstrated that some supposedly extinct species are in fact still extant in refugia and thus are in need of conservation (snails – Lee et al., 2009; Galápagos tortoise – Poulakakis et al., 2008).

Data monotypy and criteria for delimiting species

I conclude that morphological data are not relatively more informative about species limits than genetic data, and in fact that morphological data may often be rather more homoplasious than genetic data. Continued singular reliance upon, or reviewer biases requiring the use of, morphology in species delimitation owes as much to the historical roots of species delimitation (in an era with fewer and different technologies) and to biases in human sensory modalities (the term cryptic species reflects this anthropogenic bias) as it does to its purported primacy of information content. Thus, the supposed logical primacy of morphology as a type of systematic information is rather a persistent example of logical inertia.

More importantly, it is clear that no single class of data is likely to be uniformly informative about species boundaries. Although data monotypy has been vigorously advocated by both morpholo-centrists (Kelt and Brown, 2000; Dunn, 2003; Wheeler, 2004, 2008) and sequenceo-centrists (Hebert et al., 2003; Tautz et al., 2003; Blaxter, 2004; Hebert and Gregory, 2005), monotypy approaches fail to consider the inferential limitations of any monotypic data set. In view of the discussion of an HDF for evaluating species limits, it is clear that both morphological and molecular genetic data, as well as many other data types, including experimental tests of reproductive isolation and cytogenetic, behavioural, acoustic, ecological or pheromonal data, all have a place in species delimitation and phylogeny estimation (Fig 18.2). Rather, the formal consideration of multiple data types in an HDF permits statistical evaluation of data partition congruence or incongruence, and the weight of the multidimensional data set will elucidate species boundaries. Much of the acrimonious debate about cryptic diversity, an extension of this larger discussion, is made moot by the HDF approach to species delimitation (Sites and Marshall, 2003, 2004; Lee, 2004; Wiens, 2004).

To be clear, like the careless application of any powerful tool, the phylogenetic signal arising from molecular genetic data is also susceptible to a variety of issues that must be carefully considered. Molecular genetic markers can mislead if they reflect sex-biased dispersal, introgression or unanticipated recombination or selection (instead of being neutral), or if nuclear mitochondrial pseudogenes exist (Song et al., 2008). A less widely recognised issue is that genetic data can also mislead based upon limited sampling, taxonomically constrained sampling and poorly dispersed sampling. Moreover, analytical methods applied to molecular genetic data vary widely in their robustness and assumptions, meaning that poorly analysed data sets often lead to erroneous conclusions (Moritz and Cicero, 2004; DeSalle et al., 2005; Song et al., 2008).

18.4 Getting the taxonomy of cryptic diversity right

Discussions about the recognition of cryptic diversity commonly identify two distinct concerns: either that too many species or, less often, that not enough species are being recognised. These two concerns are actually related kinds of inferential error, as discussed earlier, and, both individually and synergistically, they are especially relevant to conservation.

18.4.1 Taxonomic inflation?

Critics have often charged that the use of a phylogenetic species concept (Agapow et al., 2004) as a criterion for establishing species limits simply results in conferring specific status upon geographically distinct (allopatric) demes (that may have been formally labelled as subspecies) within what really should be viewed as polytypic species (e.g. Chaitra et al., 2004; Isaac et al., 2004; Meiri and Mace, 2007; Tattersall, 2007). These and similar critiques have also often asserted that genetic data are being used in isolation, or arbitrarily. This supposed indiscriminate splitting and the consequent increases in species diversity are known as 'taxonomic inflation' (TI), which is commission of a type I error (Table 18.3). If true, these are serious concerns for taxonomy and for conservation practice. Is TI rampant, and if so, what are its causes?

Firstly, contrary to the above critics, it is important to note that the concept of TI is a phenomenon that is completely divorced from the type of data being used to delimit species, or species concepts per se; it is a type of inferential error that pertains to any system of species delimitation (Table 18.3). Thus, morphological taxonomists have also 'oversplit' many taxa, including tigers (Cracraft et al., 1998), giraffes (Brown et al., 2007), corals (Miller and Benzie, 1997), tortoises (Parham et al., 2006), and the archetypal example of Merriam's (1918) allocation of North American brown bears to 82 species.

Secondly, the charge that genetic data are conducive to this kind of 'indiscriminate splitting' is also tenuous. Molecular genetic data have often in fact served as an objective criterion that has exposed TI based upon morphological species delimitation. Thus the power of molecular genetic data to reject supposed species is a powerful counterpoint to this claim, and suggests the opposite: that when implemented in an HDF, a genetic criterion for species delimitation may actually serve to reduce type I error.

Thirdly, it has been recognised for a decade that the level of genetic distinctiveness that characterises species ranks is widely uneven across different organismal groups, which have been erected principally on the basis of morphological discontinuities (Avise and Johns, 1999; Avise and Mitchell, 2007). Species of birds are found to be far more genetically similar to each other than are species of fish or mammals, and reptile and amphibian species are even more genetically distinct, on average (Johns and Avise, 1998). There may also be unevenness due to historical differences in research effort (Harris and Froufe, 2005). For instance, tropical bird species exhibit greater interspecific genetic disparity than do their temperate counterparts (Chek et al., 2003). Avise and Johns (1999) noted that such variable taxonomic practice undermines comparative assays, such as comparative endangerment.

This unevenness of evolutionary history reflected by current taxonomy suggests either that birds exhibit higher rates of morphological to molecular evolution than other vertebrate groups (i.e. rate incongruence), or that they are taxonomically inflated (i.e. type I error has been high in avian species delimitation) relative to other vertebrates. By contrast, if birds are not oversplit, then there is much legitimate but undiscovered evolutionary diversity in the other groups (i.e. there is rampant type II error in taxonomic partitioning of non-avian vertebrates), as suggested by current rates of species discovery especially in mammals (Ceballos and Ehrlich, 2009) and amphibians (Table 18.4). Hence there is a logical and objective basis for the increase in species recognition in groups in which the current number of nominal species drastically underestimates the phylogenetic distinctiveness of many evolutionarily discrete populations.

Lastly, there is the claim that genetic data are being used in isolation or arbitrarily and that this is causing TI. As discussed earlier, it is true that some authors have argued for data monotypy in taxonomic practice, but I think that the extent of genetic data monotypy is grossly overstated. In reality, very many workers using molecular tools, including some advocates of DNA barcoding, do carefully integrate other kinds of information such as morphology, cytogenetic data, behavioural data, ecological information and so on. Only in this way can congruence or incongruence among data partitions be evaluated statistically and objectively (J. Bernardo, unpublished). For instance, Yoder et al. (2000), Heckman et al. (2006), Craul et al. (2007) and Groeneveld et al. (2009) used DNA sequences and diverse kinds of morphology to clarify species limits in lemurs, and Page et al. (2005) iteratively used DNA and morphological approaches to work out the species limits within a genus of freshwater shrimp. Numerous other examples exist (Suatoni et al., 2006; Gómez et al., 2007; Smith et al., 2008; Toews and Irwin, 2008).

Indeed, most leading evolutionary and taxonomic journals are unlikely to publish species descriptions based solely upon genetic data. Although morphological monotypy has long been the rule in species descriptions, and remains a prevalent and advocated approach (e.g. Dunn, 2003), many critics cannot tolerate the use of genetic data as diagnostic characters in species descriptions (Highton, 1990, 2000). It is high time that species delimitation based upon morphological monotypy is subjected to the same logical standard (see Bateman, Chapter 3).

18.4.2 What's in a name? Systematics, taxonomy and the taxonomy–phylogeny gap

Simpson (1945) stated: 'It is impossible to speak of the objects of any study, or to think lucidly about them, unless they are named.' Several authors have pointed out a growing gap between the pace of detection of phylogenetically distinct species and their formal taxonomic treatment (Highton, 1990, 2000; Franz, 2005). This so-called 'taxonomy-phylogeny gap' (TPG) is the failure to taxonomically recognise species, a type II error (Table 18.3); it is a logical part of the uncertainty inherent in the overall process of species delimitation (Fig 18.3). The TPG is not a new phenomenon, and it derives in part from the long-standing and sometimes tense relationship between the distinct enterprises of systematics and taxonomy. It was the basis of the lengthy discussion by contributors to the first Systematics Association volume (Huxley, 1940), and by Simpson (1943, 1945), amongst others. It has received renewed attention in the face of burgeoning analyses of evolutionary diversity using molecular systematic approaches, as well as by the advent of global conservation science and policy (e.g. the International Union for the Conservation of Nature - IUCN, 2001) in the last few decades, made all the more urgent by the pressing issue of endangerment induced by climate change (Bernardo et al., unpublished data).

Contemporary skirmishes concerning taxonomic practice and the molecular-morphology debate centre on proposals that taxonomy be based purely upon DNA sequence data (Hebert et al., 2003; Tautz et al., 2003; Blaxter, 2004; Hebert and Gregory, 2005) versus counter-arguments that DNA-based phylogenetic analyses are undermining traditional taxonomic practice, that taxonomy should not be subordinated to DNA-based methods, and that taxonomic partitioning should remain rooted in morphological characters (Ferguson, 2002; Dunn, 2003; Isaac et al., 2004; Mace, 2004; Wheeler, 2004, 2008; Meiri and Mace, 2007). It has also been ascribed to the increasing pace of species discovery due to the use of genetic data, new species concepts, and a lack of support for training of specialist taxonomists who can produce morphologically based species descriptions (Mallet and Willmott, 2003; Wheeler, 2004, 2008).

Conceptually these arguments are interesting, especially given the phylogenetic unevenness that is reflected by current taxonomy. However, in the context of conservation, in which species binomina are the currency for assessment, management and legal protection, the connection of taxonomic practice with systematic analyses is an operational, not a philosophical matter. Thus, the interplay of taxonomy and systematics has renewed significance and the hypothesised causes of the TPG require deliberate attention.

Species concepts and the TPG

There has been considerable speculation that the diversity of species concepts is grossly altering our assessments of species diversity, leading either to TI or to increasing the TPG. Related to this debate is the philosophical issue of what species are, biologically versus taxonomically. For instance, Tattersall (2007) decried increases in Madagascan lemur diversity as unreasonable applications of an arbitrary phylogenetic species concept, but ignored many other lines of evidence used in much of this work showing morphological congruence; he also did not acknowledge that genetic analyses have reduced diversity as well (e.g. Heckman et al., 2006). This logical tentativeness of the species category was acknowledged by Darwin (1859), who said:

> In short, we shall have to treat species in the same manner as those naturalists treat genera, who admit that genera are merely artificial combinations made for convenience. This may not be a cheering prospect; but we shall at least be free from the vain search for the undiscovered and undiscoverable essence of the term species.

Recent workers have echoed this view (Hendry et al., 2000; Felsenstein, 2004; but see Avise and Walker, 2000).

Agapow et al. (2004) explored how the phylogenetic species concept (PSC), which arises from phylogenetic analyses of genetic data, would impact conclusions of biodiversity studies. For a range of organisms, they estimated the differences in species number indicated by application of the PSC versus those currently recognised under the dominant BSC approach. Their conclusion, that a PSC would nearly double the number of nominal species for the groups they examined, sparked criticism and consternation, mainly because it was perceived as a cause of TI (Isaac et al., 2004). However, the relationship between TI and TPG (Table 18.3, Fig 18.3) makes clear that the opposite is true, namely that current taxonomy reflects a wide TPG. Interestingly, other workers who have examined the question of how species concept might influence taxonomy have concluded that there is relatively strong concordance between the BSC and PSC, and that current taxonomy is often remarkably robust to applications of PSC (Highton, 1990; Avise and Walker, 1999; Highton, 2000; Lefébure et al., 2006).

Figure 18.3 clarifies that these two species concepts are closely related, and differ principally in the relative evolutionary time at which a name is applied to two diverging lineages. The PSC names lineages at or shortly after they diverge from a common ancestor, whereas BSC names apply only to what I call Mayrian species: divergent lineages that are no longer reproductively compatible. Rather than thinking about these species criteria as alternative concepts, they can be thought of as lower and upper bounds at the breakpoint of speciation. Figure 18.3 also shows how the relative positions of these species concepts relate to the types of inferential error. As the lower bound, the PSC will more often tend to oversplit (i.e. have a higher type I error rate). As the upper bound, the BSC will tend to fail to recognise legitimate species differences (i.e. have a higher type II error rate).

Description reluctance

Highton (1990, 2000) identified another cause of the TPG that is not widely appreciated, which is a reluctance to formally recognise the taxonomic implications of phylogenetic analyses of genetic data. In many papers demonstrating well-supported, deep genetic differentiation of clades within a nominal species, including in cases where paraphyly is demonstrated, and cases in which corroborative lines of evidence exist, there is often an apparent reluctance to describe the lineages as species in the absence of some morphological discontinuity. Highton (2000) provided a detailed analysis of this phenomenon using allozyme data across 27 species of amphibians generated by many different laboratories. In some cases he noted the existence of corroborative data from hybridisation studies, cytogenetic or acoustic differences, yet the authors failed to recognise the cryptic forms revealed by their data.

A recent example was the surprising discovery of a large, pink land iguana in the long-studied Galápagos (Gentile et al., 2009). Nuclear and mitochondrial DNA showed that the large and morphologically distinctive form was genetically remarkably divergent from, and phylogenetically outlying to, the other Galápagos land iguanas, but the authors deliberately refrained from naming the form. Yet they made a convincing argument that the form requires specific conservation and management attention. As another example, Egge and Simons (2006) finally elevated a cryptic catfish species based upon DNA sequence data, although previous studies had shown allopatric populations to be chromosomally and allozymically (fixed differences) distinct.

I contend that this reluctance is rooted in the pervasive inertial view that some morphological difference must be identified in order to formally describe a species. It is also directly related to continued resistance to genetic data in species delimitation. This is not to say that corroborating data are not desirable; I have argued that data monotypy is a bad practice for inference of any kind and that an HDF, which admits multiple types of data, is the strongest way to infer species limits (Fig 18.2). However, there is no logical reason why morphology must be one of the types of data, although most of those who have called for an integrative approach to species delimitation insist that this should be the case (Lee, 2004; Wheeler, 2004, 2008; Wiens, 2004; Schlick-Steiner et al., 2007).

Another possible reason for this reluctance is that descriptions of species generally require not just knowledge about genealogical limits and perhaps other diagnostic characters, but knowledge of the species' distributional limits in space. This knowledge, in turn, clarifies whether genetically distinct forms come into contact (are parapatric), or even overlap in space (are sympatric), and, if so, whether they are reproductively isolated in these contact zones. Detailed knowledge of cryptic species' geographical limits is much more difficult to accumulate because it requires extensive, dense and broadly dispersed geographic sampling. Because demonstration of paraphyly does not require much sampling, the rate at which phylogenetically distinct lineages can be described is outpaced by the rate at which fieldwork to establish their geographic limits is conducted.

There are other examples of the TPG whose causes are less clear, in which multiple lines of corroborating, cytological, physiological or behavioural evidence have long existed, but distinct species were not recognised. For instance, genetic data have corroborated morphological discontinuities recognised within megafaunal species hiding in plain sight including ecologically, behaviourally and morphologically distinctive sympatric killer whales (LeDuc et al., 2008).

Does the TPG matter?

Although some evolutionary biologists have argued that the growing TPG is inconsequential (above), there are important practical issues that arise when cryptic diversity is not formally described that are especially pertinent to conservation biology (Perez and Minton, 2008). The existence of cryptic diversity transcends all the issues we have explored. Once cryptic, deeply differentiated lineages are revealed, our grasp of biodiversity of that geographical region and organismal group inevitably improves. This is true whether the lineages are formally described as species with Linnaean binomina, or are identified by compound descriptors such as the 'clade of species A inhabiting this mountain', as opposed to the 'clade of species A inhabiting that other mountain way over there'. Such descriptors belie the recognition that clade A is not a cohesive, panmictic evolutionary species, and its constituent demes have indeed been named – just not with Linnaean binomina. Finally, note that because TI and TPG are related kinds of inferential error, it should be clear that the growing TPG argues against the hypothesised growth of TI (Fig 18.3).

18.4.3 Case study. Cryptic diversity and the exponential discovery rate of amphibian diversity: taxonomic inflation or something else?

The dramatic increase in nominal species of amphibians over the last decade (Table 18.4) has been cited as an example of TI (Chaitra et al., 2004; Meiri and Mace, 2007). This increase has been occurring despite rapid population declines and extinctions, which makes the question of TI in amphibians an urgent, operational problem (Hanken, 1999; Stuart et al., 2004; Collins and Halliday, 2005; Mendelson et al., 2006; Wake and Vredenburg, 2008). The discovery rate of amphibian species over the last few decades has been explosive, with *c.* 2600 species described since 1985, or a *c.* 61–64% increase in estimated global diversity (Glaw and Köhler, 1998; Köhler et al., 2005 – Table 18.4). An important, unconsidered point is that these values are in fact underestimates, because they reflect only named species; hundreds of other species have been discovered but are not yet formally described (Highton, 2000; Meegaskumbura et al., 2002; Fouquet et al., 2007; Vieites et al., 2009).

This pace of discovery is fuelled by substantial new field effort in poorly surveyed regions, as well as detailed phylogenetic analyses of extant species utilising molecular data and evolutionary species concepts (Tilley and Mahoney, 1996). Similar causes of dramatic increases in mammalian species have also been detailed by Ceballos and Ehrlich (2009). For example, intensive analysis of Madagascan amphibians revealed hundreds of new species of frogs (Glaw and Vences, 2003; Andreone et al., 2008; Vieites et al., 2009). Intensive exploration in India and Sri Lanka revealed dozens of new caecilians (Gower et al., 2004) and hundreds of undescribed species of frogs (Meegaskumbura et al., 2002) including deeply divergent lineages meriting erection of a new anuran family (Biju and Bossuyt, 2003). Fieldwork on Guyanan tepuis similarly yielded numerous new frog species (McDiarmid and Donnelly, 2005; Means and Savage, 2007), including a new family (Heinicke et al., 2009). Other large-scale surveys of Amazonian–Guyanan frog diversity suggest minimum estimates of nearly 130 candidate species, or a *c.* 200% increase over the number described from the region surveyed (Fouquet et al., 2007). Similar work in the Tanzanian highlands and South Africa is yielding many new species (van der Meijden et al., 2005; Menegon et al., 2008).

Surprisingly, the issue of insufficient effort is not confined to remote tropical regions. For instance, biologists have been studying plethodontid salamander diversity of the southern Appalachian Mountains of eastern North America for more than two centuries, have collected hundreds of thousands of specimens and have published hundreds of papers. Yet the region continues to yield extremely geographically localised, highly divergent forms, both morphologically and otherwise, including: *Eurycea junaluska* Sever, Dundee and Sullivan, 1976 from a few streams in extreme southwestern North Carolina and southeastern Tennessee (Sever et al., 1976); *Plethodon petraeus* Wynn, Highton and Jacobs, 1988, a large, colourful, climbing salamander in northwestern Georgia (Wynn et al., 1988); an anatomically distinctive, diminutive new monotypic genus, *Urspelerpes*, phylogenetically outlying in *Eurycea*, known from a single locality in eastern Georgia (Camp et al., 2009); and a localised semi-fossorial species of *Desmognathus* from a dozen localities in northeastern Tennessee (S. G. Tilley and J. Bernardo, unpublished data). Similar discoveries have occurred in other well-studied regions such as *Plethodon asupak* Mead, Clayton, Nauman, Olsen and Pfrender, 2005 in the Klamath region of northern California (Mead et al., 2005) and the first Asian plethodontid, *Karsenia koreana* Min, Yang, Bonett, Vieites, Brandon and Wake, 2005 on the southern Korean peninsula (Min et al., 2005). Despite this explanation for the growth in amphibian species numbers, it is still possible that TI has played a role.

Köhler et al. (2005) attempted to determine the reason for taxonomic growth in Amphibia by analysing the reason for new species described between 1992 and 2003. They found that nearly 86% of the 1190 new species names added were newly discovered species, whereas only *c.* 14% were either elevated subspecies or names removed from synonomy. A further test of this idea was based on the logic that if recently named species result principally from disaggregation of species complexes, then genetic distances between new species should on average be smaller for recently described than for previously described forms. Using a large data set for Madagascan mantellid frogs, they found that average genetic distances between species discovered in the last decades are in fact significantly higher than this value computed for other multidecadal periods of the last century. This means that most new species are highly evolutionarily distinct, but previously unrecognised, not a result of disaggregation of species complexes.

18.4.4 Sampling design strategies for discovering and delimiting cryptic diversity

Elucidation of species' genealogical and geographic limits is an empirical problem. This means that the informative signals for both types of limits depend critically upon the way in which organisms are sampled from nature, before any genetic data are ever obtained. Consequently, inferences about the amount of diversity that exists, its phylogenetic structure and its partitioning across the landscape, may be inaccurate to misleading if sampling across genealogy or geography is weak. Poor sampling may be especially problematic for detecting cryptic species, which by definition are unsuspected and often highly localised. Analytical methods cannot save a study that is based on poor sampling.

Sampling design involves two factors, namely the intensity and dispersion of sampling effort. Both factors apply to each of the two dimensions of organismal sampling (geography and genealogy). Intensity refers to sample size (e.g. number of individuals) and dispersion refers to the distribution of individual samples over a continuum such as a landscape, or across genealogy. A given amount of effort (dictated by time, funding or motivation) can be allocated between intensity and dispersion in many ways. At one extreme, one might include highly dispersed samples with limited intensity at each site; or alternatively one might include relatively few sites but more individuals per site. Increasing sampling intensity generally permits increased dispersion. Once enough samples have been made to adequately cover the design space, additional effort can be allocated in a more targeted fashion depending upon the question of interest.

Careful sampling design is highly pertinent in understanding cryptic diversity and climate change. Cryptic species generally have smaller, often highly localised ranges than the more widespread forms from which they are partitioned. The likelihood of discovering such narrowly distributed (stenotopic) forms depends on sampling design. Climate change will cause changes in range occupation, so an understanding of range limits is imperative in forecasting these effects. A clear understanding of species' geographic distributions clarifies local threats and affects management decisions and legal protection. Cryptic species may differ ecologically or physiologically from their harbouring species, so to evaluate their vulnerabilities to climate change we need an accurate understanding of their genealogical limits. For these and other reasons, deliberate consideration of the dispersion of sampling effort across both genealogy and geography is vital to inform conservation efforts.

Genealogical sampling

I use the term 'genealogical sampling' instead of 'taxon sampling' to more clearly describe the goal of phylogenetic and phylogeographic analyses. Genealogical sampling also de-emphasises the role of taxonomy as a guide for sampling design, which is in fact a hypothesis that is under scrutiny in an analysis of genealogical diversity and structure (Fig 18.2).

The issue of genealogical sampling has received increasing attention in the literature concerning higher-order relationships and the performance of phylogenetic methods, including the issues of gene sampling (organellar versus nuclear markers; number of bases of sequence versus number of markers) and the issue of more bases versus more taxa (Zwickl and Hillis, 2002; Hedtke et al., 2006; Heath et al., 2008a, 2008b). Increasing attention is also being given to the issue of organismal sampling in the context of species delimitation (Templeton et al., 1995; Hedin, 1997; Sites and Crandall, 1997; Omland et al., 1999; Hedin and Wood, 2002; Funk and Omland, 2003; Morando et al., 2003; Sites and Marshall, 2003, 2004; DeSalle et al., 2005; Sinclair et al., 2005; Yoder et al., 2005). Here I specifically consider the inferential limits and inferential errors associated with different approaches to sampling populations within an HDF, especially in the context of detecting and geographically delimiting cryptic species.

Two general classes of bias occur in organism sampling, which relate respectively to genealogical and geographic dispersion. The first stems from the concept of how we define species (Fig 18.2) and might be called an inertial bias (Table 18.3) in the sampling of 'outgroup' or non-focal taxa, relative to 'ingroup' or focal taxa, guided by existing taxonomy. If we take the view that species and other taxonomic groupings (such as species complexes, sibling species etc. – Table 18.1, Fig 18.4) are hypotheses that should be evaluated using a priori criteria, then it follows that population sampling for phylogenetic and phylogeographic analysis cannot be constrained by current taxonomy.

Unfortunately inertial bias often persists in the way in which a focal group is sampled for phylogeographic and phylogenetic analysis (Funk and Omland, 2003). Studies of genetic diversity within a nominal species, or a set of supposedly closely related nominal species (e.g. a hypothesised species complex) often include very dense sampling of these focal groups, but rather sparse sampling outside the group of interest (e.g. Fig 18.4B). Such a sampling scheme presupposes that existing taxonomy is largely informative (Fig 18.2B), that is, that the nominal taxon or species complex is monophyletic or, put another way, that hypothesised species' genealogical and geographical limits are accurate. Inertial bias in sampling can thus lead to confirmation bias in hypothesis testing (Fig 18.2B, Table 18.3).

A serious concern about genealogically restricted sampling is that it may yield a highly statistically supported and apparently rigorous phylogenetic analysis of lineage diversity within the supposed species when some populations were actually paraphyletic (Fig 18.4B). If a hypothesised species or species complex comprises one or more cryptic forms, some of which are closely related, but others of which represent distantly related lineages that have ecologically converged on the same morphotype (e.g. clades $m_4(m_5 + m_6)$ and $m_3(m_1(n + m_2))$ in Fig 18.4), taxonomically guided sampling has a high likelihood of failing to detect the convergence (Fig 18.4B). This is because a phylogenetic analysis, no matter how technically sophisticated with respect to analytical methodology, cannot be informed by phylogenetic diversity that was not sampled, imparting type I error (Tables 18.2, 18.3). Lack of sampling outside the focal group (in our example, lack of samples of 'species' *d*, *e*, *f*, *g* or *h* in Fig 18.4A) prohibits formation of outgroup tree structure in a phylogenetic analysis, within which some populations of the putative species (*m*) might graft (as in Fig 18.4A), thus precluding the emergence of paraphyletic signal from the analysis. In this case, taxonomically biased sampling would recover convergently evolved cryptic species as sister taxa, implying that they are part of a species complex (i.e. a monophyletic group – Table 18.1, Fig 18.4B), whose morphological

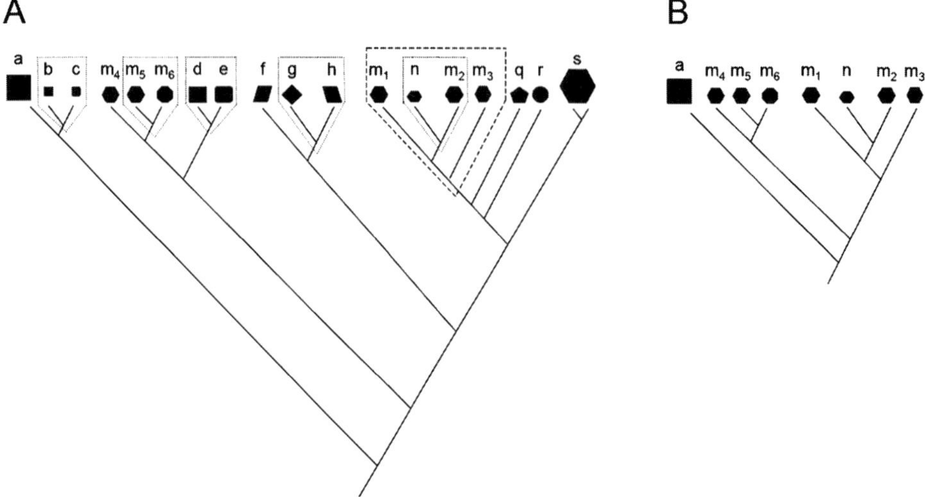

Figure 18.4 Phylogenies and species concepts. (A) A hypothetical phylogeny illustrating the phylogenetically explicit definitions of species groupings defined in Table 18.1, and some other key concepts. This hypothetical phylogeny is assumed to represent a statistically well-supported topology. The shapes illustrate the magnitude of discontinuities in size and shape used to define the current nomenclature of the sampled lineages. Different letters denote nominal species and subscripts denote population samples within nominal species. The dashed line encompasses a monophyletic species complex and dotted lines show sister species as defined in Table 18.1. Note that the m-species complex contains two morphologically similar, nominal species (*m*, *n*) and several populations of the nominal species (m_1, m_2, m_3). The paraphyly of species *m* is revealed by the nesting of species *n* within *m*. The taxonomic implications of this finding would require, for logical consistency and objectivity, either that the nominal species *n* is reduced to synonomy under *m*, or, if there are other compelling data for the specific distinctiveness of *n* (e.g. lack of ecological exchangeability with m_1, m_2 and m_3), that m_2 and m_3 be raised to specific status. Populations m_4, m_5 and m_6 of the nominal species fall well outside this monophyletic group on a widely separated part of the phylogeny, do not share a common ancestor with it, and are more closely related to morphologically disparate species (*d*, *e*, *a*, *b*, *c*) than to other morphologically similar forms. This is convergence. They do, however, comprise their own monophyletic species complex. (B) A flawed phylogeny based upon inertial sampling (i.e. guided mainly by existing taxonomy Funk and Omland, 2003) incorrectly recovers all of the *m* subpopulations as a monophyletic species complex (although it does detect that the morphologically distinct population *n* is actually a member of the *m* lineage). This false positive monophyly of *m* instigates a confirmation bias (Table 18.3). It results from poor genealogical sampling of other, morphologically distinct species (*d*, *e*, *f*, *g*, *h*) assumed to be distantly related to *m*, when in fact these other forms are more closely related to different *m* subgroups than are the *m* subgroups to each other.

or ecological similarity would then be interpreted erroneously as a result of shared ancestry and stasis (clades $m_4(m_5 + m_6)$ and $m_3(m_1(n + m_2))$ in Fig 18.4B). This leads to confirmation bias (Loehle, 1987), a type I error (Table 18.3).

Confirmation bias resulting from inertial biases in sampling for phylogenetic analysis has contributed to the conflation of the meanings of 'cryptic' and 'sibling' species (Table 18.1), that is, to the misimpression that cryptic species are necessarily closely related (e.g. Harrison, 2002; Beheregaray and Caccone, 2007). To avoid inertial bias in genealogical sampling, what is needed is a nested design comprising geographically widespread sampling of each of the congeners for a test of species boundaries, or all the families for a test of genus boundaries, and so on. Moreover, these samples must be dispersed broadly across the landscape to maximise the potential of discovering local forms and to provide ample power for tests of isolation by distance as a null expectation.

Geographic sampling

Geographically restricted or haphazard sampling is the second kind of bias in organismal sampling. While many authors recognise the need to sample multiple populations within nominal taxa, the necessity of dispersing that effort across the landscape is much less widely appreciated. Moreover, the intensity and dispersion of geographic sampling is often haphazard (e.g. biased by accessibility).

Geographically restricted sampling can decidedly mislead phylogenetic analyses and conservation decisions. An excellent example is the case of Preble's mouse, a legally protected subspecies described from a few specimens. Ramey et al. (2005) undertook a comprehensive genetic and morphological analysis of the focal form and sampled both it and numerous populations of all the other subspecies in the nominal taxon. They concluded that Preble's mouse was in fact not a genetically (or morphologically) distinct form. This conclusion provoked numerous critiques (e.g. Vignieri et al., 2006) and a new analysis by another laboratory (King et al., 2006) that concluded, using more genes and longer sequences, but much more restricted population and geographic sampling, that Ramey et al. (2005) were incorrect (Ramey et al., 2006, 2007). This debate, besides consuming a great deal of time, funds and journal space, has direct implications for future allocation of scarce conservation resources.

One issue is that cryptic species often have very small ranges that may not be predictable from physiographic features (Irwin, 2002). There are many examples from mammals (Ceballos and Ehrlich, 2009) and salamanders (Sever et al., 1976; Wynn et al., 1988; Camp et al., 2000, 2009; Anderson and Tilley, 2003). Haphazard geographic sampling or low-intensity sampling can easily miss such localised forms, resulting in type I error (Table 18.2).

A second issue related to restricted geographic sampling that is particularly relevant to conservation is that uncertainty about the geographic range of cryptic species can persist, even if compelling genealogical evidence of cryptic species is recovered. This is because it is far easier to demonstrate from a small number of samples that a particular form comprises multiple highly genetically divergent, even paraphyletic lineages, than it is to describe in detail the geographic limits of the forms and their potential contact zones or areas of overlap. While studies with restricted geographic sampling can contribute to improved species counts for large regions, such as for determining hotspots, the other conservation implications of such studies are limited. For instance, efforts to assess the relative vulnerability to climate change of a set of species require detailed information about each species' distribution on the landscape. How large are their ranges? Do the ranges of cryptic forms come into contact, or overlap, therefore permitting hybridisation? If so, do they hybridise, and how large is the hybrid zone? Hybridisation is relevant both to species recognition decisions and legal protections of populations (see Thomasset et al., Chapter 15). Are the ranges or parts of them within protected areas? What are the geopolitical entities within or across which the species occur?

A general solution to this dilemma is to saturate a landscape with sampling in an iterative approach (Morando et al., 2003). I refer to this approach as 'phylogenetically informed, iterative sampling'. The first stage of sampling should entail dispersing effort across the landscape. The second stage is informed by preliminary analyses of the first stage, which will direct more intensive effort to areas of unexpected diversity, areas in which neighbouring samples are more genetically divergent than predicted from null models (e.g. isolation by distance), or regions in which high levels of haplotype diversity are clustered. For instance, where highly dissimilar haplotypes (compared with overall diversity) are found in close proximity, subsequent dense sampling along a transect between these sites would be indicated (Tilley and Mahoney, 1996).

Sampling design is generally not given the same deliberate attention (empirical or theoretical) afforded to other decisions about analytical methods and assumptions in most published work. This is regrettable, because population sampling design determines the potential of analytical methods to reveal cryptic diversity. Poor sampling design can yield at best a limited view of standing variation, but at worst may lead to erroneous inferences about genealogical diversity and its geographical distribution, including the possibility of ecologically interacting but genetically distinctive forms. The rigour with which authors and reviewers treat the analytical approaches used in phylogeographic and phylogenetic analyses is desirable, but it can give undue credibility to statistically well-supported trees that are based on poor sampling. Because sampling design is not given deliberate attention, insidious errors arising from poor sampling often go unrecognised. Ecological journals generally require that both the design details and the analytical methods of ecological studies be laid out in the methods. A similar requirement for publication of phylogeographic and phylogenetic studies would be a constructive development.

18.5 Implications of cryptic diversity for conservation

Most large-scale conservation planning relies upon derivative indices based upon species' genealogical and geographical boundaries (e.g. Peterson and Navarro-Siguënza, 1999). As examples, species lists are used in recognising areas of importance for conservation (hotspots) that harbour high β- and γ-diversity and endemism, as well as in formal assessments of individual taxa (Fig 18.1). Accurate species delimitation can have repercussions in terms of whether geographic partitions of populations receive legal protection and appropriate management (Avise, 1989; Daugherty et al., 1990, 1994). Consequently, failure to diagnose cryptic diversity in empirical assays of species boundaries, failure to recognise it taxonomically when empirically demonstrated, or failure to consider it in conservation planning and management (Isaac et al., 2004) all have substantive implications for conservation biology in the context of climate change.

18.5.1 Taxonomy–phylogeny gap and conservation status

Although the issue of species definitions and even the system of naming lineages remains contentious, species binomina are still the basic unit for conservation biology in practice (Mace, 2004; Perez and Minton, 2008). Hence the TPG, whatever its varied causes, has distinct operational effects upon conservation. Foremost amongst these is the reality that the differential susceptibility of unnamed cryptic lineages to localised threats such as habitat loss is not contingent on whether they have been named. Furthermore, cryptic unnamed lineages lack legal standing for protection under various legal structures such as the US Endangered Species Act or the Convention on International Trade in Endangered Species (CITES).

Daugherty et al. (1990) drew attention to these problems in relation to the tuatara (an ancient reptile closely related to all lizards and snakes and endemic to New Zealand), in which long-hypothesised taxonomic divisions were ignored in its legal protection and management. This led to extinction of one evolutionarily distinctive set of populations and increased endangerment of another species. Clarification of the cryptic diversity and its formal taxonomic recognition immediately guided a new management plan (Cree and Butler, 1993), leading to aggressive management of the unrecognised species that had dwindled to a single population (Nelson et al., 2002). A similar example exists for the recently discovered cryptic species (as yet unnamed) of Galápagos tortoise that is highly divergent from all other known species, and which is critically endangered (Russello et al., 2005).

The revelation of substantial cryptic diversity within *Giraffa* (Brown et al. 2007) also illustrates both of these issues. Firstly, because giraffes are currently regarded taxonomically as a single, geographically widespread species, they are assessed by IUCN (2001) criteria as having 'lower risk' status. Secondly, the genetic analysis revealed that some unnamed cryptic species are far more endangered due to localised threats such as poaching, and that another candidate species, the West African giraffe, is geographically restricted and numbers no more than 100 individuals (Brown et al., 2007).

Daugherty et al. (1994) illustrated the scale of the TPG for New Zealand herpetofauna. About half of the 65 endemic species recognised in 1994, many of which were recently described cryptic species, are rare, threatened or endangered, and 26 species are largely or entirely restricted to offshore islands. Because these species had been undescribed, only three species had been formally protected. Reduction of the TPG by formal description of the cryptic species provoked development of a comprehensive management scheme by the New Zealand government.

In a final example, Riddle et al. (2000) demonstrated concordant cryptic diversity across a variety of supposedly widespread vertebrate species throughout the Sonoran Desert and Baja Peninsula. The concordant distributions of these cryptic species thus revealed a unique desert fauna within Baja, requiring distinct conservation attention.

18.5.2 Cryptic diversity influences estimates of global and regional diversity, estimates of endemism and other variables used in identifying hotspots

All estimates of species diversity, as well as their use in identifying geographic areas for conservation priority, are derivative indices of accurate species delimitation (Fig 18.1). The ubiquity of cryptic diversity concerns conservation biologists because the species lists used to identify biodiversity hotspots, areas of high endemism and so on are dynamic. One response has been to deliberately ignore cryptic diversity so as to stabilise species lists (Isaac et al., 2004), but the perils of doing so have been carefully detailed (Agapow, 2005; Agapow and Sluys, 2005; Padial and De la Riva, 2006).

Guilhaumon et al. (2008) considered the impact of taxonomic uncertainty (from the different perspective of the form of predictive relationships of species–area curves) on the identification of richness hotspots. They concluded that taxonomic uncertainty dramatically alters the choice of hotspots. Ceballos and Ehrlich (2009) give several examples of how underestimates of mammalian diversity revealed by the discovery of cryptic species have an impact on conservation prioritisation. For instance, they note that 'the distributions of newly discovered mammals often include large areas not considered biodiversity hotspots'.

18.5.3 Cryptic species have cryptic, smaller ranges

In 1859, Darwin wrote, 'But there is also reason to believe, that those species which are very closely allied to other species, and in so far resemble varieties, often have much restricted ranges.' A species' range is relevant to many aspects of its biology, including its conservation status and prospects for enduring climate change (Ohlemüller et al., 2008). A solid grasp of species' range limits is also of consequence to other areas of biology such as macroecology, and especially macrophysiology (Gaston et al., 2009), an approach with direct bearing upon conservation assessments. As mentioned above, species uncertainty also gives rise to range-size and distributional uncertainty, which are a critical concern for conservation biology, for several reasons.

Implications of range size per se

Perhaps the least discussed implication of cryptic diversity is that many cryptic species are more stenotopic than the more widely ranging (eurytopic), genealogically larger entity from which they were partitioned (Fig 18.5 – in salamanders, Tilley and Mahoney, 1996; Highton and Peabody, 2000; in lemurs, Yoder et al., 2000; Craul et al., 2007; and in invertebrates, Starrett and Hedin, 2007; Trontelj et al., 2009). For instance, in their analysis of the properties of 408 new mammal species described since 1993, Ceballos and Ehrlich (2009) found that the new species had significantly smaller average ranges than those of the average land mammal (81 000 km^2 versus 400 000 km^2) and 81% of the new species had ranges < 10 000 km^2. This has several distinct conservation implications.

It has long been recognised that stenotopic species are likely to be more endangered, ceteris paribus, because small ranges expose a species to a higher probability of stochastic extinction. Thus IUCN (2001) specifies range size as a criterion for assessing species' relative endangerment. As an example, a macroecological analysis of extinction-risk correlates of frogs found that range size was negatively correlated with threat status (Cooper et al., 2008).

In the specific context of climate change, restricted ranges can contribute to extinction risk in several distinct ways. A smaller range means that a species has fewer and smaller migration fronts as it tracks shifting habitat. Hence, a physiographic barrier such as a deep dry valley, a wide river or a mountain could be a more insurmountable obstacle for a narrowly distributed species (Fig 18.5A).

A related issue is that stenotopic species also have a smaller edge-to-area ratio than more eurytopic species. Thus, compared with eurytopic species, stenotopic species could suffer greater impacts from contacts with competitive or predatory species as their ranges shift, especially if nearby ecologically similar or superior competitors expand their ranges in response to climate change more quickly (Fig 18.5B). This pre-emption of suitable habitat into which the more slowly responding species would have migrated would mean

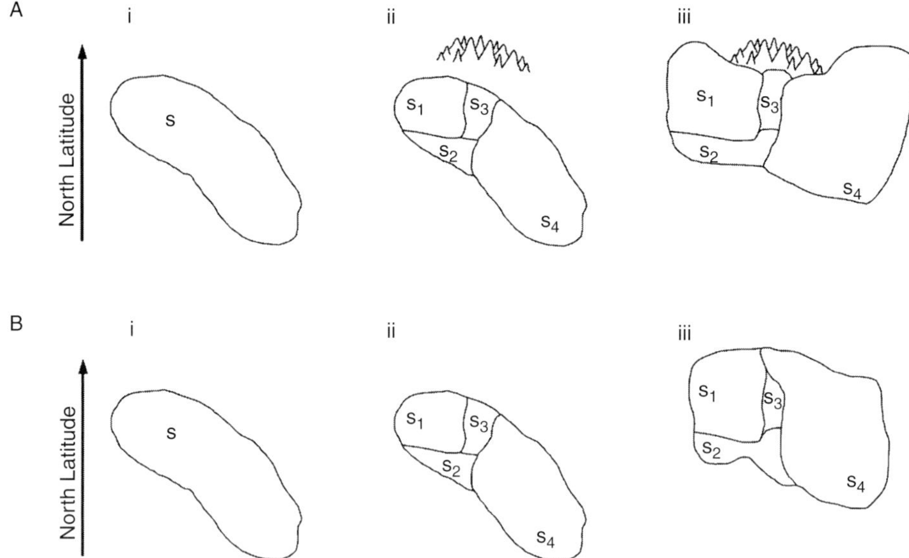

Figure 18.5 Two distinct mechanisms of increased susceptibility to climate-change-induced endangerment in stenotopic cryptic species. In both panels, (i) a single nominal species, *s*, is found to comprise (ii) four parapatric, cryptic species (s_1, s_2, s_3, s_4) with variable but smaller ranges than *s*. (A) As climate change ensues, all species track northwards, during which species s_1, s_3 and s_4 encounter a physiographic barrier (mountain range in this example). Whereas the more eurytopic species, s_1 and s_4, are wide-ranging enough to have migration fronts that are unimpeded, species s_3 has a small northern range margin, the expansion of which is entirely obstructed by the barrier. (B) As A, but species s_1 and s_4 expand more quickly than species s_3. Consequently they pre-emptively occupy favourable habitats north of the current range of s_3, which impedes any further northern range shift by s_3.

that it has to now invade an occupied habitat and endure a possibly restrictive competitive interaction (Price and Kirkpatrick, 2009), as well as the stress of climate change to which it is responding. In contrast, a species with a larger edge-to-area ratio would be more likely to have other migration fronts into unoccupied habitats.

Other consequences of stenotopy for responding to climate change relate to the association between stenotopy and climatic rarity. The observation that large numbers of species with small geographical ranges are often clustered (Orme et al., 2005; Lamoreaux et al., 2006) suggests a relationship between climatic rarity and species rarity. Our work on *Desmognathus* salamanders, for instance, has shown that montane endemics are typically stenothermal in their physiological tolerance, and stenotopic in distribution (Bernardo et al., 2007). In contrast, eurytopic congeners were also eurythermal in their tolerances.

Ohlemüller et al. (2008) used birds, plants and butterflies to evaluate three hypotheses for this coincidence of climatic and species rarity: (1) the climatic rarity hypothesis, that centres of high species rarity are associated with unusual climates; (2) the climatic relict hypothesis, that stenotopic species not only occupy rare climates but that these climates resemble historical climates; and (3) the climatic buffer hypothesis, that stenotopic species are in areas with high climatic diversity. Although there were some differences depending on the group examined, they found that small-range species occur in areas with unusual climates, that are generally higher and colder than surrounding areas. Moreover, rare species tended to occur in areas with local climatic diversity such as high elevation gradients, which could sustain populations through bouts of historic climate change. Their overall conclusion is that because climate warming will reduce the size of these climatically rare areas that sustain high levels of endemism, these areas may be extinction hotspots.

Management implications stemming from distributional uncertainty as distinct from range size

The existence of cryptic diversity means that the geographic limits of species are unclear. For species of conservation concern, this distributional uncertainty amplifies the intrinsic risk arising from the stenotopy that is characteristic of cryptic species, because management decisions are ill-informed.

In contrast, several examples show various ways in which careful geographic delimitation of cryptic lineages clarified local threats, legal protections, legal responsibilities, and informed management. *Ambystoma tigrinum* Green, 1825, an endangered salamander, comprises six highly differentiated allopatric clades. Management of the nominal species entails management of all of these evolutionarily significant components, some of which require more attention than others (Shaffer et al., 2004a). Genetic analyses of endangered desert pupfish revealed both a cryptic species and distinct lineage diversity within one species requiring distinct management needs (Echelle et al., 2000). Delimitation of cryptic diversity within Madagascan mouse lemurs clarified that several stenotopic species are potentially highly endangered by small-scale local logging activity (Yoder et al., 2000). Unsuspected cryptic lineage diversity in endangered graylings, revealed by more comprehensive geographic sampling than in previous studies, indicates a need for substantially different management plans for the species across Europe (Gum et al., 2009; Swatdipong et al., 2009). The last example is from goliath groupers, a large marine fish of subtropical and tropical waters of the Americas and western Africa. This prized food fish, long regarded as a single species, is already considered critically endangered, but it actually comprises three reproductively isolated forms that occur in distinct geopolitical waters with distinct fisheries and requiring specific management plans (Craig et al., 2008).

Understanding of species diversity and limits can also be of critical importance in genetic management of endangered species and captive breeding programmes. For instance, Kong and Li (2009) showed that a commercially important but endangered clam comprises two highly differentiated species that require separate management, and which should not be interbred in captive propagation programmes. Milinkovitch et al. (2007) showed contamination from interspecies hybridisation in a repatriation programme from captive-bred Galápagos tortoises.

Shaffer et al. (2004b) described a situation in endangered red-legged frogs (*Rana aurora* Baird and Girard, 1852 / *Rana draytonii* Baird and Girard, 1852) in which the taxonomic allocation of populations as subspecies with a broad intergradation zone, rather than as genetically isolated species, resulted in a lack of legal protection for many supposedly hybrid populations (which are not protected under the US Endangered Species Act). Detailed phylogeographic analysis, however, showed these populations to comprise genetically pure forms of one or the other species, one of which is endangered and one of which is not. Thus molecular genetic data clarified both the geographic extent of the endangered species and its genetic integrity, and thus the extent of opportunities for conservation.

Our work on green salamanders (*Aneides aeneus* Cope and Packard, 1881) has shown that a previously known disjunct that was of conservation concern is actually composed of two highly divergent cryptic species, which are not each other's closest relatives (J. Bernardo et al., unpublished). Elucidation of the two species' geographical limits clarified that they are both endangered, but to varying degrees. Application of IUCN (2001) criteria showed that one species is critically endangered, whereas the other is threatened.

Another management insight accruing from our analysis was that three different states hold two distinct species of green salamanders. Tennessee harbours two species, neither of which is endangered. Georgia harbours two species, one of which is not threatened, but the other of which is threatened. Finally, North Carolina was found to have both of the endangered forms mentioned above, one threatened and one critically endangered. In each case, the state-specific legal protections and allocation of conservation resources are usefully informed by the understanding of the geographic delimitation of the cryptic species.

A final point is that accurate delimitation of species boundaries can sometimes reveal that supposedly endangered species or subgroups are not in fact evolutionarily distinctive, and thus not deserving of either legal protection or the use of scarce conservation resources.

18.5.4 Cryptic diversity causes systematic underestimation of extinction rates

Much has been written about modern extinctions and the accelerating pace of species loss due to climate change, but there are two distinct reasons why it is rather likely that most estimates of extinction rates from climate change are too low. Cryptic diversity is relevant to both. Balmford et al. (2003) reviewed many flaws associated with extinction rate estimates, but did not discuss the impact of increasing cryptic species discovery rates (Bickford et al., 2006). If we consider the arguments above that cryptic species often have small ranges, and that small ranges make species more endangered due to climatic rarity and limited opportunities for range shifts, then it seems likely that baseline rates of extinction are higher for cryptic species than for the taxonomically well-known species described to date that have been used to estimate extinction rates.

Cryptic diversity is also an insidious error contributing to underestimation of extinction rates, because a great deal of cryptic diversity has likely been lost already without having been documented. An increase in species discovery rates is illustrated by supposedly well-studied groups such as mammals (Ceballos and Ehrlich, 2009), in which 408 new species have been discovered since 1993 (an increase of *c.* 10%). Of these, 60% are cryptic species. Similarly, the number of described amphibian species since 1993 has increased by nearly 43% (Table 18.4), even as extinction rates of amphibians are increasing to unprecedented levels (Wake and Vredenburg, 2008). The amount of cryptic diversity remaining to be discovered in other groups that are far less studied is likely to also be very high.

18.5.5 Cryptic diversity complicates efforts to assess relative endangerment of different species

It is likely that cryptic species also differ intrinsically in properties that affect their vulnerabilities to extinction induced by climate change. Accurate species delimitation, therefore, permits accurate assessments of species traits such as physiological tolerances and relative dispersal ability (e.g. Bernardo et al., 2007).

Cryptic diversity may mask physiological diversity, and hence differential susceptibility to climate change

One dimension of cryptic diversity that has not received attention in conservation is that cryptic species may be physiologically distinct (Spicer and Gaston, 1999), and these differences could influence their relative susceptibility to climate change (Bernardo et al., 2007; Chown and Gaston, 2008; Pörtner and Farrell, 2008). Although some cryptic species are revealed as allopatric populations occupying similar habitats in different areas, such as on different isolated mountains (Highton and Peabody, 2000), there is growing evidence of habitat differences among cryptic species. Examples include cryptic species of a supposed holoplanktonic copepod specialised to contrasting salinity regimes (Chen and Hare, 2008); cryptic flatworms specialised to high- versus low-energy intertidal environments (Casu and Curini-Galletti, 2006); cryptic ascidian species segregated between sheltered interior harbour sites versus more exposed littoral sites (Tarjuelo et al.,

Table 18.4 Estimates of global species diversity of amphibians and the cumulative rate of increase to the present.

Year	Minimum (maximum) estimated number of species	Minimum (maximum) cumulative % increase since 1985
1985	4003[a]	
1992	4533[b]	13.24
1993	4522[a]	
1995	4780[c]	19.41
1997	4970[d]	24.16
2003	5723[b]	42.97
2004	5605[d] (5743[e])	
2005	5828[f]	45.59
2006	5918[e]	47.84
2008	6260[e]	56.38
2009 (August)	6433[g] (6570[h])	60.70 (64.13)

[a] Hanken (1999)
[b] Köhler et al. (2005)
[c] Glaw and Köhler (1998)
[d] Vences and Wake (2007)
[e] www.iucnredlist.org/amphibians
[f] Vences and Köhler (2008)
[g] http://research.amnh.org/herpetology/amphibia
[h] http://amphibiaweb.org

2001); dramatic elevational segregation between cryptic species of tropical salamanders, with two forms occurring below 1000 m and all others occurring above 2000 m (Parra-Olea et al., 2005); and bathymetric differentiation between cryptic species of rockfish (Hyde et al., 2008).

In our work on cryptic salamanders, we have explicitly examined physiology of various cryptic forms because our prior analyses of other species showed that species' physiological differences were good predictors of both range size and genetic structure across six species (Bernardo et al., 2007). We have found striking physiological differences amongst eight cryptic species previously subsumed under three taxonomic names (J. Bernardo et al., unpublished). Based on our prior analysis showing that physiology affects both ecological occupation of the landscape and the amount of genetic exchange across the landscape, we consider these differences highly relevant to understanding the relative susceptibility of these species to climate change.

Unrecognised physiological diversity and specialisation undermines environmental niche models

The dominant approach to forecast species' responses to climate change has been to predict future ranges using environmental niche modeling (ENM), in which current distributional records are used to estimate a fundamental (biophysical) niche that is then projected onto scenarios of future environments. This epiphenomenological approach has been criticised because it disregards biotic interactions, landscape complexity and many other well-established determinants of species' ranges (Austin, 2007; Botkin et al., 2007; Beale et al., 2008; Chown and Gaston, 2008; Kearney and Porter, 2009; Rödder et al., Chapter 11).

Cryptic diversity is another, but previously unrecognised, issue that undermines ENMs, namely their implicit assumption that species delimitation is accurate. Errors accrue from unverified museum databases with inaccurate or outdated taxonomy that are used to populate ENMs. Furthermore, because cryptic species may differ fundamentally in physiological or ecological properties that are relevant to how they will respond to climate change, the existence of cryptic diversity will result in an overestimation of the biophysical range that characterises a nominal species, and therefore the overprediction of suitable habitat that is projected to exist for a nominal form under climate change.

18.6 Conclusions

The growing pace of discovery of cryptic species diversity has upset traditional taxonomic inventories and conservation thinking, provoking several pointed critiques of the very concept of cryptic species. Derivative concerns relate to the taxonomic recognition of genetically partitioned lineages, insistence upon morphological primacy and inertial species concepts, ad hoc rejection of evolutionary species concepts, and operational concerns that cryptic species discoveries will undermine conservation management and policy.

I have critically examined these concerns by casting the process of species delimitation within an HDF, which admits multiple, competing hypotheses about species limits, and multiple types of data as criteria for addressing them. I have shown that concerns about the discovery of cryptic species are really misconceptions stemming from an incomplete view of the inherent uncertainties that attend the process of species delimitation (as they do any other inferential enterprise conducted within an HDF). An inertial view of species concepts and taxonomic practice, and the perceived role that morphological data are the sine qua non criterion for species discrimination, have also contributed to these mistaken impressions of cryptic species diversity. From this analysis I conclude that critically inferred cryptic species are real evolutionary entities worthy of receiving species status and formal taxonomic recognition, a conclusion that has many implications for understanding species' vulnerabilities to climate change.

As a result, in contrast to recent critics, I argue that the detection and geographical delimitation of cryptic species are essential to sound conservation biology policy and practice, and I have provided suggestions for sampling strategies to maximise the probability of detecting them in phylogenetic analyses. The likelihood of discovering localised, cryptic, evolutionarily distinctive lineages relies upon unbiased genealogical and geographic sampling, which generally entails high-intensity sampling. Accurate delimitation of the geographic extents of natural genealogical groups is also intimately related to sampling intensity and dispersion across the landscape.

The conservation importance of recognising cryptic species is not simply that they are unrecognised diversity. Because many cryptic species are found in areas not currently considered biodiversity hotspots, their discovery may radically alter the number and distribution of conservation priority areas such as biodiversity hotspots and reserve networks, which is a sobering finding for current conservation planners. The detection and recognition of cryptic diversity is also essential to inform practical management to avoid interbreeding of distinct lineages in captive breeding programmes, to protect overall lineage diversity, and to understand local threats that may severely endanger previously unrecognised cryptic species.

Lastly, cryptic species may be relatively more endangered than many previously described species owing to their generally smaller ranges. Small ranges are already associated with endangerment, but in the context of climate change, stenotopic species may experience special threats, both because of their climatic specialisation and because the range dynamics of stenotopic versus eurytopic species are likely to differ during climate-change-induced shifts. Cryptic diversity is a reality in our efforts to understand the scope of global biodiversity, its distribution and its relative vulnerability to climate change.

Acknowledgements

I thank M. von Gordon for editorial assistance, and M. Hare, T. R. Hodkinson and S. Waldren for helpful comments. This research was supported by NSF 90-01587, DEB 94-07844 and by Cornell University. This is contribution No. 5 of the Southern Appalachian Biodiversity Institute (SABIonline.org).

References

Adams, D. C., Berns, C. M., Kozak, K.H. and Wiens, J. J. (2009). Are rates of species diversification correlated with rates of morphological evolution? *Proceedings of the Royal Society of London B*, **276**, 2729–2738.

Agapow, P. M. (2005). Species: demarcation and diversity. In *Phylogeny and Conservation*, ed. A. Purvis, J. L. Gittleman and T. M. Brooks. Cambridge: Cambridge University Press.

Agapow, P. M. and Sluys, R. (2005). The reality of taxonomic change. *Trends in Ecology and Evolution*, **20**, 278–280.

Agapow, P. M., Bininda-Emonds, O. R. P., Crandall, K. A. et al. (2004). The impact of species concept on biodiversity studies. *Quarterly Review of Biology*, **79**, 161–179.

Agnarsson, I. and Kuntner, M. (2007). Taxonomy in a changing world: seeking solutions for a science in crisis. *Systematic Biology*, **56**, 531–539.

Alizon, S., Kucera, M. and Jansen, V. A. A. (2008). Competition between cryptic species explains variations in rates of lineage evolution. *Proceedings of the National Academy of Sciences of the USA*, **105**, 12382–12386.

Allen, B., Kon, M. and Bar-Yam, Y. (2009). A new phylogenetic diversity measure generalizing the Shannon index and its application to phyllostomid bats. *American Naturalist*, **174**, 236–243.

Anderson, J. and Tilley, S. G. (2003). Systematics of the *Desmognathus ochrophaeus* complex in the Cumberland Plateau of Tennessee. *Herpetological Monographs*, **17**, 75–110.

Andreone, F., Carpenter, A. I., Cox, N. et al. (2008). The challenge of conserving amphibian megadiversity in Madagascar. *PLoS Biology*, **6**, e118.

Austin, M. (2007). Species distribution models and ecological theory: a critical assessment and some possible new approaches. *Ecological Modeling*, **200**, 1–19.

Avise, J. C. (1989). A role for molecular genetics in the recognition and conservation of endangered species. *Trends in Ecology and Evolution*, **4**, 279–281.

Avise, J. C. (1996). Introduction: the scope of conservation genetics. In *Conservation Genetics: Case Histories from Nature*, ed. J. C. Avise and J. L. Hamrick. New York, NY: Chapman and Hall, pp. 1–9.

Avise, J. C. (2000). *Phylogeography*. Cambridge, MA: Harvard University Press.

Avise, J. C. (2007). Twenty-five key evolutionary insights from the phylogeographic revolution in population genetics. In *Phylogeography of Southern European Refugia*, ed. S. Weiss and N. Ferrand. Dordrecht: Springer, pp. 7–21.

Avise, J. C. (2008). Three ambitious (and rather unorthodox) assignments for the field of biodiversity genetics. *Proceedings of the National Academy of Sciences of the USA*, **105**, 11564–11570.

Avise, J. C. and Aquadro, C. F. (1982). A comparative summary of genetic distances in the vertebrates: patterns and correlations. *Evolutionary Biology*, **15**, 151–185.

Avise, J. C. and Johns, G. C. (1999). Proposal for a standardized temporal scheme of biological classification for extant species. *Proceedings of the National Academy of Sciences of the USA*, **96**, 7358–7363.

Avise, J. C. and Mitchell, D. (2007). Time to standardize taxonomies. *Systematic Biology*, **56**, 130–133.

Avise, J. C. and Walker, D. (1999). Species realities and numbers in sexual vertebrates: perspectives from an asexually transmitted genome. *Proceedings of the National Academy of Sciences of the USA*, **96**, 9929–95.

Avise, J. C. and Walker, D. (2000). Abandon all species concepts? A response. *Conservation Genetics*, **1**, 77–80.

Baker, R. J. (1984). A sympatric cryptic species of mammal: a new species of *Rhogeessa* (Chiroptera: Vespertilionidae). *Systematic Zoology*, **33**, 178–183.

Baker, R. J. and Bradley, R. D. (2006). Speciation in mammals and the genetic species concept. *Journal of Mammalogy*, **87**, 643–662.

Balmford, A., Green, R. E. and Jenkins, M. (2003). Measuring the changing state of nature. *Trends in Ecology and Evolution*, **18**, 326–330.

Bateman, R. M. and DiMichele, W. A. (2003). Genesis of phenotypic and genotypic diversity in land plants: the present as the key to the past. *Systematics and Biodiversity*, **1**, 13–28.

Baum, D. A. and Shaw, K. L. (1995). Genealogical perspectives on the species problem. In *Monographs in Systematic Botany: Experimental and Molecular Approaches to Plant Biosystematics*, ed. P. Hoch and A. Stephenson. St Louis, MO: Missouri Botanical Garden, pp. 289–303.

Beale, C. M., Lennon, J. J. and Gimona, A. (2008). Opening the climate envelope reveals no macroscale associations with climate in European birds. *Proceedings of the National Academy of Sciences of the USA,* **105**, 14908–14912.

Beheregaray, L. and Caccone, A. (2007). Cryptic biodiversity in a changing world. *Journal of Biology*, **6**, 9.

Bell, M. A. and Foster, S. A. (1994). *The Evolutionary Biology of the Threespine Stickleback*. London: Oxford University Press.

Bernardo, J. (1994). Experimental analysis of allocation in two divergent, natural salamander populations. *American Naturalist*, **143**, 14–38.

Bernardo, J. and Agosta, S. (2003). Clinal variation in the larval life history of mountain dusky salamanders: ecological limits on foraging time and prey abundance restrict opportunities for larval growth. *Journal of Zoology*, **259**, 411–421.

Bernardo, J. and Reagan-Wallin, N. L. (2002). Plethodontid salamanders do not conform to 'general rules' for ectotherm life histories: insights from allocation models about why simple models do not make accurate predictions. *Oikos,* **97**, 398–414.

Bernardo, J. and Spotila, J. (2006). Physiological constraints on organismal response to global warming; mechanistic insights from clinally varying populations and implications for assessing endangerment. *Biology Letters*, **2**, 135–139.

Bernardo, J., Ossola, R. J., Spotila, J. and Crandall, K. A. (2007). Interspecies physiological variation as a tool for cross-species assessments of global warming-induced endangerment: validation of an intrinsic determinant of macroecological and phylogeographic structure. *Biology Letters*, **3**, 695–698.

Bickford, D., Lohman, D. J., Sodhi, N. S. et al. (2006). Cryptic species as a window on diversity and conservation. *Trends in Ecology and Evolution*, **22**,148–155.

Biju, S. D. and Bossuyt, F. (2003). New frog family from India reveals an ancient biogeographical link with the Seychelles. *Nature*, **425**, 711–714.

Blackburn, T. M. and Gaston, K. J., eds. (2003). *Macroecology: Concepts and Consequences*. British Ecological Society Annual Symposium Series. Oxford: Blackwell Science.

Blaxter, M. L. (2004). The promise of DNA taxonomy. *Philosophical Transactions of the Royal Society of London B*, **359**, 669–679.

Boissin, E., Féral, J. P. and Chenuil, A. (2008). Defining reproductively isolated units in a cryptic and syntopic species complex using mitochondrial and nuclear markers: the brooding brittle star, *Amphipholis squamata* (Ophiuroidea). *Molecular Ecology*, **17**, 1732–1744.

Botkin, D. B., Saxe, H., Araújo, M. B. et al. (2007). Forecasting the effects of global warming on biodiversity. *Bioscience*, **57**, 227–236.

Bradley, R. D. and Baker, R. J. (2001). A test of the genetic species concept: cytochrome-b sequences and mammals. *Journal of Mammalogy*, **82**, 960–973.

Brown, D. M., Brenneman, R. A., Georgiadis, N. J. et al. (2007). Extensive population genetic structure in the giraffe. *BMC Biology*, **5**, 57.

Brown, J. H. (1995). *Macroecology*. Chicago, IL: University of Chicago Press.

Bruce, R. C. (2005). Did *Desmognathus* salamanders reinvent the larval stage? *Herpetological Review*, **36**, 107–112.

Burbrink, F. T. (2001). Systematics of the North American rat snake complex (*Elaphe obsoleta*). *Herpetological Monographs*, **15**, 1–53.

Burbrink, F. T. (2002). Phylogeographic analysis of the cornsnake (*Elaphe guttata*) complex as inferred from maximum likelihood and Bayesian analyses. *Molecular Phylogenetics and Evolution*, **25**, 465–476.

Burbrink, F. T., Lawson, R. and Slowinski, J. B. (2000). Molecular phylogeography of the North American rat snake (*Elaphe obsoleta*): a critique of the subspecies concept. *Evolution*, **54**, 2107–2114.

Burns, J. M., Janzen, D. H., Hajibabaei, M., Hallwachs, W. and Hebert, P. D. N. (2008). DNA and cryptic species of skipper butterflies in the genus *Perichares* in Area de Conservacion Guanacaste, Costa Rica. *Proceedings of the National Academy of Sciences of the USA*, **105**, 6350–6355.

Camp, C. D., Marshall, J. L., Landau, K. R., Austin, R. M. and Tilley, S. G. (2000). Sympatric occurrence of two species of the two-lined salamander (*Eurycea bislineata*) complex. *Copeia*, **2000**, 572–578.

Camp, C. D., Peterman, W. E., Milanovich, J. R. et al. (2009). A new genus and species of lungless salamander (family Plethodontidae) from the Appalachian highlands of the south-eastern United States. *Journal of Zoology*, **279**, 86–94.

Casu, M. and Curini-Galletti, M. (2006). Genetic evidence for the existence of cryptic species in the mesopsammic flatworm *Pseudomonocelis ophiocephala* (Rhabditophora: Proseriata). *Biological Journal of the Linnean Society*, **87**, 553–576.

Ceballos, G. and Ehrlich, P. R. (2009). Discoveries of new mammal species and their implications for conservation and ecosystem services. *Proceedings of the National Academy of Sciences of the USA*, **106**, 3841–3846.

Chaitra, M. S., Vasudevan, K. and Shanker, K. (2004). The biodiversity bandwagon: the splitters have it. *Current Science*, **86**, 897–899.

Chek, A. A., Austin, J. D. and Lougheed, S. C. (2003). Why is there a tropical-temperate disparity in the genetic diversity and taxonomy of species? *Evolutionary Ecology Research*, **5**, 69–77.

Chen, G. and Hare, M. P. (2008). Cryptic ecological diversification of a planktonic estuarine copepod, *Acartia tonsa*. *Molecular Ecology*, **17**, 1451–1468.

Chown, S. L. and Gaston, K. J. (1999). Exploring links between physiology and ecology at macro-scales: the role of respiratory metabolism in insects. *Biological Reviews*, **74**, 87–120.

Chown, S. L. and Gaston, K. J. (2008). Macrophysiology for a changing world. *Proceedings of the Royal Society of London B*, **275**, 1469–1478.

Chown, S. L., Gaston, K.J. and Robinson, D. (2004). Macrophysiology: large-scale patterns in physiological traits and their ecological implications. *Functional Ecology*, **18**, 159–167.

Clausen, J., Keck, D. D. and Heisey, W. M. (1940). *Experimental Studies on the Nature of Species. I. Effects of Varied Environments on Western North American Plants*. Carnegie Institute of Washington Publications, 520. Washington, DC: Carnegie Institute.

Clausen, J., Keck, D. D. and Heisey, W. M. (1948). *Experimental Studies on the Nature of Species. III. Environmental Responses of Climatic Races of* Achillea. Carnegie Institute of Washington Publications, 581. Washington, DC: Carnegie Institute.

Clausen, J., Keck, D. D. and Heisey, W. M. (1958). *Experimental Studies on the Nature of Species. IV. Genetic Structure of Ecological Races*. Carnegie Institute of Washington Publications, 615. Washington, DC: Carnegie Institute.

Cohen, J. (1977). *Statistical Power Analysis for the Behavioral Sciences*. New York, NY: Academic Press.

Collins, J. P. and Halliday, T. (2005). Forecasting changes in amphibian biodiversity: aiming at a moving target. *Philosophical Transactions of the Royal Society of London B*, **360**, 309–314.

Cooper, N., Bielby, J., Thomas, G. H. et al. (2008). Macroecology and extinction risk correlates of frogs. *Global Ecology and Biogeography*, **17**, 211–221.

Cracraft, J., Feinstein, J., Vaughn, J. and Helm-Bychowski, K. (1998). Sorting out tigers (*Panthera tigris*): mitochondrial sequences, nuclear inserts, systematics, and conservation genetics. *Animal Conservation*, **1**, 139–150.

Craig, M. T., Graham, R. T., Torres, R. A. et al. (2008). How many species of goliath grouper are there? Cryptic genetic divergence in a threatened marine fish and the resurrection of a geopolitical species. *Endangered Species Research*, **7**, 167–174.

Crandall, K. A., Bininda-Emonds, O. R. P., Mace, G. M. and Wayne, R. K. (2000). Considering evolutionary processes in conservation biology. *Trends in Ecology and Evolution*, **15**, 290–295.

Crandall, K. A., Robison, H. W. and Buhay, J. E. (2009). Avoidance of extinction through nonexistence: the use of museum specimens and molecular genetics to determine the taxonomic status of an endangered freshwater crayfish. *Conservation Genetics*, **10**, 177–189.

Craul, M., Zimmermann, E., Rasoloharijaona, S., Randrianambinina, B. and Radespiel, U. (2007). Unexpected species diversity of Malagasy primates (*Lepilemur* spp.) in the same biogeographical zone: a morphological and molecular approach with the description of two new species. *BMC Evolutionary Biology*, **7**, 83.

Cree, A. and Butler, D. (1993). Tuatara recovery plan (*Sphenodon* spp.). Threatened Species Recovery Plan 9. Wellington: New Zealand Department of Conservation.

Crozier, R. H. (1997). Preserving the information content of species: genetic diversity, phylogeny and conservation worth. *Annual Review of Ecology and Systematics*, **28**, 243–268.

Crozier, R. H., Dunnett, D. J. and Agapow, P. M. (2005). Phylogenetic biodiversity assessment based on systematic nomenclature. *Evolutionary Bioinformatics Online*, **1**, 11–36.

Cruse, M., Telerant, R., Gallagher, T., Lee, T. and Taylor, J. W. (2002). Cryptic species in *Stachybotrys chartarum*. *Mycologia*, **94**, 814–822.

Darlington, C. D. (1940). Taxonomic systems and genetic systems. In *The New Systematics*, ed. J. Huxley. Oxford: Clarendon Press, pp. 137–160.

Darwin, C. R. (1859). *On the Origin of Species by Means of Natural Selection, or the Preservation of Favoured Races in the Struggle for Life*. London: John Murray.

Daugherty, C. H., Cree, A., Hay, J. M. and Thompson, M. B. (1990). Neglected taxonomy and continuing extinction of the tuatara (*Sphenodon*). *Nature*, **347**, 177–179.

Daugherty, C. H., Patterson, G. B. and Hitchmough, R. A. (1994). Taxonomic and conservation review of the New Zealand herpetofauna. *New Zealand Journal of Zoology*, **21**, 317–323.

de Queiroz, K. (1998). The general lineage concept of species, species criteria and the process of speciation. In *Endless Forms: Species and Speciation*, ed. D. Howard and S. Berlocher. Oxford: Oxford University Press, pp. 57–75.

de Queiroz, K. (1999). The general lineage concept of species and the defining properties of the species category. In *Species: New Interdisciplinary Essays*, ed. R. A. Wilson. Cambridge, Massachusetts, MA: MIT Press, pp. 49–89.

de Queiroz, K. (2005a). A unified species concept and its consequences for the future of taxonomy. *Proceedings of the California Academy of Sciences*, **56**, 196–215.

de Queiroz, K. (2005b). Different species problems and their resolution. *BioEssays*, **27**, 1263–1269.

de Queiroz, K. (2006). The PhyloCode and the distinction between taxonomy and nomenclature. *Systematic Biology*, **55**, 160–162.

de Queiroz, K. (2007a). Toward an integrated system of clade names. *Systematic Biology*, **56**, 956–974.

de Queiroz, K. (2007b). Species concepts and species delimitation. *Systematic Biology*, **56**, 879–886.

de Quieroz, K. and Donoghue, M. J. (1988). Phylogenetic systematics and the species problem. *Cladistics*, **4**, 317–338.

de Quieroz, K. and Donoghue, M. J. (1990). Phylogenetic systematics and species revisited. *Cladistics*, **6**, 83–90.

de Queiroz, K. and Gauthier, J. (1992). Phylogenetic taxonomy. *Annual Review of Ecology and Systematics*, **23**, 449–480.

Derycke, S., Remerie, T., Backeljau, T. et al. (2008). Phylogeography of the *Rhabditis* (*Pellioditis*) marina species complex: evidence for long-distance dispersal, and for range expansions and restricted gene flow in the northeast Atlantic. *Molecular Ecology*, **17**, 3306–3322.

DeSalle, R., Egan, M. G. and Siddall, M. (2005). The unholy trinity: taxonomy, species delimitation and DNA barcoding. *Philosophical Transactions of the Royal Society of London B*, **360**, 1905–1916.

Deutsch, C. A., Tewksbury, J. J., Huey, R. B. et al. (2008). Impacts of climate warming on terrestrial ectotherms across latitude. *Proceedings of the National Academy of Sciences of the USA*, **105**, 6668–6672.

Dobzhansky, T. (1937a). Genetic nature of species differences. *American Naturalist*, **71**, 404–420.

Dobzhansky, T. (1937b). *Genetics and the Origin of Species*. New York, NY: Columbia University Press.

Dobzhansky, T. (1950). Mendelian populations and their evolution. *American Naturalist*, **84**, 401.

Donoghue, M. J. (1985). A critique of the biological species concept and recommendations for a phylogenetic alternative. *The Bryologist*, **88**, 172–181.

Dubois, A. (2003). The relationships between taxonomy and conservation biology in the century of extinctions. *Comptes Rendus Biologies*, **326**, S9-S21.

Dunham, A. E., Smith, G. R. and Taylor, J. N. (1979). Evidence for character displacement in western American catostomid fishes. *Evolution*, **33**, 877–896.

Dunn, C. P. (2003). Keeping taxonomy based in morphology. *Trends in Ecology and Evolution*, **18**, 270–271.

Dytham, C. (2009). Evolved dispersal strategies at range margins. *Proceedings of the Royal Society of London B*, **276**, 1407–1413.

Echelle, A. A., van den Bussche, R. A., Malloy, T. P., Haynie, M. L. and Minckley, C. O. (2000). Mitochondrial DNA variation in pupfishes assigned to the species *Cyprinodon macularius* (Atherinomorpha: Cyprinodontidae): taxonomic implications and conservation genetics. *Copeia*, **2000**, 353–364.

Egge, J. J .D. and Simons, A. M. (2006). The challenge of truly cryptic diversity: diagnosis and description of a new madtom catfish (Ictaluridae: *Noturus*). *Zoologica Scripta*, **35**, 581–595.

Estes, S. and Arnold, S. J. (2007). Resolving the paradox of stasis: models with stabilizing selection explain evolutionary divergence on all timescales. *American Naturalist*, **169**, 227–244.

Felsenstein, J. (2004). *Inferring Phylogenies*. Sunderland, MA: Sinauer Associates.

Felsenstein, J. (2009). Phylogeny programs. http://evolution.genetics.washington.edu/phylip/software.html.

Ferguson, J. W. H. (2002). On the use of genetic divergence for identifying species. *Biological Journal of the Linnean Society*, **75**, 509–516.

Forsman, Z. H., Barshis, D. J., Hunter, C. L. and Toonen, R. J. (2009). Shape-shifting corals: molecular markers show morphology is evolutionarily plastic in *Porites*. *BMC Evolutionary Biology*, **9**, 45.

Fouquet, A., Gilles, A., Vences, M. et al. (2007). Underestimation of species richness in Neotropical frogs revealed by mtDNA analyses. *PLoS ONE*, **2**, e1109.

Franz, N. M. (2005). On the lack of good scientific reasons for the growing phylogeny/classification gap. *Cladistics*, **21**, 495–500.

Freckleton, R. P., Pagel, M. and Harvey, P. H. (2003). Comparative methods for adaptive radiations. In *Macroecology Concepts and Consequences*, ed. T. M. Blackburn and K. J. Gaston. Oxford: Blackwell, pp. 391–407.

Funk, D. J. and Omland, K. E. (2003). Species-level paraphyly and polyphyly: frequency, causes, and consequences, with insights from animal mitochondrial DNA. *Annual Review of Ecology, Evolution and Systematics*, **34**, 397–423.

Gaston, K. J. and Blackburn, T. M. (2000). *Pattern and Process in Macroecology*. Oxford: Blackwell Science.

Gaston, K. J., Chown, S. L., Calosi, P. et al. (2009). Macrophysiology: a conceptual reunification. *American Naturalist*, **174**, 595–612.

Gentile, G., Fabiani, A., Marquez, C. et al. (2009). An overlooked pink species of land iguana in the Galápagos. *Proceedings of the National Academy of Sciences of the USA*, **106**, 507–511.

Glaw, F. and Köhler J. (1998). Amphibian species diversity exceeds that of mammals. *Herpetological Review*, **29**, 11–12.

Glaw, F. and Vences, M. (2003). Introduction to amphibians. In *The Natural History of Madagascar*, ed. S. M. Goodman and J. P. Benstead. Chicago, IL: University of Chicago Press, pp. 883–898.

Gómez, A., Wright, P. J., Lunt, D. H. et al. (2007). Mating trials validate the use of DNA barcoding to reveal cryptic speciation of a marine bryozoan taxon. *Proceedings of the Royal Society of London B*, **274**, 199–207.

Good, D. A. (1994). Species limits in the genus *Gerrhonotus* (Squamata: Anguidae). *Herpetological Monographs*, **8**, 180–202.

Gower, D. J., Bhatta, G., Giri, V. et al. (2004). Biodiversity in the Western Ghats: the discovery of new species of caecilian amphibians. *Current Science*, **87**, 739–740.

Groeneveld, L. F., Weisrock, D. W., Rasoloarison, R. M., Yoder, A. D. and Kappeler, P. M. (2009). Species delimitation in lemurs: multiple genetic loci reveal low levels of species diversity in the genus *Cheirogaleus*. *BMC Evolutionary Biology*, **9**, 30.

Grube, M. and Kroken, S. (2000). Molecular approaches and the concept of species and species complexes in lichenized fungi. *Mycological Research*, **104**, 1284–1294.

Guilhaumon, F., Gimenez, O., Gaston, K. J. and Mouillot, D. (2008). Taxonomic and regional uncertainty in species-area relationships and the identification of richness hotspots. *Proceedings of the National Academy of Sciences of the USA*, **105**, 15458–15463.

Gum, B., Gross, R. and Geist, J. (2009). Conservation genetics and management implications for European grayling, *Thymallus thymallus*: synthesis of phylogeography and population genetics. *Fisheries Management and Ecology*, **16**, 37–51.

Hajibabaei, M., Janzen, D. H., Burns, J. M., Hallwachs, W. and Hebert, P. D. N. (2006). DNA barcodes distinguish species of tropical Lepidoptera. *Proceedings of the National Academy of Sciences of the USA*, **103**, 968–971.

Hanken, J. (1999). Why are there so many new amphibian species when amphibians are declining? *Trends in Ecology and Evolution,* **14**, 7–8.

Harris, D. J. and Froufe, E. (2005). Taxonomic inflation: species concept or historical geopolitical bias. *Trends in Ecology and Evolution,* **20**, 6–7.

Harrison, R. G. (2002). Species concepts. In *Encyclopedia of Evolution*, ed. M. Pagel. Oxford: Oxford University Press, pp. 1078–1083.

Harvey, P. H. and Rambaut, A. (2000). Comparative analyses for adaptive radiations. *Philosophical Transactions of the Royal Society of London B,* **355**, 1599-1606.

Heath, T. A., Zwickl, D. J., Kim, J. and Hillis, D. M. (2008a). Taxon sampling affects inferences of macroevolutionary processes from phylogenetic trees. *Systematic Biology,* **57**, 160–166.

Heath, T. A., Hedtke, S. M. and Hillis, D. M. (2008b). Taxon sampling and the accuracy of phylogenetic analyses. *Journal of Systematics and Evolution,* **46**, 239–257.

Hebert, P. D. N. and Gregory, T. R. (2005). The promise of DNA barcoding for taxonomy. *Systematic Biology,* **54**, 852–859.

Hebert, P. D. N., Cywinska, A., Ball, S. L. and de Waard, J. R. (2003). Biological identifications through DNA barcodes. *Proceedings of the Royal Society of London B,* **270**, 313–321.

Hebert, P. D. N., Penton, E. H., Burns, J. M., Janzen, D. H. and Hallwachs, W. (2004). Ten species in one: DNA barcoding reveals cryptic species in the neotropical skipper butterfly *Astraptes fulgerator. Proceedings of the National Academy of Sciences of the USA,* **101**, 14812–14817.

Heckman, K. L., Rasoazanabary, E., Machlin, E., Godfrey, L. R. and Yoder, A. D. (2006). Incongruence between genetic and morphological diversity in *Microcebus griseorufus. BMC Evolutionary Biology,* **6**, 98.

Hedin, M. (1997). Speciational history in a diverse clade of habitat-specialized spiders (Araneae: Nesticidae: *Nesticus*): inferences from geographic-based sampling. *Evolution,* **51**, 1927–1943.

Hedin, M. and Wood, D. L. (2002). Genealogical exclusivity in geographically proximate populations of *Hypochilus thorelli* Marx (Araneae, Hypochilidae) on the Cumberland Plateau of North America. *Molecular Ecology,* **11**, 1975–1988.

Hedtke, S. M., Townsend, T. and Hillis, D. M. (2006). Resolution of phylogenetic conflict in large data sets by increased taxon sampling. *Systematic Biology,* **55**, 522–529.

Heinicke, M. P., Duellman, W. E., Trueb, L. et al. (2009). A new frog family (Anura: Terrarana) from South America and an expanded direct-developing clade revealed by molecular phylogeny. *Zootaxa,* **2211**, 1–35.

Hemmerter, S., Šlapeta, J., van den Hurk, A. et al. (2007). A curious coincidence: mosquito biodiversity and the limits of the Japanese encephalitis virus in Australasia. *BMC Evolutionary Biology,* **7**, 100.

Hendry, A. P., Vamosi, S. M., Latham, S. J., Heilbuth, J. C. and Day, T. (2000). Questioning species realities. *Conservation Genetics,* **1**, 67–76.

Highton, R. (1990). Taxonomic treatment of genetically differentiated populations. *Herpetologica,* **46**, 114–121.

Highton, R. (2000). Detecting cryptic species using allozyme data. In *The*

Biology of Plethodontid Salamanders, ed. R. C. Bruce, R. G. Jaeger and L. D. Houck. New York, NY: Kluwer Academic/Plenum Publishers, pp. 215–241.

Highton, R. and Peabody, R. (2000). Geographic protein variation and speciation in salamanders of the *Plethodon jordani* and *Plethodon glutinosus* complexes in the southern Appalachian Mountains with the description of four new species. In *The Biology of Plethodontid Salamanders*, ed. R. C. Bruce, R. G. Jaeger and L. D. Houck. New York, NY: Kluwer Academic/Plenum Publishers, pp. 31–93.

Hoffmann, A. A. and Blows, M. W. (1993). Evolutionary genetics and climate change: will animals adapt to global warming? In *Biotic Interactions and Global Change*, ed. P. M. Kareiva, J. G. Kingsolver and R. B. Huey. Sunderland, MA: Sinauer Associates, pp. 165–178.

Huey, R. B. Deutsch, C., Tewksbury, J. J. et al. (2009). Climate warming puts the heat on tropical forest lizards. *Proceedings of the Royal Society of London B*, **276**, 1939–1948.

Huxley, J. S., ed. (1940). *The New Systematics.* Oxford: The Clarendon Press.

Hyde, J. R., Kimbrell, C. A., Budrick, J. E., Lynn, E. A. and Vetter, R. D. (2008). Cryptic speciation in the vermilion rockfish (*Sebastes miniatus*) and the role of bathymetry in the speciation process. *Molecular Ecology*, **17**, 1122–1136.

Irwin, D. E. (2002). Phylogeographic breaks without geographic barriers to gene flow. *Evolution*, **56**, 2383–2394.

International Union for the Conservation of Nature (I UCN) (2001). *The IUCN Red List Categories and Criteria*, version 3.1. www.iucnredlist.org/technical-documents/categories-and-criteria.

Isaac, N. J. B., Mallet, J. and Mace, G. M. (2004). Taxonomic inflation: its influence on macroecology and conservation. *Trends in Ecology and Evolution*, **19**, 464–469.

Johns, G. C. and Avise, J. C. (1998). A comparative summary of genetic distances in the vertebrates from the mitochondrial cytochrome b gene. *Molecular Biology and Evolution*, **15**, 1481–1490.

Kankare, M. and Shaw, M. R. (2004). Molecular phylogeny of *Cotesia* (Hymenoptera: Braconidae: Microgastrinae) parasitoids associated with Melitaeini butterflies (Lepidoptera: Nymphalidae: Melitaeini). *Molecular Phylogenetics and Evolution*, **32**, 207–220.

Kastin, A. J. (2006). *Handbook of Biologically Active Peptides*. New York, NY: Academic Press.

Kearney, M. and Porter, W. P. (2009). Mechanistic niche modeling: combining physiological and spatial data to predict species' ranges. *Ecology Letters*, **12**, 334–350.

Kelt, D. A. and Brown, J. H. (2000). Species as units in ecology and biogeography: are the blind leading the blind? *Global Ecology and Biogeography*, **9**, 213–217.

Kim, K. C., Brown, B. W. and Cook, E. F. (1966). A quantitative taxonomic study of the *Hoplopleura hesperomydis* complex (Anoplura, Hoplopleuridae), with notes on a posteriori taxonomic characters. *Systematic Zoology*, **15**, 24–45.

King, R. A., Tibble, A. L. and Symondson, W. O. C. (2008). Opening a can of worms: unprecedented sympatric cryptic diversity within British lumbricid earthworms. *Molecular Ecology*, **17**, 4684–4698.

King, T. L., Switzer, J. F., Morrison, C. L. et al. (2006). Comprehensive genetic analyses reveal evolutionarily distinction of a mouse (*Zapus hudsonius preblei*) proposed for delisting from the U.S. Endangered Species Act. *Molecular Ecology*, **15**, 4331–4359.

Knowlton, N. (1993). Sibling species in the sea. *Annual Review of Ecology and Systematics*, **24**, 189–216.

Knowlton N. (2000). Molecular genetic analyses of species boundaries in the sea. *Hydrobiologia*, **420**, 73–90.

Köhler, J., Vietes, D. R., Bonett, R. M. et al. (2005). New amphibians and global conservation: a boost in species discoveries in a highly endangered vertebrate group. *BioScience*, **55**, 693–696.

Kong, L. and Li, Q. (2009). Genetic evidence for the existence of a cryptic species in an endangered clam *Coelomactra antiquata*. *Marine Biology*, **156**, 1507–1515.

Lamoreux, J. F., Morrison, J. C., Ricketts, T. H. et al. (2006). Global tests of biodiversity concordance and the importance of endemism. *Nature*, **440**, 212–214.

LeDuc, R. G., Robertson, K. M. and Pitman, R. L. (2008). Mitochondrial sequence divergence among Antarctic killer whale ecotypes is consistent with multiple species. *Biology Letters*, **4**, 426–429.

Lee, M. S. Y. (2004). The molecularisation of taxonomy. *Invertebrate Systematics*, **18**, 1–6.

Lee, T., Burch, J. B., Coote, T. et al. (2009). Moorean tree snail survival revisited: a multi-island genealogical perspective. *BMC Evolutionary Biology*, **9**, 204.

Lefébure, T., Douady, C. J., Gouy, M. and Gibert, J. (2006). Relationship between morphological taxonomy and molecular divergence within Crustacea: proposal of a molecular threshold to help species delimitation. *Molecular Phylogenetics and Evolution*, **40**, 435–447.

Linnaeus, C. (1758). *Systema naturae per regna tria naturae: secundum classes, ordines, genera, species, cum characteribus, differentiis, synonymis, locis*. Stockholm.

Lipscomb, D., Platnick, N. and Wheeler, Q. (2003). The intellectual content of taxonomy: a comment on DNA taxonomy. *Trends in Ecology and Evolution*, **18**, 65–66.

Loehle, C. (1987). Hypothesis testing in ecology: psychological aspects and the importance of theory maturation. *Quarterly Review of Biology*, **62**, 397–409.

Losos, J. B., Schoener, T. W., Warheit, K. I. and Creer, D. A. (2001). Experimental studies of adaptive differentiation in Bahamian *Anolis* lizards. *Genetica*, **112–113**, 399–415.

Lubischew, A. A. (1962). On the use of discriminant functions in taxonomy. *Biometrics*, **18**, 455–477.

MacArthur, R. and Levins, R. (1967). The limiting similarity, convergence, and divergence of coexisting species. *American Naturalist*, **101**, 377–385.

Mace, G. M. (2004). The role of taxonomy in species conservation. *Philosophical Transactions of the Royal Society of London B*, **359**, 711–719.

Mace, G. M. and Purvis, A. (2008). Evolutionary biology and practical conservation: bridging a widening gap. *Molecular Ecology*, **17**, 9–19.

Malhotra, A. and Thorpe, R. S. (1997). Size and shape variation in a Lesser Antillean anole, *Anolis oculatus* (Sauria: Iguanidae) in relation to habitat. *Biological Journal of the Linnean Society*, **60**, 53–72.

Mallet, J. and Willmott, K. (2003). Taxonomy: renaissance or Tower of Babel? *Trends in Ecology and Evolution*, **18**, 57–59.

Mayer, F. and von Helversen, O. (2001). Sympatric distribution of two cryptic bat species across Europe. *Biological Journal of the Linnean Society*, **74**, 365–374.

Mayr, E. (1940). Speciation phenomena in birds. *American Naturalist*, **74**, 249–278.

Mayr, E. (1942). *Systematics and the Origin of Species*. New York, NY: Columbia University Press.

McCune, A. R. (1997). How fast is speciation: molecular, geological and phylogenetic evidence from adaptive radiations of fishes. In *Molecular Evolution and Adaptive Radiation*, ed. T. Givnish and K. Sytsma. Cambridge: Cambridge University Press, pp. 585–610.

McDiarmid, R. W. and Donnelly, M. A. (2005). The herpetofauna of the Guayana Highlands: amphibians and reptiles of the Lost World. In *Ecology and Evolution in the Tropics: a Herpetological Perspective*, ed. M. A. Donnelly, B. I. Crother, C. Guyer, M. H. Wake and M. E. White. Chicago, IL: University of Chicago Press, pp. 461–560.

McPeek, M. A. and Brown, J. M. (2007). Clade age and not diversification rate explains species richness among animal taxa. *American Naturalist*, **169**, E97–E106.

Mead, L. S., Clayton, D. R., Nauman, R. S., Olsen, D. H. and Pfrender, M. E. (2005). Newly discovered populations of salamanders from Siskiyou County, California, represent a species distinct from *Plethodon stormi*. *Herpetologica*, **61**, 158–177.

Means, D. B. and Savage, J. M. (2007). Three new malodorous rainfrogs of the genus *Pristimantis* (Anura:= Brachycephalidae) from the Wokomung Massif, in west-central Guyana, South America. *Zootaxa*, **1658**, 39–55.

Meegaskumbura, M., Bossuyt, F., Pethiyagoda, R. et al. (2002). Sri Lanka: an amphibian hotspot. Science, **298**, 379.

Meiri, S. and Mace, G. M. (2007). New taxonomy and the origin of species. *PLOS Biology*, **5**, e194,1385–1386.

Mendelson, J. R., Lips, K. R., Gagliardo, R. W. et al. (2006). Confronting amphibian declines and extinctions. *Science*, **313**, 48.

Menegon, M., Doggart, N. and Owen, N. (2008). The Nguru mountains of Tanzania, an outstanding hotspot of herpetofaunal diversity. *Acta Herpetologica*, **3**, 107–127.

Merriam, C. H. (1918). Review of the grizzly and big brown bears of North America (genus *Ursus*) with the description of a new genus, *Vetularctos*. *North American Fauna*, **41**, 1–136.

Meyer, A. (1987). Phenotypic plasticity and heterochrony in *Cichlasoma managuense* (Pisces, Cichlidae) and their implications for speciation in cichlid fishes. *Evolution*, **41**, 1357–1369.

Miles, D. B., Noecker, R., Roosenburg, W. M. and White, M. M. (2002). Genetic relationships among populations of *Sceloporus undulatus* fail to support subspecific designations. *Herpetologica*, **58**, 277–292.

Milinkovitch, M. C., Monteyne, D., Russello, M. et al. (2007). Molecular genetic analyses identify a trans-island hybrid in a repatriation program of an endangered taxon. *BMC Ecology*, **7**, 2.

Miller, K. J. and Benzie, J. A. H. (1997). No clear genetic distinction between morphological species within the coral genus *Platygyra*. *Bulletin of Marine Science*, **61**, 907–917.

Min, M. S., Yang, S. Y., Bonett, R. M. et al. (2005). Discovery of the first Asian plethodontid salamander. *Nature*, **435**, 87–90.

Molbo, D., Machado, C. A., Sevenster, J. G., Keller, L. and Herre, E. A. (2003). Cryptic species of fig pollinating wasps: implications for the evolution of the fig-wasp mutualism, sex allocation, and the precision of adaptation. *Proceedings of the National Academy of Sciences of the USA*, **100**, 5867–5872.

Morando, M., Avila, L. J. and Sites, J. W. (2003). Sampling strategies for delimiting species: genes, individuals, and populations in the *Liolaemus elongatus-kriegi* complex (Squamata; Liolaemidae) in Andean–Patagonian South America. *Systematic Biology*, **52**, 159–185.

Moritz, C. (2002). Strategies to protect biological diversity and the processes that sustain it. *Systematic Biology*, **51**, 238–254.

Moritz, C. and Cicero, C. (2004). DNA barcoding: promise and pitfalls. *PLoS Biology*, **2**, e279.

Myers, P., Lundrigan, B. L., Gillespie, B. W. and Zelditch, M. L. (1996). Phenotypic plasticity in skull and dental morphology in the prairie deer mouse (*Peromyscus maniculatus bairdii*). *Journal of Morphology*, **229**, 229–237.

Nelson, N. J., Keall, S. N., Brown, D. and Daugherty, C. H. (2002). Establishing a new wild population of Tuatara (*Sphenodon guntheri*). *Conservation Biology*, **16**, 887–894.

Ohlemüller, R., Anderson, B. J., Araújo, M. B. et al. (2008). Coincidence of climatic and species rarity: high risk to small-range species from climate change. *Biology Letters*, **4**, 568–572.

Omland, K. E. (1997). Correlated rates of molecular and morphological evolution. *Evolution*, **51**, 1381–1393.

Omland, K. E., Lanyon, S. M. and Fritz, S. J. (1999). A molecular phylogeny of the New World Orioles (*Icterus*): the importance of dense taxon sampling. *Molecular Phylogenetics and Evolution*, **12**, 224–239.

Orme, C. D. L., Davies, R. G., Burgess, M. et al. (2005). Global hotspots of species richness are not congruent with endemism or threat. *Nature*, **436**, 1016–1019.

Padial, J. M. and De la Riva, I. (2006). Taxonomic inflation and the stability of species lists: the perils of ostrich's behavior. *Systematic Biology*, **55**, 859–867.

Page, T., Choy, S. C. and Hughes, J. (2005). The taxonomic feedback loop: symbiosis of morphology and molecules. *Biology Letters*, **1**, 139–142.

Parham, J. F., Türkozan, O., Stuart, B. L. et al. (2006). Genetic evidence for premature taxonomic inflation in Middle Eastern tortoises. *Proceedings of the California Academy of Sciences*, **57**, 955–964.

Parra-Olea, G., García-París, M., Papenfuss, T. J. and Wake, D. B. (2005). Systematics of the *Pseudoeurycea bellii* species complex. *Herpetologica*, **61**, 145–158.

Perez, K. E. and Minton, R. L. (2008). Practical applications for systematics and taxonomy in North American freshwater gastropod conservation. *Journal of the North American Benthological Society*, **27**, 471–483.

Pérez-Losada, M. and Crandall, K. A. (2003). Can taxonomic richness be used as a surrogate for phylogenetic distinctness indices for ranking areas for conservation? *Animal Biodiversity and Conservation*, **26**, 77–84.

Peterson, A. T. and Navarro-Siguënza, A. G. (1999). Alternate species concepts as bases for determining priority conservation areas. *Conservation Biology*, **13**, 427–431.

Pfenniger, M. and Schwenk, K. (2007). Cryptic animal species are homogeneously distributed among taxa and biogeographical regions. *BMC Evolutionary Biology*, **7**, 6.

Platt, J. R. (1964). Strong inference. *Science*, **146**, 347–353.

Pörtner, H. O. and Farrell, A. P. (2008). Physiology and climate change. *Science*, **322**, 690–692.

Poulakakis, N., Glaberman, S., Russello, M. et al. (2008). Historical DNA analysis reveals living descendants of an extinct species of Galápagos tortoise. *Proceedings of the National Academy of Sciences of the USA*, **105**, 15464–15469.

Price, T. (1997). Correlated evolution and independent contrasts. *Philosophical Transactions of the Royal Society of London B*, **352**, 519–529.

Price, T. and Kirkpatrick, M. (2009). Evolutionarily stable range limits set by interspecific competition. *Proceedings of the Royal Society of London B*, **276**, 1429–1434.

Pringle, A., Baker, D. M., Platt, J. L. et al. (2005). Cryptic speciation in the cosmopolitan and clonal human pathogenic fungus *Aspergillus fumigatus*. *Evolution*, **59**, 1886–1899.

Quesada, H., Posada, D., Caballero, A., Morán, P. and Rolán-Alvarez, E. (2007). Phylogenetic evidence for multiple sympatric ecological diversification in a marine snail. *Evolution*, **61**, 1600–1612.

Ramey, R. R., Liu, H. P., Epps, C. W., Carpenter, L. and Wehausen, J. D. (2005). Genetic relatedness of the Preble's meadow jumping mouse (*Zapus hudsonius preblei*) to nearby subspecies of *Z. hudsonius* as inferred from variation in cranial morphology, mitochondrial DNA, and microsatellite DNA: implications for taxonomy and conservation. *Animal Conservation*, **8**, 329–346.

Ramey, R. R., Wehausen, J. D., Liu, H. P., Epps, C. W. and Carpenter, L. (2006). Response to Vignieri et al. (2006): Should hypothesis testing or selective post hoc interpretation of results guide the allocation of conservation effort? *Animal Conservation*, **9**, 244–247.

Ramey, R. R., Wehausen, J. D., Liu, H. P., Epps, C. W. and Carpenter, L. M. (2007). How King et al. (2006) define an 'evolutionarily distinct' subspecies: a response. *Molecular Ecology*, **16**, 3518–3521.

Riddle, B. R. and Hafner, D. J. (1999). Species as units of analysis in ecology and biogeography: time to take the blinders off. *Global Ecology and Biogeography*, **8**, 433–441.

Riddle B. R., Hafner, D. J., Alexander, L.F. and Jaeger, J. R. (2000). Cryptic vicariance in the historical assembly of a Baja California Peninsular Desert biota. *Proceedings of the National Academy of Sciences of the USA*, **97**, 14438–14443.

Robalo, J. I., Sousa-Santos, C., Cabral, H., Castilho, R. and Almada, V. C. (2009). Genetic evidence fails to discriminate between *Macroramphosus gracilis* Lowe 1839 and *Macroramphosus scolopax* Linnaeus 1758 in Portuguese waters. *Marine Biology*, **156**, 1733–1737.

Roca, A. L., Georgiadis, N., Pecon-Slattery, J. and O'Brien, S. J. (2001). Genetic evidence for two species of elephant in Africa. *Science*, **293**, 1473–1477.

Roy, V., Demanche, C., Livet, A. and Harry, M. (2006). Genetic differentiation in the soil-feeding termite *Cubitermes* sp. *affinis subarquatus*: occurrence of cryptic species revealed by nuclear and mitochondrial markers. *BMC Evolutionary Biology*, **6**, 102.

Russello, M. A., Glaberman, S., Gibbs, J. P. et al. (2005). A cryptic taxon of Galápagos tortoises in conservation peril. *Biology Letters*, **1**, 287–290.

Schlick-Steiner, B. C., Seifert, B., Stauffer, C. et al. (2007). Without morphology cryptic species stay in taxonomic crypsis following discovery. *Trends in Ecology and Evolution*, **22**, 391–392.

Schluter, D. (2000a). *The Ecology of Adaptive Radiation*. Oxford: Oxford University Press.

Schluter, D. (2000b). Ecological character displacement in adaptive radiation. *American Naturalist*, **156**, S4–S16.

Sechrest, W., Brooks, T. M., da Fonseca, G. A. B. et al. (2002). Hotspots and the conservation of evolutionary history. *Proceedings of the National Academy of Sciences of the USA*, **99**, 2067–2071.

Sever, D. M., Dundee, H. A. and Sullivan, C. D. (1976). A new *Eurycea* (Amphibia: Plethodontidae) from southwestern North Carolina. *Herpetologica*, **32**, 26–29.

Shaffer, H. B., Pauly, G. B., Oliver, J. C. and Trenham, P. C. (2004a). The molecular phylogenetics of endangerment: cryptic variation and historical phylogeography of the California tiger salamander, *Ambystoma californiense. Molecular Ecology*, **13**, 3033–3049.

Shaffer, H. B., Fellers, G. M., Voss, S. R., Oliver, J. C. and Pauly, G. B. (2004b). Species boundaries, phylogeography, and conservation genetics of the red-legged frog (*Rana aurora/draytonii)* complex. *Molecular Ecology*, **13**, 2667–2677.

Siddall, M. E., Trontelj, P., Utevsky, S. Y., Nkamany, M. and Macdonald, K. S. (2007). Diverse molecular data demonstrate that commercially available medicinal leeches are not *Hirudo medicinalis. Proceedings of the Royal Society of London B*, **274**, 1481–1487.

Simpson, G. G. (1943). Criteria for genera, species and subspecies in zoology and paleontology. *Annals of the New York Academy of Science*, **44**, 145–178.

Simpson, G. G. (1945). The principles of classification and a classification of mammals. *Bulletin of the American Museum of Natural History*, **85**, 1–350.

Sinclair, E. A., Pérez-Losada, M. and Crandall, K. A. (2005). Molecular phylogenetics for conservation biology. In *Phylogeny and Conservation*, ed. A. Purvis, J. L. Gittleman and T. Brooks. Cambridge: Cambridge University Press, pp. 19–56.

Sites, J. W. and Crandall, K. A. (1997). Testing species boundaries in biodiversity studies. *Conservation Biology*, **11**, 1289–1297.

Sites, J. W. and Marshall, J. C. (2003). Species delimitation: a Renaissance issue in systematic biology. *Trends in Ecology and Evolution*, **18**, 462–470.

Sites, J. W. and Marshall, J. C. (2004). Operational criteria for delimiting species. *Annual Review of Ecology, Evolution, and Systematics*, **35**, 199–229.

Smith, M. A., Woodley, N. E., Janzen, D. H., Hallwachs, W. and Hebert, P. D. N. (2006). DNA barcodes reveal cryptic host-specificity within the presumed polyphagous members of a genus of parasitoid flies (Diptera: Tachinidae). *Proceedings of the National Academy of Sciences of the USA*, **103**, 3657–3662.

Smith, M. A., Rodriguez, J. J., Whitfield, J. B. et al. (2008). Extreme diversity of tropical parasitoid wasps exposed by iterative integration of natural history, DNA barcoding, morphology, and collections. *Proceedings of the National Academy of Sciences of the USA*, **105**, 12359–12364.

Song, H., Buhay, J. E., Whiting, M. F. and Crandall, K. A. (2008). Many species in one: DNA barcoding overestimates the number of species when nuclear mitochondrial pseudogenes are coamplified. *Proceedings of the National Academy of Sciences of the USA*, **105**, 13486–13491.

Spicer, J. I. and Gaston, K. J. (1999). *Physiological Diversity and Its Ecological Implications*. Oxford: Blackwell Science.

Starrett, J. and Hedin, M. (2007). Multilocus genealogies reveal multiple cryptic species and biogeographic complexity in the California turret spider *Antrodiaetus riversi* (Mygalomorphae, Antrodieatidae). *Molecular Ecology*, **16**, 583–604.

Stillman, J. H. (2002). Physiological tolerance limits in intertidal crabs. *Integrative and Comparative Biology*, **42**, 790–796.

Stillman, J. H. (2003). Acclimation capacity underlies climate change susceptibility. *Science*, **301**, 65.

Stillman, J. H. (2004). A comparative analysis of plasticity of thermal limits in porcelain crabs across latitudinal and intertidal zone clines. *International Congress Series*, **1275**C, 267–275.

Strathmann, R. R., Fenaux, L. and Strathmann, M. F. (1992). Heterochronic developmental plasticity in larval sea urchins and its implications for evolution of nonfeeding larvae. *Evolution*, **46**, 972–986.

Stuart, B. L., Inger, R. F. and Voris, H. K. (2006). High level of cryptic species diversity revealed by sympatric lineages of Southeast Asian forest frogs. *Biology Letters*, **2**, 470–474.

Stuart, S. N., Chanson, J. S., Cox, N. A. et al. (2004). Status and trends of amphibian declines and extinctions worldwide. *Science*, **306**, 1783–1786.

Suatoni, E., Vicario, S., Rice, S., Snell, T. and Caccone, A. (2006). An analysis of species boundaries and biogeographic patterns in a cryptic species complex: the rotifer – *Brachionus plicatilis. Molecular Phylogenetics and Evolution*, **41**, 86–98.

Swatdipong, A., Vasemägi, A., Koskinen, M. T., Piironen, J. and Primmer, C. R. (2009). Unanticipated population structure of European grayling in its northern distribution: implications for conservation prioritization. *Frontiers in Zoology*, **6**, 6.

Tarjuelo, I., Posada, D., Crandall, K. A., Pascual, M. and Turon, X. (2001). Cryptic species of *Clavelina* (Ascidiacea: Aplousobranchiata,

Polycitoridae) in two different habitats: harbours and rocky littoral zones in the northwestern Mediterranean. *Marine Biology*, **139**, 455–462.

Tattersall, I. (2007). Madagascar's lemurs: cryptic diversity or taxonomic inflation? *Evolutionary Anthropology*, **16**, 12–23.

Tautz, D., Arctander, P., Minelli, A., Thomas, R. H. and Vogler, A. P. (2003). A plea for DNA taxnomy. *Trends in Ecology and Evolution*, **18**, 70–74.

Templeton, A. R., Routman, E. and Phillips, C. A. (1995). Separating population structure from history: a cladistic analysis of the geographical distribution of mitochondrial DNA haplotypes in the tiger salamander. *Ambystoma tigrinum. Genetics*, **140**, 767–782.

Templeton, A. R., Robertson, R. J., Brisson, J. and Strasburg, J. (2001). Disrupting evolutionary processes: the effect of habitat fragmentation on collared lizards in the Missouri Ozarks. *Proceedings of the National Academy of Sciences of the USA*, **98**, 5426–5432.

Thomas, C. D., Cameron, A., Green, R. E. et al. (2004). Extinction risk from climate change. *Nature*, **427**, 145–148.

Thorpe, J. P. (1982). The molecular clock hypothesis: biochemical evolution, genetic differentiation and systematics. *Annual Review of Ecology and Systematics*, **13**, 139–168.

Thorpe, J. P., Smartt, J. and Allcock, A. L. et al. (1995). Genetic diversity as a component of biodiversity. In *Global Biodiversity Assessment*, ed. V. H. Heywood and R. T. Watson. Cambridge: Cambridge University Press, pp. 57–87.

Tilley, S. G. (1981). A new species of *Desmognathus* (Amphibia: Caudata: Plethodontidae) from the southern Appalachian Mountains. *Occasional Papers of the Museum of Zoology, University of Michigan*, **695**, 1–23.

Tilley, S. G. and Mahoney, M. J. (1996). Patterns of genetic differentiation in salamanders of the *Desmognathus ochrophaeus* complex *(Amphibia: Plethodontidae). Herpetological Monographs*, **10**, 1–42.

Tilley, S. G., Merritt, R. B., Wu, B. and Highton, R. (1978). Genetic differentiation in salamanders of the *Desmognathus ochrophaeus* complex. *Evolution*, **32**, 93–115.

Toda, M., Hikida, T. and Ota, H. (2001). Discovery of sympatric cryptic species within *Gekko hokouensis* (Gekkonidae: Squamata) from the Okinawa Islands, Japan, by use of allozyme data. *Zoologica Scripta*, **30**, 1–11.

Todd, P. A. (2008). Morphological plasticity in scleractinian corals. *Biological Reviews*, **83**, 315–337.

Toews, D. P. L. and Irwin, D. E. (2008). Cryptic speciation in a Holarctic passerine revealed by genetic and bioacoustic analyses. *Molecular Ecology*, **17**, 2691–2705.

Toft, C. A. and Shea, P. J. (1983). Detecting community-wide patterns: estimating power strengthens statistical inference. *American Naturalist*, **122**, 618–625.

Trewick, S. A. (1998). Sympatric cryptic species in New Zealand Onychophora. *Biological Journal of the Linnean Society*, **63**, 307–329.

Trontelj, P. and Fiser, C. (2009). Cryptic species diversity should not be trivialised. *Systematics and Biodiversity*, **7**, 1–3.

Trontelj, P., Douady, C. J., Fiser, C. et al. (2009). A molecular test for cryptic diversity in ground water: how large are the ranges of macro-stygobionts? *Freshwater Biology*, **54**, 727–744.

van der Meijden, A., Vences, M., Hoegg, S. and Meyer, A. (2005). A previously unrecognized radiation of ranid frogs in southern Africa revealed by nuclear and mitochondrial DNA sequences. *Molecular Phylogenetics and Evolution*, **37**, 674–685.

Vences, M. and Köhler, J. (2008). Global diversity of amphibians (Amphibia) in freshwater. *Hydrobiologia*, **595**, 569–580.

Vences, M. and Wake, D. B. (2007). Speciation, species boundaries and phylogeography of amphibians. In *Amphibian Biology, Vol. 6, Systematics*, ed. H. H. Heatwole and M. Tyler. Chipping Norton, Australia: Surrey Beatty & Sons, pp. 2613–2669.

Vieites, D. R., Wollenberg, K. C., Andreone, F. et al. (2009). Vast underestimation of Madagascar's biodiversity evidenced by an integrative amphibian inventory. *Proceedings of the National Academy of Sciences of the USA*, **106**, 8267–8272.

Vignieri, S. N, Hallerman, E. M., Bergstrom, B. J. et al. (2006). Mistaken view of taxonomic validity undermines conservation of an evolutionarily distinctive mouse: A response to Ramey et al. *Animal Conservation*, **9**, 237–243.

Wake, D. B. and Vredenburg, V. T. (2008). Are we in the midst of the sixth mass extinction? A view from the world of amphibians. *Proceedings of the National Academy of Sciences of the USA*, **105**, 11466–11473.

Wheeler, Q. D. (2004). Taxonomic triage and the poverty of phylogeny. *Philosophical Transactions of the Royal Society of London B*, **359**, 571–583.

Wheeler, Q. D. (2008). Undisclosed thinking: morphology and Hennig's unfinished revolution. *Systematic Entomology*, **33**, 2–7.

Whittaker, R. H. (1972). Evolution and measurement of species diversity. *Taxon*, **21**, 213–251.

Wiens, J. J. (2004). The role of morphological data in phylogeny reconstruction. *Systematic Biology*, **53**, 653–661.

Wiens, J. J. and Penkrot, T. (2002). Delimiting species using DNA and morphological variation and discordant species limits in spiny lizards (*Sceloporus*). *Systematic Biology*, **51**, 69–91.

Will, K. W. and Rubinoff, D. (2004). Myth of the molecule: DNA barcodes for species cannot replace morphology for identification and classification. *Cladistics*, **20**, 47–55.

Wullschleger, E. B. and Jokela, J. (2002). Morphological plasticity and divergence in life-history traits between two closely related freshwater snails, *Lymnaea ovata and Lymnaea peregra*. *Journal of Molluscan Studies*, **68**, 1–5.

Wynn, A. H., Highton, R. and Jacobs, J. F. (1988). A new species of rock-crevice dwelling *Plethodon* from Pigeon Mountain, Georgia. *Herpetologica*, **44**, 135–143.

Yoder, A. D., Rasoloarison, R. M., Goodman, S. M. et al. (2000). Remarkable species diversity in Malagasy mouse lemurs (Primates, *Microcebus*). *Proceedings of the National Academy of Sciences of the USA*, **97**, 11325–11330.

Yoder, A. D., Burns, M. M. and Genin, F. (2002). Molecular evidence of reproductive isolation in sympatric sibling species of mouse lemurs. *International Journal of Primatology*, **23**, 1335–1343.

Yoder, A. D., Olson, L. E., Hanley, C. et al. (2005). A multidimensional approach for detecting species patterns in Malagasy vertebrates. *Proceedings of the National Academy of Sciences of the USA*, **102**, 6587–6594.

Zwickl, D. J. and Hillis, D. M. (2002). Increased taxon sampling greatly reduces phylogenetic error. *Systematic Biology*, **51**, 588–589.

19

Climate change and Cyperaceae

D. A. Simpson

Royal Botanic Gardens, Kew, UK

C. Yesson

Institute of Zoology, Zoological Society of London, UK

A. Culham

School of Biological Sciences and *The Walker Institute for Climate Change, University of Reading, UK*

C. A. Couch

Royal Botanic Gardens, Kew, UK

A. M. Muasya

Department of Botany, University of Cape Town, South Africa

Climate Change, Ecology and Systematics, ed. Trevor R. Hodkinson, Michael B. Jones, Stephen Waldren and John A. N. Parnell. Published by Cambridge University Press.

Abstract

Cyperaceae (sedges) are a monocotyledenous angiosperm plant family with over 5300 species. Despite their global importance, few, if any, climate change studies have been carried out on, or with, Cyperaceae. However, they may be a model family on which to base such work. They are of economic, ethnobotanical, conservation and environmental importance, and a wide range of resources for Cyperaceae is available. Examples are given of where Cyperaceae may win or lose in the climate change stakes. Taxa with C_4 photosynthetic pathways, such as *Cyperus rotundus* ('the world's worst weed'), *C. longus* and members of *Cyperus* sect. *Arenarii*, are potential winners that could considerably extend their distributions. Niche modelling results are presented showing the predicted areas of climatic suitability for *C. rotundus* (globally) and *C. longus* (British Isles) in 2050. Furthermore, historical distribution data are presented that show the northward range expansion of *C. longus* in Britain during the last 100 years. The chapter highlights the threat of climate change to endemic taxa with restricted distributions, such as *Carex* spp., *Isolepis* spp., *Khaosokia caricoides* and *Mapania* spp. These appear particularly vulnerable, although, as yet, there is no direct evidence of climate change threatening or eliminating taxa.

19.1 Introduction

Cyperaceae (sedges) are a monocotyledenous angiosperm plant family with 106 genera and 5387 species (Govaerts et al., 2007). They are placed in the order Poales and have a superficial similarity to Poaceae (grasses). Both families have much reduced flowers and are primarily wind-pollinated. Phylogenetic studies using DNA sequencing show that Cyperaceae have a sister-group relationship with Juncaceae (rushes) (Muasya et al., 1998; Chase et al., 2006) while Poaceae are more distantly related. Grasses include some of the world's most impor tant crops, such as rice (*Oryza sativa* L.), wheat (*Triticum* L. spp.) and maize (*Zea mays* L.). Grasses and sedges combined include some of the world's worst weeds, such as alang-alang (*Imperata cylindrica* (L.) Beauv.; Poaceae), purple nutsedge (*Cyperus rotundus* L.; Cyperaceae) and yellow nutsedge (*Cyperus esculentus* L.; Cyperaceae), giving the Poales top place in the league of economic plants. Despite their global importance, few, if any, climate change studies have been carried out on, or with, Cyperaceae. The aim of this chapter is to demonstrate that Cyperaceae may be a model family on which to base climate change work, and to give some examples of where Cyperaceae may win or lose in the climate change stakes.

19.2 Why are Cyperaceae important?

Cyperaceae are the seventh largest family in the angiosperms and third largest in the monocotyledons, so there is a wide range of material from which to choose for study. They are more or less cosmopolitan in distribution and are absent only from the Antarctic mainland. They occur through a wide range of altitudes from sea level up to 5000 m in the Himalayas, and are present in a broad range of habitats, from high Arctic tundra through to tropical forest, and to seasonally wet grasslands. They can be major or even dominant components of many plant communities. For example, in the UK, many mire, heath and wetland communities recognised by the National Vegetation Classification (Rodwell, 2006) have Cyperaceae as key taxa, such as *Carex elata* swamp and *Scirpus cespitosus – Eriophorum vaginatum* blanket mire (Rodwell, 1991, 1995), while in East Africa, the well-known papyrus (*Cyperus papyrus* L.) can totally dominate the margins of lakes and wetlands forming an impenetrable, floating mass of plants (Haines and Lye, 1983).

Cyperaceae have significant economic and ethnobotanical importance. Nearly 10% of the family is put to use by humans (Simpson and Inglis, 2001; Simpson, 2008) with the focus of use in the tropics. They can make an important contribution to local and regional economies. For example, *Actinoscirpus grossus* (L.f) Goetgh. and D. A. Simpson and *Cyperus corymbosus* L. have been major contributors to the rural economy of northeastern Thailand, where they are cultivated and woven into matting and basketry (Simpson, 1992), although recent personal observations by the first author suggest this is now in decline. *Cyperus corymbosus* is similarly used in south India (Amalraj, 1991).

Cyperaceae also have conservation and environmental importance. They are major or even dominant components of wetland habitats. Wetland species attract wildlife by providing food and shelter, and the decline of sedge species within these types of habitats is a useful indicator of potential habitat damage. In terms of ecosystem services they can play a particular role in the maintenance and improvement of water quality. Constructed wetlands, artificial marshes or swamps created for anthropogenic discharge such as wastewater, stormwater, runoff or sewage treatment in various parts of the world have included Cyperaceae species such as *Baumea articulata* (R. Br.) S. T. Blake, *Bolboschoenus fluviatilis* (Torr.) Soják, *Cyperus involucratus* Rottb. and *Schoenoplectus californicus* (C. A. Mey.) Soják, amongst others. For example, water treatments using Cyperaceae have demonstrated up to 92% removal of total nitrogen (Tanner, 1996) and significant sequestration of metals such as copper (Murray-Gulde et al., 2005).

19.3 Resources available

Resources available on the family are wide-ranging. There is a large number of specimens of Cyperaceae available in the world's herbaria. A particularly comprehensive collection is available at the Royal Botanic Gardens, Kew, UK, where research on Cyperaceae has been undertaken since the 1880s and the herbarium has *c.* 120 000 specimens covering all genera and nearly all species (www.kew.org/science/directory/teams/MonocotsII/background.html). Large and important herbarium collections of Cyperaceae are also available at the New York Botanical Garden (USA), Missouri Botanical Garden (USA), Muséum National d'Histoire Naturelle (France), Institute of Botany, Chinese Academy of Sciences (China) and the Royal Botanic Gardens Sydney (Australia), amongst many others. Overall, these build into a vast databank of information about Cyperaceae, which can increasingly be accessed online, such as via the Herbarium Catalogue at Kew (http://apps.kew.org/herbcat/).

Other resources based on herbarium specimens include the World Checklist of Cyperaceae (Govaerts et al., 2007), a listing of all accepted names of taxa and synonyms in Cyperaceae with distributional information down to Biodiversity Information Standards Level 3 of the International Working Group on Taxonomic Databases for Plant Sciences (TDWG – Brummitt et al., 2001). Checklists for Level 3 geographical units can be built in the online version (Govaerts et al., 2009). Distribution data from herbarium specimens could be important in understanding past and present distributions of Cyperaceae, which in turn may be linked to climate change data.

Preliminary conservation status assessments for species, using the International Union for Conservation of Nature categories and criteria (IUCN, 2001), can be made using distribution data. This is an exercise which, to date, has hardly been undertaken for Cyperaceae, but which is urgently needed. For example, we have completed assessments for Thailand using CATS GIS software tools (Willis et al., 2003 – www.kew.org/gis/projects/cats), which show that 40 species out of a total of 250 in Thailand are in the Vulnerable (VU), Endangered (EN) or Critically Endangered (CR) categories (C. A. Couch and D. A. Simpson, unpublished). Another 15 are in the Near Threatened (NT) category. Such assessments serve to highlight taxa that may be particularly vulnerable to climate change and where an accelerated programme of conservation assessment is required.

An understanding of the classification of Cyperaceae and relationships between the taxa underpins any biological work on the family. Classifications are traditionally based on the use of morphological data to delimit groups. However, one of the inherent problems of using such data in Cyperaceae is their reduced structure, especially in the inflorescence, which makes it difficult to work out homologies within characters. Many earlier workers also saw evolution in the family as a linear series of reductions in floral complexity. This has often followed the pattern of treating forest-dwelling genera with a complex floral structure (e.g. *Scirpodendron* and *Mapania*) as 'primitive', and genera of predominantly open habitats with a simplified structure, such as *Carex*, as 'advanced' (Kern, 1974). DNA-based phylogenetic trees have become available over the past decade or so (Muasya et al., 1998, 2000, 2009; Simpson et al., 2007); these have helped to shed light on a number of anomalies, and different patterns of relationship are now coming to light. For example, there is strong evidence that Cyperaceae comprises just two subfamilies, Cyperoideae and Mapanioideae (Simpson et al., 2007; Muasya et al., 2009), rather than the four (Caricoideae, Cyperoideae, Mapanioideae and Sclerioideae) previously recognised (Bruhl, 1995; Goetghebeur, 1998). We now also understand that *Carex*, the largest genus in Cyperaceae, together with allied genera which, for many years, were placed as an isolated 'advanced group' in their own subfamily or even separate family Kobresiaceae (Gilly, 1952), are closely related to tribe Scirpeae. This relationship was hinted at from a morphological standpoint by Dahlgren et al. (1985) and is clearly supported by molecular data (Simpson et al., 2007).

While some problems may have been resolved through molecular work, others have arisen, such as the placement of *Cyperus* species with either the C_3 and C_4 photosynthetic pathway into different clades despite the lack of unambiguous characters separating the C_3 and C_4 species. Despite progress in the phylogenetics and taxonomy of Cyperaceae at tribal level, there is still much more work to do at lower taxonomic ranks. Preliminary attempts at dating phylogenetic trees that include Cyperaceae have also been made (Bremer, 2002) and a recent study including most lineages of Cyperaceae show a stem age of the family at about 87 million years ago (Besnard et al., 2009). The recent discovery of *Volkeria messelensis* S. Y. Smith, Collinson, D. A. Simpson, Rudall, Marone and Stampanone (Smith et al., 2009), a 48-million-year-old fossil mapanioid sedge, should help to give more accuracy to the date estimations of major Cyperaceae clades.

Finally, there is a range of expertise in the family based in many parts of the world. Cyperologists have demonstrated an excellent level of cooperation over recent years. Cyperaceae symposia have been held at the Monocots II, III and IV conferences in 1998, 2003 and 2008, respectively, and at the International Botanical Congress in Vienna, Austria, in 2005. The Sedges 2002 Symposium was held at Delaware State University in the USA in 2002 (Naczi and Ford, 2008). These regular gatherings have ensured that our understanding of the family has advanced and have been the impetus for the molecular phylogenetic work indicated above (e.g. Muasya et al., 2000, 2009; Simpson et al., 2007).

19.4 Climate change: examples of potential winners and losers in Cyperaceae

19.4.1 Potential winner: C_4 photosynthetic pathway and *Cyperus rotundus* – the world's worst weed

Two patterns of vegetative anatomy are present in Cyperaceae, one with radiate chlorenchyma and green chlorenchymatous sheaths within the vascular bundles, the other with non-radiate chlorenchyma and vascular bundles surrounded by a sheath of colourless cells (Bruhl and Wilson, 2007). These differences are correlated with the presence, in the former, of the C_4 photosynthetic pathway. Estimates of the number of Cyperaceae species with the C_4 pathway vary between 27% (Sage et al., 1999) and 34% (Bruhl and Wilson, 2007). Within the family the C_4 pathway has evolved at least four times (Soros and Bruhl, 2000; Christin et al., 2008), giving rise to the 'chlorocyperoid' (C_4 members of *Cyperus* and allied genera), 'eleocharoid' (*Eleocharis*), 'fimbristyloid' (*Fimbristylis* and allied genera) and 'rhynchosporoid' (*Rhynchospora*) anatomical types.

C_4 plants can have an advantage over C_3 plants under conditions of water stress, high temperatures and high light intensity (Hopkins and Hüner, 2004; Bryson and Carter, 2008; Bouchenak-Khelladi and Hodkinson, Chapter 7); thus they are better adapted to hotter, drier climates. Genera of Cyperaceae with the C_4 pathway are particularly found in seasonally wet or drought-prone habitats within the tropics, exhibiting the greatest species diversity within these regions. This is exemplified by *Cyperus*, which has the greatest species diversity in eastern and southern Africa, regions which are frequently prone to periodic drought, with associated high temperature and light intensity.

Cyperaceae contain a number of taxa that are highly aggressive and potent weeds. Terry (2001) estimated this number to be 230 species, while Simpson (2008) categorised 168. Among the most widespread and competitive are *Cyperus difformis* L., *C. esculentus*, *C. iria* L., *C. haspan* L., *C. rotundus* and *Fimbristylis dichotoma* (L.) Vahl. All are pantropical and cause very serious problems for agriculture and horticulture (Terry, 2001; Bryson and Carter, 2008; Simpson, 2008) throughout their ranges. Four of these species are C_4: *C. esculentus*, *C. iria*, *C. rotundus* and *F. dichotoma*.

Cyperus rotundus is the most troublesome and has often been called the world's worst weed (Holm et al., 1977; Terry, 2001), because it causes major problems in 52 crops in 92 countries, mostly in the tropics and subtropics (Parsons and Cuthbertson, 1992; Bromilow, 1995). It spreads by underground rhizomes and tubers; there may be up to 500 plants per m^2 and there may be 53 000 tubers per m^3 of soil (Parsons and Cuthbertson, 1992). Indeed, so successful does this method of perennation appear to be that *C. rotundus* seldom sets fruit (Kern, 1974). It is resistant to many common herbicides and costs the world millions of dollars per year in lost production and control measures (Tuor and Froud-Williams, 2002).

With climate change, the potential for *C. rotundus* to move into new regions, especially those likely to become hotter and drier, is a serious cause for concern. We undertook a preliminary investigation of this potential by developing a niche model to determine areas climatically suitable for *C. rotundus* in 2050. The model was built with openModeller v.1.0.7 (http://openmodeller.sourceforge.net) using the 'GARP with best subsets – new openModeller implementation' algorithm, with the default settings (AUC = 0.83). Distribution data for *C. rotundus* were gathered from digitised herbarium records accessed from the Global Biodiversity Information Facility (www.gbif.org). Climate data were as used by Yesson and Culham (2006; Chapter 12) with present-day climate data from the Climate Research Unit (CRU) CL1 data set and future climate data based on the Hadley A2c model. The future model accounts for a moderate increase in carbon dioxide (CO_2).

Figure 19.1A shows current climatic conditions favourable to *C. rotundus*. There is a particular concentration in tropical, subtropical and warm temperate regions, especially where there is a hot dry period during the year, which would favour plants with the C_4 pathway. Figure 19.1B shows climatic conditions predicted for 2050. The overall pattern is similar, with a focus on present-day tropical to warm temperate regions. However, there are some notable differences. For example, conditions become more favourable further north in Europe up to the English Channel and North Sea coasts, with a possible spread into southern England. This would match with modelled change to hotter drier summers in northern Europe. *Cyperus rotundus* could therefore present a threat to agriculture in parts of Europe where it does not currently occur or where it is non-competitive with crops at the present time. There is also some predicted spread northwards in the USA and around the southern coasts of Australia. In contrast, conditions are predicted to become less favourable in other places, notably a large swathe of South America south of the Amazon basin, much of India, parts of Thailand and Indochina, and much of inland Australia.

These data must be treated with caution, as the distribution points focus only on data currently available through GBIF and are not an accurate representation of the worldwide distribution of *C. rotundus* (see Culham and Yesson, Chapter 10); nor are the models able to account fully for the change in competitiveness of C_4 and C_3 plants in elevated CO_2 conditions, where the physiological advantage of C_4 photosynthesis may be less than under current conditions.

19.4.2 Potential winner: *Cyperus longus* – a C_4 species in the UK

Cyperus longus L. is widely distributed from Europe to southern Africa, in the Middle East and in parts of Asia. It is present in the UK at the northern limits of its range (Collins et al., 1988). It is found in marshy pond margins, often near the

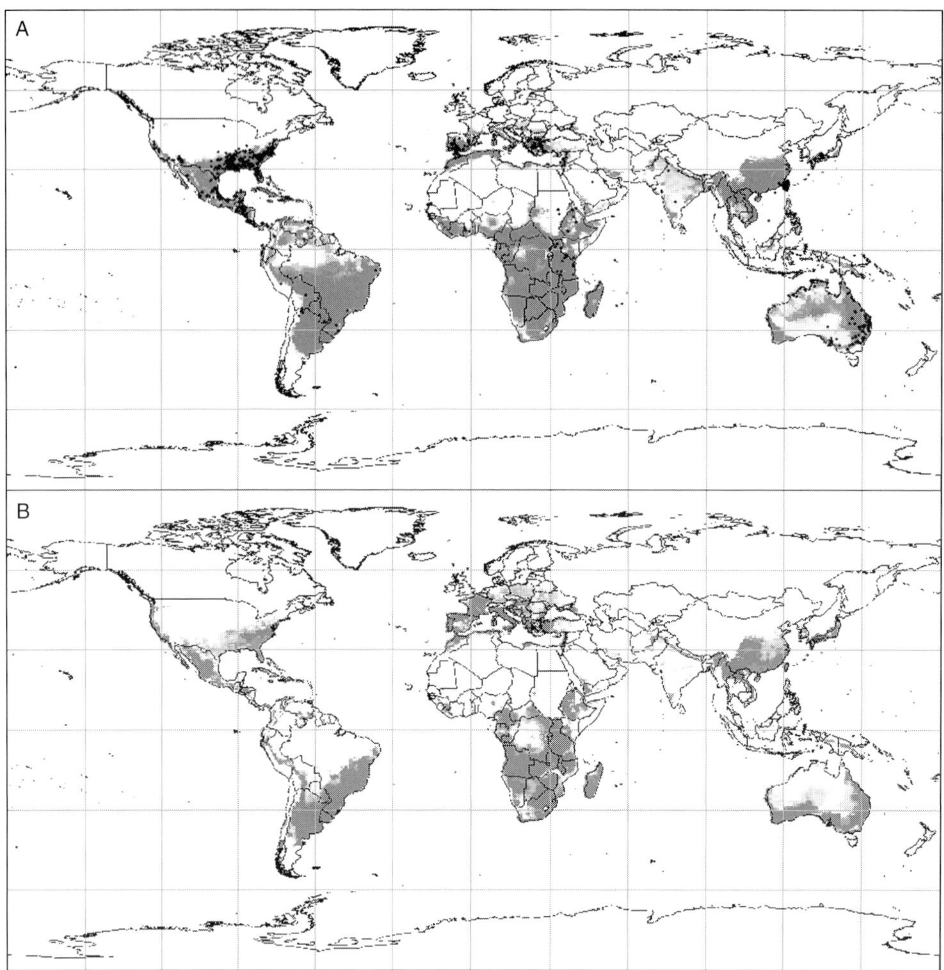

Figure 19.1 Areas climatically suitable for *Cyperus rotundus*, as determined by a GARP niche model within (A) present-day climate and (B) predicted climate for 2050. Present-day distribution suitability based on the CRU CL1 model and future distribution based on the Hadley A2c model. Darker shaded areas indicate higher suitability. Black dots show current distribution points. Global climate data are of 0.25 degree resolution and the overlying grid is of 30 degree squares.

coast, but in recent years it has become established at inland sites, where it can show vigorous rhizome growth (Jermy et al., 2007). *Cyperus longus* is one of the few native British C_4 species, and the physiology of *C. longus* under low-temperature conditions has been the subject of a number of studies (e.g. Jones et al., 1981; Collins and Jones, 1986, 1988). Such studies suggest that leaf extension is very slow at lower temperatures, which restricts leaf canopy development and is a factor in determining the outcome of competition between *C. longus* and other species. However, small differences in relative growth rate at temperatures of 10–30 °C result in large differences in foliage area between lower and higher temperatures, suggesting that the plant can better compete at higher temperatures.

Figure 19.2 shows the distribution of *C. longus* in the UK from pre-1930 to the present day. Before 1930 the species was confined to southern England, with one record as far north as Lincolnshire. In subsequent years there was an increase in the number of records together with a distributional spread northwards. By the 1970–86 period it had reached Northumberland, and in the 1987–99 period there were two records from Scotland. The most recent map (2000+) shows one record from Scotland and a number of records from northeastern and northwestern England, so it appears to be firmly established in these regions.

The northward spread is coincident with an increase in the mean annual temperature across the UK. For pre-1930 the mean annual temperature was *c.* 8.5 °C, for the 1930–69 period it was *c.* 9 °C and for 2000+ it was *c.* 9.5 °C (Hulme et al., 2002). We also prepared a GARP niche model for *C. longus* in the UK and Ireland to show areas of suitability for the growth of this species under present-day conditions and predicted conditions in 2050, again using data based on the Hadley A2c model (Fig 19.3). The data show that, at the present time, conditions in the UK are highly suitable up to *c.* 55° N, whereas by 2050 there is a shift to *c.* 56° N, with large areas of northern England and southern Scotland becoming highly suitable.

At this stage it is uncertain whether the shift of *C. longus* further north can be directly attributable to the increase in temperature; further studies are necessary. The situation is made even more complex by the fact that most of the recent records are introductions rather than natural occurrences (Jermy et al., 2007). *Cyperus longus* is frequently grown in gardens as a bog or pond-side plant, and it is possible that many of these records represent material of garden origin. Whatever the form of spread, it is nevertheless apparent that the plant is now occurring much further north than it has done in the past. If this C_4 species can spread in such a way then the possibilities of other, more pernicious taxa, such as *C. rotundus* and *C. esculentus*, getting a foothold in the UK and Ireland would seem to be increasing.

19.4.3 Potential winner: *Cyperus* sect. *Arenarii*

Cyperus sect. *Arenarii* Kunth ex Jaub. and Spach. is a group of 25 species (e.g. *Cyperus conglomeratus* Rottb., *C. crassipes* Vahl – Väre and Kukkonen, 2005) distributed from the Mediterranean region to India and Sri Lanka and throughout Africa. Many of the species are psammophytic, being adapted to hot, dry conditions, especially deserts and sandy areas on coasts. They have a number of characteristics that help to reduce water loss, including a thick cuticle all over the plant, bluish-green coloration and narrow, canaliculate leaves. In common with other Cyperaceae they have a reduced inflorescence structure. A characteristic feature

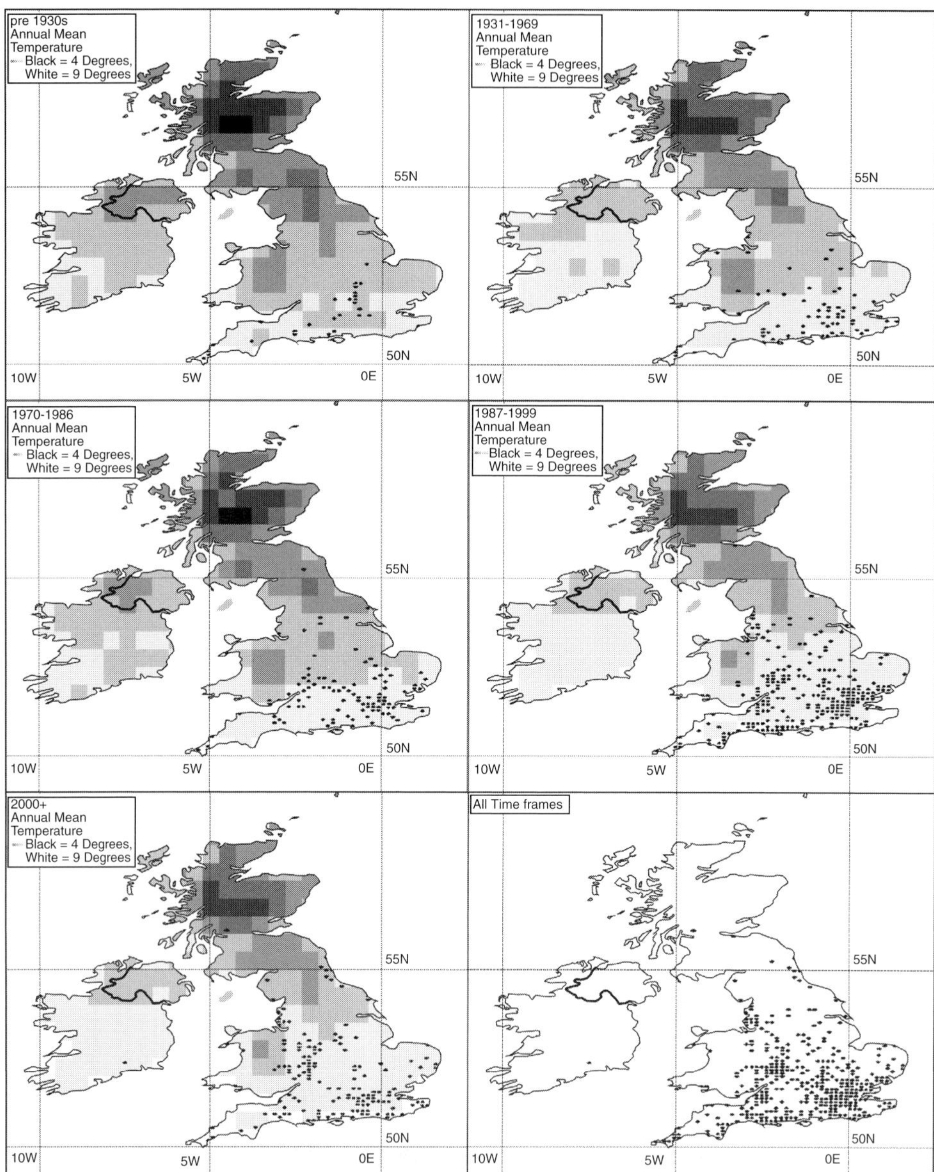

Figure 19.2 Changes in the UK distribution of *Cyperus longus* over the past century. Distribution data from the Botanical Society of the British Isles (www.bsbimaps.org.uk/atlas). The 'All Timeframes' map summarises all the distribution points. Gridded shading indicates annual mean temperature data for the relevant time period (except for 2000+, where data are not available, and therefore 1990s data are shown). Shading is in increments of 1 °C (black = 4 °C, white = 9 °C). Data taken from the CRU TS 2.1 global time series data sets (Mitchell and Jones, 2005). Summary of all distribution data shown in all timeframes.

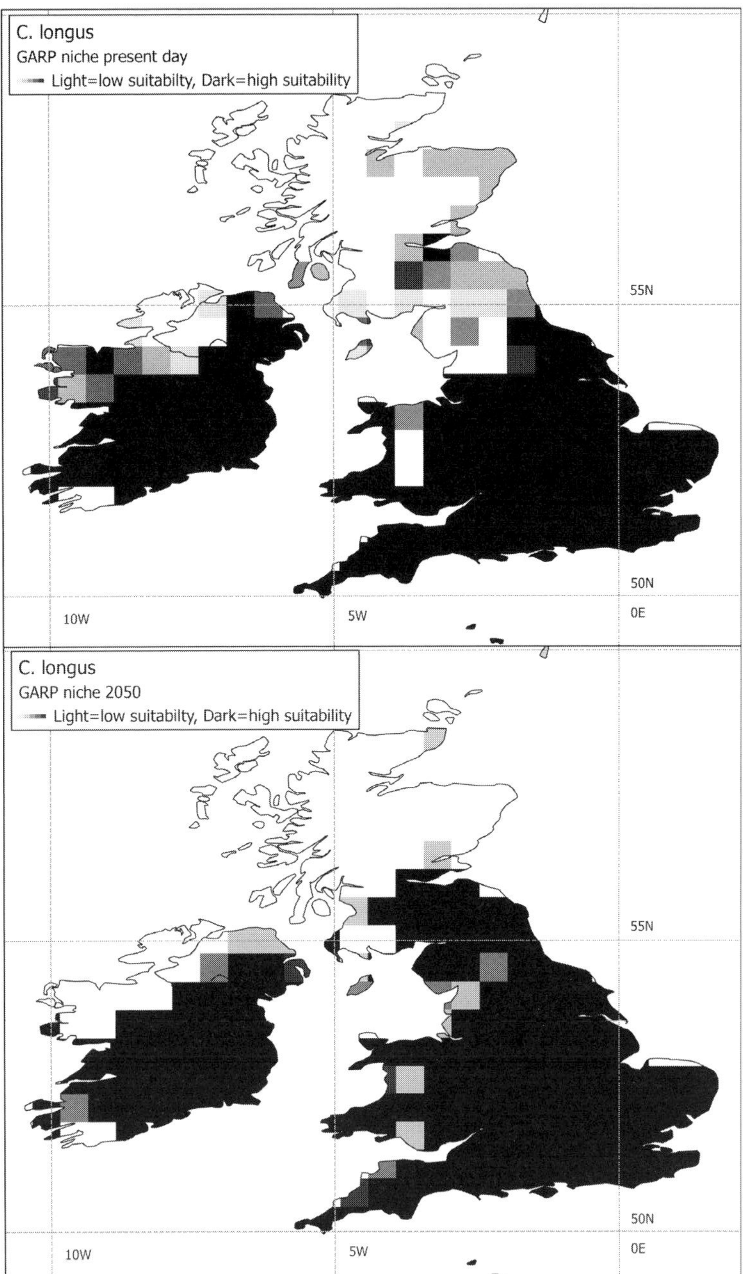

Figure 19.3 Areas climatically suitable for *Cyperus longus* in the UK and Ireland. Areas determined by a GARP niche model within present-day climate and predicted climate for 2050, trained with distribution data from GBIF using present-day climate data. This model was projected into the predicted climate for 2050 (scenario A2c). Darker shaded areas indicate higher suitability.

of the perennial species is the long creeping rhizome, which is often densely tomentose. The rhizome may be deeply buried in sand, and the tomentum probably helps to reduce water loss and aid survival in conditions of extreme aridity (Väre and Kukkonen, 2005). Väre and Kukkonen (2005) noted that some species are able to penetrate into the most arid parts of the Sahara where little else survives. The four taxa analysed to date for photosynthetic pathway are all C_4. All members of this group would probably show range extensions if hotter, drier conditions began to prevail outside their current distribution, for example in parts of southern Europe. Interestingly, some members of the group, especially *C. conglomeratus*, are recognised for their sand-binding ability, and are used for that purpose in parts of the Middle East (Simpson and Inglis, 2001). There could be an increasing use of this and other related species in areas that become increasingly prone to desertification through climate change. A niche modelling approach, based on the distribution data of the taxa in *Cyperus* sect. *Arenarii*, would again be a worthwhile exercise to predict future distribution patterns.

19.4.4 Potential losers

There is a wide range of taxa that may be under threat from climate change. Endemic taxa with restricted distributions appear particularly vulnerable, and there are numerous examples of such taxa among the Cyperaceae. It should be noted that, at this stage, we have no direct evidence of climate change threatening or eliminating taxa, and the examples given below are intended to highlight potential problems.

Mapania is a pantropical genus of *c.* 80 species and is a characteristic group of the herb layer of ever-wet tropical forests. Simpson (1992) noted that they occur where little light penetrates through the forest canopy, mostly 'in areas where the soil is particularly damp, muddy or peaty or in swampy depressions or by the side of pools or streams'. Many species are local endemics with highly restricted distributions; just a few species are widespread: for example *Mapania cuspidata* (Miq.) Uttien, which occurs throughout Southeast Asia. Whether widespread or restricted in distribution they are seldom abundant in any one locality. Engelbrecht et al. (2007) observed that drought has a limiting effect on tropical plant distributions and that in tropical rainforests global warming could result in loss of diversity and perhaps species extinction. Drier conditions could, therefore, accelerate the decline, and possible extinction, of many *Mapania* species throughout the tropics. Of course, other factors could be involved. Simpson (1992) noted that some of the local species may already be extinct through forest clearance, as they seem unable to tolerate higher light and lower humidity levels. The survival of *Mapania* spp. may also depend on factors such as dispersal ability of the species and niche suitability, but it is not yet known how well *Mapania* could migrate with changing climate or adapt to new conditions. Further work is necessary.

Khaosokia caricoides D. A. Simpson, Chayam. and J. Parn. was first discovered in Khao Sok National Park, Surat Thani Province, Thailand, in 2001 (Simpson et al., 2005). So far it has only been recorded from one locality, and its habitat (crevices on sunny limestone cliffs) is highly unusual for Cyperaceae. It occurs in an area once covered by rainforest which was flooded to produce a reservoir. Prior to that the species would have been inaccessible, growing high up on the limestone cliffs. However, the formation of the reservoir has allowed the plant to be reached by boat. It could be present in similar limestone habitats elsewhere in Thailand and Malaysia that are so far inaccessible to collectors. Such habitats are already threatened by human activity, especially limestone extraction. Human activity is unlikely to threaten the known locality for *K. caricoides*, because the reservoir offers protection, but other populations, should they exist, could be put under pressure.

Figure 19.4 shows current and predicted annual mean temperatures for Thailand based on the same climate models used in the niche model for *C. rotundus*, above. The locality for *K. caricoides* is indicated by 'A'. The data suggest that

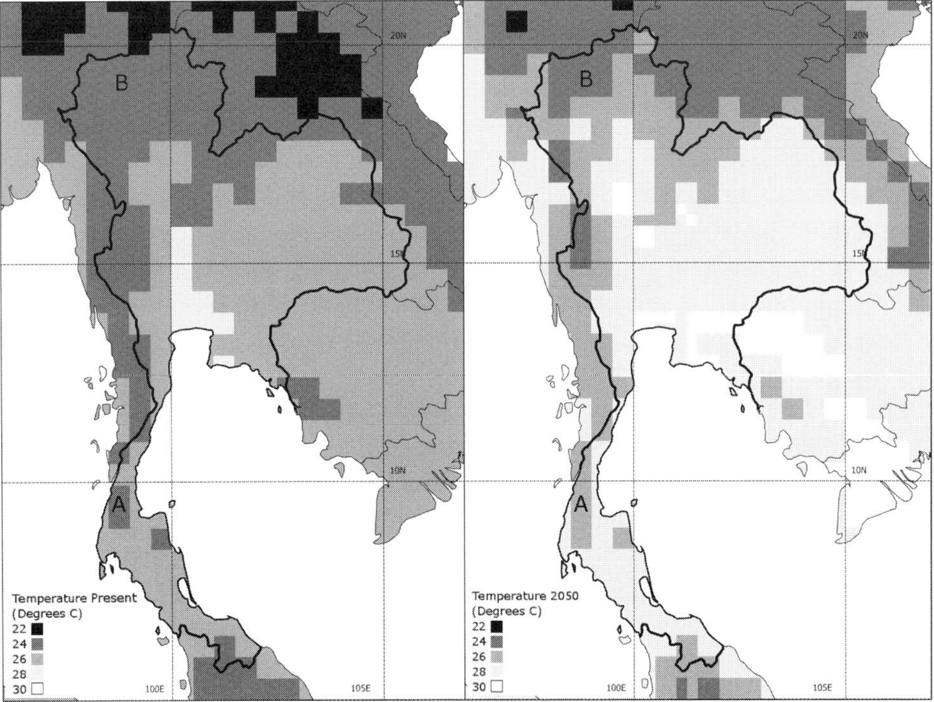

Figure 19.4 Annual mean temperature for Thailand. Climate data for present and future predictions as for Fig 19.1. A, only known locality of *Khaosokia caricoides*. B, *Carex phyllocaula*, endemic to Doi Chiang Dao Mountain.

temperatures will increase at this locality by 2 °C by 2050. Yusuf and Francisco (2009) indicated that Surat Thani is one of the provinces in Thailand vulnerable to the effects of multiple climate change hazards including drought and flood, while Boonprakrob and Hattirat (2006) suggested that Khao Sok National Park will be adversely affected by climate change, based on the UK 89 climate model. Moreover, neighbouring Krabi Province will be affected by a 10% decrease in total annual rainfall by 2033 (WWF, 2008) due to a predicted shortening of the annual monsoon season over peninusular Thailand by four weeks. Unfortunately we know next to nothing about the ecological requirements of *K. caricoides*, but, as far as we can tell, rain falling directly onto the cliffs, together with any associated runoff, is the only way the plant obtains water. Should the climate become drier at Khao Sok, *K. caricoides* could be affected by the reduced amount of rainfall it would receive. Simpson et al. (2005) also demonstrated it to be a C_3 species, which suggests it could be less tolerant of hotter, drier conditions. Once again, further studies are necessary.

Many species of the large and cosmopolitan genus *Carex* have restricted distributions. Several species in Thailand have been shown to have an IUCN conservation status of VU to CR (C. A. Couch and D. A. Simpson, unpublished data). One such example is *Carex phyllocaula* Nelmes, which is classified as EN (based on Area of Occupancy – IUCN, 2001) and is endemic to Doi Chiang Dao, a mountain in northern Thailand. It is found in thickets and open grassy ground at altitudes of 1500–2100 m and is only known from a few specimens. Figure 19.4 shows the locality for *C. phyllocaula*, marked as 'B' on the map. Present-day and predicted temperatures are the same. However, there is evidence for a predicted long-term decrease in September rainfall over the Indochina peninsula, which includes northern Thailand (Takahashi et al., 2008). This may have a longer-term effect on the vegetation on the mountain, and suitable habitat for *C. phyllocaula* may be reduced, affecting the plant's survival. Other rare species in Thailand could be similarly affected.

Other mountain endemic species are known worldwide. For example, three species of *Isolepis* R. Br., *I. keniaensis* Lye, *I. kilimanjarica* R. W. Haines and Lye and *I. ruwenzoriensis* R. W. Haines and Lye are respectively endemic to Mount Kenya, Mount Kilimanjaro and Mount Ruwenzori in East Africa (Haines and Lye, 1983; Muasya and Simpson, 2002). They only occur at high altitudes (3650–4350 m) and are known from just one specimen each, an indication of their rarity. All are also reported to occur in alpine bogs. Climate change, especially changes in rainfall pattern and/or higher temperatures, could again affect the composition of the vegetation at these altitudes, perhaps by changing species composition or encouraging scrub or forest to move to higher altitudes in the mountains.

19.5 Where next?

There is much work to be done in order to obtain a comprehensive picture of how Cyperaceae will be affected by climate change. Indeed, we are just at the start of gaining an understanding in this area. We must focus on the distributions of taxa and their controlling environmental variables. Some distributions, especially for taxa that are common weeds, are well understood and there are many records available, either in the literature or from herbarium specimens. Much less is known about those with restricted distributions; are they truly rare, or is their perceived rarity an artefact due to insufficient collecting? We also need to consider linking past and present distributions to different climate change models to see how the taxa would be affected under various scenarios. The preliminary niche modelling work with *Cyperus rotundus* described above has shown the potential use of such techniques. We need a better understanding of the evolution of characters and traits to determine how taxa might respond to climate change. The occurrence of C_3 and C_4 photosynthesis in Cyperaceae needs particular study, as the potential for C_4 species to extend their distributions and for the weedy taxa to wreak havoc on agricultural systems in temperate regions is high.

In all this work, taxonomy and systematics will play a vital role. Both will help to provide basic data about taxa, whether on characters and traits or on distributions and rarity, as well as a better understanding of the relationships between taxa and character evolution. Such research will be needed to underpin planning for agricultural systems and food supplies in Europe and around the world, for planning the conservation of biodiversity at global and local scales, and for discovering which species may have new uses in the stabilisation of soils and protection from erosion of water catchment areas.

References

Amalraj, V. A. (1991). Cultivated sedges of S. India for mat weaving industry. *Journal of Economic and Taxonomic Botany*, **14**, 629–631.

Besnard, G., Muasya, A. M., Russier, F. et al. (2009). Phylogenomics of C_4 photosynthesis in sedges (Cyperaceae): multiple appearances and genetic convergence. *Molecular Biology and Evolution*, **26**, 1909–1919.

Boonprakrob, K. and Hattirat, S. (2006). *Crisis or Opportunity: Climate Change and Thailand*. Bangkok: Greenpeace Southeast Asia.

Bremer, K. (2002). Gondwanan evolution of the grass alliance of families (Poales). *Evolution*, **56**, 1374–1387.

Bromilow, C. (1995). *Problem Plants of South Africa*. Arcadia: Briza Publications.

Bruhl, J. J. (1995). Sedge genera of the world: relationships and a new classification of Cyperaceae.

Australian Systematic Botany, **8**, 125–305.

Bruhl, J. J. and Wilson, K. (2007). Towards a comprehensive survey of C_3 and C_4 photosynthetic pathways in Cyperaceae. *Aliso*, **23**, 99–148.

Brummitt, R. K., Pando, F., Hollis, S. et al. (2001). *World Geographic Scheme for Recording Plant Distributions. Plant Taxonomic Database Standards No. 2*, 2nd edn. Pittsburgh, PA: Hunt Institute for Botanical Documentation.

Bryson, C. T. and Carter, R. (2008). The significance of Cyperaceae as weeds. In *Sedges: Uses, Diversity, and Systematics of the Cyperaceae*, ed. R. F. C. Naczi and B. A. Ford. Monographs in Systematic Botany from the Missouri Botanical Garden. St Louis, MO: Missouri Botanical Garden Press, pp. 15–101.

Chase, M. W., Fay, M. F., Devey, D. S. et al. (2006). Multigene analyses of monocot relationships: a summary. *Aliso*, **22**, 63–75.

Christin, P. A., Salamin, N., Muasya, A. M. et al. (2008). Genetic convergence exposes determinants of Rubisco higher efficiency in C_4 plants. *Molecular Biology and Evolution*, **25**, 2361–2368.

Collins, R. P. and Jones, M. B. (1986). The seasonal pattern of growth and production of a temperate C4 species, *Cyperus longus*. *Journal of Experimental Botany*, **37**, 1823–1835.

Collins, R. P. and Jones, M. B. (1988). The effects of temperature on leaf growth in *Cyperus longus*, a temperate C4 species. *Annals of Botany*, 61, 355–362.

Collins, R. P., McNally, S. F., Simpson, D. A. and Jones, M. B. (1988). Infraspecific variation of *Cyperus longus* L. in Europe. *New Phytologist*, **110**, 279–289.

Dahlgren, R. M. T., Clifford, H. T. and Yeo, P. F. (1985). *The Families of the Monocotyledons*. Berlin: Springer-Verlag.

Engelbrecht, B. M. J., Comita, L. S., Condit, R. et al. (2007). Drought sensitivity shapes species distribution patterns in tropical forests. *Nature*, **447**, 80.

Gilly, G. L. (1952). Phylogenetic development of the inflorescence and generic relationships in the Kobresiaceae. *Iowa State College Journal of Science*, **26**, 210–212.

Goetghebeur, P. (1998). Cyperaceae. In *The Families and Genera of Vascular Plants 4*, ed. K. Kubitzki, H. Huber, P. J. Rudall, P. S. Stevens and T. Stützel. Berlin: Springer-Verlag, pp. 141–190.

Govaerts, R., Simpson, D. A., Bruhl, J. et al. (2007). *World Checklist of Cyperaceae*. London: Royal Botanic Gardens, Kew.

Govaerts, R., Simpson, D. A., Bruhl, J. et al. (2009). *World Checklist of Cyperaceae*. London: Royal Botanic Gardens, Kew. www.kew.org/wcsp/monocots.

Haines, R. W. and Lye, K. (1983). *Sedges and Rushes of East Africa*. Nairobi: East African Natural History Society.

Holm, L. G., Plucknett, D. L., Pancho, J. V. and Herberger, J. P. (1977). *The World's Worst Weeds*. Honolulu, HI: University Press of Hawaii.

Hopkins, W. G. and Hüner, N. P. A. (2004). *Introduction to Plant Physiology*, 3rd edn. New York, NY: Wiley.

Hulme, M., Jenkins, G. J., Lu, X. et al. (2002). *Climate Change Scenarios for the United Kingdom: the UKCIP02 Scientific Report*. Norwich: Tyndall Centre for Climate Change Research, University of East Anglia.

International Union for the Conservation of Nature (I UCN) (2001). *The IUCN Red List Categories and Criteria*, version

3.1. www.iucnredlist.org/technical-documents/categories-and-criteria.

Jermy, A. C., Simpson, D. A., Foley, M. J. Y. and Porter, M. (2007). *Sedges of the British Isles*, 3rd edn. BSBI Handbook 1. London: Botanical Society of the British Isles.

Jones, M. B., Hannon, G. E. and Coffey, M. D. (1981). C_4 photosynthesis in *Cyperus longus* L., a species occurring in temperate climates. *Plant, Cell and Environment*, **4**, 161–168.

Kern, J. H. (1974). Cyperaceae. In *Flora Malesiana*, Vol. 1, ed. C. G. G. J. van Steenis. Leiden: Noordhoff, pp. 435–753.

Mitchell, T. D. and Jones, P. D. (2005). An improved method of constructing a database of monthly climate observations and associated high-resolution grids. *International Journal of Climatology*, **25**, 693–712.

Muasya, A. M. and Simpson, D. A. (2002). A revision of the genus *Isolepis* (Cyperaceae). *Kew Bulletin*, **57**, 257–362.

Muasya, A. M., Simpson, D. A., Culham, A. and Chase, M. W. (1998). An assessment of suprageneric phylogeny in Cyperaceae using *rbcL* sequence data. *Plant Systematics and Evolution*, **211**, 257–271.

Muasya, A. M., Bruhl, J. J., Simpson, D. A., Culham, A. and Chase, M. W. (2000). Suprageneric phylogeny of Cyperaceae. In *Monocots: Systematics and Evolution*, ed. K. L. Wilson and D. Morrison. Melbourne: CSIRO Publishing, pp. 593–601.

Muasya, A. M., Simpson, D. A., Verboom, G. A. et al. (2009). Phylogeny of Cyperaceae based on DNA sequence data: current progress and future prospects. *Botanical Review*, **75**, 52–66.

Murray-Gulde, C. L., Huddleston, G. M., Garber, K. V. and Rodgers, J. H. (2005). Contributions of *Schoenoplectus californicus* in a constructed wetland system receiving copper contaminated wastewater. *Water, Air and Soil Pollution*, **163**, 355–378.

Naczi, R. F. C. and Ford, B. A., eds. (2008). *Sedges: Uses, Diversity, and Systematics of the Cyperaceae*. St Louis, MO: Missouri Botanical Garden Press.

Parsons, W. T. and Cuthbertson, E. G. (1992). *Noxious Weeds of Australia*. Melbourne/Sydney: Inkata Press.

Rodwell, J. S., ed. (1991). *British Plant Communities, Volume 2. Mires and Heaths*. Cambridge: Cambridge University Press.

Rodwell, J. S., ed. (1995). *British Plant Communities, Volume 4. Aquatic Communities, Swamps and Tall-herb Fens*. Cambridge: Cambridge University Press.

Rodwell, J. S. (2006). *National Vegetation Classification: Users' Handbook*. Peterborough: Joint Nature Conservation Committee.

Sage, R. F., Li, M. and Monson, R. K. (1999). The taxonomic distribution of C_4 photosynthesis. In *C_4 Plant Biology*, ed. R. F. Sage and R. K. Monson. San Diego, CA: Academic Press, pp. 551–584.

Simpson, D. A. (1992). Notes on Cyperaceae in north-eastern Thailand. *Cyperaceae Newsletter*, **10**, 10–12.

Simpson, D. A. (2008). Frosted curls to tiger nuts: ethnobotany of Cyperaceae. In *Sedges: Uses, Diversity, and Systematics of the Cyperaceae. Monographs in Systematic Botany from the Missouri Botanical Garden*, ed. R. F. C. Naczi and B. A. Ford. St Louis, MO: Missouri Botanical Garden Press, pp. 1–14.

Simpson, D. A. and Inglis C. A. (2001). Cyperaceae of economic, ethnobotanical and horticultural importance: a checklist. *Kew Bulletin*, **56**, 257–360.

Simpson, D. A., Muasya, A. M., Chayamarit, K. et al. (2005). *Khaosokia caricoides*, a new genus and species of Cyperaceae from Thailand. *Botanical Journal of the Linnean Society*, **149**, 357–364.

Simpson, D. A., Muasya, A. M., Alves, M. et al. (2007). Phylogeny of Cyperaceae based on DNA sequence data: a new *rbcL* analysis. *Aliso*, **23**, 72–83.

Smith, S. Y., Collinson, M. E., Simpson, D. A. et al. (2009). Elucidating the affinities and habitat of ancient, widespread Cyperaceae: *Volkeria messelensis* gen. et sp. nov., a fossil mapanioid sedge from the Eocene of Europe. *American Journal of Botany*, **96**, 1581–1593.

Soros, C. L. and Bruhl, J. J. (2000). Multiple evolutionary origins of C_4 photosynthesis in the Cyperaceae. In *Monocots: Systematics and Evolution*, ed. K. L. Wilson and D. Morrison. Melbourne: CSIRO Publishing, pp. 629–636.

Takahashi, H. G., Yoshikane, T., Hara, M. and Yasunari, T. (2008). High-resolution regional climate simulations of the long-term decrease in September rainfall over Indochina. *Atmospheric Science Letters*, **10**, 14–18.

Tanner, C. C. (1996). Plants for constructed wetland treatment systems: a comparison of the growth and nutrient uptake of eight emergent species. *Ecological Engineering*, **7**, 59–83.

Terry, P. J. (2001). The Cyperaceae: still the world's worst weeds? In *The World's Worst Weeds*, ed. C. R. Riches. Symposium Proceedings No. 77. Farnham: British Crop Protection Council, pp. 3–18.

Tuor, F. A. and Froud-Williams, R. J. (2002). Interaction between purple nutsedge, maize and soybean. *International Journal of Pest Management*, **48**, 65–71.

Väre, H. and Kukkonen, I. (2005). Seven new species of *Cyperus* (Cyperaceae) section *Arenarii* and one new combination and typification. *Annales Botanici Fennici*, **42**, 473–483.

Willis, F., Moat, J. and Paton, A. (2003). Defining a role for herbarium data in Red List assessments: a case study of *Plectranthus* from eastern and southern tropical Africa. *Biodiversity and Conservation*, **12**, 1537–1552.

WWF (2008). *Climate Change Impacts in Krabi Province, Thailand*. Bangkok: WWF and SEA-START. http://assets.panda.org/downloads/thailand_full_final_report.pdf.

Yesson, C. and Culham, A. (2006). A phyloclimatic study of *Cyclamen*. *BMC Evolutionary Biology*, **6**, 72.

Yusuf, A. A. and Francisco, H. A. (2009). *Climate Change Vulnerability Mapping for Southeast Asia*. Singapore: Economy and Environment Program for Southeast Asia.

20

An interdisciplinary review of climate change trends and uncertainties: lichen biodiversity, arctic–alpine ecosystems and habitat loss

C. J. Ellis and R. Yahr

Royal Botanic Garden Edinburgh, UK

Abstract

We provide an overview of trends and uncertainties emerging from the growing field of climate change and biodiversity research using lichens as a study group. Problems in understanding the implications of global change for lichens are relevant to other groups comprising subdominant species such as algae, mosses and liverworts. Ecological study of lichens represents a diverse range of the ascomycete fungi, which have adopted a strategy in symbiosis with an inhabitant autotrophic partner. In general lichens may be considered 'stress tolerators', although contrasting lichens encompass a range of life histories with respect to reproduction, dispersal and habitat specialisation. Lichens typically occupy microhabitats nested within a larger-scale habitat mosaic and are relatively little studied compared to

Climate Change, Ecology and Systematics, ed. Trevor R. Hodkinson, Michael B. Jones, Stephen Waldren and John A. N. Parnell. Published by Cambridge University Press.

vascular plants and animals. We examine two main themes: (1) the direct effect of climate warming on lichens with respect to arctic–alpine ecosystems; and (2) the indirect effect of climate change on lichens resulting from interaction with other environmental factors. Within this framework we discuss the current limits to bioclimatic modelling, the role of molecular ecology in climate change studies, species interactions, and opportunities for conservation in the face of climate change uncertainty. We draw on research from across geographic regions, with several focused examples referring to lichens in Britain and Ireland, which have the advantage of being among the best-explored lichen floras in the world.

20.1 Introduction

Lichens represent a fungal life-history strategy in which a heterotrophic fungal species forms a functional association with an inhabitant autotrophic partner, i.e. with a green alga and/or a cyanobacterium, collectively known as the 'photobiont' (Hale, 1983). Lichenisation is thought to have evolved in the fungi at least 400 million years ago (Taylor et al., 1995), arising independently multiple times within the basidiomycetes and ascomycetes (Gargas et al., 1995). However, 98% of lichenised fungi are ascomycetes, and it has been suggested that lichenisation represents the evolutionary ancestral state of most of the diversity of filamentous ascomycetes (Lutzoni et al., 2001), with the multiple subsequent losses of lichenisation explaining an intermixed phylogenetic pattern of lichenised, non-lichenised and lichenicolous fungi (Lutzoni et al., 2001; Grube and Winka, 2002; Guiedan et al., 2008).

A majority of the biomass and the gross morphology of the lichen thallus is typically formed of fungal tissue that is modified according to function (Sanders, 2001). The structure of fungal tissues serves to protect the photosynthetic partner (e.g. chemical defences against herbivory or ultraviolet (UV) light damage – Solhaug et al., 2003; Gauslaa, 2005), and to harvest carbohydrates from the photobiont to support fungal growth and reproduction (Honegger, 1991; Palmqvist, 2000). The scientific name of the 'lichen' refers to the lichenised fungus: different lichens are in fact different fungal species, and the fungal species may be associated with the same or different algal or cyanobacterial species from a wide range of phylogenetic lineages. Taxonomic authorities throughout follow Smith et al. (2009) for lichen fungi and Stace (1997) for vascular plants.

20.1.1 Reproduction and dispersal

Reproduction in lichens can be either clonal, with both partners dispersing together, or via sexual reproduction of the fungus. While contained in the lichenised relationship the photobiont is limited to reproducing asexually (Bubrick, 1988), and clonal reproduction and dispersal of both the fungal and photobiont partner together occurs via asexual propagules (e.g. isidia or soredia). Where sexual reproduction of the fungus occurs, dispersal is by fungal spores. Colonising spores are typically required to reassociate with a photosynthetic partner in order to form a lichen, although a few fungal genera can codisperse their sexual spores with algal cells (Gueidan et al., 2007). In addition, there are a few examples where a fungus may have the capacity to adopt both a lichenised and an alternate trophic strategy, e.g. saprotrophic (Wedin et al., 2004). Sexual systems in ascomycetes can be heterothallic (effectively outcrossing, requiring the meeting of two different mating types) or homothallic (effectively selfing, in which case a single thallus can produce spores by itself). Breeding systems of lichen fungi have been little studied, but cases of both heterothallism (Zoller et al., 1999; Murtagh et al., 2000; Scherrer et al., 2005; Seymour et al., 2005a) and homothallism are known (Scherrer et al., 2005).

Contrasting reproductive modes are expected to have consequences for the evolutionary adaptation of lichens to environmental change. Both in theory and in empirical studies, sexual reproduction results in higher genetic diversity (Fahselt, 1989; Hageman and Fahselt, 1990; Zoller et al., 1999), and potentially more frequent long-distance dispersal by spores, although this may be balanced against the requirement to locate a suitable photobiont partner. Asexual reproduction may cause reduced genetic diversity, build-up of deleterious mutations and less frequent long-distance dispersal, but is possibly a more effective short-distance dispersal mechanism, in part because of the guaranteed availability of a photobiont (Seymour et al., 2005b). In cases where a single lichen species exhibits both sexual and asexual strategies, investment in one or the other reproductive mode may be partitioned according to ecological circumstances (Hestmark, 1992).

20.1.2 Lichen physiology and ecological success

Lichens lack a complex vascular system and are poikilohydric (i.e. with internal water relations linked strongly to ambient climatic conditions). Nevertheless, they are extremely successful ecologically, and the lichenised relationship demonstrates important emergent physiological properties. Lichens are highly desiccation-tolerant with active metabolic processes to counter the effect of extreme or prolonged dryness (Kranner et al., 2008). They are able to hydrate and reactivate rapidly during short periods of exposure to water (e.g. as dew, fog, rain or snowmelt), maintaining a positive carbon balance under conditions that may be limiting for the growth of vascular plants: e.g. in fog deserts (Lange et al., 2006, 2007) or in subzero temperatures in dry polar environments (Kappen, 1990, 2000; Schroeter et al., 1994). Lichens also efficiently capture limiting nutrients from atmospheric sources (e.g. nitrogen and phosphorus) and can be independent of soil nutrient cycling (Crittenden, 1989; Ellis et al., 2004). Accordingly, lichens may be considered 'stress-tolerators' (Grime, 1977) or 'extremophiles', colonising habitats where there is an absence of competition from vascular plants, or hitching a ride on the top-competitors as epiphytes and leaf-dwelling species.

20.1.3 Lichens and climate change

Lichens are ecologically successful in 'stressed' environments and microhabitats. As a functional group they occur ubiquitously, from tropical forests to the polar regions, from the littoral seashore to the summits of the world's highest mountains, and from desert conditions to aquatic freshwater environments. However, considering individual species employing this fungal strategy, certain lichens appear to be globally cosmopolitan and to occupy a relatively wide range of habitats, while others are geographically restricted and may be considered microhabitat specialists.

An assessment of the threat from climate change therefore needs to be carefully tailored to lichen species that invoke some expectation for climatic sensitivity. In this regard we focus on species that show biogeographic restriction putatively explained by present-day climate (i.e. species with a distinctive arctic–alpine, or montane, bioclimatic range). For these species we provide a summary of the evidence base relating to climate change sensitivity and critique this evidence by focusing on the interaction of lichen species with their biotic and abiotic environment. We also explore the role of climate change as an additional larger-scale factor operating alongside an amalgam of multiple known drivers of lichen biodiversity change. In this context the indirect consequences of climate change are potentially spread across a wider spectrum of species from many biogeographic regions. Relatively little research has been conducted to explicitly address the consequences of climate change for lichen diversity, and this chapter therefore draws on evidence from a range of disciplines to provide a current summary. We also summarise potential avenues of future research that may provide greater insight into lichen sensitivity to climate change.

20.2 Arctic–alpine species and the direct threat of climate warming

In many assessments of climatic sensitivity, attention has focused on arctic and alpine environments, as such ecosystems are expected to be directly threatened by global warming. Lichens form important components of these environments in terms of vegetation structure and function (Longton, 1988). Arctic–alpine environments thus provide an important case study for the examination of the impacts of climate change.

Species which are adapted to arctic–alpine environments are thought to be severely threatened by climate change; range shifts and phenological change have been observed across the taxonomic spectrum (Parmesan, 2006). There is evidence that many components of the arctic–alpine vascular plant flora evolved during a late Tertiary cooling period in earth's geologic history (Tribsch and Stuessy, 2003; Brochmann and Brysting, 2008), with genetic diversity structured in part by climatic variation during the late Pliocene and Pleistocene (Comes and Kadereit, 2003; Brochmann and Brysting, 2008). Given the putative evolutionary adaptation of arctic–alpine species to habitat regimes that originated during periods characterised by low global temperatures, continued climate warming might be expected to pose a direct threat to the arctic–alpine flora. This threat has been assessed using predictive models, which unequivocally demonstrate a loss of bioclimatic space for arctic–alpine species (e.g. Thuiller et al., 2005; Hamann and Wang, 2006; Rehfeldt et al., 2006; Trivedi et al., 2008a). The projected impact of climate warming is supported by observational data to demonstrate a shifting balance between the declining occurrence or abundance of arctic–alpine species (Klanderud and Birks, 2003; Lesica and McCune, 2004) and an expansion of species from lower latitudes or altitudes (Grabherr et al., 1994; Sturm et al., 2001; Kullman 2002; Lenoir et al., 2008).

20.2.1 Climatic sensitivity and lichen biogeography

In making a preliminary assessment of climatic sensitivity and the potential impacts of climate change on a regional flora, extreme caution must be exercised with respect to biogeographic context. Lichens are often significant components of the vegetation in arctic and alpine habitats, where the growth of tall-stature plants is limited by environmental stress. In continental Antarctica, for example, *c.* 386 lichen species have been recorded, compared to only two vascular plants and 111 mosses (Øvstedal and Lewis Smith, 2001; Ochyra et al., 2008). On this basis, one might expect climate change to pose a generic threat to arctic–alpine and Antarctic lichens. However, the Antarctic lichen flora includes many species which appear to be globally cosmopolitan; for example, species such as *Amandinea punctata* (Hoffm.) Coppins and Scheid., *Buellia aethalea* (Ach.) Th. Fr., *Candelariella vitellina* (Hoffm.) Müll. Arg., *Lecanora polytropa* (Hoffm.) Rabenh., *Parmelia saxatilis* (L.) Ach. and *Rhizocarpon geographicum* (L.) DC. are common both in Antarctica and in the more genteel surroundings of British churchyards (Øvstedal and Lewis Smith, 2001; Dobson, 2005).

Accurate biogeographic data are necessary to identify candidate species that might be considered threatened, though such data are often lacking for lichens, which remain poorly recorded across many regions of the world. As one example of a potential candidate group threatened by global warming, the *Neuropogon*-type

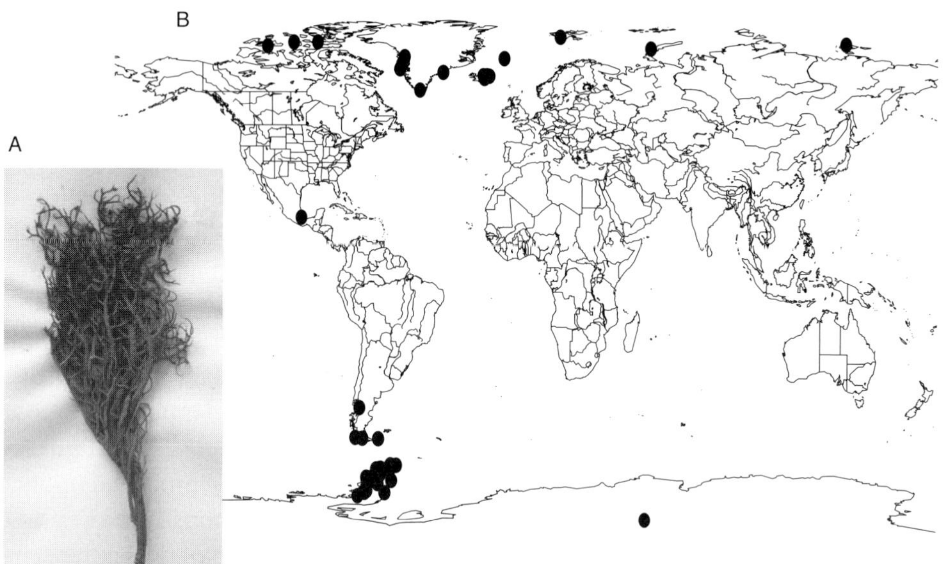

Figure 20.1 (A) Specimen of *Usnea aurantiaco-atra*, an example of an *Usnea* with *Neuropogon*-type traits. (B) Worldwide distribution of the *Neuropogon*-type *Usnea* spp., drawn from data presented by Walker (1985). The bipolar distribution belongs to *Usnea sphacelata* R. Br.

species in the genus *Usnea* meet the criteria of biogeographic restriction and specific adaptation (Fig 20.1A). Formerly treated as a separate genus, subgenus or section within *Usnea*, the *Neuropogon*-type *Usnea* species are characterised by a distributional centre in the Antarctic and subantarctic regions, with outliers at high altitudes (Fig 20.1B). However, molecular evidence has shown that members of the *Neuropogon* group are polyphyletic within the genus *Usnea* (Wirtz et al., 2006; Seymour et al., 2007). This indicates that morphological characters common to the group (e.g. a dark pigmented cortex) may be independently evolved adaptations to harsh polar or high-altitude environments (Wirtz et al., 2006). As individual species, *Usnea* spp. with *Neuropogon*-type traits and global biogeographic restriction to arctic–alpine environments, appear to be strong candidates when assessing threats from climate change.

Caveats to species-scale data

However, in the case of lichens, caution may be required when assessing climate change threat based on bioclimatic models. These models rely on a backbone of distribution or abundance data presented at the species level (Fig 20.1B). However, numerous cases of cryptic or poorly understood intraspecific diversity have been documented in lichens (e.g. LaGreca, 1999; Kroken and Taylor, 2001; Myllys et al.,

2001; Crespo et al., 2002; Buschbom and Mueller, 2006), suggesting that hidden genetic diversity may control adaptation to arctic–alpine environments (Sonesson et al., 1992). Such diversity has been documented in the ostensibly cosmopolitan lichen species *Xanthoria elegans* Link Th. Fr., whereby considerable genetic structure and differences in physiological traits (i.e. relative growth rate) were recorded in common garden experiments on a range of *X. elegans* ecotypes taken from differing climatic settings (Murtagh et al., 2002). These results for *X. elegans* suggest that lichen climatic response assessed at a species scale may provide an incomplete picture of the biological consequences of climate change, as potential adaptive traits relevant to climate impacts may be hidden below the widely used concept of the lichen morphospecies. As a consequence, species-scale assessments of climate change threat may sometimes be of restricted value for taxonomically difficult groups such as lichens.

Furthermore, threatened species have the potential to adapt to climate change (Franks et al., 2007) and in-situ adaptation may provide a buffer against population extinction. Limitations to understanding climate change impacts on lichen diversity, i.e. resulting from cryptic diversity coupled with the potential for evolutionary adaptation, may be tackled by synthesising genomic and environmental data (Holderegger et al., 2008), enabling an additional focus on adaptive as well as neutral genetic variation. However, combined genomic environment analyses specifically relevant to climate change adaptation are currently lacking for lichens. Since it is presently impossible to predict which populations support the genetic potential for in-situ evolution and survival under a future climatic scenario, or have sufficient gene flow between populations to enable the spread of favourable alleles, we must seek alternative methods to assess threat, and to complement and extend taxonomic species-scale data.

Evolutionary adaptation to climate change

Although adaptation in lichens may be expected to be complicated by their symbiotic status, we suggest several reasons why this concern may be overestimated. Lichen photobionts are widely dispersed as soredia (e.g. aerially – Tormo et al., 2001) and, as such, lichen photobionts appear to be geographically widespread and possibly cosmopolitan (Piercey-Normore and DePriest, 2001; O'Brien et al., 2005). In Antarctic lichens, a single *Nostoc* strain was detected both free-living and in five different lichen fungi (Wirtz et al., 2003), while four species of *Umbilicaria* appear to lack specificity for particular green algal associations (Romeike et al., 2002). Both of these studies suggest repeated re-establishment and colonisations by fungi and photobionts. This apparent evolutionary disassociation of lichen fungi and their photobionts has been demonstrated repeatedly, suggesting that algal switching and reassociations are frequent (Blaha et al., 2006; Yahr et al., 2006; Nelsen and Gargas, 2008) and that trophic strategies may vary depending on ecological setting (Gassmann and Ott, 2000; Schaper and Ott, 2003; Wedin et al., 2004). Furthermore, the frequent occurrence of strictly sexual reproduction by lichen fungi necessitates symbiotic reassociation on short timescales, such as has been observed for early establishment of lichens on rocks (e.g. Clayden, 1998). Therefore, we focus on what is known about the factors affecting evolutionary responses in the fungal partner. The link between observed diversity of neutral genetic markers and fitness (Reed and Frankham, 2003; Leimu et al., 2006) allows the evolution of adaptation in lichens to be tentatively assessed here with respect to genetic variation and correlated life-history traits (e.g. Hamrick and Godt, 1996).

Genetic variation

Standing genetic variation is assumed to relate to fitness and evolutionary potential (Ellstrand and Elam, 1993; Reed and Frankham, 2003), with population genetic variation providing a favourable evolutionary starting point for adaptation to changing conditions. Several studies of lichen population genetics have suggested regional differentiation (e.g. Högberg et al., 2002; Walser et al., 2005), but population dynamics are generally poorly understood and it is still too early to make broad generalisations (Werth, 2010). Instead, relevant genetic variation may be assessed indirectly, with respect to population size and key life-history traits.

Population size is a frequent predictor of genetic variation (Leimu et al., 2006). Large range size and population stability are expected to be correlated with higher genetic variation (partly due to their relationship with population size), relative to populations characterised by smaller ranges or frequent fluctuations. It has been argued for the circumpolar boreal/arctic lichen *Porpidia flavicunda* (Ach.) Gowan that populations have steadily and slowly advanced and receded in a continuous vegetation belt and that populations appear not to have experienced genetic bottlenecks in response to major climatic fluctuations during the Pleistocene (Buschbom, 2007). This hypothesis should be tested for other arctic species, but it may provide a positive outlook for future population changes, since many arctic and boreal lichens probably do have relatively large range sizes in the northern hemisphere. Similarly, the widely disjunct populations of *Cavernularia hultenii* Degel. from North America and Europe share a core of ancestral genotypes and were inferred to have arisen from a once continuous belt of evergreen forest during a warmer climate (Printzen et al., 2003). Repeated population bottlenecks reduce population size and genetic diversity, but have not yet been tested explicitly in lichens (but see Worth, 2010).

Lichen life-history traits expected to correlate with high genetic variation include their sessile nature, hypothesised to lead to local adaptations (as in plants – Antonovics, 1972) and their long life spans, allowing multiple cohorts to coexist, each of which might have experienced different selective pressures at establishment (Hamrick et al., 1981). In the arctic/boreal map lichens (*Rhizocarpon* spp.), apothecial production has been documented in thalli as small as 2 mm (Clayden, 1998), and thalli probably can live hundreds of years, often growing at less than 1 mm per year in cold climates (Bradwell and Armstrong, 2007).

Life-history traits which correlate directly with gene flow, such as mode of reproduction and breeding system, are also expected to be important predictors of genetic variation (Hamrick and Godt, 1996) and these vary among lichen species. In 31 of 36 studied cases in nature (crosses are so far impossible experimentally), lichen fungi have outcrossing (heterothallic) breeding systems (Culberson et al., 1988, 1993; Zoller et al., 1999; Murtagh et al., 2000; Honegger et al., 2004; Scherrer et al., 2005; Seymour et al., 2005a; Honegger and Zippler, 2007), a strategy for generating novel genotypes and relatively higher genetic diversity than that expected for selfing (homothallism). Still, some of these species may be facultatively heterothallic, permitting selfing as a bet-hedging strategy or for dealing with harsh environmental conditions (Murtagh et al., 2000). It has been suggested that sexual reproduction (via selfing or outcrossing) is the predominant strategy where environmental conditions are extreme in polar and arctic–alpine habitats (Fahselt et al., 1989; Seymour et al., 2005b). Although this might not be expected, given the risks associated with having to reassociate with a suitable photobiont, it has been observed that lichen fungi are able to temporarily associate with ecologically suboptimal photobionts (Ott, 1987; Schaper and Ott, 2003). Furthermore, asexual species have to face their own genetic risks, including the lack of standing genetic variation and low effective population sizes.

Dispersal

Dispersability is a key limiting factor in a species' response to climate change, but several studies have provided convincing evidence for repeated dispersal by lichens across large geographical areas. For example, lichen communities are significantly better predicted by wind connectivity than by geographic distance in the southern hemisphere (Muñoz et al., 2004), and repeated colonisations of Antarctica have been suggested for species of *Umbilicaria* (Romeike et al., 2002) and *Cladonia* (Myllys et al., 2003). Furthermore, the presence of disjunct mountain-top populations of arctic–alpine species probably attests to their long-distance dispersal potential (Galloway and Aptroot, 1995), as has been demonstrated for the arctic–alpine species *Flavocetraria cucullata* (Bellardi) Kärnefelt and A. Thell and *F. nivalis* (L.) Kärnefelt and A. Thell (Geml et al., 2010). In this study, the finding of identical sequences across the northern hemisphere was attributed to transoceanic migration (gene flow) rather than shared ancestral genotypes. However, in other species, dispersal limitation has been suggested at distances of less than a few kilometres (Dettki et al., 2000; Sillett et al., 2000; Hilmo and Såstad, 2001; Cassie and Piercey-Normore, 2003).

Theoretical predictions of gene flow and evolution may also inform us of the ability of lichens to adapt to changing climates. Firstly, only small amounts of gene flow are required to propagate advantageous genetic changes (Rieseberg and Burke, 2001), because of the large effect of selection on the spread of such changes (Slatkin, 1976). Secondly, large and relatively rapid changes in species' distributions have been observed historically and can be seen to be governed by rare, long-distance events (Neubert and Caswell, 2000) as predicted by theoretical studies (Slatkin, 1976). In the boreal and montane lichen *Letharia vulpina* (L.) Hue, long-distance dispersal from a North American source has been suggested to explain the genetically depauperate marginal European populations (Högberg et al., 2002), although the alternative explanation of bottlenecks in the European range and glacial refugia in North America would produce the same disparity in genetic diversity (Werth, 2010).

20.2.2 The current evidence base

Notwithstanding the important caveats outlined above with respect to species-scale data and the need for improved genetic information, candidate species with putative climatic sensitivity have been identified based on biogeographic, ecological and population parameters. Using data from three compatible approaches – predictive modelling, experimentation and observation (monitoring) – evidence from different studies in different regions can be combined to provide a preliminary assessment of lichen sensitivity to climate change.

Consistent with large-scale bioclimatic projections for arctic–alpine species and communities in Europe and North America (Thuiller et al., 2005; Hamann and Wang, 2006; Rehfeldt et al., 2006), projections for the British Isles indicate a particular threat to the montane flora (Berry et al., 2002; Trivedi et al., 2008a, 2008b). This projected threat has been extended to British lichens, identifying a loss of bioclimatic space for northern and montane species and an expansion of 'southern' species (Ellis et al., 2007a). The caveats associated with bioclimatic models have been discussed at length (e.g. Hampe, 2004; Guisan and Thuiller, 2005; Heikkinen et al., 2006), although, with appropriate circumspection, predictive modelling at least indicates the potential for the reorganisation of the British lichen flora under standard Intergovernmental Panel on Climate Change (IPCC) climate change scenarios (Nakicenovic, 2000; Ellis et al., 2007a, 2007b). Importantly, bioclimatic modelling indicating a loss of montane species (e.g. arctic–alpine elements of the British lichen flora, such as *Alectoria nigricans* (Ach.) Nyl., *Flavocetraria nivalis* and *Thamnolia vermicularis* (Sw.) Ach. Ex Schaer) is consistent with emerging trends in the European flora towards a regional increase in southern warm-temperate species, or species with tropical affinities, and a decrease in boreal species (van Herk et al., 2002; Aptroot and van Herk, 2007). This projected and observational evidence is also broadly consistent with experimental evidence in the functional ecology of arctic and alpine habitats, which provides a process-based understanding of climate change response.

Experimental support for a threat to montane lichens

The balance of predictive and observational evidence indicates a particular threat to montane and terricolous lichens (Aptroot and van Herk, 2007; Ellis et al., 2007a) largely driven by competition with vascular plants. Relative to plants, the success of lichens in arctic–alpine environments is in part due to several peculiar ecological traits. For example, terricolous 'mat-forming' lichens have an apical canopy, formed by vertical indefinite growth, that efficiently sequesters nutrients from atmospheric precipitation (Crittenden, 1989, 1991; Ellis et al., 2004). Additionally, mat-forming lichens efficiently recycle nutrients from structurally intact senescent tissue to support growth, maintaining high nutrient productivity and mean residence times (Ellis et al., 2005; Kytöviita and Crittenden, 2007). Nevertheless, there are constraining limits to the size and growth rate of lichens, and, being independent of the soil system, mat-forming lichens are often dependent upon the microhabitat of a structural matrix, provided either by intertwining with other mat-forming species and/or within a vascular plant canopy of low growth rate and reduced stature. Lichen-rich terricolous communities are thus maintained by the absence of strong interspecific competition, and may therefore be susceptible to vegetation change driven by climate warming. Climate warming is expected to alter plant community dynamics directly, such as through increased air and soil temperatures, and indirectly through more rapid nutrient cycling and the improved availability of limiting soil nutrients (Chapin et al., 1995; Robinson et al., 1998; Shaver and Jonasson, 1999; Walker et al., 2006). Experimental research to simulate the effects of climate warming (increased temperature and improved soil nutrient status) has demonstrated an increase in canopy height and the dominance of more competitive tall-stature species, causing a decline in abundance of small-stature species, including terricolous lichens (Chapin et al., 1995; Press et al., 1998; Cornelissen et al., 2001; Graglia et al., 2001; Hollister et al., 2005; Klanderud and Totland, 2005; Walker et al., 2006; Klanderud, 2008).

Problems with downscaling regional trends

The evidence base for climate change threat to arctic–alpine lichens at a local or regional scale, as described above, appears relatively robust. This includes shifts in bioclimatic space for montane species that are consistent with emerging observational evidence, and experimental results from arctic–alpine habitats, which provide a functional basis for threat to terricolous montane lichen communities. Nevertheless, problems arise in scaling down the general impacts of regional climate change to focus on local vegetation patterns. This is exemplified by lichen-rich montane heathland in the British mountains.

The Cairngorm Mountains in northeast Scotland include examples of ground-layer lichen vegetation which in terms of composition and physical structure provide an important and geographically outlying example of tundra heath (Fig 20.2A; cf. Ahti and Oksanen, 1990; Fryday, 2001a). However, in the Cairngorms, the low heath vegetation and local dominance of lichens is maintained by limiting climatic conditions including high wind speeds. High wind speeds are a frequent occurrence in the British mountains, and several key habitats locally dominated by lichens provide examples of wind-driven vegetation. The effect of wind speeds may be direct, through physical erosion of the canopy (Metcalfe, 1950; Burges, 1951), or indirect, through the redistribution of snow cover (Watt and Jones, 1948; Poore and McVean, 1957). Consistent with the potential role of snow cover in controlling the vegetation, British terricolous lichens have been described as chionophobous (snow hating), occurring within a low-growing ericaceous canopy that is blown clear of protective winter snow by strong winds (Gilbert and Fox, 1985). Perhaps the best example of wind-driven vegetation is provided by low-growing *Calluna* heath patterned by the wind into a series of stripes (Fig 20.2B). The prevailing wind imposes directionality to the formation of vegetation stripes, and drives a dynamic process in which unidirectional spread is maintained along a leeward front. Ericaceous shoots are sheltered by the bulk of the plant, spreading into areas of eroded soil, while the windward edge is senescent (Watt, 1947; Rodwell, 1991). Terricolous lichens establish at a point within this cycle at which the ericaceous shrubs are reduced in vigour, before their complete senescence (Watt, 1947; Metcalfe, 1950). This description of wind-driven vegetation from the Cairngorms is just one example of a physiognomy that is otherwise found in suitable localities from different floras throughout the world (Walton, 1922; Barrow et al., 1968; Burke et al., 1989).

In the Cairngorms, areas in which the stature of the vascular plant vegetation is climatically limited provide suitable habitat for low-growing terricolous montane lichens, such as *Alectoria* spp., *Cetraria islandica* (L.) Ach. and *Flavocetraria nivalis*. A recent detailed analysis examined the ecology of the mat-forming terricolous lichen species *Alectoria ochroleuca* (Hoffm.) A. Massal. in areas of wind-driven heath on Creagan Gorm (Fig 20.3A) (Ellis et al., unpublished). The results of this study demonstrate two important features with respect to wind-driven vegetation. Firstly, there is a negative exponential relationship between the canopy height of ericaceous shrubs, such as *Arctostaphylos uva-ursi* (L.) Spreng., *Calluna vulgaris* (L.) Hull and *Vaccinium myrtillus* L., and the amount of bare and unvegetated ground (Fig 20.3B). This trend captures a relationship between wind clipping of

A

B

Figure 20.2 (A) Lichen-rich montane heath in the Cairngorm Mountains of northeast Scotland. (B) In areas of wind exposure the vegetation is patterned into a series of stripes.

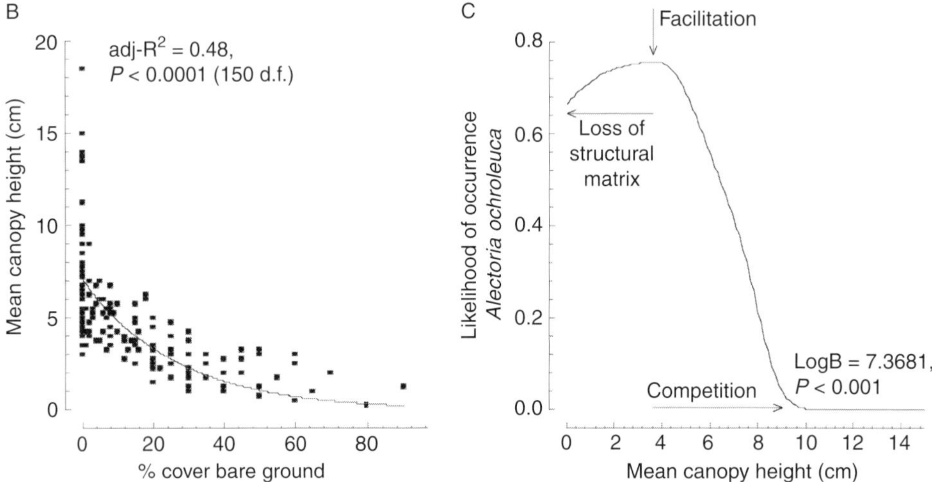

Figure 20.3 Terricolous lichen *Alectoria ochroleuca*. (A) *A. ochroleuca* growing in a canopy of *Arctostaphylos*. (B) The exponential relationship between mean height of the ericaceous canopy in 1 m^2 plots, and per cent cover of bare ground. (C) Unimodal response of *A. ochroleuca* to mean canopy height, annotated to highlight inferred ecological processes: loss of a physical matrix, facilitation and competition. The response was modelled using non-parametric multiplicative regression with a randomisation test to calculate *P* (McCune, 2006): sensitivity = 1.49, tolerance = 0.913 (5%).

the canopy and destruction of the vegetation by frost heave and high wind speeds (cf. King, 1960). Secondly, the occurrence of *Alectoria ochroleuca* within the vegetation demonstrates a skewed unimodal response, with peak occurrence at a mean canopy height of *c.* 5 cm.

Bearing in mind the ecology of mat-forming lichens, described in detail above, we interpret this pattern for *A. ochroleuca* as a response to the vascular plant canopy (Fig 20.3C). An optimum canopy height may therefore equate to a facilitation effect, providing a structural matrix and equitable microhabitat in which establishment is possible and where the branches of *A. ochroleuca* are anchored. As the canopy decreases in height and the amount of bare ground increases, physical support and protection will be diminished, and the occurrence of *A. ochroleuca* will decline. However, as the height of the vascular canopy increases, *A. ochroleuca* becomes subject to the competitive effect of overshading. Plant interaction is thus inferred to provide contrasting mechanisms operating to shape the lichen response, from a loss of facilitation and exposure to high wind speeds to competitive effects. This response pattern is consistent with the 'stress-gradient hypothesis': an ecological trade-off in the balance of facilitation and competition along a gradient in environmental stress (Bertness and Callaway, 1994; Maestre et al., 2009). The stress-gradient hypothesis has been confirmed as an important process structuring arctic–alpine plant communities (Callaway et al., 2002), and, as suggested here, may incorporate physical shelter effects related to plant architecture (Carlsson and Callaghan, 1991). However, because for the Cairngorm lichen vegetation these subtly changing interaction patterns are driven in part by wind speed, the results also point to significant uncertainty in downscaling climate predictions. In terms of climate change projections relevant to lichen-rich heath (e.g. Berry et al., 2002; Ellis et al., 2007a; Trivedi et al., 2008a, 2008b) the modelled response of certain British montane species may be inaccurately represented if their local occurrence is primarily controlled not only by temperature (and precipitation) but also by wind patterns. Bioclimatic projections rarely incorporate wind patterns, which are notoriously difficult to model and are considered to be extremely uncertain (Barnett et al., 2006; Christensen et al., 2007).

These preliminary results for *A. ochroleuca* have been extended to other sites and comparable lichen species (Crabtree and Ellis, 2010), though need to be confirmed by experimentation. They nevertheless suggest that the downscaling from regional temperature/precipitation gradients to local climatic effects presents a severe problem in understanding the smaller-scale response for the many species, such as lichens, whose distribution is critically dependent on microhabitat availability. In addition, consistent with studies demonstrating the modifying effects of species interaction on the climatic response (Klanderud, 2005; Suttle et al., 2007), the nested relationship of subdominant mat-forming lichen species with respect to the vascular plant canopy adds additional complexity to the interpretation and reliability of climate change projections. This multiplicative framework includes not only the individual response of each species to regional climate change, but also species interactions (Brooker, 2006; Brooker et al., 2008) and additional environmental drivers such as microclimate (i.e. local topographic effects on wind speeds, snow-lie, soil temperatures etc.), pollution and grazing (Fryday, 2001b; Britton and Fisher, 2006).

20.3 Climate change and the amalgam of environmental threat

Clearly, scientific concern stemming from climate change impacts is wide-ranging, integrating not just arctic–alpine habitats but species and communities from many contrasting biogeographic zones (Walther et al., 2002; Rosenzweig et al., 2008). This wider concern, which implicates climate change (specifically, climate warming) in potential biodiversity loss, has to be compared against previous climatic variability during geological history. Integrating a functional relationship between biodiversity and spatial/temporal gradients in the temperature and water regime (Currie and Paquin, 1987; Kalmar and Currie, 2006; Willis et al., 2007), past climate shifts have been cited as an evolutionary driver of biodiversity (Jaramillo et al., 2006; MacDonald et al., 2008). In this sense, if all other factors were equal, climate change per se might be considered a driver, although not necessarily a direct threat to biodiversity on macroevolutionary timescales. The threat of human-induced climate change stems from its temporal scale and its interaction with a variety of additional factors, which limit the opportunity for species to respond (migrate, acclimate or adapt) to large-scale global change (Travis, 2003). The additional effect of massive habitat loss and fragmentation spreads the burden of climate change impacts across many more biological realms than the directly threatened arctic–alpine vegetation.

Habitat loss and fragmentation are amongst the most important drivers contributing to the indirect threat of climate change. Based on metapopulation theory (Hanski, 1999), the epiphyte response to habitat fragmentation is controlled by available habitat (i.e. the extent and suitability of habitat in a landscape) and dispersal limitation (i.e. the distance between habitat patches relative to a dispersal kernel –Snäll et al., 2005; Löbel et al., 2006a, 2006b). Habitat fragmentation results in fewer opportunities for interpatch colonisation for dispersal-limited species (Sillett et al., 2000; Öckinger et al., 2005), combined with a reduction in the range and availability of suitable habitat (Gignac and Dale, 2005). In lichens, there is an ongoing debate about the balance between dispersal and establishment limitation in structuring metapopulations at contrasting scales (Werth et al., 2006). However, establishment limitation may be expected to have the same general impact in shifting the balance of colonisation and extinction dynamics in a fragmented landscape, with an effective reduction in the extent and availability of habitat. Spatial analysis suggests that in managed and fragmented landscapes lichen populations may become increasingly aggregated (isolated) and spatially restricted to relatively fewer high-quality habitat patches (Gu et al., 2001; Johansson and Ehrlén, 2003). Thus, species facing a double jeopardy of both habitat specialisation and dispersal (or establishment) limitation will be especially susceptible to climate change. Certain lichen epiphyte species are likely to find themselves trapped in isolated habitat patches (e.g. old-growth forest patches in a 'sea' of farmland), with reduced opportunity to colonise stepping stones between disjunct high-quality habitat, and therefore a reduced chance of migration in response to a changing environment.

20.3.1 Population genetic consequences

Population genetic consequences of climate change incorporate changes in range, either via a change in range size leading to shifting population size, or change in range position. Species responding to changes in climate are expected to migrate along latitudinal or altitudinal gradients from their existing ranges, where they may be challenged by lower population genetic variation and weak connectivity at range margins (Eckert et al., 2008), and by more limited habitat extents (e.g. with upward shifts in altitudinal zones). However, selection for increased dispersability may become stronger, as jumps to newly available habitats are increasingly favoured (Parmesan, 2006). Dispersal has been examined empirically in the model lichen fungus *Lobaria pulmonaria* (L.) Hoffm., where vegetative propagules have maximum dispersal on the scale of only a few hundred metres (Walser, 2004), but where high rates of gene flow are observed among woodland patches over the scale of several kilometres, presumably by spores (Werth et al., 2007). Studies of other lichens have similarly suggested high rates of gene flow at the scale of several to tens of kilometres (Werth and Sork, 2008; Lättman et al., 2009). Furthermore, studies using PCR detection of fungus-specific DNA markers in snow samples detected *L. pulmonaria* from localities where it is currently absent, suggesting that dispersal does not limit its populations (Walser et al., 2001; Werth et al., 2006). Instead, using transplants of vegetative propagules, it appears that habitat quality and its effects on establishment limit the local distribution of the species (Werth et al., 2006), a factor that may be expected to change with changing climate. On a regional level, geographically separated populations are genetically differentiated (Walser et al., 2005; Werth et al., 2007), suggesting that populations may respond individualistically to large and rapid bioclimatic range shifts. A case-by-case approach will likely be wise for other lichen fungi, as both continental-scale population differentiation (e.g. Printzen et al., 2003; Palice and Printzen, 2004) and gene flow at the scale of hundreds of kilometres (e.g. Baloch and Grube, 2009) have been demonstrated.

20.3.2 Alleviating climate change impacts

The emergence of human-induced climate change as a threat to biodiversity adds further ecological complexity to an existing framework of anthropogenic pressure. However, while governments grapple with reducing greenhouse gas emissions, and accept that some degree of climate warming is now inevitable, there are opportunities to ease climate change impacts on biodiversity by reducing additional stresses. These opportunities for mitigation exist irrespective of uncertainty in the climatic response of biological groups.

Habitat restoration: the case of oceanic lichens

The northwest British Isles represents one of the most important regions in Europe for temperate rainforest epiphytes. Analogous communities with a preponderance of cyanobacterial lichens (James et al., 1977; Green and Lange, 1991; Ellyson and Sillett, 2003) exist in northwest North America, Chile, New Zealand and Tasmania, and comparable epiphyte communities may once have been widespread along the Atlantic coastline of Europe. However, owing to the impacts of deforestation and pollution, oceanic European epiphyte communities are now extremely restricted, with the west coast of Scotland providing a premier European example of intact cool-temperate rainforest epiphytes. The composition of temperate rainforest epiphyte communities has been explained by limits to cyanobacterial photosynthesis and nitrogen fixation, e.g. a requirement for liquid water (Lange et al., 1986, 1993) and moderate temperatures (MacFarlane and Kershaw, 1977; Antoine, 2004). Such lichen communities are also explained phylogeographically, i.e. many species in temperate rainforests appear to be outlying representatives of genera with tropical affinities (Ellis et al., 2009). While these epiphyte communities appear to reach a zenith in mild and consistently humid oceanic climates, it is possible that component species were once more widespread across lowland continental Europe (Rose, 1988). In support of this suggestion, many such species occur in habitats of varying quality under optimal oceanic conditions, but become increasingly restricted to old-growth habitat along a gradient of increasing continentality (Ellis and Coppins, 2007). Thus, in relatively continental eastern Scotland, epiphyte species which are common in oceanic western Scotland are applied as indicators of ecological continuity (Coppins and Coppins, 2002). The observation of changing habitat specificity with climatic setting is an intriguing one, and explains the suggestion that certain 'oceanic' species may once have been far more widespread than during the present day (Rose, 1988). For example, 5000 years ago, all forest was functionally old-growth forest.

Attempts to project the response of British oceanic epiphyte species have failed to reach a consensus, indicating either an increase in overall diversity, possibly owing to warmer climatic conditions (Ellis and Coppins, 2007), or contrasting responses (including negative impacts) for individual oceanic species (Ellis et al., 2007a, 2009). However, studies to integrate the effects of climate and habitat quality (i.e. extent of old-growth woodland) unequivocally demonstrate the context-dependent interaction between climatic setting and habitat, and the potential role of habitat quality in offsetting climate change impacts (Ellis and Coppins, 2007; Ellis et al., 2009). These integrative studies suggest that regeneration of native woodland habitat targeted to certain climatic regions, increasing extent and connectivity, may provide an effective buffer against regional climate change, ensuring some degree of certainty to biodiversity protection and countering uncertainty in the species' response to future climate change.

The pollution regime

Pollution may be invoked as amongst the most important drivers of habitat quality for lichen epiphytes. Lichens have been used accordingly as bioindicators for a variety of pollutants, including sulphur dioxide (Hawksworth and Rose, 1970) and nitrogen (van Herk, 1999; van Herk et al., 2003). Pollution indices have, in turn, been criticised for neglecting to account for underlying differences between lichen communities in response to climatic sensitivity over relatively small spatial scales (Ellis and Coppins, 2006). The relative importance of climate and pollution in controlling lichen community structure is symptomatic of a key issue in conservation, in which potentially confounded relationships between climate and additional drivers, such as pollution and habitat loss, have yet to be resolved.

However, recent evidence concerning the respective contribution of three large-scale drivers – pollution, climatic setting and woodland history – to epiphyte community composition clearly implicates pollution as a relatively strong regional driver of community structure (Ellis and Coppins, 2009). Exemplified for lichen epiphytes, the implication is that even with large-scale habitat recreation, the forest network over large regions would be unavailable for colonisation by lichens because pollution creates a blanket form of habitat loss operating regardless of the actual availability of trees as a substratum. Beginning in the late eighteenth and early nineteenth centuries, large-scale toxic pollution resulting from the industrial burning of fossil fuels (e.g. SO_2, NO_x), and associated acidification of substrata, decimated lichen diversity within and downwind of industrial centres in the British Isles and Europe (Coppins et al., 2001; van Dobben et al., 2001). As these pollutants began to decline following targeted emission controls (Woodin, 1989), pollution-sensitive lichens have begun to recolonise areas of Britain from which they had previously suffered extinction (Rose and Hawksworth, 1981; Seaward, 1998). Reducing pollution loads to below threshold levels effectively creates habitat newly available for colonisation by lichen epiphytes (forest patches in a landscape that was formerly polluted), providing a strong benefit in terms of habitat availability and connectivity. Evidence following declining levels of SO_2 pollution and associated acidification suggests that previously polluted habitats may regain suitability for colonisation relatively quickly (in 10–100 years – Rose and Hawksworth, 1981; Seaward, 1998). However, a recent decline in levels of SO_2 and acid rain from burning fossil fuels has been accompanied by a steep rise in N pollution from intensive agriculture (Woodin, 1989; Stevens et al., 2004). Current evidence suggests that, at a landscape scale, one pollution regime (SO_2) has been replaced by another (N), and scenarios for N pollution point to a rising trend beyond 2100. These scenarios of N pollution include a worrying increase not only in Europe but also in biodiversity hotspots such as Central America, India and southeast Asia (Millennium Ecosystem Assessment, 2005).

20.4 Conclusions and future directions

Research on lichen biodiversity, evolutionary ecology and climate change is in its infancy. For many regions of the world the lichen flora is entirely unknown, and vulnerable species in unexplored regions (e.g. in the alpine zones of tropical mountains) may face extinction as a consequence of climate change before they are ever recorded. In contrast, a minority of regions have been well studied with respect to predictive modelling (e.g. the British Isles and parts of North America), integrating both simple bioclimatic models (Ellis et al., 2007a, 2007b), species–habitat interactions (Ellis and Coppins, 2007; Ellis et al., 2009) and multiple landscape-scale drivers (Jovan and McCune, 2005; Geiser and Neitlich, 2007; Ellis and Coppins, 2009). This modelling approach is supported by experimental physiology to indicate lichen sensitivity to the climatic regime (Lange et al., 1986, 1993; del Prado and Sancho, 2007), although physiological studies also indicate opportunities for acclimation (Schofield et al., 2003; Lange and Green, 2005), including the potential for photobiont switching in response to environmental setting (Piercey-Normore, 2006; Yahr et al., 2006). Physiological experimentation and molecular ecological studies used to explain species distributions would provide a powerful platform for predictive research incorporating species' functional response. Such an approach would facilitate better recognition of important ecological processes currently absent from bioclimatic models, such as dispersal and evolutionary adaptation, non-equilibrium range patterns, or feedback mechanisms enforcing contrasting stable states. Future bioclimatic research would thus benefit from the use of functional response models, as an extension to the species presence/absence approach (cf. Woodward and Beerling, 1997; Morin and Thuiller, 2009). Species interactions nested at scales beneath the regional climate and large-scale habitat are critically important, and may fundamentally alter species' climatic response when downscaling from regional to small-scale patterns. These interactions may necessitate the consideration of additional microclimatic variables, such as wind patterns and snow-lie, as well as a range of smaller-scale interacting factors, such as land management, grazing and pollution regime. Research examining the importance of microclimatic setting and species interactions (e.g. a facilitation–competition trade-off) is likely to provide important new insight into the response of lichen biodiversity to climate change. This work should extend to the consideration of trophic interactions, e.g. grazers whose distribution under a changing climate may have consequences for epiphyte community structure (Asplund and Gauslaa, 2008; Gauslaa, 2008). In addition, research should examine in more detail the metapopulation response to climate change: i.e. the spatial arrangement of habitat patches of varying quality, with respect to likelihoods of dispersal and establishment, and spatially explicit changes in model parameters under climate change scenarios. The metapopulation approach would benefit from targeted molecular research to examine population processes in lichens, also integrating studies on cryptic speciation and the potential for in-situ adaptation. Ultimately, the best ecological models will provide an uncertain predictive framework for conservation strategy, and projections should be confirmed or refuted, and recalibrated, by applying taxonomic expertise in fieldwork and direct long-term monitoring.

References

Ahti, T. and Oksanen, J. (1990). Epigeic lichen communities of taiga and tundra regions. *Vegetatio*, **86**, 39–70.

Antoine, M. E. (2004). An ecophysiological approach to quantifying nitrogen fixation by *Lobaria oregana*. *Bryologist*, **107**, 82–87.

Antonovics, J. (1972). Population dynamics of the grass *Anthoxanthum odoratum* on a zinc mine. *Journal of Ecology*, **60**, 351–367.

Aptroot, A. and van Herk, C. M. (2007). Further evidence of the effects of global warming on lichens, particularly those with *Trentepohlia* phycobionts. *Environmental Pollution*, **146**, 293–298.

Asplund, J. and Gauslaa, Y. (2008). Mollusc grazing limits growth and early development of the old forest lichen *Lobaria pulmonaria* in broadleaved deciduous forests. *Oecologia*, **155**, 93–99.

Baloch, E. and Grube, M. (2009). Pronounced genetic diversity in tropical epiphyllous lichen fungi. *Molecular Ecology*, **18**, 2185–2197.

Barnett, C., Hossell, J., Perry, M., Procter, C. and Hughes, G. (2006). *A Handbook of Climate Trends Across Scotland*. Edinburgh: Scotland and Northern Ireland Forum for Environmental Research.

Barrow, M. D., Costin, A. B. and Lake, P. (1968). Cyclical changes in an Australian Fjaeldmark community. *Journal of Ecology*, **56**, 89–96.

Berry, P. M., Dawson, T. P., Harrison, P. A. and Pearson, R. G. (2002). Modelling potential impacts of climate change on the bioclimatic envelope of species in Britain and Ireland. *Global Ecology and Biogeography*, **11**, 453–462.

Bertness, M. D. and Callaway, R. (1994). Positive interactions in communities. *Trends in Ecology and Evolution*, **9**, 191–193.

Blaha, J., Baloch, E. and Grube, M. (2006). High photobiont diversity in symbioses of the euryoecious lichen *Lecanora rupicola* (Lecanoraceae, Ascomycota). *Biological Journal of the Linnean Society*, **88**, 283–293.

Bradwell, T. and Armstrong, R. A. (2007). Growth rates of *Rhizocarpon geographicum* lichens: a review with new data from Iceland. *Journal of Quaternary Science*, **22**, 311–320.

Britton, A. J. and Fisher, J. M. (2006). Interactive effects of nitrogen deposition, fire and grazing on diversity and composition of low-alpine prostrate *Calluna vulgaris* heathland. *Journal of Applied Ecology*, **44**, 125–135.

Brochmann, C. and Brysting, A. K. (2008). The Arctic: an evolutionary freezer? *Plant Ecology and Diversity*, **1**, 181–195.

Brooker, R. W. (2006). Plant–plant interactions and environmental change. *New Phytologist*, **171**, 271–284.

Brooker, R. W., Maestre, F. T., Callaway, R. M. et al. (2008). Facilitation in plant communities: the past, the present, and the future. *Journal of Ecology*, **96**, 18–34.

Bubrick, P. (1988). Methods for cultivating lichens and isolated bionts. In *Handbook of Lichenology*, Vol. 3, ed. M. Galun. Boca Raton, FL: CRC Press, pp. 127–138.

Burges, A. (1951) The ecology of the Cairngorms. III. The *Empetrum–Vaccinium* zone. *Journal of Ecology*, **39**, 271–284.

Burke, I. C., Reiners, W. A. and Olson, R. K. (1989). Topographic control of vegetation in a mountain big sagebrush steppe. *Vegetatio*, **84**, 77–86.

Buschbom, J. (2007). Migration between continents: geographical structure and long-distance gene flow in *Porpidia flavicunda*. *Molecular Ecology*, **16**, 1835–1846.

Buschbom, J. and Mueller, G. M. (2006). Testing 'species pair' hypotheses: evolutionary processes in the lichen-forming species complex *Porpidia flavocoerulescens* and *Porpidia melinodes*. *Molecular Biology and Evolution*, **23**, 574–586.

Callaway, R. M., Brooker, R. W., Choler, P. et al. (2002). Positive interactions among alpine plants increase with stress. *Nature*, **417**, 844–848.

Carlsson, B. Å. and Callaghan, T. V. (1991). Positive plant interactions in tundra vegetation and the importance of shelter. *Journal of Ecology*, **79**, 973–983.

Cassie, D. M. and Piercey-Normore, M. D. (2008). Dispersal in a sterile lichen-forming fungus, *Thamnolia subuliformis* (Ascomycotina: Icmadophilaceae). *Botany*, **86**, 751–762.

Chapin, F. S., Shaver, G. R., Giblin, A. E., Nadelhoffer, K. J. and Laundre, J. A. (1995). Responses of arctic tundra to experimental and observed changes in climate. *Ecology*, **76**, 694–711.

Christensen, J. H., Hewitson, B., Busuioc, A. et al. (2007). Regional climate projections. In *Climate Change 2007: the Physical Science Basis. Contribution of Working Group I to the Fourth Assessment Report of the Intergovernmental Panel on Climate Change*, ed. S. Solomon, D. Qin, M. Manning et al. Cambridge: Cambridge University Press, pp. 847–940.

Clayden, S. (1998). Thallus initiation and development in the lichen *Rhizocarpon lecanorinum*. *New Phytologist*, **139**, 685–695.

Comes, H. P. and Kadereit, J. W. (2003). Spatial and temporal patterns in the evolution of the flora of the European Alpine system. *Taxon*, **52**, 451–462.

Coppins, A. M. and Coppins, B. J. (2002). *Indices of Ecological Continuity for Woodland Epiphytic Lichen Habitats in the British Isles*. London: British Lichen Society.

Coppins, B. J., Hawksworth, D. L. and Rose, F. (2001). Lichens. In *The Changing Wildlife of Great Britain and Ireland*, ed. D. L. Hawksworth. London: Taylor and Francis.

Cornelissen, J. H. C., Callaghan, T. V., Alatalo, J. M. et al. (2001). Global change and arctic ecosystems: is lichen decline a function of increases in vascular plant biomass? *Journal of Ecology*, **89**, 984–994.

Crabtree, D. and Ellis, C. J. (2010). Species interaction and response to wind-speed alter the impact of projected temperature change in a montane ecosystem. *Journal of Vegetation Science*, **21**, 744–760.

Crespo, A., Carmen Molina, M., Blanco, O. et al. (2002). rDNA ITS and β-tubulin gene sequence analyses reveal two monophyletic groups within the cosmopolitan lichen *Parmelia saxatilis*. *Mycological Research*, **106**, 788–795.

Crittenden, P. D. (1989). Nitrogen relations of mat-forming lichens. In *Nitrogen, Phosphorus and Sulphur Utilization by Fungi*, ed. L. Boddy, R. Marchant and C. J. Read. Cambridge: Cambridge University Press.

Crittenden, P. D. (1991). Ecological significance of necromass production in mat-forming lichens. *The Lichenologist*, **23**, 323–331.

Culberson, C. F., Culberson, W. L. and Johnson, A. (1988). Gene flow in lichens. *American Journal of Botany*, **75**, 1135–1139.

Culberson, W. L., Culberson, C. F. and Johnson, A. (1993). Speciation in lichens of the *Ramalina siliquosa* complex (Ascomycotina, Ramalinaceae): gene flow and reproductive isolation. *American Journal of Botany*, **80**, 1472–1481.

Currie, D. J. and Paquin, V. (1987). Large-scale biogeographical patterns of species richness of trees. *Nature*, **329**, 326–327.

Del Prado, R. and Sancho, L. G. (2007). Dew as a key factor for the distribution pattern of the lichen species *Teloschistes lacunosus* in the Tabernas Desert (Spain). *Flora*, **202**, 417–428.

Dettki, H., Klintberg, P. and Esseen, P. A. (2000). Are epiphytic lichens in young forests limited by local dispersal? *Ecoscience*, **7**, 317–325.

Dobson, F. (2005). *Lichens: an Illustrated Guide to British and Irish Species*. Slough: Richmond Publishing.

Eckert, C. G., Samis, K. E. and Lougheed, C. (2008). Genetic variation across species' geographical ranges: the central-marginal hypothesis. *Molecular Ecology*, **17**, 1170–1188.

Ellis, C. J. and Coppins, B. J. (2006). Contrasting functional traits maintain lichen epiphyte diversity in response to climate and autogenic succession. *Journal of Biogeography*, **33**, 1643–1656.

Ellis, C. J. and Coppins, B. J. (2007). Changing climate and historic-woodland structure interact to control species diversity of the '*Lobarion*' epiphyte community in Scotland. *Journal of Vegetation Science*, **18**, 725–734.

Ellis, C. J. and Coppins, B. J. (2009). Quantifying the role of multiple landscape-scale drivers controlling epiphyte composition and richness in a conservation priority habitat (juniper scrub). *Biological Conservation*, **142**, 1291–1301.

Ellis, C. J., Crittenden, P. D. and Scrimgeour, C. M. (2004). Soil as a potential source of nitrogen for mat-forming lichens. *Canadian Journal of Botany*, **82**, 145–149.

Ellis, C. J., Crittenden, P. D., Scrimgeour, C. M. and Ashcroft, C. J. (2005). Translocation of 15N indicates nitrogen recycling in the mat-forming lichen *Cladonia portentosa*. *New Phytologist*, **168**, 423–434.

Ellis, C. J., Coppins, B. J., Dawson, T. P. and Seaward, M. R. D. (2007a). Response of British lichens to climate change scenarios: trends and uncertainties in the projected impact for contrasting biogeographic groups. *Biological Conservation*, **140**, 217–235.

Ellis, C. J., Coppins, B. J. and Dawson, T. P. (2007b). Predicted response of the lichen epiphyte *Lecanora populicola* to climate change scenarios in a clean-air region of northern Britain. *Biological Conservation*, **135**, 396–404.

Ellis, C. J., Yahr, R. and Coppins, B. J. (2009). Local extent of old-growth woodland modifies epiphyte response to climate change. *Journal of Biogeography*, **36**, 302–313.

Ellstrand, N. C. and Elam, D. R. (1993). Population genetic consequences of small population size: implications for plant conservation. *Annual Review Ecology and Systematics*, **24**, 217–242.

Ellyson, W. J. T. and Sillett, S. C. (2003). Epiphyte communities on sitka spruce in an old-growth redwood forest. *Bryologist*, **106**, 197–211.

Fahselt, D. (1989). Enzyme polymorphism in sexual and asexual umbilicate lichens from Sverdrup Pass, Ellesmere Island, Canada. *Lichenologist*, **21**, 279–285.

Fahselt, D., Maycock, P. and Wong, P. Y. (1989). Reproductive modes of lichens in stressful environments in central Ellesmere Island, Canadian high arctic. *The Lichenologist*, **21**, 343–353.

Franks, S. J., Sim, S. and Weis, A. E. (2007). Rapid evolution of flowering time by an annual plant in response to a climate fluctuation. *Proceedings of the National Academy of Sciences of the USA*, **104**, 1278–1282.

Fryday, A. M. (2001a). Phytosociology of terricolous lichen vegetation in the Cairngorm Mountains, Scotland. *The Lichenologist*, **33**, 331–351.

Fryday, A. M. (2001b). Effects of grazing animals on upland/montane lichen vegetation in Great Britain. *Botanical Journal of Scotland*, **53**, 1–19.

Galloway, D. J. and Aptroot, A. (1995). Bipolar lichens: a review. *Cryptogamic Botany*, **5**, 184–191.

Gargas, A., DePriest, P. T., Grube, M. and Tehler, A. (1995). Multiple origins of lichen symbiosis in fungi suggested by SSU rDNA phylogeny. *Science*, **268**, 1492–1495.

Gassmann, A. and Ott, S. (2000). Growth strategy and the gradual symbiotic interactions of the lichen *Ochrolechia frigida*. *Plant Biology*, **2**, 368–378.

Gauslaa, Y. (2005). Lichen palatability depends on investments in herbivore defence. *Oecologia*, **143**, 94–105.

Gauslaa, Y. (2008). Mollusc grazing may constrain the ecological niche of the old forest lichen *Pseudocyphellaria crocata*. *Plant Biology*, **10**, 711–717.

Geiser, L. H. and Neitlich, P. N. (2007). Air pollution and climate gradients in western Oregon and Washington indicated by epiphytic macrolichens. *Environmental Pollution*, **145**, 203–218.

Geml, J., Kauff, F., Brochmann, C. and Taylor, D. L. (2010). Surviving climate changes: high genetic diversity and transoceanic gene flow in two arctic-alpine lichens, *Flavocetraria cucullata* and *F. nivalis* (Parmeliaceae, Ascomycota). *Journal of Biogeography*, **37**, 1529–1542.

Gignac, L. D. and Dale, M. R. T. (2005). Effects of fragment size and habitat heterogeneity on cryptogam diversity in the low-boreal forest of western Canada. *Bryologist*, **108**, 50–66.

Gilbert, O. L. and Fox, B. W. (1985). Lichens of high ground in the Cairngorm Mountains, Scotland. *The Lichenologist*, **17**, 51–66.

Grabherr, G., Gottfried, M. and Pauli, H. (1994). Climate effects on mountain plants. *Nature*, **369**, 448.

Graglia, E., Jonasson, S., Michelsen, A. et al . (2001). Effects of environmental perturbations on abundance of subarctic plants after three, seven and ten years of treatments. *Ecography,* **24**, 5–12.

Green, T. G. A. and Lange, O. L. (1991). Ecophysiological adaptations of the lichen genera *Pseudocyphellaria* and *Sticta* to south temperate rainforests. *The Lichenologist*, **23**, 267–282.

Grime, J. P. (1977). Evidence for the existence of three primary strategies in plants and its relevance to ecological and evolutionary theory. *American Naturalist*, **111**, 1169–1194.

Grube, M. and Winka, K. (2002). Progress in understanding the evolution and classification of lichenized ascomycetes. *Mycologist*, **16**, 67–76.

Gu, W. D., Kuusinen, M., Konttinen, T. and Hanski, I. (2001). Spatial pattern in the occurrence of the lichen *Lobaria pulmonaria* in managed and virgin boreal forests. *Ecography*, **24**, 139–150.

Gueidan, C., Roux, C. and Lutzoni, F. (2007). Using a multigene phylogenetic analysis to assess generic delineation and character evolution in Verrucariaceae (Verrucariales, Ascomycota). *Mycological Research*, **111**, 1145–1168.

Gueidan, C., Ruibal Villaseñor, C., de Hoog et al. (2008). A rock-inhabiting ancestor for mutualistic and pathogen-rich fungal lineages. *Studies in Mycology*, **61**, 111–119.

Guisan, A. and Thuiller, W. (2005). Predicting species distribution: offering more than simple habitat models. *Ecology Letters*, **8**, 993–1009.

Hageman, C. and Fahselt, D. (1990). Multiple enzyme forms as indicators of functional sexuality in the lichen *Umbilicaria vellea*. *The Bryologist*, **93**, 389–394.

Hale, M. E. (1983). *The Biology of Lichens*. London: Edward Arnold.

Hamann, A. and Wang, T. (2006). Potential effects of climate change on ecosystem and tree species distribution in British Columbia. *Ecology*, **87**, 2773–2786.

Hampe, A. (2004). Bioclimatic envelope models: what they detect and what they hide. *Global Ecology and Biogeography*, **13**, 469–476.

Hamrick, J. L. and Godt, M. J. W. (1996). Effects of life history traits on genetic diversity in plant species. *Philosophical Transactions of the Royal Society of London B,* **351**, 1291–1298.

Hamrick, J. L., Milton, J. B. and Linhart, Y. B. (1981). *Levels of Genetic Variation in Trees: Influence of Life History Characteristics*. Gen. Tech. Rep.

PSW-GTR-48. Berkeley, CA: Pacific Southwest Forest and Range Exp. Stn, Forest Service, U.S. Department of Agriculture, pp. 35–41.

Hanski, I. (1999). *Metapopulation Ecology*. Oxford: Oxford University Press.

Hawksworth, D. L. and Rose, F. (1970). Qualitative scale for estimating sulphur dioxide air pollution in England and Wales using epiphytic lichens. *Nature*, **227**, 145–148.

Heikkinen, R. K., Luoto, M., Araújo, M. B. et al. (2006). Methods and uncertainties in bioclimatic envelope modelling under climate change. *Progress in Physical Geography*, **30**, 751–777.

Hestmark, G. (1992). Sex, size, competition and escape: strategies of reproduction and dispersal in *Lasallia pustulata* (Umbilicariaceae, Ascomycetes). *Oecologia*, **92**, 305–312.

Hilmo, O. and Såstad, S. M. (2001). Colonization of old-forest lichens in a young and an old boreal *Picea abies* forest: an experimental approach. *Biological Conservation*, **102**, 251–259.

Högberg, N., Kroken, S., Thor, G. and Taylor, J. W. (2002). Reproductive mode and genetic variation suggest a North American origin of European *Letharia vulpina*. *Molecular Ecology*, **11**, 1191–1196.

Holderegger, R., Herrmann, D., Poncet, B. et al. (2008). Land ahead: using genome scans to identify molecular markers of adaptive radiation. *Plant Ecology and Diversity*, **1**, 273–283.

Hollister, R. D., Webber, P. J. and Bay, C. (2005). Plant response to temperature in northern Alaska: implications for predicting vegetation change. *Ecology*, **86**, 1562–1570.

Honegger, R. (1991). Functional aspects of the lichen symbiosis. *Annual Review of Plant Physiology and Plant Molecular Biology*, **42**, 553–578.

Honegger, R. and Zippler, U. (2007). Mating systems in representatives of Parmeliaceae, Ramalinaceae and Physciaceae. *Mycological Research*, **111**, 424–432.

Honegger, R., Zippler, U., Gansner, H. and Scherrer, S. (2004). Mating systems in the genus *Xanthoria* (lichen-forming ascomycetes). *Mycological Research*, **108**, 480–488.

James, P. W., Hawksworth, D. L. and Rose, F. (1977). Lichen communities in the British Isles: a preliminary conspectus. In *Lichen Ecology*, ed. M. R. D. Seaward. London: Academic Press.

Jaramillo, C., Rueda, M. J. and Mora, G. (2006). Cenozoic plant diversity in the Neotropics. *Science*, **311**, 1893–1896.

Johansson, P. and Ehrlén, J. (2003). Influence of habitat quantity, quality and isolation on the distribution and abundance of two epiphytic lichens. *Journal of Ecology*, **91**, 213–221.

Jovan, S. and McCune, B. (2005). Air-quality bioindication in the Greater Central Valley of California, with epiphytic macrolichen communities. *Ecological Applications*, **15**, 1712–1726.

Kalmar, A. and Currie, D. J. (2006). A global model of island biogeography. *Global Ecology and Biogeography*, **15**, 72–81.

Kappen, L. (1990). *Usnea sphacelata*, its role in the vegetation and its possible growth capacity on Bailey Peninsula, Wilkes Land. *Bibliotheca Lichenologica*, **38**, 277–289.

Kappen, L. (2000). Some aspects of the general success of lichens in Antarctica. *Antarctic Science*, **12**, 314–324.

King, R. B. (1960). Vegetation destruction in the sub-alpine and alpine zones of the Cairngorm Mountains. *Scottish Geographical Magazine*, **87**, 103–115.

Klanderud, K. (2005). Climate change effects on species interactions in an alpine plant community. *Journal of Ecology*, **93**, 127–137.

Klanderud, K. (2008). Species-specific responses of an alpine plant community under simulated environmental change. *Journal of Vegetation Science*, **19**, 363–372.

Klanderud, K. and Birks, H. J. B. (2003). Recent increases in species richness and shifts in altitudinal distributions of Norwegian mountain plants. *The Holocene*, **13**, 1–6.

Klanderud, K. and Totland, Ø. (2005). Simulated climate change altered dominance hierarchies and diversity of an alpine biodiversity hotspot. *Ecology*, **86**, 2047–2054.

Kranner, I., Beckett, R., Hochman, A. and Nash, T. H. (2008). Desiccation-tolerance in lichens: a review. *The Bryologist*, **111**, 576–593.

Kroken, S. and Taylor, J. W. (2001). A gene genealogical approach to recognize phylogenetic species boundaries in the lichenized fungus *Letharia*. *Mycologia*, **93**, 38–53.

Kullman, L. (2002). Rapid recent range-margin rise of tree and shrub species in the Swedish Sacndes. *Journal of Ecology*, **90**, 68–77.

Kytöviita, M. M. and Crittenden, P. D. (2007). Growth and nitrogen relations in the mat-forming lichens *Stereocaulon paschale* and *Cladonia stellaris*. *Annals of Botany*, **100**, 1537–1545.

LaGreca, S. (1999). A phylogenetic evaluation of the *Ramalina americana* chemotype complex (Lichenized Ascomycota, Ramalinaceae) based on rDNA ITS sequence data. *The Bryologist*, **102**, 602–618.

Lange, O. L. and Green, T. G. A. (2005). Lichens show that fungi can acclimate their respiration to seasonal changes in temperature. *Oecologia*, **142**, 11–19.

Lange, O. L., Kilian, E. and Ziegler, H. (1986). Water vapour uptake and photosynthesis of lichens: performance differences in species with green and blue-green algae as phycobionts. *Oecologia*, **71**, 104–110.

Lange, O. L., Büdel, B., Meyer, A. and Kilian, E. (1993). Further evidence that activation of net photosynthesis by dry cyanobacterial lichens requires liquid water. *The Lichenologist*, **25**, 175–189.

Lange, O. L., Green, T. G. A., Melzer, B., Meyer, A. and Zellner, H. (2006). Water relations and CO_2 exchange of the terrestrial lichen *Teloschistes capensis* in the Namib fog desert: measurements during two seasons in the field and under controlled conditions. *Flora*, **201**, 268–280.

Lange, O. L., Green, T. G. A., Meyer, A. and Zellner, H. (2007). Water relations and carbon dioxide exchange of epiphytic lichens in the Namib fog desert. *Flora*, **202**, 479–487.

Lättman, H., Lindblom, L., Mattsson, J. E. et al. (2009). Estimating the dispersal capacity of the rare lichen *Cliostomum corrugatum*. *Biological Conservation*, **142**, 1870–1878.

Leimu, R., Mutikainen, P., Koricheva, J. and Fischer, M. (2006). How general are positive relationships between plant population size, fitness and genetic variation? *Journal of Ecology*, **94**, 942–952.

Lenoir, J., Gégout, J. C., Marquet, P. A., de Ruffray, P. and Brisse, H. (2008).

A significant upward shift in plant species optimum elevation during the 20th century. *Science*, **320**, 1768–1771.

Lesica, P. and McCune, B. (2004). Decline of arctic–alpine plants at the southern margin of their range following a decade of climatic warming. *Journal of Vegetation Science*, **15**, 679–690.

Löbel, S., Snäll, T. and Rydin, H. (2006a). Metapopulation processes in epiphytes inferred from patterns of regional distribution and local abundance in fragmented landscapes. *Journal of Ecology*, **94**, 856–868.

Löbel, S., Snäll, T. and Rydin, H. (2006b). Species richness patterns and metapopulation processes: evidence from epiphyte communities in boreo-nemoral forests. *Ecography*, **29**, 169–182.

Longton, R. E. (1988). *Biology of Polar Bryophytes and Lichens*. Cambridge: Cambridge University Press.

Lutzoni, F., Pagel, M. and Reeb, V. (2001). Major fungal lineages are derived from lichen symbiotic ancestors. *Nature*, **411**, 937–940.

MacDonald, G. M., Bennett, K. D., Jackson, S. T. et al. (2008). Impacts of climate change on species, populations and communities: palaeobiogeographical insights and frontiers. *Progress in Physical Geography*, **32**, 139–172.

MacFarlane, J. D. and Kershaw, K. A. (1977). Physiological–environmental interactions in lichens, IV. Seasonal changes in the nitrogenase activity of *Peltigera canina* var. *praetextata* and *P. canina* var. *rufescens*. *New Phytologist*, **79**, 403–408.

Maestre, F. T., Callaway, R. M., Valladares, F. and Lortie, C. J. (2009). Refining the stress-gradient hypothesis for competition and facilitation in plant communities. *Journal of Ecology*, **97**, 199–205.

McCune, B. (2006). Non-parametric habitat models with automatic interactions. *Journal of Vegetation Science*, **17**, 819–830.

Metcalfe, G. (1950). The ecology of the Cairngorms, part II. The mountain Callunetum. *Journal of Ecology*, **38**, 46–74.

Millennium Ecosystem Assessment (2005). *Ecosystems and Human Well- being: Biodiversity Synthesis*. Washington, DC: World Resources Institute.

Morin, X. and Thuiller, W. (2009). Comparing niche- and process-based models to reduce prediction uncertainty in species range shifts under climate change. *Ecology*, **90**, 1301–1313.

Muñoz, J., Felicisimo, A. M., Cabezas, F., Burgaz, A. R. and Martinez, I. (2004). Wind as a long-distance dispersal mechanism in the Southern Hemisphere. *Science*, **304**, 1144–1147.

Murtagh, G. J., Dyer, P. S. and Crittenden, P. D. (2000). Sex and the single lichen. *Nature*, **404**, 564.

Murtagh, G. J., Dyer, P. S., Furneaux, P. A. and Crittenden, P. D. (2002). Molecular and physiological diversity in the bipolar lichen-forming fungus *Xanthoria elegans*. *Mycological Research*, **106**, 1277–1286.

Myllys, L., Lohtander, K. and Tehler, A. (2001). Beta-tubulin, ITS and group I intron sequences challenge the species pair concept in *Physcia aipolia* and *P. caesia*. *Mycologia*, **93**, 335–343.

Myllys, L., Stenroos, S. and Thell, A. (2003). Phylogeny of bipolar *Cladonia arbuscula* and *C. mitis* (Lecanorales, Eusacomycetes). *Molecular Phylogenetics and Evolution*, **27**, 58–69.

Nakicenovic, N. (2000). *Special Report on Emissions Scenarios*. IPCC III. Cambridge: Cambridge University Press.

Nelsen, M. P. and Gargas, A. (2008). Dissociation and horizontal transmission of co-dispersed lichen symbionts in the genus *Lepraria* (Lecanorales: Stereocaulaceae). *New Phytologist*, **177**, 264–275.

Neubert, M. G. and Caswell, H. (2000). Dispersal and demography: calculation and sensitivity analysis of invasion speed for structured populations. *Ecology*, **81**, 1613–1628.

O'Brien, H., Miadlikowska, J. and Lutzoni, F. (2005). Assessing host specialization in symbiotic Cyanobacteria associated with four closely related species of the lichen fungus *Peltigera*. *European Journal of Phycology*, **40**, 363–378.

Ochyra, R., Lewis Smith, R. I. and Bednarck-Ochyra, H. (2008). *Illustrated Moss Flora of Antarctica*. Cambridge: Cambridge University Press.

Öckinger, E., Niklasson, M. and Nilsson, S. G. (2005). Is local distribution of the epiphytic lichen *Lobaria pulmonaria* limited by dispersal capacity of habitat quality? *Biodiversity and Conservation*, **14**, 759–773.

Ott, S. (1987). Reproductive strategies in lichens. *Bibliotheca Lichenologica*, **25**, 81–93.

Øvstedal, D. O. and Lewis Smith, R. I. (2001). *Lichens of Antarctica and South Georgia*. Cambridge: Cambridge University Press.

Palice, Z. and Printzen, C. (2004). Genetic variability in tropical and temperate populations of *Trapeliopsis glaucolepidea*: Evidence against long-range dispersal in a lichen with disjunct distribution. *Mycotaxon*, **90**, 43–54.

Palmqvist, K. (2000). Carbon economy in lichens. *New Phytologist*, **148**, 11–36.

Parmesan. C. (2006). Ecological and evolutionary responses to recent climate change. *Annual Review of Ecology, Evolution, and Systematics*, **37**, 637–669.

Piercey-Normore, M. D. (2006). The lichen-forming ascomycete *Evernia mesomorpha* associates with multiple genotypes of *Trebouxia jamesii*. *New Phytologist*, **169**, 331–344.

Piercey-Normore, M. D. and DePriest, P. T. (2001). Algal switching among lichen symbioses. *American Journal of Botany*, **88**, 1490–1498.

Poore, M. E. D. and McVean, D. N. (1957). A new approach to Scottish mountain vegetation. *Journal of Ecology*, **45**, 401–439.

Press, M. C., Potter, J. A., Burke, M. J. W., Callaghan, T. V. and Lee, J. A. (1998). Responses of a subarctic dwarf shrub community to simulated environmental change. *Journal of Ecology*, **86**, 315–327.

Printzen, C., Ekman, S. and Tønsberg, T. (2003). Phylogeography of *Cavernularia hultenii*: evidence of slow genetic drift in a widely disjunct lichen. *Molecular Ecology*, **12**, 1473–1486.

Reed, D. H. and Frankham, R. (2003). Correlation between fitness and genetic diversity. *Conservation Biology*, **17**, 230–237.

Rehfeldt, G. E., Crookston, N. L., Warwell, M. V. and Evans, J. S. (2006). Empirical analysis of plant community relationships for the western United States. *International Journal of Plant Geography*, **167**, 1123–1150.

Rieseberg, L. H. and Burke, J. M. (2001). The biological reality of species: gene flow, selection, and collective evolution. *Taxon*, **50**, 47–67.

Robinson, C. H., Wookey, P. A., Lee, J. A., Callaghan, T. V. and Press, M. C. (1998). Plant community responses to simulated environmental change at a High Arctic polar semi-desert. *Ecology*, **79**, 856–866.

Rodwell, J. S., ed. (1991). *British Plant Communities, Volume 2. Mires and Heaths*. Cambridge: Cambridge University Press.

Romeike, J., Friedl, T., Helms, G. and Ott, S. (2002). Genetic diversity of algal and fungal partners in four species of *Umbilicaria* (lichenized ascomycetes) along a transect of the Antarctic peninsula. *Molecular Biology and Evolution*, **19**, 1209–1217.

Rose, C. I. and Hawksworth, D. L. (1981). Lichen recolonization in London's cleaner air. *Nature*, **289**, 289–292.

Rose, F. (1988). Phytogeographical and ecological aspects of *Lobarion* communities in Europe. *Botanical Journal of the Linnean Society*, **96**, 69–79.

Rosenzweig, C., Karoly, D., Vicarelli, M. et al. (2008). Attributing physical and biological impacts to anthropogenic climate change. *Nature*, **453**, 353–357.

Sanders, W. B. (2001). Lichens: the interface between mycology and plant morphology. *Bioscience*, **51**, 1025–1035.

Schaper, T. and Ott, S. (2003). Photobiont selectivity and interspecific interactions in lichen communities. I. Culture experiments with the mycobiont *Fulgensia bracteata*. *Plant Biology*, **5**, 441–450.

Scherrer, S., Zippler, U. and Honegger, R. (2005). Characterisation of the mating-type locus in the genus *Xanthoria* (lichen-forming ascomycetes, Lecanoromycetes). *Fungal Genetics and Biology*, **42**, 976–988.

Schofield, S. C., Campbell, D. A., Funk, C. and MacKenzie, T. D. B. (2003). Changes in macromolecular allocation in nondividing algal symbionts allow for photosynthetic acclimation in the lichen *Lobaria pulmonaria*. *New Phytologist*, **159**, 709–718.

Schroeter, B., Green, T. G. A., Kappen, L. and Seppelt, R. D. (1994). Carbon dioxide exchange at subzero temperatures: field measurements on *Umbilicaria aprina* in Antarctica. *Cryptogamic Botany*, **4**, 233–241.

Seaward, M. R. D. (1998). Time-space analysis of the British lichen flora, with particular reference to air quality surveys. *Folia Cryptogamica Estonica*, **32**, 85–96.

Seymour, F. A., Crittenden, P. D., Dickinson, M. J. et al. (2005a). Breeding systems in the lichen-forming fungal genus *Cladonia*. *Fungal Genetics and Biology*, **42**, 554–563.

Seymour, F. A., Crittenden, P. D. and Dyer, P. S. (2005b). Sex in the extremes: lichen-forming fungi. *Mycologist*, **19**, 51–58.

Seymour, F. A., Crittenden, P. D., Wirtz, N. et al. (2007). Phylogenetic and morphological analysis of Antarctic lichen-forming *Usnea* species in the group *Neuropogon*. *Antarctic Science*, **19**, 71–82.

Shaver, G. R. and Jonasson, S. (1999). Response of Arctic ecosystems to climate change: results of long-term field experiments in Sweden and Alaska. *Polar Research*, **18**, 245–252.

Sillett, S. C., McCune, B., Peck, J. E., Rambo, T. R. and Ruchty, A. (2000). Dispersal limitations of epiphytic lichens result in species dependent on old-growth forests. *Ecological Applications*, **10**, 789–799.

Slatkin, M. (1976). The rate of spread of an advantageous allele in a subdivided population. In *Population Genetics and Ecology*, ed. S. Karlin and E. Nevo. New York, NY: Academic Press, pp. 767–780.

Smith, C. W., Aptroot, A., Coppins, B. J. et al., eds. (2009). *The Lichens of Great Britain and Ireland*. Slough: Richmond Publishing Co.

Snäll, T., Pennanen, J., Kivistö, L. and Hanski, I. (2005). Modelling epiphyte metapopulation dynamics in a dynamic forest landscape. *Oikos*, **109**, 209–222.

Solhaug, K. A., Gauslaa, Y., Nybakken, L. and Bilger, W. (2003). UV-induction of sun-screening pigments in lichens. *New Phytologist*, **158**, 91–100.

Sonesson, M., Schipperges, B. and Carlsson, B. Å. (1992). Seasonal patterns of photosynthesis in alpine and subalpine populations of the lichen *Nephroma arcticum*. *Oikos*, **65**, 3–12.

Stace, C. (1997). *New Flora of the British Isles*, 2nd edn. Cambridge: Cambridge University Press.

Stevens, C. J., Dise, N. B., Mountford, J. O. and Gowing, D. J. (2004). Impact of nitrogen deposition on the species richness of grasslands. *Science*, **303**, 1876–1879.

Sturm, M., Racine, C. and Tape, K. (2001). Increasing shrub abundance in the Arctic. *Nature*, **411**, 546.

Suttle, K. B., Thomsen, M. A. and Power, M. E. (2007). Species interactions reverse grassland responses to changing climate. *Science*, **315**, 640–642.

Taylor, T. N., Hass, H., Remy, W. and Kerp, H. (1995). The oldest fossil lichen. *Nature*, **378**, 244.

Thuiller, W., Lavorel, S., Araújo, M. B., Sykes, M. T. and Prentice, I. C. (2005). Climate change threats to plant diversity in Europe. *Proceedings of the National Academy of Sciences of the USA*, **102**, 8245–8250.

Tormo, R., Recio, D., Silva, I. and Muñoz, A. F. (2001). A quantitative investigation of airborne algae and lichen soredia obtained from pollen traps in south-west Spain. *European Journal of Phycology*, **36**, 385–390.

Travis, J. M. J. (2003). Climate change and habitat destruction: a deadly anthropogenic cocktail. *Proceedings of the Royal Society of London B*, **270**, 467–473.

Tribsch, A. and Stuessy, T. F. (2003). Evolution and phylogeography of arctic and alpine plants in Europe: introduction. *Taxon*, **52**, 415–416.

Trivedi, M. R., Morecroft, M. D., Berry, P. M. and Dawson, T. P. (2008a). Potential effects of climate change on plant communities in three montane nature reserves in Scotland, UK. *Biological Conservation*, **141**, 1665–1675.

Trivedi, M. R., Berry, P. M., Morecroft, M. D. and Dawson, T. P. (2008b). Spatial scale affects bioclimate model projections of climate change impacts on mountain plants. *Global Change Biology*, **14**, 1089–1103.

van Dobben, H. F., Wolterbeek, H. T., Wamelink, G. W. W. and ter Braak, C. J. F. (2001). Relationship between epiphytic lichens, trace elements and gaseous atmospheric pollutants. *Environmental Pollution*, **112**, 163–169.

van Herk, C. M. (1999). Mapping of ammonia pollution with epiphytic lichens in the Netherlands. *The Lichenologist*, **31**, 9–20.

van Herk, C. M., Aptroot, A. and van Dobben, H. F. (2002). Long-term monitoring in the Netherlands suggests that lichens respond to global warming. *The Lichenologist*, **34**, 141–154.

van Herk, C. M., Mathijssen, E. A. M. and de Zwart, D. (2003). Long distance nitrogen air pollution effects on lichens in Europe. *The Lichenologist*, **35**, 347–359.

Walker, J. (1985). The lichen genus *Usnea* subgenus *Neuropogon*. *Bulletin of the British Museum of Natural History*, **13**, 1–130.

Walker, M. D., Wahren, C. H., Hollister, R. D. et al. (2006). Plant community responses to experimental warming across the tundra biome. *Proceedings of the National Academy of Sciences of the USA*, **103**, 1342–1346.

Walther, G. R., Post, E., Convey, P. et al. (2002). Ecological responses to recent climate change. *Nature*, **416**, 389–395.

Walser, J. C. (2004). Molecular evidence for limited dispersal of vegetative propagules in the epiphytic lichen *Lobaria pulmonaria*. *American Journal of Botany*, **91**, 1273–1276.

Walser, J. C., Zoller, S., Buchler, U. and Scheidegger, C. (2001). Species-specific detection of *Lobaria pulmonaria* (lichenized ascomycete) diaspores in litter samples trapped in snow cover. *Molecular Ecology*, **10**, 2129–2138.

Walser, J. C., Holderegger, R., Gugerli, F., Hoebee, S. and Scheidegger, C. (2005). Microsatellites reveal regional population differentiation and isolation in *Lobaria pulmonaria*, an epiphytic lichen. *Molecular Ecology*, **14**, 457–467.

Walton, J. (1922). A Spitsbergen salt marsh: with observations on the ecological phenomena attendant on the emergence of land from the sea. *Journal of Ecology*, **19**, 109–121.

Watt, A. S. (1947). Pattern and process in the plant community. *Journal of Ecology*, **35**, 1–22.

Watt, A. S. and Jones, E. W. (1948). The ecology of the Cairngorms. I. The environment and altitudinal zonation of the vegetation. *Journal of Ecology*, **36**, 283–304.

Wedin, M., Döring, H. and Gilenstam, G. (2004). Saprotrophy and lichenization as options for the same fungal species on different substrata: environmental plasticity and fungal lifestyles in the *Stictis-Conotrema* complex. *New Phytologist*, **164**, 459–465.

Werth, S. (2010). Population genetics of lichen-forming fungi: a review. *The Lichenologist*, **42**, 499–520.

Werth, S. and Sork, V. L. (2008). Local genetic structure in a North American epiphytic lichen, *Ramalina menziesii* (Ramalinaceae). *American Journal of Botany* **95**, 568–576.

Werth, S., Wagner, H. H., Gugerli, F. et al. (2006). Quantifying dispersal and establishment limitation in a population of an epiphytic lichen. *Ecology*, **87**, 2037–2046.

Werth, S., Gugerli, F., Holderegger, R. et al. (2007). Landscape-level gene flow in *Lobaria pulmonaria*, an epiphytic lichen. *Molecular Ecology*, **16**, 2807–2815.

Willis, K. J., Kleczkowski, A., New, M. and Whittaker, R. J. (2007). Testing the impact of climatic variability on European plant diversity: 320,000 years of water-energy dynamics and its long-term influence on plant taxonomic richness. *Ecology Letters*, **10**, 673–679.

Wirtz, N., Lumbsch, H. T., Green, T. G. A. et al. (2003). Lichen fungi have low cyanobiont selectivity in maritime Antarctica. *New Phytologist*, **160**, 177–183.

Wirtz, N., Printzen, C., Sancho, L. G. and Lumbsch, H. T. (2006). The phylogeny and classification of Neuropogon and *Usnea* (Parmeliaceae, Ascomycota) revisited. *Taxon*, **55**, 367–376.

Woodin, S. J. (1989). Environmental effects of air pollution in Britain. *Journal of Applied Ecology*, **26**, 749–761.

Woodward, F. I. and Beerling, D. J. (1997). The dynamics of vegetation change: health warnings for equilibrium 'dodo' models. *Global Ecology and Biogeography Letters*, **6**, 413–418.

Yahr, R., Vilgalys, R. and DePreist, P. T. (2006). Geographic variation in algal partners of *Cladonia subtenuis* (Cladoniaceae) highlights the dynamic nature of the lichen symbiosis. *New Phytologist*, **171**, 847–860.

Zoller, S., Lutzoni, F. and Scheidegger, C. (1999). Genetic variation within and among populations of the threatened lichen *Lobaria pulmonaria* in Switzerland and implications for its conservation. *Molecular Ecology*, **8**, 2049–2059.

21

Climate change and oceanic mountain vegetation: a case study of the montane heath and associated plant communities in western Irish mountains

R. L. Hodd and M. J. Sheehy Skeffington

Botany and Plant Science, National University of Ireland, Galway, Ireland

Abstract

Plant communities in montane regions are useful for studying the potential effects of climate change. Many mountain species have affinities with colder climates and may not survive local temperature rises. Although Irish mountains are not of high altitude and are influenced by the tempering effect of the Atlantic Ocean, they support some species of arctic-montane affinity. In Ireland, the climate termed hyperoceanic, with its constant moisture and mild temperatures, prevails on western mountains. There it benefits the growth of bryophyte communities, which are more abundant due to higher cloud cover and precipitation as well as lower evapotranspiration. As these bryophyte communities occur up to *c*. 1000 m, alongside the arctic-montane higher plant species, they can be complementary

Climate Change, Ecology and Systematics, ed. Trevor R. Hodkinson, Michael B. Jones, Stephen Waldren and John A. N. Parnell. Published by Cambridge University Press.

as climate change indicators, as they respond differently to such change. There is little systematic information on the distribution of these scarce montane plant communities. Their distribution on the mountains of the west of Ireland is being mapped, and data are being gathered on the local climate of selected mountains. This will supply useful case-study material for climate change modelling, specifically providing information on regions that have little precise climatic information and on plant communities that are likely to be very vulnerable to aspects of climate change.

21.1 Introduction

It is thought that many species may be unlikely to adapt as fast as the current accelerated rate of climate change or that predicted for the future (Berry et al., 2005). There are projected to be major changes in ecosystem structure and function, species' ecological interactions and shifts in geographical ranges of species, with predominantly negative consequences for biodiversity and ecosystem goods and services (IPCC, 2007). It is therefore important to focus on the potential effects of such climate change on plant and animal communities. Some communities will be more susceptible than others to climate change, partly due to their geographical location; they may not be able to migrate to regions with a compatible climate, either because their dispersal is limited or because the suitable regions are too far away or non-existent (Walther et al., 2002).

Islands and mountain tops present these difficult geographical challenges to the communities that inhabit them. For example, Walther et al. (2005) concluded that many alpine species have shifted their range upwards in the past few decades, a trend that has accelerated greatly since 1985. This upward shift of many species leads, in turn, to increased competition for plants specifically adapted to higher mountain environments, which also become restricted in their range, as they may lose already limited climate space (Gottfried et al., 1999). In order to combat the effects of current rapid climate change, organisms must be able to adapt quickly to survive in their changing environment (Visser, 2008). However, many alpine and montane vascular plant species, along with numerous oceanic bryophyte species, are poor competitors and have limited dispersal ability, and are therefore unable to adapt rapidly to changes in their environment (Körner, 2003; Porley and Hodgetts, 2005).

In attempting to evaluate the effects of climate change on biodiversity, it is therefore useful to select habitats that have specific, often localised climatic conditions. Plants provide the habitat and food for other species and, because they are not usually individually mobile, they can be good indicators of long-term environmental change (Donnelly et al., 2004). Plant communities of mountain tops are in an extreme habitat and are vulnerable to change, specifically to higher temperatures, since many species have an arctic–montane/arctic–alpine distribution, occurring in regions where temperatures are low. The potential changes in climatic conditions that have been predicted for the near future (IPCC, 2007) may therefore result in corresponding changes in plant and animal communities. However, in order to interpret these changes in vegetation, it is important to understand the current climate and its influence on the present vegetation.

Ireland, on the western fringe of Europe, has a climate that is predominantly influenced by the Atlantic Ocean and therefore represents another extreme in climate. The mountains in western Ireland, being the first landfall for weather systems, are particularly subject to constant high humidity (Sweeney, 1997). Certain plant communities, particularly of bryophytes (mosses and liverworts), live alongside the arctic–montane communities, but are dependent on the more equable oceanic climate of the region. Both communities are therefore potentially useful indicators of climate change.

Climate change is projected to have a major impact on the phenology, physiology, distribution, species interactions and community composition of plants in western Europe (Walther, 2003; Donnelly et al., Chapter 8). Indeed, studies over the past 30 years provide evidence that warmer temperatures have affected these characteristics in organisms (Walther et al., 2002; Jones et al., 2006). Knowledge of vegetation patterns and distribution provides a baseline for detecting future changes in these patterns and, although species respond individually to changes in local climate, they interact at community level. Therefore, studies of plant and animal communities will give an overview of the interactive effects of climate change (Duckworth et al., 2000). The potential effects of climate change on the vegetation in oceanic and coastal montane areas have been little studied compared to those of continental areas (Fosaa et al., 2004).

21.2 Climate

21.2.1 The oceanic climate of western Ireland

The climatic conditions of western Ireland, which are very different from those even of the midlands and east coast of the country, are among the principal factors affecting the composition of the vegetation in the region. As Ireland is located on the Atlantic fringe of Europe, it has a highly oceanic climate. Oceanicity is a term used to describe conditions of temperature, humidity and other factors that result from maritime influences and alter the ecology and environment of oceanic areas (Crawford, 2000). An oceanic climate does not have extreme high or low temperatures or a high temperature lapse rate, but does have high cloudiness, high and frequent rainfall and humidity, low vapour pressure saturation deficits and often very high wind speeds (Ratcliffe, 1968; Grace, 1997). The degree of oceanicity varies locally, and there is a gradient of increasing oceanicity from east to west, within the British–Irish Isles (Ratcliffe, 1968; Brown et al., 1993). Due to the topography of Ireland, where the uplands are largely coastal, there is a high climatological contrast between the oceanically influenced maritime margins and the relatively continental interior of the country (Sweeney, 1997). Climate change is likely to lead to an overall increase in oceanicity in western European areas that already have an oceanic climate (Crawford, 2000).

Oceanicity can be defined by a number of meteorological methods, most commonly using the mean annual temperature range and other associated meteorological and geographical factors. According to the majority of methods of calculating oceanicity, such as Conrad's index of continentality ($K = [1.7A/\sin(\varphi + 10)] - 14$, in which K = index of continentality; A = average annual temperature range; φ = latitude – Conrad, 1946), western Ireland, along with the Faroe Islands and western Scotland, is classified as hyperoceanic (Crawford, 2000). Kirkpatrick and Rushton (1990) calculated the degree of oceanicity of a number of sites in northern Ireland, using Conrad's index of continentality and Kotilainen's index of oceanicity ($K = Ndt/100\Delta$, in which N = precipitation in millimetres per year; dt = number of days with mean temperature between 0 and 10 °C; Δ = difference between mean temperature of warmest and coldest months – Godske, 1944), showing that the most oceanic areas were in the western montane and coastal parts of that area of Ireland studied. Averis et al. (2004) also calculated an oceanicity index, based on the mean annual number of wet days (> 1 mm of rain) divided by the range of monthly mean temperatures in °C. An isoline of 20 for this index separates off the western fringe of Ireland, including the western mountain ranges, from the rest of the island (Fig 21.1).

Temperatures in Ireland are relatively high for its latitude, due to the warming influence of the Gulf Stream ocean current. As a result there is little frost or snow during the winters, especially in western coastal areas. However, due to the cooling effect of the ocean in summer, temperatures do not rise as high as in more continental areas at similar or higher latitudes (Sweeney, 1997). Temperatures have a very small range across Ireland between summer and winter and are strongly influenced by the oceanic climate, especially in the west. The mean annual temperature in Ireland is about 9 °C (www.met.ie); average daily January temperatures in Kerry, in the southwest, range from 6 to 6.5 °C, while those for July range from 15 to 15.5 °C (Rohan, 1986). The difference in mean monthly temperature in Connemara, in western Galway, between the warmest and coldest month is 8 °C, which is very small compared to elsewhere in Britain and Ireland (Horsfield et al., 1991). Growth is low below an air temperature of 5 °C, and the number of days for the grass growing season in the west of Ireland ranges from

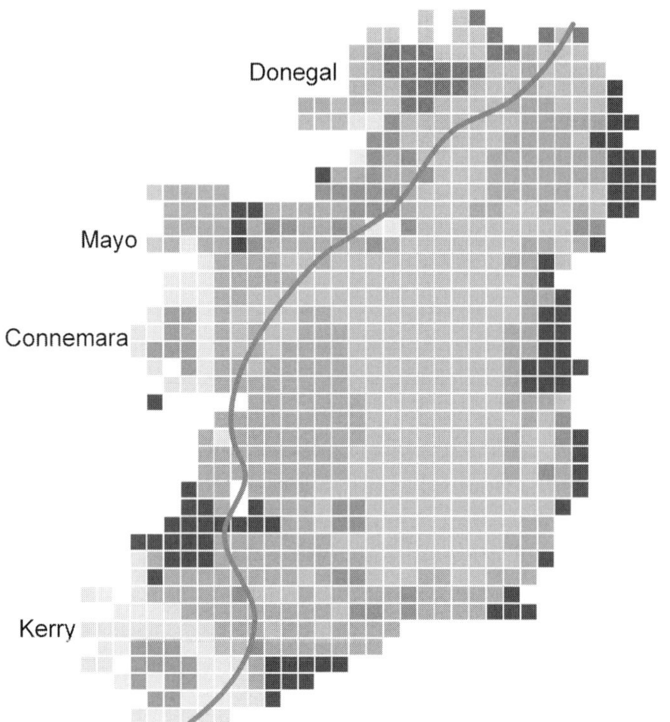

Figure 21.1 Map of Irish bioclimatic regions, with regions defined by the MONARCH project (Berry et al., 2005). Each group of coloured squares represents a different bioclimatic region. The mountains of Kerry and Connemara are within a distinct bioclimatic region, different from that of the mountains of Donegal. The bold line represents a value of 20 on the index of climatic oceanicity (after Averis et al., 2004). This is calculated as the mean annual number of wet days (> 1 mm of rain) divided by the range of monthly mean temperatures in °C. *See colour plate section.*

330 days in the extreme southwest to 300 days along the Atlantic coast north to Mayo, but on the Donegal coast in the north of Ireland it drops to 270 days (Collins and Cummins, 1996).

Rainfall and humidity are also strongly influenced by the proximity to the Atlantic Ocean. The average annual rainfall in the west of Ireland is generally between 1000 and 1250 mm, rising to over 2000 mm in mountainous areas (Carruthers, 1998). In general, most areas of western Ireland have over 180 rain days (days on which more than 1 mm of rain falls) per year, with mountainous areas having over 220 rain days per year (Ratcliffe, 1968). The frequent rainfall contributes to a highly humid atmosphere, especially in mountain areas of the west. The average relative humidity in the west is between 70% and 90% (Whilde, 1994; Carruthers, 1998). The high humidity is further enhanced by high levels of cloudiness, especially in the west, with complete cloud cover over Ireland for more than 50% of the year (www.met.ie).

Evapotranspiration rates are low in western Ireland due to a combination of high rainfall, humidity, cloudiness and low sunshine. Rainfall exceeds evaporation, so the atmosphere is constantly moist (Whilde, 1994). This is despite the fact that the oceanic areas of the British–Irish Isles are among the windiest areas in the world (Grace and Unsworth, 1988). However, the intensity and frequency of strong winds varies greatly across Ireland; the northwestern coastal areas are the windiest, with an average of over 60 days with gales per year (Sweeney, 1997).

21.2.2 The climate of oceanic mountains

Ireland is a mainly lowland country, with 5% of its land surface above 300 m in altitude and 0.25% of the land surface above 600 m (Pochin Mould, 1976). Despite the scarcity of land above 600 m, there are mountain ranges in most coastal areas, especially in the west. The highest point in the country, Corrán Tuathail, in the Macgillycuddy's Reeks of Co. Kerry, is 1041 m in altitude.

The climatic conditions that occur on the mountain summits differ in a number of ways from the climate at their base. Temperatures decrease with an increase in altitude; in oceanic areas there is a lapse rate of about 0.8–1 °C per 100 m of altitude gained (Crawford, 2000). This is significantly higher than the lapse rate in continental mountain ranges, which is about 0.65 °C per 100 m altitude gained (Nagy et al., 2005), because of the increased exposure and cloudiness in oceanic areas.

Precipitation is higher in mountain areas than in lowland areas. For example, the mean annual precipitation (1961–90) for the summit of Mangerton Mountain in Co. Kerry (altitude 808 m) was 3230 mm, whereas, for the same period, it was 1430 mm at Valentia Island (altitude 9 m), less than 50 km west of Mangerton (Sweeney, 1997). More precipitation falls in the form of snow and hail at higher altitudes than in the lowlands, but snowfall is not a significant component of the precipitation regime in western Ireland, even at higher latitudes or altitudes (Sweeney, 1997).

Humidity is also higher on mountain summits, particularly in areas close to the sea, as areas above 500–600 m are covered by cloud for most of the time. This increased cloud cover also results in a reduction in the amount of sunlight that mountain areas are exposed to (Barry, 1992). Comparing Ben Nevis in Scotland (1344 m) with Fort William at its base (53.6 m), for the years 1884–1903, Tansley (1949) listed mean rainfall on the summit (4084 mm) as more than double that at Fort William (2002 mm), temperatures much cooler (mean July: 5 °C versus 14 °C) and mean sunshine incidence much lower (17% versus 31%). On the other hand, wind speed increases with altitude, with high winds frequent at higher altitudes. However, wind speed is more strongly influenced by topography than by altitude (Barry, 1992).

21.2.3 Mountain microclimate and topography

As mountains are topographically complex, a range of topoclimatic and microclimatic conditions can occur within a small area of a single mountain depending on altitude, aspect, slope and other factors (Barry, 1992). These factors, in tandem with the overall climate, geology, soils and land use, determine the species and communities that will grow on a particular area of a mountain (Barry, 1992). Factors such as temperature, wind and cloud cover vary not only with altitude but also with aspect. South-facing slopes receive more sunshine than north-facing ones (Barry, 1992) and are therefore drier and warmer than those facing north, where, in turn, less evapotranspiration takes place, resulting in higher humidity (Ratcliffe, 1968). Exposure is greatest on summits, ridge tops and cols, as wind speeds are higher and surface temperatures are lower in these places than on the slopes. The complex effects of high winds and low surface temperatures in exposed places can result in drier but colder conditions than in more sheltered areas (Grace and Unsworth, 1988).

Variations in the steepness of slopes also affect the microclimate, as steep slopes are generally drier than shallower slopes, where moisture is able to accumulate. The presence of cliffs and corries alters the conditions on the slopes below the cliffs, as water running off the cliffs accumulates at the cliff base (Tansley, 1949; Hodd, 2007; Hodd and Sheehy Skeffington, in preparation). There is also less sunlight in corries, especially those which face north, which are also sheltered from strong winds. These factors all result in less evapotranspiration and, therefore, higher humidity in corries than on surrounding slopes (Hodd, 2007; Hodd and Sheehy Skeffington, in preparation).

21.3 Vegetation

21.3.1 The vegetation of hyperoceanic mountain areas of northwestern Europe

Western Ireland, western Scotland and the Faroe Islands all share a hyperoceanic climate (Crawford, 2000), and these areas have many plant communities in common. The mountains of these areas are of similar altitude (maximum altitude *c.* 800–1350 m), so the main gradient between them is one of latitude. In all these areas, the montane vegetation is dominated, to varying degrees, by *Calluna* heath, moss heath, grassland and blanket bog. The vegetation of all three regions is highly modified by human activities, such as grazing, burning and deforestation (Baxter and Thompson, 1995; Bleasdale and Sheehy Skeffington, 1995; Fosaa, 2001).

As the Faroes are located considerably further north than Ireland or Scotland, they have a colder and more severe climate. The vegetation of the Faroe Islands is dominated mainly by grassland at all altitudes, and heath up to an altitude of about 400 m. At higher altitudes, in the alpine zone (up to 800–900 m), fell fields and snow beds are frequent, with *Racomitrium* moss-heath prevalent on many summits (Fosaa, 2001). *Calluna vulgaris – Erica cinerea* heath, a vegetation type that is common in a wide range of situations in more southerly oceanic areas, is restricted in distribution on the Faroes, as it is at its northern limit here (Fosaa, 2001).

The vegetation of the Scottish Highlands shares some similarities with the Faroe Islands, with dwarf shrub heath giving way to moss heath and grassland at higher altitudes and some snow-bed vegetation present on the higher peaks (Tansley, 1949; McVean and Ratcliffe, 1962; Averis et al., 2004). Due to a combination of lower altitude, and especially latitude, snow-bed and associated cold-climate vegetation types are completely absent from the western Irish mountains. There are, however, many similarities in the vegetation of the heaths, bogs and grassland of these areas.

21.3.2 The vegetation of the western Irish mountains

The mountains of western Ireland are predominantly covered by blanket bog, rough pasture and heathland (Whilde, 1994; Carruthers, 1998). Blanket bog vegetation is widespread throughout the Irish mountains, especially on shallow slopes, where deep peat is able to accumulate. Peatlands are strongly reliant on the occurrence of oceanic conditions, as they require high rainfall and low evapotranspiration rates to form (Taylor, 1983; Mitchell and Ryan, 2001). Grassland also occurs throughout the Irish mountains, often on more basic rocks and where grazing or exposure prevents or eliminates the growth of larger shrub species, such as *Calluna vulgaris* (L.) Hull (Bleasdale and Sheehy Skeffington, 1995; McKee et al.,1998).

Many Irish mountains have abundant rocky habitats, mainly in the form of screes and cliffs. The scree slopes are generally colonised by vegetation from surrounding stands of grassland and heath and can support rare oceanic bryophyte species such as *Scapania ornithopodioides* (With.) Waddell and *Mastigophora woodsii* (Hook.) Nees (Hodd, 2007; Hodd and Sheehy Skeffington, in preparation). The cliffs support fragmentary vegetation, and where conditions of geology and aspect are suitable, a number of rare arctic–montane species occur (Praeger, 1934; Roden, 1986).

Although the mountains of Ireland are low in relation to the majority of mountain ranges in other parts of Europe, the higher areas of these mountain ranges could be classified as alpine. Alpine areas are defined as those above the potential tree line (the highest altitude at which trees > 3 m in height form distinct groups – Körner, 2003). However, in Britain and Ireland, it is more appropriate to refer to areas above the potential tree line as montane, and areas above the limits of enclosed farmland as submontane (Ratcliffe and Thompson, 1988).

At Ireland's latitude, between 51° and 54° north, the tree line would be expected to be well over 1000 m (Körner, 2003), higher than most of its peaks.

However, due to the influence of the oceanic climate, the potential tree line on the western Irish mountains is depressed, as a consequence of increased cloudiness and exposure (Crawford, 2000). Grace (1997) suggests that the factors that control the tree line are different in oceanic areas than in areas with a continental climate. Factors such as frost stress and temperature limitation, which limit the tree line in continental areas (Körner, 2003), are ameliorated by the mild oceanic climate. But in Britain and Ireland wind is likely to be the main controlling factor in limiting tree growth at high altitude (Tansley, 1949; Grace, 1997). Firstly, there is a wind cooling effect, which brings the vegetation temperature down close to that of the air. The second factor is wind blasting, where small particles lifted from soil and rocks abrade the plant surfaces; and the third factor is the mechanical response of the plant structure to strong winds (Grace, 1997). Wind pruning of exposed coastal trees curtails their growth (Crawford, 2005), as the salt in these winds inhibits the meristematic growth on the seaward side (Doutt, 1941).

The highest surviving natural forests in Ireland occur up to an altitude of 330 m (Cross, 2006), but the natural tree line would be located at a higher altitude than this. It is not easy to define the potential tree line, as anthropogenic impacts, through grazing, burning and felling of natural forests, coupled with the hyperoceanic climate, have led to a major reduction in the extent of Irish woodland (Crawford, 2005), with any woodland near the potential location of the tree line having been eradicated. Once the woodland was felled, it was difficult for it to become re-established, due to leaching, impoverishment and waterlogging of the soils, and it was replaced by encroaching bog and heath (Mitchell and Ryan, 2001).

In Britain, it has been estimated that the former Holocene tree line varies from less than 500 m in the northwest Scottish Highlands and islands to around 800 m in the Cairngorms and Pennines (Birks, 1988). There are no equivalent estimates for Ireland, but Cross (2006) suggests that montane *Betula pubescens* Ehrh. forest, in mosaic with bog and heath, without any anthropogenic impact, would occur in Ireland up to an altitude of 700 m. Whilde (1994) notes that the tree line in Connemara is at 450 m, but the original source of this estimate is not clear. Taking into account differences in latitude and the hyperoceanic climate of western Ireland, it is likely that the potential natural tree line on the western Irish mountains would be, at its maximum, between 550 and 650 m, perhaps lower than 500 m in northwest Donegal (colder and more wind) and lower again in areas exposed directly to the influence of the ocean. As exposure is the main controlling factor in the position of the natural tree line in oceanic regions, it is likely that the altitude of the tree line would fluctuate greatly, depending on the degree of exposure, ranging from close to sea level in very exposed areas to relatively high altitudes in sheltered valleys (Poore and McVean, 1957).

21.3.3 Montane heath in the west of Ireland

Much of the vegetation of the mountains of western Ireland can be classified as montane heath. In the strictest sense, the term heath is used to describe only areas that are dominated by shrubby, ericaceous species, such as *Calluna* (Gimingham, 1972). However, in a broader sense, especially in Britain, heath is used to describe any community occurring on acidic, podsolised soils and can include lichen, moss and rush heath, among others (Rodwell, 1992; Averis et al., 2004). There is no clear delineation between lowland, montane and alpine heath, so montane heath in Ireland will here be taken to mean any heath community (with or without *Calluna*) occurring above an altitude of about 400 m. Montane heath has been shown to be very sensitive to climate change, and many species are not likely to be able to migrate to areas of suitable climate (Harrison et al., 2001).

There is a variety of heath communities present on the western Irish mountains, which are similar in many instances to those described for Britain. The majority of montane heath vegetation in Ireland grows on peat, so, in many instances, the heath communities intergrade into upland blanket bog. Heath is generally distinguished from blanket bog by the depth of peat (greater than 1 m in blanket bog – Rodwell, 1991), the presence of more ericoid species and less *Sphagnum* moss. Wet and dry heath are also distinguished by peat depth (greater than 50 cm in wet heath) and the species composition, with less *Sphagnum* and fewer sedge (Cyperaceae) species in dry heath (Rodwell, 1991). However, in practice, it is often difficult to distinguish clear boundaries between plant communities in extremely oceanic areas (Fosaa, 2001). Plant communities that would be described as dry heath may occur on peat greater than 50 cm deep, and wet heath type communities may grow on relatively shallow peat.

Montane heath communities in the west of Ireland are dominated by dwarf shrubs, bryophytes, rushes (Juncaceae) and sedges. The most frequent type of montane heath is characterised by the dwarf shrub *Calluna vulgaris*, which occurs in most montane heath vegetation types and at all altitudes (Gimingham, 1972). On slopes at lower altitudes and in sheltered corries free from overgrazing, *Calluna* can form a canopy of up to 1 m, under which a luxuriant understory of bryophytes often occurs (Fig 21.2). This can consist of either large pleurocarpous mosses or a rare liverwort community (hepatic mats), where conditions are right (Averis et al., 2004). Other dwarf shrubs often associated with *Calluna vulgaris* include *Vaccinium myrtillus* L., *Empetrum nigrum* L. and, mainly in northern areas of the country, the arctic-montane species *Arctostaphylos uva-ursi* (L.) Sprengel and *Vaccinium vitis-idaea* L. The dwarf willow (*Salix herbacea* L.) is also frequent on Irish mountain tops (Fig 21.3). *Calluna vulgaris* grows up to an altitude of over 1000 m, but at altitudes over about 600–700 m generally does not

Figure 21.2 Northern hepatic mat vegetation. Relatively ungrazed *Calluna vulgaris* heath, on Errigal, Co. Donegal, Ireland, supporting northern hepatic mat vegetation. Photo: R. Hodd. *See colour plate section.*

grow taller than 5–10 cm, and is often subdominant to other species, depending on the degree of exposure and grazing (Tansley, 1949; McKee et al., 1998). Heaths of central and eastern Irish mountains tend to be more frequently dominated by *C. vulgaris* and, though bryophytes can be abundant, they are less species-rich and are mostly mosses (Tansley, 1949; Moore, 1960).

Grass, rush and sedge heaths occur at all altitudes in Ireland, although it could be argued that these communities should not be strictly classified as heath vegetation (*sensu* Gimingham, 1972), but rather as grassland. However, these communities are often derived from ericaceous heath communities as a result of overgrazing, and are usually dominated by the grass *Nardus stricta* L. and the rush *Juncus squarrosus* L. (Bleasdale and Sheehy Skeffington, 1995). It has been found by the first author that, although the majority of grass and rush heaths are relatively species-poor, some stands of rush heath may contain a relatively large number of bryophyte species. Stands of this vegetation are most frequent in the southwest of Ireland. Sedge heath, dominated by *Carex bigelowii* Torr. ex Schwein. is occasional in Ireland; it is mainly a community of snow beds in Scotland (McVean and Ratcliffe, 1962; Averis et al., 2004). Vegetation dominated by *Carex bigelowii* occurs sparingly on mountain summits throughout Ireland.

Figure 21.3 *Salix herbacea*, a widespread arctic–montane species, on Dooish, Co. Donegal, Ireland. Photo: R. Hodd. *See colour plate section.*

Bryophyte heath is a highly oceanic vegetation type (thriving in the almost continuously moist climate), and, as such, is frequent in the mountains of western Ireland, but is not abundant in central and eastern parts of the island. The most widespread and well-documented type of bryophyte heath is *Racomitrium* heath, which is dominated by the large acrocarpous moss *R. lanuginosum* (Hedw.) Brid. (Fig 21.4). It occurs mainly at higher altitudes, on exposed summits and ridges (Averis et al., 2004), and usually grows in association with *Calluna*. Another type of bryophyte heath, defined by Mhic Daeid (1976), is dominated by the liverwort *Herbertus aduncus* (Dicks.) Gray subsp. *hutchinsiae* (Gottsche) R. M. Schust. without a tall *Calluna vulgaris* canopy. This community occurs mainly, but not exclusively, on east- to north-facing slopes and corries. It often contains many oceanic species, where it becomes known as a mixed northern hepatic mat.

21.3.4 Arctic–montane species

Arctic–montane species are those that are distributed mainly north of the arctic timber line and in moderate- to high-altitude areas further south. They differ from arctic–alpine species in that they can be found below the tree line in southern areas, and can grow at sea level in oceanic areas (Dahl, 1998). In comparison to the Scottish Highlands, and especially continental mountain ranges, Ireland has

Figure 21.4 Montane heath, dominated by the moss *Racomitrium lanuginosum*, on the Slieve Mish mountains, Co. Kerry, Ireland. Photo: R. Hodd. *See colour plate section.*

a small and much reduced arctic–montane flora, due to a combination of the low altitude of the Irish mountains and the lack of suitability of oceanic mountains in general for the growth of many arctic–montane plant species (Crawford, 2008). In general the Irish arctic–montane flora is restricted to either north-facing montane cliffs and corries, with some calcareous bedrock present (Roden, 1986), or montane heath on mountain summits and ridges.

These species are relicts that are believed to have lasted through the previous ice age (Pearsall, 1950; Mitchell and Ryan, 2001), and the majority only grow in areas with little or no plant competition or grazing and with a suitable aspect, altitude and geology. Therefore, these species grow in scattered localities throughout the west of Ireland (Preston et al., 2002) and are often restricted to isolated patches where all of the conditions are suitable for their growth. Although there are a number of relatively large areas of high ground in coastal areas of eastern Ireland, especially in Wicklow and the Mourne Mountains, fewer arctic–montane species occur here than in the west of Ireland. This is possibly because the eastern Irish mountain ranges are less montane than mountain ranges in the west of Ireland, as they are less oceanic. Therefore, the exposure is less severe, resulting in increased competition from more vigorous species. Hart (1891) also noted the lack of suitable habitat in the Wicklow Mountains. The potential reduced distribution and loss of arctic–montane species may also be of use as indicators of climate change in Ireland (Donnelly et al., 2004).

In general, more arctic–montane species occur in the northwest of Ireland than elsewhere in the country (Praeger, 1950). These species also occur in greater abundance and at lower altitudes in the northwest of the country (Hart, 1891), suggesting that temperature has a major control on the distribution of these species in Ireland, as has been recorded in other parts of Europe (Dahl, 1998). However, geology also has a strong influence on the distribution of arctic–montane species, as there is more suitable montane ground on calcareous bedrock in the northwest of Ireland than in other montane parts of Ireland (Praeger, 1950). Competition and disturbance also limit the distribution of arctic–montane and alpine species, as these species, in response to relatively harsh montane conditions, are generally very slow-growing and low in habit (Körner, 2003), so have a low tolerance of changes in their environment or invasion of species of more temperate (micro)climates (Crawford, 2008). Indeed, many may only set seed on occasion (Hart, 1891). The arctic–montane bryophyte communities are poorly represented in Ireland, but Watson (1925) lists over 30 bryophyte species as occurring almost exclusively above 600 m in Britain. Such species are adapted to extremes in climate and also may not frequently reproduce sexually (Watson, 1925).

21.3.5 Mixed northern hepatic mats

Leafy liverworts (or hepatics) are particularly thin-leaved and weak plants that require almost constant moisture to grow and photosynthesise, and are dormant in the absence of water (Porley and Hodgetts, 2005). Therefore, liverworts abound only in the wettest regions of the world. Western Ireland, and the southwest in particular, is a European diversity hotspot for liverwort species, especially those which are restricted in Europe to the oceanic areas of Macaronesia, Ireland, Scotland, the Faroe Islands and western Norway. Due to the equable climate of southwestern Ireland, species of both southern and northern Atlantic distribution grow in close proximity to each other, with southern species, such as *Adelanthus decipiens* (Hook.) Mitt. and the fern *Trichomanes speciosum* Willd., able to grow at relatively high altitudes, and northern species, such as *Scapania nimbosa* Taylor ex Lehm. and *Bazzania pearsonii* Steph., descending to relatively low altitudes at some sites in Kerry (Ratcliffe, 1968).

One distinct liverwort community is the mixed northern hepatic mat, or liverwort heath, a community of large leafy liverwort species of highly oceanic distribution (Ratcliffe, 1968). It is confined primarily to western Scotland and western Ireland, with some outlying, less species-rich, stands in Wales, northern England, the Faroe Islands and southern Norway (Averis, 1994). Many of the species of this community have their European stronghold in these hyperoceanic areas and are of disjunct distribution, growing elsewhere only in northwestern North America, the Himalayas and China (Averis, 1994). Liverwort heath communities, similar in composition to Irish and Scottish hepatic mats, have been recorded from various parts of the Himalayas and other Asian mountain ranges, such as by Long (2008) on the Gaoligong Shan mountain on the Sino-Burmese border.

Within Ireland, this community grows only in upland areas of Donegal, Mayo, Connemara and Kerry (Hill et al., 1991; Holyoak, 2003). Its distribution is controlled primarily by two factors, climate and topography. Hepatic mats require constant high humidity and are therefore confined mainly to areas with more than 220 rain days per year (Ratcliffe, 1968; Hobbs, 1988). However, topography is equally important, as hepatic mats favour north- to east-facing slopes, which are cooler and more shaded than other aspects, as well as receiving less sunlight, resulting in an increase in atmospheric humidity (Averis, 1992). The reduction in sunlight also results in an increase in the growth and competitiveness of bryophyte species (Poore and McVean, 1957).

These communities form pure mats, rich in bryophytes, in a number of habitats in mountain areas (Fig 21.5). In Scotland, hepatic mats most frequently occur in *Calluna vulgaris* - *Vaccinium myrtillus* - *Sphagnum capillifolium* heath (Averis et al., 2004) under a moderately open canopy of tall heather. Stands of hepatic mat

Figure 21.5 Northern hepatic mat vegetation on Errigal, Co. Donegal, with *Bazzania pearsonii* ssp. *hutchinsiae* and *Scapania ornithopodioides* prominent. Photo: R. Hodd. *See colour plate section.*

vegetation, of varying species richness, also occur on rocky, grassy slopes and inaccessible cliffs in corries, as well as occasionally in woodland (Ratcliffe, 1968). In Irish stands of hepatic mat vegetation, the first author has observed that in Donegal hepatic mat vegetation usually grows under a canopy of *Calluna vulgaris*, whereas in Kerry it occurs almost exclusively on steep rocky slopes, with little or no *Calluna* present.

Hepatic mats are particularly vulnerable to any disturbances to their habitat or changes in conditions, as, apart from having specific climatic requirements, they also have low dispersal ability, being restricted mainly to vegetative reproduction (Rothero, 2003). Only one of the principal liverwort species of this community, *Anastrophyllum donnianum* (Hook.) Steph., which does not occur in Ireland, has been known to produce sporophytes in Britain, on one occasion (Averis, 1992). Currently, the main threats to hepatic mats in Scotland and Ireland are burning, grazing (Porley and Hodgetts, 2005; Holyoak, 2006) and climate change.

The effects of overgrazing on hepatic mats are clearly visible in many parts of western Ireland, especially Connemara, where sheep overstocking has led, at one of the few European sites for the rare liverwort *Adelanthus lindenbergianus* (Lehm.) Mitt., to the complete loss of *Calluna* cover and nearly all of the hepatic mat vegetation (Holyoak, 2006). The hepatic mat vegetation is in better condition in parts of Kerry, Mayo and Donegal, but degradation due to grazing has occurred in all these areas, to varying extents. Climate change is a threat to the survival of hepatic mat vegetation, as hepatic mats are highly dependent on suitable climatic conditions for growth (Averis, 1992; Porley and Hodgetts, 2005).

21.3.6 Human impact on mountain vegetation in Ireland

In assessing climate effects on vegetation, it is important to consider other environmental impacts, especially anthropogenic ones, that might mask climatic effects. The main human impact on montane vegetation in Ireland is through the grazing of sheep. Increases in the number of sheep over the past 25 years have led to overgrazing in many parts of the Irish uplands, especially in Connemara and Mayo (Bleasdale and Sheehy Skeffington, 1995; Sheehy Skeffington et al., 1996; Geerling et al., 2002). However, the intensity of the overgrazing is less severe in other parts of the country, where there is less livestock, such as in Kerry (Hodd, 2007; Hodd and Sheehy Skeffington, in prep.).

The severity of grazing varies depending on topography and soils, with less grazing generally occurring on steeper slopes and cliffs. Overgrazing can lead to changes in vegetation type and cover, with, for example, grassland replacing heath. In severe cases it can result in the erosion of the majority of vegetation and soil from a slope (Bleasdale, 1998). Rare bryophytes of the mixed northern hepatic mat community are particularly vulnerable to the effects of overgrazing, and many stands of hepatic mat have been severely damaged and destroyed in

Ireland (Holyoak, 2006). Forestry is planted over large areas of the uplands in western Ireland, but not usually above 400–500 m in altitude. Other human impacts are minor, with leisure activities such as hill walking and rock climbing restricted to a few of the more popular mountains (MacGowan and Doyle, 1998; Hodd, 2007).

Nitrogen deposition, due to various human activities such as fossil fuel combustion and intensive agriculture, has been shown to have a major effect on the composition of upland vegetation, as mountain areas are particularly vulnerable to N deposition (Britton et al., 2005). At high levels of N deposition, montane heath can be replaced by grassland (Leith et al., 1999). Oceanic bryophyte species, especially the frequent and important *Racomitrium lanuginosum*, are particularly vulnerable to N deposition (Pearce et al., 2003). However, as western Ireland is located far from any major industrial centres and upwind of them, given the prevailing winds, the effects of N deposition are less pronounced.

21.4 The conservation status of montane habitats and species in Ireland

Heath, in general, is of conservation significance, and nine different heath habitats are listed in the European Union Habitats Directive (European Communities, 1992). Northern Atlantic wet heaths, alpine heaths and European heaths are present in Ireland, but the characteristics listed for these (European Commission, 1999) do not fully correspond to those of the upland oceanic heaths of the west of Ireland, especially as the former are all characterised by small shrubs. Because of the location of Ireland on the hyperoceanic fringe of western Europe, the vegetation of the montane heaths is different from that of more continental mountains and, as such, is rare and of high conservation importance.

Since the arctic–montane species in Ireland are on the edge of their climatic range, they may disappear with climate change. This is particularly likely since they are on mountain summits and cannot migrate further up in altitude. They may shift their distribution to the more northerly mountains in Ireland, but this depends on their dispersal ability (Sætersdal and Birks, 1997). Many vascular species listed (Curtis and McGough, 1988) as potentially threatened by climate change are montane.

Hepatic mats specifically, and oceanic bryophyte communities in general, are confined to western Scotland and Ireland and are therefore of international importance, yet they have little protective legislation (Holyoak, 2006). Ireland, as a European stronghold for hepatic mat vegetation, has a special responsibility for their conservation. The Habitats Directive Manual (European Commission, 1999) does not refer to any bryophyte community under the heath category, and therefore hepatic mats and *Racomitrium* heath still require international recognition. However, many mountains in Ireland are designated as Special Areas of Conservation (SACs) under the Habitats Directive for their rare or unusual plant communities, and therefore receive some protection and international recognition. High-altitude plant communities in particular need to be maintained in good condition to maximise their resilience to climate change (Trivedi et al., 2008). However, their conservation is hampered by heavy grazing in many uplands (Sheehy Skeffington et al., 1996; Bleasdale, 1998; Geerling et al., 2002). The inclusion of plant and animal species in a red data list (Curtis and McGough, 1988; see also www.botanicgardens.ie/herb/census/threatnd.htm) highlights their vulnerability to any habitat change, and currently about 14 arctic-montane vascular plants are listed as endangered. Only 18 species of bryophyte are listed in the Irish Flora (Protection) Order (Government of Ireland, 1999), and work is still ongoing to create a reliable red list of bryophytes. This is partly due to a lack of in-depth knowledge of the distribution/abundance of rare species, including montane bryophytes, a situation that has been improved by recent survey work (Holyoak, 2006).

21.5 Potential climate change and its effects on western Irish montane vegetation

21.5.1 Temperature effects on oceanic montane vegetation

It is predicted that the climate of western Ireland will change in a number of ways as a result of global climate change. In Ireland as a whole, temperatures are predicted to rise by up to 3–4 °C by the end of the century (McGrath and Lynch, 2008). This rise is expected to be greater in the south and east of the country and, by the middle of the century, winters in the north of Ireland are projected to become similar to the winters experienced in Cork and Kerry in the southwest during the period 1960–91 (Sweeney and Fealy, 2002). These rises in temperature may result in a reduction in the frequencies of frost, especially in oceanic mountain areas. As the length of the growing season is controlled primarily by the occurrence of frost, plant growth may be able to continue year round (Pepin, 1997).

The projected rises in temperature and associated changes may have a number of effects on the biodiversity of montane habitats in western Ireland. Montane species are particularly vulnerable to rises in temperature, as they are adapted in a number of ways to colder conditions (Trivedi et al., 2008). The most likely plants to be affected are the arctic-montane vascular species, due to their adaptation to cold conditions and low growth rates. Temperature rises may have an indirect, rather than a direct, effect on these species (Sætersdal and Birks, 1997). Rises in temperature may cause lowland species to expand their range to higher altitudes, thus coming into direct competition with arctic-montane species (Grabherr et al.,

1994; Gottfried et al., 1999; Walther et al., 2005). As these species are limited in range and slow-growing, often reproducing vegetatively, their ability to compete or adjust their range to higher altitudes is likely to be minimal. Many are stress tolerators and are poor competitors (*sensu* Grime, 1979). A number of oceanic bryophyte communities are also likely to be affected, including *Racomitrium* heath and hepatic mat vegetation, which may lose cover and growth space as a result of increased competition from vascular plants (Porley and Hodgetts, 2005; Trivedi et al., 2008).

The possible drier summers may also inhibit bryophyte community growth for a longer period of the year. The changes in competitive balance may lead to an upward altitudinal shift in the distribution of montane vegetation, resulting in possible major changes in plant community structure and a significant loss of climate space for many species (Berry et al., 2003). Temperature rises and warmer winters also affect the phenology of plants (Jones et al., 2006), leading to early bud burst. This could be especially detrimental to plant growth and survival in mountain areas, where relatively cold, severe periods can still occur into spring.

21.5.2 Rainfall changes and effects on montane oceanic vegetation

Precipitation in Ireland is expected to increase overall, with larger variations between seasons. By the end of the century, rainfall in Ireland is expected to increase by about 15–20% in winter and autumn, but to decrease in the summer by up to 25% (Sweeney and Fealy, 2002; McGrath and Lynch, 2008). It is the decrease in summer precipitation that is likely to result in the greatest changes for oceanic vegetation. Even the smallest change in summer water balance can change the conditions for plant growth, with peatlands being regarded as particularly vulnerable to change (Jones et al., 2006).

The mixed northern hepatic mat bryophytes, which have very specific climatic requirements, are also likely to be vulnerable to decreases in summer precipitation. Although species of the hepatic mat community have been shown to be moderately resistant to prolonged drought (Averis, 1994), liverworts require abundant moisture to grow and survive in the long term. In the event of prolonged drought, the liverworts' growing season would be interrupted and their already limited dispersal capacity is likely to become further reduced. This would make them more vulnerable to competition from other species, and any disturbance, such as that provided by grazing, would have an even more detrimental effect when combined with climate change.

Temperature rises may also negatively affect hepatic mats, as the species are mainly of northern and montane distribution and are not tolerant of very high temperatures (Averis, 1994). A combination of decreases in summer precipitation and increases in temperatures may result in a marked contraction of the range of hepatic mat vegetation, especially at the edges of its range, on more open slopes and in southwestern Ireland.

21.5.3 Other potential impacts on montane biodiversity

Due to the uncertainty involved with applying climate models, especially to oceanic and montane areas (Coll et al., 2005; Trivedi et al., 2008), it is likely that changes other than those discussed above may occur. It is possible that the proximity of Ireland to the Atlantic Ocean may have a buffering effect, with the ocean to a certain extent mitigating the effects of climate change. Climate change may also result in an increase in oceanicity (Crawford, 2000), which may be favourable to the growth and survival of Atlantic bryophyte species. Precipitation changes may not occur as projected, due to a large amount of uncertainty in modelling patterns of precipitation, particularly in topographically diverse oceanic areas (Coll et al., 2005).

Temperature changes may also not occur as projected, as some models predict a weakening of the Gulf Stream, which keeps winter temperatures high along the Atlantic coast of Europe (Sweeney, 1997). If temperatures were to drop as a consequence of the weakening of the Gulf Stream, certain species, especially those of arctic-montane distribution, may migrate downwards to lower altitudes, leading to an expansion in range (Fosaa et al., 2004). Even in the event of temperatures rising in the west of Ireland, lowland species may not be able to colonise mountain summits and ridges. In oceanic areas especially, temperature is not the only limiting factor for plant growth on mountains. Wind speed also limits the spread of lowland species up mountains, so the plant communities of mountain tops and ridges may not change significantly, as many lowland species would be unable to cope with high winds and exposure (Sætersdal and Birks, 1997). Although there are substantial uncertainties involved in projecting the exact impact of changes in the climate on the plant communities of the mountains of western Ireland, it is likely that the composition and distribution of these communities will undergo great change in response to the changing climate.

21.6 Conclusions

This chapter has reviewed current knowledge of climate change and the montane heath communities of western Ireland. As for many species and their communities, knowledge of the exact current distribution is not complete. Our current research on montane heath communities in the west of Ireland is focusing on mapping them, with reference to arctic–montane vascular plants and to oceanic bryophyte communities. This systematic collection of distribution data will provide specific data to inform ongoing modelling to project the effects of climate change.

Climate (air temperature, relative humidity, rainfall, solar radiation, wind speed and direction, as well as soil humidity and temperature) is also being monitored above 600 m on two mountains, one in Kerry, the other in Donegal, which have different bioclimatic characteristics, to acquire information on the climatic range of oceanic mountain tops in Ireland. A series of data loggers measuring temperature and humidity have also been placed at different aspects of selected mountains in the same counties. Such data do not exist for Irish mountains, and they should provide more precise information on the climatic requirements of these rare and vulnerable communities. The maps will provide specific baseline data on the distribution and occurrence of these communities that can be monitored in future years. If conditions can otherwise be maintained favourable for these communities, any changes in their distribution may be directly attributable to climate change.

Since conservation actions are difficult without precise projection information, it is recommended that best conservation practice be applied in any habitats that contain potentially vulnerable plant or animal species (Ellis and Good, 2005). Most of the sites under study are SACs, and some are also in national parks. The precise mapping of the plant communities will inform conservation managers of the location of the rarer ones, and conservation action, such as preventing overgrazing and visitor access, will at least aim to maintain a maximum area for the occurrence of these communities. Monitoring of any changes in their distribution may help predict any long-term climate change effects. Furthermore, monitoring the migration of species up mountains will enable projections to be made concerning the degree of threat that such species will pose to these vulnerable communities.

Acknowledgements

We would like to acknowledge the funding supplied by the Irish Research Council for Science, Engineering and Technology (IRCSET) and the National Parks and Wildlife Service (NPWS) for this project. Thanks are due to Caitriona Douglas and Neil Lockhart (NPWS), David Bourke (TCD), Michael O'Connell and Pat O'Rafferty (NUI Galway), Bob Crawford (St Andrew's) and John Coll (NUI Maynooth), for providing advice, literature and technical assistance. We also thank Pam Berry for providing bioclimatic data from the MONARCH project.

References

Averis, A. M. (1992). Where are all the hepatic mat liverworts in Scotland? *Botanical Journal of Scotland,* **46**,191–198.

Averis, A. M. (1994). The ecology of an Atlantic liverwort community. Unpublished PhD thesis, University of Edinburgh.

Averis, A. M., Averis, A. B. G., Birks, H. J. B . et al. (2004). *An Illustrated Guide to British Upland Vegetation.* Peterborough: Joint Nature Conservation Committee.

Barry, R. G. (1992). *Mountain Weather and Climate.* London: Routledge.

Baxter, C. and Thompson, D. B. A. (1995). *Scotland: Land of Mountains.* Grantown-on-Spey: Colin Baxter.

Berry, P. M., Dawson, T. P., Harrison, P. A., Pearson, R. and Butt, N. (2003). The sensitivity and vulnerability of terrestrial habitats and species in Britain and Ireland to climate change. *Journal of Nature Conservation,* **11**, 15–23.

Berry, P. M., Harrison, P. A., Dawson, T. P. and Walmsley, C. A. (2005). *Climate Change and Nature Conservation in the UK and Ireland: Modelling Natural Resource Responses to Climate Change (MONARCH2).* Oxford: UK Climate Impacts Programme.

Birks, H. J. B. (1988). Long-term ecological change in the British uplands. In *Ecological Change in the Uplands,* ed. M. B. Usher and D. B. A. Thompson. British Ecological Society Special Publication 7. Oxford: Blackwell Scientific Publications, pp. 37–56.

Bleasdale, A. (1998). Overgrazing in the west of Ireland: assessing solutions. In *Towards a Conservation Strategy for the Bogs of Ireland,* ed. G. O 'Leary and F. Gormley. Dublin: Irish Peatland Conservation Council, pp. 67–76.

Bleasdale, A. and Sheehy Skeffington, M. (1995). The upland vegetation of north-east Connemara in relation to sheep grazing. In *Irish Grasslands,* ed. D. W. Jeffrey, M. B. Jones and J. H. McAdam. Dublin: Royal Irish Academy, pp. 110–124.

Britton, A. J., Fisher, J. and Baillie, G. (2005). The effects of nitrogen deposition on mountain vegetation: the importance of geology. In *Mountains of Northern Europe: Conservation, Management, People and Nature,* ed. D. B. A. Thompson, M. F. Price and C. A. Galbraith. Edinburgh: The Stationery Office, pp. 121–125.

Brown, A., Birks, H. J. B. and Thompson, D. B. A. (1993). A new biogeographical classification of the Scottish uplands. II. Vegetation–environment relationships. *Journal of Ecology,* **81**, 231–251.

Carruthers, T. (1998). *Kerry: A Natural History.* Cork: Collins Press.

Coll, J., Gibb, S. W. and Harrison, S. J. (2005). Modelling future climates in the Scottish Highlands: an approach integrating local climatic variables and regional climatic model outputs. In *Mountains of Northern Europe: Conservation, Management, People and Nature,* ed. D. B. A. Thompson, M. F. Price and C. A. Galbraith. Edinburgh: The Stationery Office, pp. 103–120.

Collins, J. F. and Cummins, T. (1996). *Agroclimatic Atlas of Ireland.* Dublin: Agmet.

Conrad, V. (1946). Usual formulas of continentality and their limits of validity. *Transactions of the American Geophysical Union*, **27**, 663–664.

Crawford, R. M. M. (2000). Ecological hazards of oceanic environments: Tansley Review 114. *New Phytologist*, **147**, 257–281.

Crawford, R. M. M. (2005). Trees by the sea: advantages and disadvantages of oceanic climates. *Biology and Environment: Proceedings of the Royal Irish Academy*, **105B**, 129–139.

Crawford, R. M. M. (2008). *Plants at the Margin: Ecological Limits and Climate Change*. Cambridge: Cambridge University Press.

Cross, J. R. (2006). The potential natural vegetation of Ireland. *Biology and Environment: Proceedings of the Royal Irish Academy*, **106B**, 65–116.

Curtis, T. G. F. and McGough, H. N. (1988). *The Irish Red Data Book. 1. Vascular Plants*. Dublin: Government Publications.

Dahl, E. (1998). *The Phytogeography of Northern Europe (British Isles, Fennoscandia and Adjacent Areas)*. Cambridge: Cambridge University Press.

Donnelly, A., Jones, M. B. and Sweeney, J. (2004). A review of indicators of climate change for use in Ireland. *International Journal of Biometeorology*, **49**, 1–12.

Doutt, J. K. (1941). Wind pruning and salt spray as factors in ecology. *Ecology*, **22**, 195–196.

Duckworth, J. C., Bunce, R. G. H. and Malloch, A. J. C. (2000). Vegetation gradients in Atlantic Europe: the use of existing phytosociological data in preliminary investigations on the potential effects of climate change on British vegetation. *Global Ecology and Biodiversity*, **9**, 187–199.

Ellis, N. E. and Good, J. E. G. (2005). Climate change and effects on Scottish and Welsh ecosystems: a conservation perspective. In *Mountains of Northern Europe: Conservation, Management, People and Nature*, ed. D. B. A. Thompson, M. F. Price and C. A. Galbraith. Edinburgh: The Stationery Office, pp. 99–102.

European Commission (1999). *Interpretation Manual of European Union Habitats*. EUR 15/2. Brussels: European Commission, DG Environment.

European Communities (1992). Council Directive 92/43/EEC on the conservation of natural habitats and of wild fauna and flora. *Official Journal* L 206, 22/07/1992. http://ec.europa.cu/cnvironmcnt/nature/legislation/habitatsdirective.

Fosaa, A. M. (2001). A review of plant communities of the Faroe Islands. *Fróðskaparrit*, **48**, 41–54.

Fosaa, A. M., Sykes, M. T., Lawesson, J. E. and Gaard, M. (2004). Potential effects of climate change on plant species in the Faroe Islands. *Global Ecology and Biogeography*, **13**, 427–437.

Geerling G., van Gestel C., Sheehy Skeffington, M. et al. (2002). Blanket bog degradation in river catchments in the west of Ireland. In *Application of Geographic Information Systems and Remote Sensing in River Studies*, ed. R. S. E. W. Leuven, I. Poudevigne and R. M. Teeuw. Leiden: Backhuys Publishers, pp. 25–40.

Gimingham, C. H. (1972). *Ecology of Heathlands*. London: Chapman and Hall.

Godske, C. L. (1944). The geographical distribution in Norway of certain indices of humidity and oceanicity.

Bergens Museums Arbeider Naturvitenskapelige række, **8**.

Gottfried, M., Pauli, H., Reiter, K. and Grabherr, G. (1999). A fine-scale predictive model for changes in species distribution patterns of high mountain plants induced by climate warming. *Diversity and Distribution*, **5**, 241–251.

Government of Ireland (1999). *Flora Protection Order*. S.I. No. 94 of 1999. Dublin: Government of Ireland.

Grabherr, G., Gottfried, M. and Pauli, H. (1994). Climate effects on mountain plants. *Nature*, **369**, 448.

Grace, J. (1997). The oceanic tree-line and the limit for tree growth in Scotland. *Botanical Journal of Scotland*, **49**, 223–236.

Grace, J. and Unsworth, H. (1988). Climate and microclimate of the uplands. In *Ecological Change in the Uplands*, ed. M. B. Usher and D. B. A. Thompson. British Ecological Society Special Publication 7. Oxford: Blackwell Scientific Publications, pp. 137–150.

Grime, J. P. (1979). *Plant Strategies and Vegetation Processes*. New York, NY: John Wiley and Sons.

Harrison, P. A., Berry, P. M. and Dawson, T. P., eds. (2001). *Climate Change and Nature Conservation in Britain and Ireland: Modelling Natural Resource Responses to Climate Change (The MONARCH Project)*. Oxford: UK Climate Impacts Programme.

Hart, H. C. (1891). On the range of flowering plants and ferns on the mountains of Ireland. *Proceedings of the Royal Irish Academy 3rd Series I*, **4**, 512–570.

Hill, M. O., Preston, C. D. and Smith, A. J. E. (1991). *Atlas of the Bryophytes of Britain and Ireland. Vol. 1. Liverworts (Hepaticae and Anthocerotae)*. Colchester: Harley Books.

Hobbs, A. M. (1988). Conservation of leafy liverwort-rich *Calluna vulgaris* heath in Scotland. In *Ecological Change in the Uplands*, ed. M. B. Usher and D. B. A. Thompson. British Ecological Society Special Publication 7. Oxford: Blackwell Scientific Publications, pp. 339–343.

Hodd, R. L. (2007). *The Vegetation and Ecology of the Scree Slopes of the Macgillycuddy's Reeks, Co. Kerry*. BSc thesis, Department of Botany, NUI, Galway.

Holyoak, D. T. (2003). *The Distribution of Bryophytes in Ireland*. Dinas Powys: Broadleaf Books.

Holyoak, D. T. (2006). Progress towards a species inventory for conservation of bryophytes in Ireland. *Biology and Environment: Proceedings of the Royal Irish Academy*, **106B**, 225–236.

Horsfield, D., Averis, A., Averis, B. and Kinnes, L. (1991). *The Vegetation of Connemara in Relation to Plant Communities of Great Britain*. CSD Report No. 1253. Peterborough: Nature Conservancy Council.

Intergovernmental Panel on Climate Change (IPCC) (2007). *Climate Change 2007: Synthesis Report. Contribution of Working Groups I, II and III to the Fourth Assessment Report of the Intergovernmental Panel on Climate Change*, ed. R. K. Pachauri and A. Reisinger. Geneva: IPCC.

Jones, M .B., Donnelly, A. and Albanito, F. (2006). Responses of Irish vegetation to future climate change. *Biology and Environment: Proceedings of the Royal Irish Academy*, **106B**, 323–334.

Kirkpatrick, A. H. and Rushton, B. S. (1990). The oceanicity/continentality of the climate of the north of Ireland. *Weather*, **45**, 322–326.

Körner, C. (2003). *Alpine Plant Life: Functional Plant Ecology of High Mountain Ecosystems*, 2nd edn. Berlin: Springer.

Leith, I. D., Hicks, W. K., Fowler, D. and Woodin, S. J. (1999). Differential responses of UK upland plants to nitrogen deposition. *New Phytologist*, **141**, 277–289.

Long, D. G. (2008). The Gaoligong Shan mountains of the Sino-Burmese border. *Field Bryology*, **96**, 28–38.

MacGowan, F. and Doyle, G. (1998). Vegetation of Atlantic blanket bogs damaged by tourist trampling in the west of Ireland. In *Towards a Conservation Strategy for the Bogs of Ireland*, ed. G. O 'Leary and F. Gormley. Dublin: Irish Peatland Conservation Council, pp. 159–166.

McGrath, R. and Lynch, P., eds. (2008). *Ireland in a Warmer World: Scientific Predictions of the Irish Climate in the 21st Century*. Dublin: Community Climate Change Consortium for Ireland (C_4I).

McKee, A. M., Bleasdale, A. J. and Sheehy Skeffington, M. (1998). The effects of different grazing pressures on the above-ground biomass of vegetation in the Connemara uplands. In *Towards a Conservation Strategy for the Bogs of Ireland*, ed. G. O 'Leary and F. Gormley. Dublin: Irish Peatland Conservation Council, pp. 177–188.

McVean, D. N. and Ratcliffe, D. A. (1962). *Plant Communities of the Scottish Highlands*. London: Her Majesty's Stationery Office.

Mhic Daeid, E. C. (1976). A phytosociological and ecological study of the vegetation of peatlands and heaths of the Killarney valley. Unpublished PhD thesis, Trinity College Dublin.

Mitchell, G. F. and Ryan, M. (2001). *Reading the Irish Landscape*. Dublin: Town House.

Moore, J. J. (1960). A re-survey of the vegetation of the district lying south of Dublin (1905–1956). *Proceedings of the Royal Irish Academy*, **61B**, 1–36.

Nagy, L., Grabherr, G., Körner, C. and Thompson, D. B. A., eds. (2005). *Alpine Biodiversity in Europe: an Introduction*. Peterborough: Joint Nature Conservation Committee.

Pearce, I. S. K., Woodin, S. J. and van der Wal, R. (2003). Physiological and growth response of the montane bryophyte *Racomitrium lanuginosum* to atmospheric nitrogen deposition. *New Phytologist*, **160**, 145–155.

Pearsall, W. H. (1950). *Mountains and Moorlands*. London: Collins.

Pepin, N. C. (1997). Scenarios of future climate change: effects on frost occurrence and severity in the maritime uplands of northern England. *Geografiska Annaler*, **79A**, 121–137.

Pochin Mould, D. D. C. (1976). *The Mountains of Ireland*. Dublin: Gill and Macmillan.

Poore, M. E. D. and McVean, D. N. (1957). A new approach to Scottish mountain vegetation. *Journal of Ecology*, **45**, 401–439.

Porley, R. and Hodgetts, N. (2005). *Mosses and Liverworts*. New Naturalist Library. London: HarperCollins.

Praeger, R. L. (1934). *The Botanist in Ireland*. Wakefield: EP Publishing.

Praeger, R. L. (1950). *Natural History of Ireland: a Sketch of its Flora and Fauna*. London: Collins.

Preston, C. D., Pearman, D. A. and Dines, T. D., eds. (2002). *New Atlas of the British and Irish Flora*. Oxford: Oxford University Press.

Ratcliffe, D. A. (1968). An ecological account of Atlantic bryophytes in the British Isles. *New Phytologist*, **67**, 365–439.

Ratcliffe, D. A. and Thompson, D. B. A. (1988). The British uplands: their ecological character and international significance. In *Ecological Change in the Uplands*, ed. M. B. Usher and D. B. A. Thompson. British Ecological Society Special Publication 7. Oxford: Blackwell Scientific Publications, pp. 9–36.

Roden, C. M. (1986). A survey of the flora of some mountain ranges in the west of Ireland. *Irish Naturalists Journal*, **22**, 52–59.

Rodwell, J. S., ed. (1991). *British Plant Communities, Volume 2. Mires and Heaths*. Cambridge: Cambridge University Press.

Rodwell, J. S., ed. (1992). *British Plant Communities, Volume 3. Grassland and Montane Plant Communities.* Cambridge: Cambridge University Press.

Rohan, P. K. (1986). *The Climate of Ireland.* Dublin: The Stationery Office.

Rothero, G. (2003). Bryophyte conservation in Scotland. *Botanical Journal of Scotland*, **55**, 17–26.

Sætersdal, M. and Birks, H. J. B. (1997). A comparative ecological study of Norwegian mountain plants in relation to possible future climate change. *Journal of Biogeography*, **24**, 127–152.

Sheehy Skeffington, M. J., Bleasdale, A. and McKee A. M. (1996). Research in the Connemara uplands: vegetation changes and peat erosion (sustainability and sheep grazing in the Connemara uplands). In *Seeking a Partnership Towards Managing Ireland's Uplands*, ed. D. Hogan and A. Phillips. Galway: Irish Uplands Forum, pp. 143–148.

Sweeney, J. (1997). Ireland. In *Regional Climates of the British Isles*, ed. D. Wheeler and J. Mayes. London: Routledge, pp. 254–275.

Sweeney, J. and Fealy, R. (2002). A preliminary investigation of future climate scenarios for Ireland. *Biology and Environment: Proceedings of the Royal Irish Academy*, **102B**, 121–128.

Tansley, A. G. (1949). *The British Islands and their Vegetation*. Cambridge: Cambridge University Press.

Taylor, J. A. (1983). The peatlands of Great Britain and Ireland. In *Ecosystems of the World 4B. Mires: Swamp, Bog, Fen and Moor*, ed. A. J. P. Gore. Amsterdam: Elsevier, pp. 1–46.

Trivedi, M. R., Morecroft, M. D., Berry, P. M. and Dawson, T. P. (2008). Potential effects of climate change on plant communities in three montane nature reserves in Scotland, UK. *Biological Conservation*, **141**, 1665–1675.

Visser, M. E. (2008). Keeping up with a warming world: assessing the rate of adaptation to climate change. *Proceedings of the Royal Society of London B*, **275**, 649–659.

Walther, G. R. (2003). Plants in a warmer world. *Perspectives in Plant Ecology, Evolution and Systematics*, **6**, 169–185.

Walther, G. R., Post, E., Convey, P. et al. (2002). Ecological responses to recent climate change. *Nature*, **416**, 389–395.

Walther, G. R., Beissner, S. and Burga, C. A. (2005). Trends in upward shift of Alpine plants. *Journal of Vegetation Science*, **16**, 541–548.

Watson, W. (1925) The bryophytes and lichens of arctic-alpine vegetation. *Journal of Ecology*, **XIII** (1), 1–26.

Whilde, T. (1994). *The Natural History of Connemara*. London: Immel Publishing.

Index

Systematics Association Publications

1. Bibliography of Key Works for the Identification of the British Fauna and Flora, 3rd edition (1967)†
   Edited by G. J. Kerrich, R. D. Meikie and N. Tebble
2. The Species Concept in Palaeontology (1956)†
   Edited by P. C. Sylvester-Bradley
3. Function and Taxonomic Importance (1959)†
   Edited by A. J. Cain
4. Taxonomy and Geography (1962)†
   Edited by D. Nichols
5. Speciation in the Sea (1963)†
   Edited by J. P. Harding and N. Tebble
6. Phenetic and Phylogenetic Classification (1964)†
   Edited by V. H. Heywood and J. McNeill
7. Aspects of Tethyan Biogeography (1967)†
   Edited by C. G. Adams and D. V. Ager
8. The Soil Ecosystem (1969)†
   Edited by H. Sheals
9. Organisms and Continents through Time (1973)*
   Edited by N. F. Hughes
10. Cladistics: A Practical Course in Systematics (1992)‡
    P. L. Forey, C. J. Humphries, I. J. Kitching, R. W. Scotland, D. J. Siebert and D. M. Williams
11. Cladistics: The Theory and Practice of Parsimony Analysis, 2nd edition (1998)‡
    I. J. Kitching, P. L. Forey, C. J. Humphries and D. M. Williams

---

† Published by the Systematics Association (out of print)
* Published by the Palaeontological Association in conjunction with the Systematics Association
‡ Published by Oxford University Press for the Systematics Association

Systematics Association Special Volumes

1. The New Systematics (1940)[a]
   Edited by J. S. Huxley (reprinted 1971)
2. Chemotaxonomy and Serotaxonomy (1968)*
   Edited by J. C. Hawkes
3. Data Processing in Biology and Geology (1971)*
   Edited by J. L. Cutbill
4. Scanning Electron Microscopy (1971)*
   Edited by V. H. Heywood
5. Taxonomy and Ecology (1973)*
   Edited by V. H. Heywood
6. The Changing Flora and Fauna of Britain (1974)*
   Edited by D. L. Hawksworth
7. Biological Identification with Computers (1975)*
   Edited by R. J. Pankhurst
8. Lichenology: Progress and Problems (1976)*
   Edited by D. H. Brown, D. L. Hawksworth and R. H. Bailey
9. Key Works to the Fauna and Flora of the British Isles and Northwestern Europe, 4th edition (1978)*
   Edited by G. J. Kerrich, D. L. Hawksworth and R. W. Sims
10. Modern Approaches to the Taxonomy of Red and Brown Algae (1978)*
    Edited by D. E. G. Irvine and J. H. Price
11. Biology and Systematics of Colonial Organisms (1979)*
    Edited by C. Larwood and B. R. Rosen
12. The Origin of Major Invertebrate Groups (1979)*
    Edited by M. R. House
13. Advances in Bryozoology (1979)*
    Edited by G. P. Larwood and M. B. Abbott
14. Bryophyte Systematics (1979)*
    Edited by G. C. S. Clarke and J. G. Duckett
15. The Terrestrial Environment and the Origin of Land Vertebrates (1980)*
    Edited by A. L. Panchen

16. Chemosystematics: Principles and Practice (1980)*
    Edited by F. A. Bisby, J. G. Vaughan and C. A. Wright
17. The Shore Environment: Methods and Ecosystems (2 volumes) (1980)*
    Edited by J. H. Price, D. E. C. Irvine and W. F. Farnham
18. The Ammonoidea (1981)*
    Edited by M. R. House and J. R. Senior
19. Biosystematics of Social Insects (1981)*
    Edited by P. E. House and J.-L. Clement
20. Genome Evolution (1982)*
    Edited by G. A. Dover and R. B. Flavell
21. Problems of Phylogenetic Reconstruction (1982)*
    Edited by K. A. Joysey and A. E. Friday
22. Concepts in Nematode Systematics (1983)*
    Edited by A. R. Stone, H. M. Platt and L. F. Khalil
23. Evolution, Time and Space: The Emergence of the Biosphere (1983)*
    Edited by R. W. Sims, J. H. Price and P. E. S. Whalley
24. Protein Polymorphism: Adaptive and Taxonomic Significance (1983)*
    Edited by G. S. Oxford and D. Rollinson
25. Current Concepts in Plant Taxonomy (1983)*
    Edited by V. H. Heywood and D. M. Moore
26. Databases in Systematics (1984)*
    Edited by R. Allkin and F. A. Bisby
27. Systematics of the Green Algae (1984)*
    Edited by D. E. G. Irvine and D. M. John
28. The Origins and Relationships of Lower Invertebrates (1985)†
    Edited by S. Conway Morris, J. D. George, R. Gibson and H. M. Platt
29. Infraspecific Classification of Wild and Cultivated Plants (1986)†
    Edited by B. T. Styles
30. Biomineralization in Lower Plants and Animals (1986)†
    Edited by B. S. C. Leadbeater and R. Riding
31. Systematic and Taxonomic Approaches in Palaeobotany (1986)†
    Edited by R. A. Spicer and B. A. Thomas
32. Coevolution and Systematics (1986)†
    Edited by A. R. Stone and D. L. Hawksworth
33. Key Works to the Fauna and Flora of the British Isles and Northwestern Europe, 5th edition (1988)†
    Edited by R. W. Sims, P. Freeman and D. L. Hawksworth
34. Extinction and Survival in the Fossil Record (1988)†
    Edited by G. P. Larwood
35. The Phylogeny and Classification of the Tetrapods (2 volumes) (1988)†
    Edited by M. J. Benton

36. Prospects in Systematics (1988)†
Edited by J. L. Hawksworth

37. Biosystematics of Haematophagous Insects (1988)†
Edited by M. W. Service

38. The Chromophyte Algae: Problems and Perspective (1989)†
Edited by J. C. Green, B. S. C. Leadbeater and W. L. Diver

39. Electrophoretic Studies on Agricultural Pests (1989)†
Edited by H. D. Loxdale and J. den Hollander

40. Evolution, Systematics and Fossil History of the Hamamelidae (2 volumes) (1989)†
Edited by P. R. Crane and S. Blackmore

41. Scanning Electron Microscopy in Taxonomy and Functional Morphology (1990)†
Edited by D. Claugher

42. Major Evolutionary Radiations (1990)†
Edited by P. D. Taylor and G. P. Larwood

43. Tropical Lichens: Their Systematics, Conservation and Ecology (1991)†
Edited by G. J. Galloway

44. Pollen and Spores: Patterns and Diversification (1991)†
Edited by S. Blackmore and S. H. Barnes

45. The Biology of Free-Living Heterotrophic Flagellates (1991)†
Edited by D. J. Patterson and J. Larsen

46. Plant–Animal Interactions in the Marine Benthos (1992)†
Edited by D. M. John, S. J. Hawkins and J. H. Price

47. The Ammonoidea: Environment, Ecology and Evolutionary Change (1993)†
Edited by M. R. House

48. Designs for a Global Plant Species Information System (1993)†
Edited by F. A. Bisby, G. F. Russell and R. J. Pankhurst

49. Plant Galls: Organisms, Interactions, Populations (1994)†
Edited by M. A. J. Williams

50. Systematics and Conservation Evaluation (1994)†
Edited by P. L. Forey, C. J. Humphries and R. I. Vane-Wright

51. The Haptophyte Algae (1994)†
Edited by J. C. Green and B. S. C. Leadbeater

52. Models in Phylogeny Reconstruction (1994)†
Edited by R. Scotland, D. I. Siebert and D. M. Williams

53. The Ecology of Agricultural Pests: Biochemical Approaches (1996)**
Edited by W. O. C. Symondson and J. E. Liddell

54. Species: The Units of Diversity (1997)**
Edited by M. F. Claridge, H. A. Dawah and M. R. Wilson

55. Arthropod Relationships (1998)**
Edited by R. A. Fortey and R. H. Thomas

56. Evolutionary Relationships Among Protozoa (1998)[**]
Edited by G. H. Coombs, K. Vickerman, M. A. Sleigh and A. Warren

57. Molecular Systematics and Plant Evolution (1999)[††]
Edited by P. M. Hollingsworth, R. M. Bateman and R. J. Gornall

58. Homology and Systematics (2000)[††]
Edited by R. Scotland and R. T. Pennington

59. The Flagellates: Unity, Diversity and Evolution (2000)[††]
Edited by B. S. C. Leadbeater and J. C. Green

60. Interrelationships of the Platyhelminthes (2001)[††]
Edited by D. T. J. Littlewood and R. A. Bray

61. Major Events in Early Vertebrate Evolution (2001)[††]
Edited by P. E. Ahlberg

62. The Changing Wildlife of Great Britain and Ireland (2001)[††]
Edited by D. L. Hawksworth

63. Brachiopods Past and Present (2001)[††]
Edited by H. Brunton, L. R. M. Cocks and S. L. Long

64. Morphology, Shape and Phylogeny (2002)[††]
Edited by N. MacLeod and P. L. Forey

65. Developmental Genetics and Plant Evolution (2002)[††]
Edited by Q. C. B. Cronk, R. M. Bateman and J. A. Hawkins

66. Telling the Evolutionary Time: Molecular Clocks and the Fossil Record (2003)[††]
Edited by P. C. J. Donoghue and M. P. Smith

67. Milestones in Systematics (2004)[††]
Edited by D. M. Williams and P. L. Forey

68. Organelles, Genomes and Eukaryote Phylogeny (2004)[††]
Edited by R. P. Hirt and D. S. Horner

69. Neotropical Savannas and Seasonally Dry Forests: Plant Diversity, Biogeography and Conservation (2006)[††]
Edited by R. T. Pennington, G. P. Lewis and J. A. Rattan

70. Biogeography in a Changing World (2006)[††]
Edited by M. C. Ebach and R. S. Tangney

71. Pleurocarpous Mosses: Systematics & Evolution (2006)[††]
Edited by A. E. Newton and R. S. Tangney

72. Reconstructing the Tree of Life: Taxonomy and Systematics of Species Rich Taxa (2006)[††]
Edited by T. R. Hodkinson and J. A. N. Parnell

73. Biodiversity Databases: Techniques, Politics, and Applications (2007)[††]
Edited by G. B. Curry and C. J. Humphries

74. Automated Taxon Identification in Systematics: Theory, Approaches and Applications (2007)[††]
Edited by N. MacLeod

75. Unravelling the Algae: The Past, Present, and Future of Algal Systematics (2008)[††]
    Edited by J. Brodie and J. Lewis
76. The New Taxonomy (2008)[††]
    Edited by Q. D. Wheeler
77. Palaeogeography and Palaeobiogeography: Biodiversity in Space and Time (in press)[††]
    Edited by P. Upchurch, A. McGowan and C. Slater

---

[a] Published by Clarendon Press for the Systematics Association
* Published by Academic Press for the Systematics Association
† Published by Oxford University Press for the Systematics Association
** Published by Chapman & Hall for the Systematics Association
†† Published by CRC Press for the Systematics Association